Ohio State University Mathematical Research Institute Publications 11

Editors: Gregory R. Baker, Walter D. Neumann, Karl Rubin

Ohio State University
Mathematical Research Institute Publications

1 Topology '90, *B. Apanasov, W. D. Neumann, A. W. Reid, L. Siebenmann* (Eds.)
2 The Arithmetic of Function Fields, *D. Goss, D. R. Hayes, M. I. Rosen* (Eds.)
3 Geometric Group Theory, *R. Charney, M. Davis, M. Shapiro* (Eds.)
4 Groups, Difference Sets, and the Monster, *K. T. Arasu, J. F. Dillon, K. Harada, S. Sehgal, R. Solomon* (Eds.)
5 Convergence in Ergodic Theory and Probability, *V. Bergelson, P. March, J. Rosenblatt* (Eds.)
6 Representation Theory of Finite Groups, *R. Solomon* (Ed.)
7 The Monster and Lie Algebras, *J. Ferrar, K. Harada* (Eds.)
8 Groups and Computation III, *W. M. Kantor, Á. Seress* (Eds.)
9 Complex Analysis and Geometry, *J. D. McNeal* (Ed.)
10 Codes and Designs, *K. T. Arasu, Á. Seress* (Eds.)

Automorphic Representations, L-Functions and Applications: Progress and Prospects

Proceedings of a conference honoring
Steve Rallis
on the occasion of his 60th birthday

The Ohio State University
March 27–30, 2003

Editors

James W. Cogdell
Dihua Jiang
Stephen S. Kudla
David Soudry
Robert Stanton

Walter de Gruyter · Berlin · New York

Editors

James W. Cogdell
Department of Mathematics, Oklahoma State University, Stillwater, OK 74078, USA
Department of Mathematics, The Ohio State University, Columbus, OH 43210-1174, USA

Dihua Jiang
School of Mathematics, University of Minnesota, Minneapolis, MN 55455, USA

Stephen S. Kudla
Math Department, University of Maryland, College Park, MD 20742-4015, USA

David Soudry
School of Mathematical Sciences, Sackler Faculty of Exact Sciences, Tel Aviv University, Ramat-Aviv, 69978 Israel

Robert Stanton
Department of Mathematics, The Ohio State University, Columbus, OH 43210-1174, USA

Series Editors

Gregory R. Baker
Department of Mathematics, The Ohio State University, Columbus, OH 43210-1174, USA

Karl Rubin
Department of Mathematics, University of California, Irvine, CA 92697-3875, USA

Walter D. Neumann
Department of Mathematics, Columbia University, New York, NY 10027, USA

Mathematics Subject Classification 2000:
primary: 11F70, 22E05, 22E50, 11−06, 22−06; *secondary:* 11F67, 11F72, 11F27, 11F33

Keywords:
automorphic representations, representation theory of *p*-adic groups, Rankin-Selberg L-functions, relative trace formula

∞ Printed on acid-free paper which falls within the guidelines of the ANSI
to ensure permanence and durability.

Library of Congress Cataloging-in-Publication Data

Automorphic representations, L-functions, and applications : progress and prospects : proceedings of a conference honoring Steve Rallis on the occasion of his 60th birthday, the Ohio State University, March 27−30, 2003 / edited by James W. Cogdell ... [et al.].
p. cm. − (Ohio State University Mathematical Research Institute publications ; 11)
ISBN-13: 978-3-11-017939-2 (hardcover : acid-free paper)
ISBN-10: 3-11-017939-3 (hardcover : acid-free-paper)
1. L-functions − Congresses. 2. Automorphic forms − Congresses. 3. Representations of groups − Congresses. I. Rallis, Stephen, 1942− II. Cogdell, James W., 1953− III. Series.
QA246.A98 2005
512.7'3−dc22

2005025652

ISSN 0942-0363
ISBN-13: 978-3-11-017939-2
ISBN-10: 3-11-017939-3

Bibliographic information published by Die Deutsche Bibliothek

Die Deutsche Bibliothek lists this publication in the Deutsche Nationalbibliografie;
detailed bibliographic data is available in the Internet at <http://dnb.ddb.de>.

Printed in Germany.
Typeset using the authors' TEX files: M. Pfizenmaier, Berlin.
Printing and binding: Hubert & Co. GmbH & Co. KG, Göttingen.
Cover design: Thomas Bonnie, Hamburg.

Preface

This volume is the proceedings of the conference on Automorphic Representations, L-functions and Applications: Progress and Prospects, held at the Department of Mathematics of The Ohio State University, March 27–30, 2003, in honor of the 60th birthday of Steve Rallis.

The term proceedings is used here in the sense that these 14 contributed papers reflect many of the main themes and directions of the conference. Among the topics covered are:

Rankin–Selberg L-functions (Bump, Ginzburg–Jiang–Rallis, Lapid–Rallis)

the relative trace formula (Jacquet, Mao–Rallis)

automorphic representations (Gan–Gurevich, Ginzburg–Rallis–Soudry)

representation theory of p-adic groups (Baruch, Kudla–Rallis, Mœglin, Cogdell–Piatetski-Shapiro–Shahidi)

p-adic methods (Harris–Li–Skinner, Vigneras)

and arithmetic applications (Chinta–Friedberg–Hoffstein).

The continuing vigor and diversity of research on automorphic representations and their applications to arithmetic are clearly reflected here. Also reflected are the depth and breadth of Rallis's influence. His vision and energy have been a remarkable source of inspiration for many other researchers. We hope that he will enjoy the harvest of results contained in this volume.

We gratefully acknowledge the financial support of the National Science Foundation, the Institute for Mathematics Applications (IMA), the Office of the Vice President for Research of The Ohio State University, and the Mathematics Research Institute of the Department of Mathematics of The Ohio State University. Also, we want to recognize the skilled staff of the Department, Ms. Karen Blessing, Ms. Denise Clark and Ms. Marilyn Radcliff, whose professional expertise made possible the smooth functioning at the conference.

The Organizing Committee:

James Cogdell
Dihua Jiang
Steve Kudla
David Soudry
Robert J. Stanton

Table of contents

Bessel functions for $GL(n)$ over a p-adic field

Ehud Moshe Baruch

Abstract. We attach Bessel functions to generic representations of $GL_n(F)$ where F is a p-adic field and show that they are given locally by orbital integrals.

1. Introduction and main results

Let F be a non-archimedean local field. In [3] we attached Bessel functions to every generic representation of a quasi-split reductive group over F using a distribution approach similar to Harish-Chandra's approach for the character functions. In the present paper we attach Bessel functions to generic representations of $GL_n(F)$ using a Whittaker integral method similar to the one in [6],[13],[1], [5] and generalizing the results in [4]. As in [4] we show that these Bessel functions are given locally by orbital integrals. Hence it follows from [12] that they have an asymptotic expansion in terms of the Jacquet-Ye germs.

Acknowledgments. I thank J. Cogdell, H. Jacquet, I. Piatetski-Shapiro, and S. Rallis for sharing their insight with me and their constant encouragement.

1.1. Main results We state here our main theorems. We shall only consider here the main Bessel function of a representation which is the one attached to the open Bruhat cell. Other Bessel functions are described in Section 8.

Let $G = GL_n(F)$ and let B be the Borel subgroup of upper triangular matrices, A the subgroup of diagonal matrices and N the subgroup of upper unipotent matrices. Let ψ be a non-degenerate character of N. Let $\mathbb{W} = N(A)/A$ be the Weyl group where $N(A)$ is the normalizer of A. We identify $\mathbb{W}$ with the set of permutation matrices in $N(A)$. This set is also identified with S_n, the symmetric group on n-letters in a natural way. Let

$$w_0 = \begin{pmatrix} & & & 1 \\ & & 1 & \\ & \cdot & & \\ 1 & & & \end{pmatrix} \tag{1}$$

be the longest Weyl element in $\mathbf{W}$. Let (π, V) be an irreducible admissible representation of G. We say that π is generic if there exists a nonzero functional $L : V \to C$ such that

$$L(\pi(n)v) = \psi(n)L(v) \quad n \in N,\ v \in V.$$

It is well known that such a functional is unique up to scalar multiples. We call this functional a ψ Whittaker functional. Now define

$$W_v(g) = L(\pi(g)v) \quad v \in V,\ g \in G. \tag{2}$$

and let G act on the space of these functions by right translations. That is, if $g_1 \in G$ and W is a function on G then we define

$$(\rho(g_1)W)(g) = W(gg_1), \quad g \in G. \tag{3}$$

Then the map $v \to W_v$ gives a realization of π on a space of Whittaker functions satisfying

$$W(ng) = \psi(n)W(g) \quad n \in N,\ g \in G.$$

We denote this space by $\mathcal{W}(\pi, \psi)$. In Section 3 we define the subspace $\mathcal{W}^0(\pi, \psi)$ of $\mathcal{W}(\pi, \psi)$. In the case where π is supercuspidal we have that $\mathcal{W}^0(\pi, \psi) = \mathcal{W}(\pi, \psi)$. Let $\alpha_1, \dots, \alpha_{n-1}$ be the positive roots realized as functions on A (See (19)). Let $M > 0$ be a constant and let

$$A^M = A^M(w_0) = \{a \in A : \alpha_i(a) < M, i = 1, 2, \dots, n-1\}. \tag{4}$$

Our first main theorem is the following.

Theorem 1.1. *Let $W \in \mathcal{W}^0(\pi, \psi)$ and M a positive constant. Then the function*

$$(a, n) \mapsto W(aw_0n)$$

defined on the set $A^M \times N$ is compactly supported in N. That is, if $W(aw_0n) \neq 0$ and $a \in A^M, n \in N$ then n is in some compact set independent of a.

Since A^M cover A as $M \to \infty$ we get the following corollary.

Corollary 1.2. *Let π be a supercuspidal representation of G and $W \in \mathcal{W}(\pi, \psi)$ a Whittaker function associated to π. Fix $g \in Bw_0B$. Then the function*

$$n \mapsto W(gn)$$

is compactly supported in N.

Proof. Write $g = n_1 a w_0 n_2$ and choose M large enough such that $a \in A^M$. Since $W(gn) = \psi(n_1)W(aw_0n_2n)$ the result follows from Theorem 1.1 □

This result allows us to define Bessel functions for supercuspidal representations (See Section 6). In order to treat all irreducible admissible representations we will need the following result which allows us to move from $\mathcal{W}(\pi, \psi)$ to $\mathcal{W}^0(\pi, \psi)$.

Theorem 1.3. *Let $W \in \mathcal{W}(\pi, \psi)$. There exists a compact open subgroup $N_0 = N_0(W)$ of N such that the function $W_{N_0,\psi} \in \mathcal{W}^0(\pi, \psi)$.*

Here $W_{N_0,\psi}$ is defined by

$$W_{N_0,\psi}(g) = \int_{N_0} W(gn)\psi^{-1}(n)dn.$$

Corollary 1.4.

$$\mathcal{W}^0(\pi, \psi) \neq \{0\}.$$

Proof. Let $W \in \mathcal{W}(\pi, \psi)$ be such that $W(e) \neq 0$. Then $W_{N_0,\psi}(e) \neq 0$ for every compact open subgroup N_0 in N. □

Let $N_1 \subset N_2 \subset N_3 \subset \ldots$ be a filtration of N with compact open subgroups N_i, $i = 1, 2, \ldots$, such that $N = \bigcup_{i=1}^{\infty} N_i$. We denote this filtration by $\mathcal{N}$. Let $f : N \to \mathbb{C}$ be a locally constant function.

Definition 1.5.

$$\int_N^{\mathcal{N}} f(n)dn = \lim_{m\to\infty} \int_{N_m} f(n)dn$$

if this limit exists.

Corollary 1.6. *Let $\mathcal{N} = \{N_i, i > 0\}$ be a filtration of N as above. Let $g \in Bw_0B$ and $W \in \mathcal{W}(\pi, \psi)$. Then*

$$\int_N^{\mathcal{N}} W(gn)\psi^{-1}(n)dn$$

is convergent, and the value is independent of the choice of filtration $\mathcal{N}$.

Proof. Let $N_0 = N_0(W)$ be a compact open subgroup of N as in Theorem 1.3. There exists an integer M such that $N_0 \subset N_m$ for all $m > M$. Let $m > M$. We have

$$\frac{1}{\mathrm{vol}(N_0)} \int_{N_m} W_{N_0,\psi}(gn)\psi^{-1}(n)dn \tag{5}$$

$$= \frac{1}{\mathrm{vol}(N_0)} \int_{N_m} \int_{N_0} W(gn_1n_2)\psi(n_1n_2)dn_1dn_2.$$

Applying Fubini and changing variables $n = n_1n_2$ we get that the last integral is the same as

$$\int_{N_m} W(gn)\psi^{-1}(n)dn.$$

By Theorem 1.1 the function $n \mapsto W_{N_0,\psi}(gn)$ is compactly supported in N, hence we can take the limit $m \to \infty$ in (5) to get the value

$$\frac{1}{\mathrm{vol}(N_0)} \int_{N} W_{N_0,\psi}(gn)\psi^{-1}(n)dn.$$

It is clear that this value is independent of the filtration $\mathcal{N}$. □

Let $g \in Bw_0B$ and define the linear functional $L_g : V \to \mathbb{C}$ by

$$L_g(v) = \int_N^{\mathcal{N}} W_v(gn)\psi^{-1}(n)dn.$$

It is easy to see that L_g is a Whittaker functional, hence it follows from the uniqueness of Whittaker functionals that there exists a scalar $j_{\pi,\psi}(g)$ such that

$$L_g(v) = j_{\pi,\psi}(g)L(v) \tag{6}$$

for all $v \in V$. We call $j_\pi = j_{\pi,\psi}$ the Bessel function of π. (See Section 8 for other Bessel functions). The Bessel function j_π is defined on the set Bw_0B and we will prove that it is locally constant there. It is clear that $j_\pi(g)$ satisfies

$$j_\pi(n_1gn_2) = \psi(n_1n_2)j_\pi(g), \quad n_1, n_2 \in N, \, g \in Bw_0B,$$

hence it is determined by its values on the set Aw_0. As is the case with the character of the representation [**7**], the Bessel function j_π is expected to be locally integrable on G. Harish-Chandra's proof of the local integrability of the character depended on certain relations between the asymptotics of the character and certain orbital integrals. In this paper we establish that the asymptotics of j_π are the same as the asymptotics of certain orbital integrals which were studied by Jacquet and Ye [**12**]. We now describe the relation between the Bessel functions and orbital integrals.

Let $C_c^\infty(G)$ be the space of locally constant and compactly supported functions on G. Let Z be the center of G and let ω be a quasi character on Z.

For $\phi \in C_c^\infty(G)$ and $g \in Bw_0B$ we define the orbital integral (see **[12]**, (6))

$$J_{\psi,\omega}(g,\phi) = \int_{N\times Z\times N} \phi(n_1zgn_2)\psi^{-1}(n_1n_2)\omega^{-1}(z)dn_1dn_2dz.$$

It follows from **[12]** that this integral converges absolutely and defines a locally constant function on Bw_0B.

Theorem 1.7. *Let π be an irreducible admissible representation of G with central character ω_π. Let $x \in G$. There exists a neighborhood U_x of x in G and a function $\phi \in C_c^\infty(G)$ such that*

$$j_{\pi,\psi}(g) = J_{\psi,\omega_\pi}(g,\phi)$$

for all $g \in U_x$.

Remark 1.8. Since $j_{\pi,\psi}$ and J_{ψ,ω_π} are only defined on Bw_0B we are really asserting the equality on $Bw_0B \cap U_x$. Another option is to define these functions as having value zero outside of Bw_0B in which case the equality above does hold. In any case, the equality is true up to a set of measure zero.

Corollary 1.9. *If $g \mapsto J_{\psi,\omega_\pi}(g,\phi)$ is locally integrable as a function on G for every $\phi \in C_c^\infty(G)$ then $j_{\pi,\psi}$ is locally integrable on G.*

Hence the question of local integrability of the Bessel function reduces to the question of the local integrability of the orbital integral.

Our paper is divided as follows. In Section 2 we introduce some notations including roots, weights and Bruhat ordering. In Section 3 we study some cones and dual bases in a Euclidean space. These will be applied later for different bases of roots and weights. In Section 4 we describe the method of proof used for our main results and prove a result which is needed later using this method. In Section 5 we prove a more general version of Theorem 1.1. In Section 6 we define Bessel functions for supercuspidal representations. In Section 7 we prove Theorem 1.3. In Section 8 we define Bessel functions for general generic representations, including Bessel functions attached to other Weyl elements. In Section 9 we prove Theorem 1.7 and in Section 10 we indicate how to generalize our results to simply laced groups.

2. Notations and preliminaries

Let F be a non-archimedean local field. Let O be the ring of integers in F and let P be the maximal ideal in F. Let ϖ be a generator of P. We denote by

$|x|$ the normalized absolute value of $x \in F$. Let $q = |O/P|$ be the order of the residue field of F. Then $|\varpi| = q^{-1}$. Let $G = GL_n(F)$ and let A be the group of diagonal matrices in G consisting of matrices of the form

$$d(a_1, a_2, ..., a_n) = \begin{pmatrix} a_1 & & & & \\ & a_2 & & & \\ & & \cdot & & \\ & & & \cdot & \\ & & & & a_n \end{pmatrix}.$$

We let

$$Z = Z(G) = \{d(a, a, \ldots, a) : a \in F^*\}. \tag{7}$$

Let $X(A) = \mathrm{Hom}_F(A, F)$ be the group of rational homomorphisms. Then each $\alpha \in X(A)$ is of the form

$$\alpha(d(a_1, a_2, \ldots, a_n)) = a_1^{k_1} a_2^{k_2} \ldots a_n^{k_n}$$

with $k_1, k_2, \ldots, k_n \in \mathbf{Z}$. We view $X(A)$ as a group under addition where the addition is defined by

$$(\alpha + \beta)(a) = \alpha(a)\beta(a), \quad \alpha, \beta \in X(A), a \in A. \tag{8}$$

We let $|X| = X(A) \otimes \mathbf{R}$. Then we shall identify $|X|$ with the vector space of functions $|\alpha| = |\alpha|_{\lambda_1, \lambda_2, \ldots, \lambda_n}$ from A to $\mathbf{R}$ of the form

$$|\alpha|(d(a_1, a_2, \ldots, a_n)) = |a_1|^{\lambda_1} |a_2|^{\lambda_2} \ldots |a_n|^{\lambda_n} \tag{9}$$

where $\lambda_1, \lambda_2, \ldots, \lambda_n \in \mathbf{R}$. Here addition is defined as in (8) and scalar multiplication is defined by

$$(\lambda|\alpha|)(a) = (|\alpha|(a))^\lambda, \quad |\alpha| \in |X|, a \in A, \lambda \in \mathbf{R}.$$

We define an inner product on $|X|$ by

$$< \alpha_{\lambda_1, \lambda_2, \ldots, \lambda_n}, \alpha_{\gamma_1, \gamma_2, \ldots, \gamma_n} >= \sum_{i=1}^{n} \lambda_i \gamma_i. \tag{10}$$

For $i, j \in \{1, 2, \ldots, n\}$, $i \neq j$ we let $\alpha_{i,j} : A \to F$ be the functions defined by

$$\alpha_{i,j}(d(a_1, a_2, \ldots, a_n)) = \frac{a_i}{a_j}$$

and $|\alpha|_{i,j}(a) = |\alpha_{i,j}(a)|$. Let $\Phi = \{\alpha_{i,j}\}$. Then $|\Phi| = \{|\alpha|_{i,j}\}$ is a root system in $|X|$. We have that $\Phi = \Phi^+ \cup \Phi^-$ where $\Phi^+ = \{\alpha_{i,j} : i < j\}$ is the set of positive roots and Φ^- is the set of negative roots. Let $E_{i,j}$ be the matrix whose (i,j)th entry is 1 and all other entries are zero. For $\alpha = \alpha_{i,j} \in \Phi$ and for $b \in F$ we let

$$x_\alpha(b) = x_{i,j}(b) = I + bE_{i,j}$$
$$h_\alpha(b) = h_{i,j}(b) = bE_{i,i} - b^{-1}E_{j,j}.$$

For each $\alpha \in \Phi$ we let $N_\alpha = \{x_\alpha(b) : b \in F\}$. Let $\mathbf{W} = N(A)/A$ be the Weyl group of G. We shall identify $\mathbf{W}$ with S_n, the symmetric group. In particular if $\sigma \in S_n$ then we let w_σ be the associated permutation matrix. In particular, $w_{(i,j)}$ is the permutation matrix having 1s in the (i,j) and (j,i) entries and in the (k,k) entries for $k \neq i,j$. $\mathbf{W}$ acts on Φ and $|\Phi|$ in a natural way. We have that if $i \neq j$ then

$$a x_\alpha(b) a^{-1} = x_\alpha(\alpha(a)b), \quad a \in A, b \in F, \tag{11}$$
$$x_\alpha(b) x_{-\alpha}(-b^{-1}) x_\alpha(b) = w_\alpha h_\alpha(b), \quad b \in F, \tag{12}$$
$$w x_\alpha(b) w^{-1} = x_{w(\alpha)}(b), \quad w \in \mathbf{W}, b \in F. \tag{13}$$

Also, if $\alpha, \beta \in \Phi$ and $\alpha \neq \pm\beta$ then

$$x_\alpha(b_1) x_\beta(b_2) = x_{\alpha+\beta}(\epsilon b_1 b_2) x_\beta(b_2) x_\alpha(b_1) \tag{14}$$

where $\epsilon = \pm 1$ and $x_{\alpha+\beta}(r) = e$ if $\alpha + \beta \notin \Phi$. Let N be the subgroup of upper unipotent matrices. Then every $n \in N$ can be written uniquely in the form

$$n = \prod_{i>j} x_{i,j}(b_{i,j})$$

where $b_{i,j} \in F$ and the order of the product is fixed. (Any fixed order is fine).

2.1. Roots, and weights The root system $|\Phi|$ spans a subspace $|V|$ of $|X|$ given by

$$|V| = \{|\alpha|_{\lambda_1,\lambda_2,\ldots,\lambda_n} : \lambda_1 + \lambda_2 + \ldots + \lambda_n = 0\}. \tag{15}$$

Then $\Delta = \{|\alpha|_{i,i+1} : i = 1, \ldots, n-1\}$ is a basis for $|V|$ consisting of simple roots. If $\mathcal{B}$ is a basis for $|V|$ then we denote by $\mathcal{B}^*$ the dual basis (up to positive scalars) with respect to (10). In particular, the fundamental weights $\lambda_1, \ldots, \lambda_{n-1}$ give Δ^*, the basis dual to Δ where

$$\lambda_1 = |\alpha|_{n-1,-1,-1,\ldots,-1}, \quad \lambda_2 = |\alpha|_{n-2,n-2,-2,-2,\ldots,-2}, \quad \lambda_{n-1} = |\alpha|_{1,1,\ldots,1,1-n}. \tag{16}$$

We write $\alpha_i = |\,\alpha|_{i,i+1}$. Then

$$\Delta = \{\alpha_1, \ldots, \alpha_{n-1}\} \tag{17}$$

and

$$\Delta^* = \{\lambda_1, \ldots, \lambda_{n-1}\}. \tag{18}$$

Notice that we have chosen λ_i so that $< \alpha_i, \lambda_j >= 0$ if $i \neq j$ and that $< \alpha_i, \lambda_i >> 0$. We now recall the three different notations that we have for the simple roots for future reference:

$$\alpha_i = |\alpha|_{i,i+1} = |\alpha|_{0,\dots,0,1,-1,0,\dots,0}. \tag{19}$$

Remark 2.1. It is easy to see that if $\alpha \in \Phi$ is a negative root and if $r \in F$ is such that $|r| \geq D$ for some constant $D > 0$ then there exists a constant $C = C_D > 0$ such that

$$\lambda(h_\alpha(r)) < C \tag{20}$$

for all $\lambda \in \Delta^*$. Moreover, if $|r| \geq 1$ then

$$\lambda(h_\alpha(r)) \leq 1. \tag{21}$$

Each positive root $\alpha \in \Phi$ can be written as a positive integral combination of simple roots, that is,

$$\alpha = \sum_{i=1}^{n-1} c_i \alpha_{i,i+1}$$

with c_i is a non negative integer. We define the height of the positive root α to be

$$\text{height}(\alpha) = \sum_{k=1}^{n-1} c_k.$$

If α is a negative root then we define height(α) = height$(-\alpha)$. It is easy to check that for $j > i$

$$\text{height}(\alpha_{i,j}) = j - i.$$

2.2. Bruhat ordering For each $\alpha \in \Phi$ we let $w_\alpha \in \mathbb{W}$ be the reflection associated with α. That is, $w_{\alpha_{i,j}} = w_{(i,j)}$ Since $\mathbb{W}$ is generated by the simple reflections $w_{(i,i+1)}$ it follows that each $w \in \mathbb{W}$ can be written as a product of simple reflections. Let $w \in \mathbb{W}$ and write

$$w = w_{\beta_1} w_{\beta_2} \cdots w_{\beta_l}, \quad \beta_i \in \Delta,\ i = 1, \dots, k. \tag{22}$$

If (22) is a minimal expression for w, then we define

$$l(w) = l, \tag{23}$$
$$S(w) = \{\beta_1, \dots, \beta_l\} \subseteq \Delta. \tag{24}$$

It is well known (see **[10]**) that $l(w)$ and $S(w)$ are independent of the the minimal decomposition (22). We define the Bruhat partial order on $\mathbb{W}$ by

$w' \le w \iff w'$ can be written as a subexpression of w, i.e,

$$w' = w_{\beta_{i_1}} w_{\beta_{i_2}} \cdots w_{\beta_{i_t}} \quad 1 \le i_1 < i_2 < \ldots < i_t \le l.$$

This Bruhat ordering does not depend on the choice of minimal expression in (22) (see [10]).

It is clear that if $w_1 \le w_2$ then $S(w_1) \subseteq S(w_2)$. It is well known that w_0 is the longest Weyl element in $\mathbf{W}$ and that $w_0 \ge w$ for all $w \in \mathbf{W}$. Also, by ([10] 5.9, example 3) we have that

$$w_1 \le w_2 \Leftrightarrow w_1 w_0 \ge w_2 w_0.$$

It will be convenient to use the following notation:

Definition 2.2.

$$S^0(w) = S(ww_0).$$

It follows from the above discussion that

$$w_1 \le w_2 \Rightarrow S^0(w_1) \supseteqq S^0(w_2). \tag{25}$$

We define

$$S^-(w) = \{\alpha \in \Phi^+ : w(\alpha) < 0\}, \quad S^+(w) = \{\alpha \in \Phi^+ : w(\alpha) > 0\}. \tag{26}$$

Let S be a subset of simple roots, that is, $S \subset \Delta$. Let $\Phi(S) \subset \Phi$ be the set of roots in Φ which are in the span of S. It is well known that $\Phi(S)$ is itself a root system. We say that a root in $\Phi(S)$ has support in S. We let $\mathbf{W}_S$ be the Weyl group associated with S and we identify $\mathbf{W}_S$ as the subgroup of $\mathbf{W}$ generated by the simple reflections w_α, $\alpha \in S$. We let w_S be the longest Weyl element in $\mathbf{W}_S$.

Let $w = w_{i,i+1}$ be a simple reflection. It is easy to see that w sends $\alpha = \alpha_{i,i+1}$ to $-\alpha$ and that α permutes all the other positive roots. The following lemma is well known.

Lemma 2.3.
(a) w permutes the positive roots which do not have support in $S(w)$.
(b) $S^-(w) \subset \Phi(S(w))$.
(c) If $\alpha \in \Phi(S(w))$ then $w(\alpha) \in \Phi(S(w))$.

Proof. (a) Write w as a minimal product of simple reflections. It is clear from the above remark on the simple reflections $w_{i,i+1}$ that each simple reflection in the decomposition of w permutes the positive roots with support not in S. Hence w also permutes the positive roots with support not in S.
(b) If α is positive and $w(\alpha)$ is negative then it follows from part (a) that α is supported on S.
(c) Since $w \in \mathbf{W}_S$ and $\alpha \in \Phi(S)$ it is clear that $w(\alpha) \in \Phi(S)$. □

Let $S^0(w)$ be defined as in (2.2).

Corollary 2.4. *Let $\alpha \in \Phi^+$ be such that $w(\alpha) > 0$ then $w(\alpha) \in \Phi(S^0(w))$.*

Proof. Let $w' = ww_0$. Since $w_0^2 = e$ we have $w'(w_0(\alpha)) = ww_0w_0(\alpha) = w(\alpha) > 0$. Since $w_0(\alpha) < 0$, it follows from Lemma 2.3 (b) that $w_0(\alpha) \in \Phi(S(w'))$. It follows from Lemma 2.3 (c) that $w'(w_0(\alpha)) \in \Phi(S(w'))$. Since $w'w_0 = w$ we get that $w(\alpha) \in \Phi(S(w')) = \Phi(S(ww_0)) = \Phi(S^0(w))$. □

Corollary 2.5. *Let $\alpha \in \Phi^+$ be such that $w(\alpha) < 0$. Let $w_1 = ww_\alpha$ then $w(\alpha) \in \Phi(S^0(w_1))$.*

Proof. We have that $w_1(\alpha) = -w(\alpha) > 0$. Since $w_1(\alpha) > 0$ it follows from Corollary 2.4 that $w_1(\alpha) \in \Phi(S^0(w_1))$. □

2.3. Bruhat decomposition

We define

$$N_w^- = \prod_{\alpha \in S^-(w)} N_\alpha, \quad N_w^+ = \prod_{\alpha \in S^+(w)} N_\alpha. \tag{27}$$

It is well known that $|S^-(w)| = l(w)$ and that $N = N_w^+ N_w^-$. The Bruhat decomposition of G is given by

$$G = \bigcup_{w \in \mathbf{W}} BwB.$$

Moreover, we have, $BwB = NAwN_w^-$ with uniqueness of expression. That is, every $g \in BwB$ can be uniquely written in the form $g = n_1awn_2$ with $n_1 \in N$, $a \in A$ and $n_2 \in N_w^-$. Hence, if $S^-(w) = \{\alpha_1, \ldots, \alpha_l\}$ then every $g \in BwB$ can be written uniquely in the form

$$g = nawx_{\alpha_1}(r_1) \cdots x_{\alpha_l}(r_l)$$

with $a \in A$,$n \in N$ and $r_1, \ldots, r_l \in F$. The following lemma will play a crucial role in the proofs of our main results in this paper. It supplies us with a tool to move from smaller Bruhat cells to larger Bruhat cells and to cover G in an inductive way.

Lemma 2.6. *Let $w \in \mathbf{W}$ and assume that $S^-(w) = \{\alpha_1, \ldots, \alpha_l\}$. Assume also that height(α_i) $\geq$ height(α_{i+1}) for $i = 1, \ldots, l-1$. Let $g \in BwB$ and assume that*

$$g = nawx_{\alpha_1}(r_1) \cdots x_{\alpha_t}(r_t)$$

with $t \leq l$. Assume also that $r_t \neq 0$. Let

$$g_1 = gx_{-\alpha_t}(-1/r_t).$$

Then $g_1 \in Bw_1B$ *with* $w_1 = ww_{\alpha_t}$ *and in particular* $w_1 < w$. *Moreover, if* $g_1 = n_1a_1w_1n_2$ *is the unique decomposition of* g_1 *with* $n_1 \in N$, $a_1 \in A$ *and* $n_2 \in N^-_{w_1}$ *then*

$$a_1 = ah_{w(\alpha_t)}(r_t).$$

Proof. By (12) we have $x_{\alpha_t}(r_t)x_{-\alpha_t}(-1/r_t) = w_{\alpha_t}h_{\alpha_t}(r_t)x_{\alpha_t}(-r_t)$. Hence

$$g_1 = nawx_{\alpha_1}(r_1)\cdots x_{\alpha_{t-1}}(r_{t-1})w_{\alpha_t}h_{\alpha_t}(r_t)x_{\alpha_t}(-r_t) \tag{28}$$

Since $w_{\alpha_t} = w^{-1}_{\alpha_t}$ it follows from (13) that

$$w^{-1}_{\alpha_t}x_{\alpha_i}(r_i)w_{\alpha_t} = x_{w_{\alpha_t}(\alpha_i)}(r_i).$$

Since height$(\alpha_i) \geq$ height(α_t) for $i < t$ we have that $w_{\alpha_t}(\alpha_i) > 0$. (To see this, write $\alpha_t = \alpha_{a,b}$, $\alpha_i = \alpha_{c,d}$ and $w_{\alpha_t} = w_{a,b}$. Since height$(\alpha_i) \geq$ height(α_t) it follows that $d - c > b - a$. The claim now follows from a case by case computation.) Hence by conjugating $w_{\alpha_t}h_{\alpha_t}$ across the expression in (28) we get that

$$g = na_1w_1n_3$$

with a_1 and w_1 as defined above and $n_3 \in N$. Hence it is clear that $g \in Bw_1B$. To get the unique form of g we decompose $n_3 = n_3^+n_3^-$ with $n_3^+ \in N^+_{w_1}$ and $n_3^- \in N^-_{w_1}$. We can move n_3^+ across a_1w_1 by conjugating to get the unique form of g_1. It is clear that a_1 gives the required torus part. □

3. Bases and cones in a Euclidean space

We recall some facts about dual bases and cones in a Euclidean space. We shall apply these facts to the base Δ of $|V|$ defined in Section 2. Let E be an m dimensional vector space over $\mathbb{R}$ equipped with an inner product $< u, v >$. If $S = \{v_1, ..., v_r\}$ is a set of linearly independent vectors in E then we let $S^* = \{v_1^S, \ldots, v_r^S\}$ where v_i^S, $i = 1, \ldots, r$, is in the linear subspace spanned by S and is determined by the following equations

$$< v_i^S, v_j >= \delta_{i,j}, \quad j = 1, \ldots, r. \tag{29}$$

Remark 3.1. In most cases we will be satisfied by finding a vector w_i in the linear span of S satisfying $< w_i, v_j >= 0$ for $i \neq j$ and $< w_i, v_i >> 0$. It is clear that w_i is a positive multiple of v_i^S and we will not bother normalizing w_i.

Remark 3.2. ([9], pg. 72 ex.7 and ex.8) Write $v_i^S = \sum c_{i,j} v_j$. Then $c_{i,i} > 0$. Moreover, if S is an obtuse set of vectors, that is, $< v_i, v_j > \leq 0$ for $i \neq j$ then $c_{i,j} \geq 0$ and $< v_i^S, v_j^S > \geq 0$ for every i and j.

Let $\Delta = \{v_1, \ldots, v_m\}$ be a fixed base of E. For $v \in \Delta$ we denote $v^* = v^\Delta$. Let $S \subset \Delta$.

Definition 3.3. We denote by $\Delta(S)$ the set of m vectors where we replaced v with v^* when $\alpha \notin S$, That is, $\Delta(S) = \{w_1, w_2, \ldots, w_m\}$ is given by

$$w_i = \begin{cases} v_i & \text{if } v_i \in S; \\ v_i^* & \text{if } v_i \notin S. \end{cases}$$

Lemma 3.4.

(a) $\Delta(S)$ is a basis for E.

(b) The dual basis $\Delta(S)^* = \{u_1, \ldots, u_m\}$ is given up to positive scalar multiplications (see Remark 3.1) by

$$u_i = \begin{cases} v_i^S & \text{if } v_i \in S; \\ v_i^{S \cup \{v_i\}} & \text{if } v_i \notin S. \end{cases} \tag{30}$$

Proof. (a) Assume that $w = \sum c_i w_i = 0$ where $c_i \in \mathbf{R}$. Write $w = x_1 + x_2$ where $x_1 = \sum_{v_i \in S} c_i v_i$ and $x_2 = \sum_{v_j \notin S} c_j v_j^*$. It is clear that $< x_1, x_2 >= 0$ and since $w = 0$ we have $< x_1, w >=< x_2, w >= 0$. Hence $< x_1, x_1 >= 0$ and $< x_2, x_2 >= 0$ so $x_1 = x_2 = 0$. Since both Δ and Δ^* form a basis we get that $c_i = 0$ for all i.

(b) Assume $v_i \in S$ and let $u_i = v_i^S$. Then u_i is in the linear span of S hence $(u_i, v_j^*) = 0$ for all $v_j \notin S$. By definition $(u_i, v_j) = 0$ for all $v_j \in S$, $v_j \neq v_i$ and $(u_i, v_i) = 1$

If $v_i \notin S$ then $u_i = v_i^{S \cup \{v_i\}}$ is in the linear span of $S \cup \{v_i\}$. So $(u_i, v_j^*) = 0$ for all $v_j \notin S \cup \{v_i\}$. If $v_j \in S$ then by definition $(u_i, v_j) = 0$.

Since $\Delta(S)$ is a basis, and $u_i \neq 0$ we must have $< u_i, v_i^* > \neq 0$. If we write $u_i = cv_i + \sum_{v_j \in S} c_j v_j$ for $c, c_j \in \mathbf{R}$ then we have that $< u_i, v_i^* >= c$ and by Remark 3.2, $c > 0$. ☐

3.1. Polyhedral cones Let S be a finite set of vectors in E. We define the cones

$$C(S) = \{\sum_{v \in S} c_v v : \ c_v \geq 0, \},$$

$$C^*(S) = \{u \in E : < u, v > \geq 0, v \in S\}.$$

If S is minimal then S is called a basis for $C(S)$. It is a well known theorem that these two representations of polyhedral cones are equivalent, that is, for every S there exist finite sets $T_1, T_2 \subset E$ such that $C(S) = C^*(T_1)$ and $C^*(S) = C(T_2)$. When $S = \Delta$ is a basis of E, this theorem is easy to prove and is summarized in the following lemma:

Lemma 3.5.

$$C(\Delta) = C^*(\Delta^*), \quad C^*(\Delta) = C(\Delta^*).$$

We now assume that $\Delta = \{v_1, \dots, v_m\}$ is an obtuse base of E, that is,

$$< v_i, v_j > \leq 0, \quad i \neq j.$$

Notice that our base Δ of $|V|$ of simple roots defined in (17) is obtuse.

By Remark 3.2 it follows that if Δ is an obtuse base then Δ^* is an acute base and

$$C(\Delta^*) \subseteq C(\Delta). \tag{31}$$

We will also need the following Lemma:

Lemma 3.6. *Let Δ be an obtuse base of E and let $S \subset \Delta$. Then*

$$C^*(S \cup \Delta^*) = C^*(\Delta(S)) = C(\Delta(S)^*).$$

Proof. The second equality follows from the fact that $\Delta(S)$ is a basis for E (see Lemma 3.4 (a)) and from Lemma 3.5. Here we do not need Δ to be an obtuse basis.

Since $S \cup \Delta^* \supseteq \Delta(S)$ it follows that

$$C^*(S \cup \Delta^*) \subseteq C^*(\Delta(S)).$$

To finish the proof will show that

$$C(\Delta(S)^*) \subseteq C^*(S \cup \Delta^*).$$

Let $\Delta = \{v_1, ..., v_m\}$ and $\Delta(S)^* = \{u_1, ..., u_m\}$ where by (30)

$$u_i = \begin{cases} v_i^S & \text{if } v_i \in S; \\ v_i^{S \cup \{v_i\}} & \text{if } v_i \notin S. \end{cases}$$

Let $u \in C(\Delta(S)^*)$. Then $u = \sum c_i u_i$ with $c_i \geq 0$ for all i. To show that $u \in C^*(S \cup \Delta^*)$ it is enough to show that $< u_i, x > \geq 0$ for all $x \in S \cup \Delta^*$.

(i) Assume that $v_i \in S$, hence $u_i = v_i^S$.

If $x = v_j \in S$ then

$$< u_i, x > = < v_i^S, v_j > = \delta_{i,j} \geq 0.$$

If $x = v_j^* \in \Delta^*$ and $v_j \notin S$ then

$$< u_i, x >=< v_i^S, v_j^* >= 0$$

since v_i^S is in the span of S. If $x = v_j^* \in \Delta^*$ and $v_j \in S$ then we write $u_i = v_i^S = \sum_{v_t \in S} d_t v_t$. Since S is a set of linearly independent obtuse vectors it follows from Remark 3.2 that $d_i \geq 0$. Hence

$$< u_i, x >=< v_i^S, v_j^* >= \sum_{v_t \in S} d_t < v_t, v_j^* >= d_j < v_j, v_j^* > \geq 0.$$

(ii) Assume $u_i = v_i^{S \cup \{v_i\}}$ where $v_i \notin S$. Similar arguments as above will show that $< u_i, x > \geq 0$ for all $x \in S \cup \Delta^*$ hence we are done. □

4. Method of proof

The main method of proof in this paper is to use the Bruhat decomposition for a cell by cell analysis of the functions that we are interested in. It is important to understand how the Bruhat decomposition compares with the Iwasawa decomposition.

We present an explicit method of obtaining such information which follows a simple pattern. The idea is to analyze the Bruhat cells inductively going from the closed cell up to the open cell. The induction is on the height of the respective Weyl element. Another induction takes place inside an individual cell where we "peel" the root groups one by one. For this process we shall appeal repeatedly to Lemma 2.6 which allows us to obtain information on a larger cell from a smaller cell.

The main results in this paper are proved using this method. In this section we illustrate the method by proving a result that we will need later. This result is probably known to experts. For the case of $GL_3(F)$ see ([**4**], Section 3).

4.1. Iwasawa decomposition Let $K = GL_n(O)$. It is well known that

$$G = NAK.$$

For every $|\alpha| \in |X|$ we extend $|\alpha|$ (see [**11**]) to G by defining

$$|\alpha|(g) = |\alpha|(a) \tag{32}$$

where $g = nak$, $n \in N, a \in A, k \in K$, is an Iwasawa decomposition of g. It is easy to see that $|\alpha|$ is independent of the choice of decomposition.

Recall that $\Delta^* = \{\lambda_1, \ldots, \lambda_n\}$ is the set of fundamental weights where $\lambda_i = |\alpha|_{n-i,\ldots,n-i,i,\ldots,i}$ (See (9).) We view λ_i as a function on G as above. The main theorem of this section is the following:

Theorem 4.1. *Let* $\lambda \in \Delta^*$ *and* $w \in \mathbf{W}$. *Then*

$$\lambda(wn) \le 1$$

for every $n \in N_w^-$.

Remark 4.2. Since every $n \in N$ can be written in the form $n = n_+ n_-$ for $n_+ \in N_w^+$ and $n_- \in N_w^-$ it follows that

$$wn = wn_+n_- = n_+^w wn_-$$

where $n_+^w = w^{-1}n_+w \in N$. Hence $\lambda(wn) = \lambda(wn_-)$. It follows that the statement in the above Theorem is equivalent to the statement $\lambda(wn) \le 1$ for all $n \in N$ and $\lambda \in \Delta^*$.

Proof. We will proceed with two inductions. The first induction is on $l(w)$.

$l(w) = 0$:

In this case $w = e$, $N_w^- = \{e\}$. Hence, $wn = e$. Since $\lambda(e) = 1$ we are done.

Now let $w \in \mathbf{W}$ be such that $l(w) > 0$ and assume that the Theorem is true for all $w_1 \in \mathbf{W}$ such that $l(w_1) < l(w)$.

We order the roots in

$$S^-(w) = \{\alpha \in \Phi^+ : w(\alpha) < 0\} = \{\alpha_1, \ldots, \alpha_l\}$$

as in Lemma 2.6 so that height$(\alpha_i) \ge$ height(α_{i+1}) for $i = 1, \ldots, l-1$. If $n \in N_w^-$ then we can write

$$n = x_{\alpha_1}(r_1)x_{\alpha_2}(r_2)\cdots x_{\alpha_t}(r_t) \tag{33}$$

with $r_1, \ldots, r_t \in F$ and $t \le l$. Notice that we can always take $t = l$ at the cost of having the last r_is being zeroes. However, we are interested in having t as small as possible. We would like to prove that $\lambda_i(wn) \le 1$. We shall do so by induction on t.

Since w is a permutation matrix whose entries are 0 or 1 it follows that $w \in K$. Hence if $t = 0$ then $wn = w$ and $\lambda_i(w) = 1$. So assume that n is of the form (33) with $t > 0$ and assume that the Theorem is true for $t-1$. We divide into two cases.

If $|r_t| \le 1$ then $x_{\alpha_t}(r_t) \in K$ hence

$$\lambda(wn) = \lambda(wx_{\alpha_1}(r_1)\cdots x_{\alpha_{t-1}}(r_{t-1})x_{\alpha_t}(r_t)) = \lambda(wx_{\alpha_1}(r_1)\cdots x_{\alpha_{t-1}}(r_{t-1})).$$

Hence we can use our second induction assumption to conclude that $\lambda(wn) \le 1$.

If $|r_t| \geq 1$ then $x_{-\alpha_t}(-r_t^{-1}) \in K$. Hence we have

$$\lambda(wn) = \lambda(wnx_{-\alpha_t}(-r_t^{-1})) = \lambda(wx_{\alpha_1}(r_1) \cdots x_{\alpha_t}(r_t)x_{-\alpha_t}(-r_t^{-1})).$$

By Lemma 2.6 we have that $wnx_{-\alpha_t}(-r_t^{-1}) = n_1 a_1 w_1 n_2$ with $w_1 < w_2$, $n_1, n_2 \in N$ and $a_1 = h_{w(\alpha_t)}(r_t)$. Hence

$$\lambda(wn) = \lambda(h_{w(\alpha_t)}(r_t))\lambda(w_1 n_2).$$

By Remark 2.1, we have $\lambda(h_{w(\alpha_t)}(r_t)) \leq 1$. By our first induction assumption we have $\lambda(w_1 n_2) \leq 1$. Hence we get the result. □

5. Spaces of Whittaker functions

In this section we define a subspace of the space of Whittaker functions on G and prove some properties of this subspace. In particular we prove Theorem 5.7 which asserts that certain functions on unipotent subgroups are compactly supported. This is one of our main theorems in this paper.

5.1. Whittaker functions Let ψ_F be a character of F and assume that ψ_F is identically one on O and nontrivial on P^{-1}. For a unipotent matrix $y \in N$ we set

$$\psi(y) = \psi_F(y_{1,2} + y_{2,3} + \ldots + y_{n-1,n}), \tag{34}$$

where $y_{i,j}$ are the entries of y. We let $\mathcal{W} = \mathcal{W}(G, \psi)$ be the set of functions $W : G \to C$ such that W is smooth on the right and

$$W(ng) = \psi(n)W(g), \quad n \in N, g \in G.$$

Examples of such functions are Whittaker functions associated with generic representations of G. Other examples are given by projecting compactly supported and locally constant functions to this space as follows.

$$W_f(g) = \int_N f(ng)\psi^{-1}(n)dn, \quad f \in C_c^\infty(G).$$

we shall study the space of such functions $\{W_f : f \in C_c^\infty(G)\}$ in Section 9.

For every $|\alpha| \in |X|$ we extend $|\alpha|$ to G as in (32).

For $w \in \mathbf{W}$ we let $S^0(w)$ be the set of simple roots defined in (2.2). That is, $S^0(w) = S(ww_0)$, where $S(ww_0)$ is defined in (24).

Definition 5.1. Let $\mathcal{W} = \mathcal{W}(G,\psi)$ be the space of Whittaker functions defined above. We define $\mathcal{W}^0 = \mathcal{W}^0(G,\psi) \subset \mathcal{W}$ to be the set of functions $W \in \mathcal{W}$ such that for every $w \in \mathbf{W}$ and every $\alpha \in S^0(w)$ there exist positive constants $D_\alpha < E_\alpha$ such that if $g \in BwB$ then

$$W(g) \neq 0 \Longrightarrow D_\alpha < |\alpha|(g) < E_\alpha, \quad \alpha \in S^0(w). \tag{35}$$

In other words, $W \in \mathcal{W}^0$ if for each $w \in \mathbf{W}$ and each $\alpha \in S^0(w)$ the support of W in BwB has bounded image under α.

Remark 5.2. The condition $|\alpha|(g) < E_\alpha$, $\alpha \in S^0(w)$ that appears in (35) is redundant since by **[11]** the support of W is contained in the set $\{g : |\alpha|(g) < C, \alpha \in \Delta\}$ for some positive number C. Moreover, if W is a Whittaker function in the Whittaker model of a supercuspidal representation of G then it follows from **[11]** that W is compactly supported mod NZ. Hence it follows that W satisfies condition (35) for every $\alpha \in \Delta$ and every $g \in G$ and in particular $W \in \mathcal{W}^0$.

Definition 5.3. Let $\alpha \in \Delta$. We define the sets

$$X_{C_1,C_2}(\alpha) = \{g \in G \mid C_1 < |\alpha|(g) < C_2\},$$
$$A_{C_1,C_2}(\alpha) = \{a \in A \mid C_1 < |\alpha|(a) < C_2\}$$

and the sets

$$X_{C_1,C_2} = \bigcap_{\alpha\in\Delta} X_{C_1,C_2}(\alpha), \quad A_{C_1,C_2} = \bigcap_{\alpha\in\Delta} A_{C_1,C_2}(\alpha).$$

Lemma 5.4. *Let $\alpha \in \Delta$, $C_1 < C_2$ positive numbers and R a compact set in G. Then*
(a) There exist constants $C_1' < C_2'$ such that

$$X_{C_1,C_2}(\alpha)R \subset X_{C_1',C_2'}(\alpha).$$

(b) Let Y be a subset of G and assume that for every $y \in Y$ there exists $r \in R$ such that $yr \in X_{C_1,C_2}(\alpha)$. Then there exist positive constants $C_1' < C_2'$ such that $Y \subset X_{C_1',C_2'}(\alpha)$.

Proof. (a) We can write $X_{C_1,C_2}(\alpha) = NA_{C_1,C_2}(\alpha)K$. It is clear that $|\alpha|(X_{C_1,C_2}(\alpha)R) = |\alpha|(A_{C_1,C_2}(\alpha))|\alpha|(KR)$. Since KR is a a compact set in G and $|\alpha|$ is continuous the result follows.
(b) Let $y \in Y$ and let $y = n_0a_0k_0$ be an Iwasawa decomposition for y. If $r \in R$ then $|\alpha|(yr) = |\alpha|(y)||\alpha|(k_0r)$. Since KR is a compact set, there exist positive constants $D_1 < D_2$ such that $D_1 < |\alpha(k_0r)| < D_2$ for all $k_0 \in K$ and $r \in R$. By our assumption, there exists $r_0 \in R$ such that $C_1 < |\alpha|(yr_0) < C_2$. Hence $C_1/D_2 < |\alpha|(y) < C_2/D_1$ and we can choose $C_1' = C_1/D_2$ and $C_2' = C_2/D_1$. □

Corollary 5.5.
(a) The set $\mathcal{W}^0$ *is invariant under right translations by* B*, i.e, if* $W \in \mathcal{W}^0$ *then for every* $b \in B$*,* $W_b \in \mathcal{W}^0$ *where* $W_b(g) = W(gb)$*.*
(b) W^0 *is invariant under right integration by compact open subset of closed subgroups of* B*, i.e, if* H *is a closed subgroup of* B *and* $X \subset H$ *is open and compact in* H *then for every* $W \in \mathcal{W}^0$*,* $W_X \in \mathcal{W}^0$ *where* $W_X(g) = \int_X W(gh)dh$*.*

Proof. (a) We take R to be the Singleton, $R = \{b^{-1}\}$, where $b \in B$. By Lemma 5.4, $X_{C_1,C_2}(\alpha)b^{-1} \subset X_{C_1',C_2'}(\alpha)$. Thus if W restricted to the set BwB is supported on $X_{C_1,C_2}(\alpha) \cap BwB$ then W_b restricted to BwB will be supported on the set $X_{C_1',C_2'}(\alpha) \cap BwB$.
(b) Since W is smooth on the right, W_X is a finite linear combination of W_{b_i} for some $b_i \in B$, hence (b) follows from (a). □

For each $w \in W$ we let $\Delta(S^0(w))$ be the basis of $|V|$ defined in (3.3) and let $\Delta(S^0(w))^*$ be the dual basis (up to scalars) that we fixed in Lemma 3.4 (b). Let M be a positive constant. We define the cone $A^M(w) \subset A$ to be

$$A^M(w) = \{a \in A : |\beta|(a) < M, \quad \text{for all } \beta \in \Delta(S^0(w))^*\}.$$

Lemma 5.6. *Let* $w_1, w \in \mathbf{W}$ *and* $M > 0$*. If* $w_1 < w$ *then there exists a constant* $M_1 > 0$ *such that*

$$A^M(w) \subset A^{M_1}(w_1).$$

Proof. By (25) and Lemma 3.6

$$C(\Delta(S^0(w_1))^*) \subseteq C(\Delta(S^0(w))^*).$$

Hence every $\lambda \in \Delta(S^0(w_1))^*$ can be written as a non-negative linear combination of elements in $\Delta(S^0(w))^*$. Thus there exists a constant $M_1 > 0$ such that $|\lambda|(a) < M_1$ for all $a \in A^M(w)$ and $\lambda \in \Delta(S^0(w_1))^*$. □

Our first main theorem of this paper is the following:

Theorem 5.7. *Let* $W \in \mathcal{W}^0$ *and* M *a positive constant. Then the function*

$$(a, n) \mapsto W(awn)$$

defined on the set $A^M(w) \times N_w^-$ *is compactly supported in* N_w^-*. That is, if* $W(awn) \neq 0$ *and* $a \in A^M(w), n \in N_w^-$ *then* n *is in some compact set independent of* a*.*

Note that if $w = w_0$ then $S^0(w) = \emptyset$ hence $\Delta(S^0(w)) = \Delta^*$ and $\Delta(S^0(w))^* = \Delta$. It follows that $A^M(w) = A^M$ as defined in (4). Since $N_w^- = N$ in that case, Theorem 1.1 follows from the above Theorem.

Proof. Our proof will use a double induction argument as in the proof of Theorem 4.1. We begin by induction on $l(w)$.

$l(w) = 0$: That is, $w = e$.

In this case, $N_w^- = \{e\}$ and there is nothing to prove. Now let $w \in \mathbf{W}$ and assume the Theorem is true for all $w_1 \in \mathbf{W}$ such that $l(w_1) < l(w)$.

We order the roots in

$$S^-(w) = \{\alpha_1, \ldots, \alpha_l\}$$

as in Lemma 2.6 so that height$(\alpha_i) \geq$ height(α_{i+1}) for $i = 1, \ldots, l-1$. If $n \in N_w^-$ then we can write

$$n = x_{\alpha_1}(r_1)x_{\alpha_2}(r_2)\cdots x_{\alpha_t}(r_t) \tag{36}$$

with $r_1, \ldots, r_t \in F$ and $t \leq l$. Here we will use induction on t. The precise induction statement is the following: Fix $W \in \mathcal{W}^0$ and M a positive constant. Let $n \in N_w^-$ be written as in (36) and $a \in A^M(w)$. If

$$W(awn) \neq 0$$

then there exists a constant $C = C(W, w, M) > 0$ independent of a such that $|r_i| \leq C$ for $i = 1, \ldots, t$.

Assume $t = 1$. Then we can write $n = x_{\alpha_1}(r_1)$. We assume $W(awn) \neq 0$ with $a \in A^M(w)$. Since W is smooth on the right, there exists $D > 0$ so that

$$W(gx_{-\alpha_1}(-r_1^{-1})) = W(g)$$

for every $r_1 \in F$ such that $|r_1| > D$ and every $g \in G$. Hence if $|r_1| > D$ then

$$W(awn) = W(awx_{\alpha_1}(r_1)x_{-\alpha_1}(-r_1^{-1})) \neq 0.$$

By Lemma 2.6 we have that $awx_{\alpha_1}(r_1)x_{-\alpha_1}(-r_1^{-1})) = n_1a_1w_1n_2$ with $w_1 = ww_{\alpha_1}$, $a_1 = ah_{w(\alpha_1)}(r_1)$ and $n_1, n_2 \in N$. More precisely, it is easy to see that in this case

$$awx_{\alpha_1}(r_1)x_{-\alpha_1}(-r_1^{-1})) = n_3ah_{w(\alpha_1)}(r_1)w_1$$

for some $n_3 \in N$. Set

$$\alpha_0 = w(\alpha_1).$$

We get that $W(awn) \neq 0$ implies that

$$W(ah_{\alpha_0}(r_1)w_1) \neq 0.$$

Set $S_1 = S^0(w_1)$. Since $w_1 \in K$ and since $W \in \mathcal{W}^0$ we have that for every $\beta \in C(S_1)$ there exists a positive constant D_β such that

$$\beta(ah_{w(\alpha_1)}(r_1)) \geq D_\beta. \tag{37}$$

Since $a \in A^M(w)$ and since $w_1 < w$ it follows from Lemma 5.6 that there exists $M_1 > 0$ such that $a \in A^{M_1}(w_1)$. Hence for every $\gamma \in C(\Delta(S_1)^*) = C^*(\Delta(S_1))$ there exists $E_\gamma > 0$ depending only on M_1 and γ such that

$$\gamma(a) \leq E_\gamma. \tag{38}$$

By Corollary 2.5 we have that $\alpha_0 \in \Phi(S_1)$. Since α_0 is negative it follows that

$$\alpha_0 = \sum_{\alpha \in S_1} c_\alpha \alpha \tag{39}$$

with $c_\alpha \leq 0$ for all $\alpha \in S_1$. If $\gamma \in C^*(\Delta(S_1))$ then it follows from the definition of $\Delta(S_1)$ that

$$< \gamma, \alpha > \geq 0, \quad \text{for all } \alpha \in S_1.$$

Hence it follows from (39) that for all $\gamma \in C^*(\Delta(S_1))$

$$< \gamma, \alpha_0 > \leq 0.$$

Since $C^*(\Delta(S_1)) = C(\Delta(S_1)^*)$ contains the basis $\Delta(S_1)^*$ and since $\alpha_0 \neq 0$ it follows that there exists $\gamma_0 \in C(\Delta(S_1)^*)$ such that

$$< \gamma_0, \alpha_0 > < 0. \tag{40}$$

Let $\Delta(S_1)^* = \{\gamma_1, \ldots, \gamma_{n-1}\}$ where γ_i is defined by (30). Then we can write

$$\gamma_0 = \sum d_i \gamma_i$$

with $d_i \geq 0$. Since $\alpha_0 \in \Phi(S_1)$ it follows that $< \alpha_0, \gamma_i > = 0$ for all i such that $\alpha_i \notin S_1$. Hence we can (and will) assume that $d_i = 0$ for i such that $\alpha_i \notin S_1$. (That is, we are replacing γ_0 with $\tilde{\gamma}_0 = \sum_{\alpha_i \in S} d_i \gamma_i$. It is clear that $\tilde{\gamma}_0 \in C(\Delta(S_1)^*)$ and that $< \tilde{\gamma}_0, \alpha_0 > < 0$).

Since S_1 is an obtuse set we get that $\gamma_i = \alpha_i^{S_1}$ is in $C(S_1)$ for all i such that $\alpha_i \in S_1$. Hence $\gamma_0 \in C(S_1)$. Hence by (37)

$$\gamma_0(ah_{\alpha_0)}(r_1)) \geq D_{\gamma_0}.$$

Since

$$\gamma_0(ah_{\alpha_0}(r_1)) = \gamma_0(a)\gamma_0(h_{\alpha_0}(r_1))$$

and since $\gamma_0(a) \leq E_{\gamma_0}$ by (38) we get that

$$\gamma_0(h_{\alpha_0}(r_1)) \geq \frac{D_{\gamma_0}}{E_{\gamma_0}}.$$

Write $\gamma_0 = |\alpha|_{\lambda_1,\ldots,\lambda_n}$ (see (9)) and $\alpha_0 = |\alpha|_{i,j}$ (see (19)). Then by (40)

$$< \gamma_0, \alpha_0 > = < |\alpha|_{\lambda_1,\ldots,\lambda_n}, |\alpha|_{i,j} > = \lambda_i - \lambda_j < 0.$$

On the other hand

$$\gamma_0(h_{\alpha_0}(r_1)) = \gamma_0(h_{i,j}(r_1)) = |r_1|^{\lambda_i-\lambda_j} \geq \frac{D_{\gamma_0}}{E_{\gamma_0}}.$$

Hence there exists $C > 0$ depending on W,w,α_1 and M but not on $a \in A^M(w)$ such that

$$W(awx_{\alpha_1}(r_1)) \neq 0 \Rightarrow |r_1| \leq C.$$

We now prove the general case. Let $t > 1$. Assume that our second induction hypothesis holds for $t-1$. Let $S^-(w) = \{\alpha_1, \ldots, \alpha_l\}$ with height$(\alpha_i) \geq$ height(α_{i+1}) and let $n \in N_w^-$ be of the form

$$n = x_{\alpha_1}(r_1)x_{\alpha_2}(r_2)\cdots x_{\alpha_t}(r_t).$$

Let $a \in A^M(w)$ and assume that

$$W(awn) \neq 0.$$

Let $D > 0$ be such that if $|r_t| \geq D$ then

$$W(gx_{-\alpha_t}(-r_t^{-1})) = W(g)$$

for all $g \in G$. Assume $|r_t| \geq D$. Then

$$W(awn) = W(awnx_{-\alpha_t}(-r_t^{-1})) \neq 0.$$

Let $g_1 = awnx_{-\alpha_1}(-r_t^{-1})$. and $\alpha_0 = w(\alpha_t)$. Then by Lemma 2.6

$$g_1 = n_1ah_{\alpha_0}(r_t)w_1n_2$$

with $n_1 \in N$, $w_1 = ww_{\alpha_t}$ and $n_2 \in N_{w_1}^-$. Since $W(g_1) \neq 0$ we get that

$$W(ah_{\alpha_0}(r_t)w_1n_2) \neq 0$$

with $l(w_1) < l(w)$. We wish to invoke our first induction assumption for w_1. Notice that we have assumed that $a \in A^M(w)$ and that $|r_t| \geq D$. To use the induction assumption we need to show that $ah_{\alpha_0}(r_t) \in A^{M_2}(w_1)$ for some constant $M_2 > 0$ which depends only on D and M. By Lemma 5.6, $A^M(w) \subset A^{M_1}(w_1)$ for some constant $M_1 > 0$.

Let $S_1 = S^0(w_1)$. Let $\gamma \in \Delta(S_1)^*$. By the same arguments as in the $t = 1$ case we have that

$$< \gamma, \alpha_0 > \leq 0.$$

Hence, (see $t = 1$ case) $\gamma(h_{\alpha_0}(r_t)) = |r_t|^p$ with $p \leq 0$. It follows that

$$\gamma(h_{\alpha_0}(r_t)) \leq D^p.$$

Now since

$$\gamma(ah_{\alpha_0}(r_t)) = \gamma(a)\gamma(h_{\alpha_0}(r_t)) \leq M_1D^p$$

it follows that there exists $M_2 > 0$ such that $ah_{\alpha_0}(r_t) \in A^{M_2}(w_1)$ for all $a \in A^M(w)$ and $|r_t| \geq D$.

Now it follows from our induction hypothesis that n_2 is inside a compact set in N_w^- independent of a and r_t.

Since $W \in \mathcal{W}^0$ it follows that for every $\alpha \in S_1$ there exist positive constants $D_\alpha < E_\alpha$ such that

$$D_\alpha \leq \alpha(ah_{\alpha_0}(r_t)w_1n_2) \leq E_\alpha.$$

Since n_2 is in a fixed compact set it follows from Lemma 5.4 (a) that there exist positive constants $D'_\alpha \leq E'_\alpha$ such that

$$D'_\alpha \leq \alpha(ah_{\alpha_0}(r_t)) \leq E'_\alpha.$$

Hence if $\beta \in C(S_1)$ there exists a positive constant D'_β such that

$$D'_\beta \leq \beta(ah_{\alpha_0}(r_t)).$$

The proof now follows word for word the $t = 1$ case. That is, we find $\gamma_0 \in C^*(\Delta(S_1)) \cap C(S_1)$ such that $< \gamma_0, \alpha_0 > < 0$. Since $a \in A^{M_1}(w_1)$ there exists $M_{\gamma_0} > 0$ such that

$$\gamma_0(a) \leq M_{\gamma_0}.$$

By (37)

$$\gamma_0(ah_{\alpha_0}(r_t)) = \gamma_0(a)\gamma_0(h_{\alpha_0}(r_t)) \geq D'_{\gamma_0}$$

hence

$$\gamma_0(h_{\alpha_0}(r_t)) \geq \frac{D'_{\gamma_0}}{M_{\gamma_0}}.$$

Since $< \gamma_0, \alpha_0 > < 0$ we have that $\gamma_0(h_{\alpha_0}(r_t)) = |r_t|^p$ with $p < 0$ hence $|r_t|$ is bounded.

To summarize, we have just proved that if

$$W(awn) \neq 0$$

with $a \in A^M(w)$ and n written as in (36) then there exists $C_t > 0$ independent of a and $r_1, \ldots, r_{t-1}$ so that $|r_t| \leq C_t$.

It remains to prove that $r_1, \ldots, r_{t-1}$ are also bounded. Consider the space $\{\rho(x_{\alpha_t}(r_t))W \ : \ |r_t| \leq C_t\}$ where

$$\rho(x_{\alpha_t}(r_t))W)(g) = W(gx_{\alpha_t}(r_t)).$$

Since W is smooth on the right it follows that this space is spanned by a finite number of functions $W_1, \ldots, W_p$. By Corollary 5.5 (b), each such function is in $\mathcal{W}^0$. Let $a \in A^M(w)$ and $n' = x_{\alpha_1}(r_1) \cdots x_{\alpha_{t-1}}(r_{t-1})$. It follows from our induction assumption on t that for each such function W_i there exists a constant A_i such that

$$W_i(awn') \neq 0 \implies |r_j| < A_i, \ j = 1, \ldots, t-1. \tag{41}$$

Let $A = \max\{A_1, \ldots, A_p\}$. Then it is clear that (41) holds with A replacing A_i. We now write

$$W(awn) = W(awx_{\alpha_1}(r_1) \cdots x_{\alpha_{t-1}}(r_{t-1})x_{\alpha_t}(r_t))$$
$$= c_1(r_t)W_1(awx_{\alpha_1}(r_1) \cdots x_{\alpha_{t-1}}(r_{t-1})) + \ldots + c_p(r_t)W_l(awx_{\alpha_1}(r_1) \cdots x_{\alpha_{t-1}}(r_{t-1}))$$

for every r_t such that $|r_t| \leq C_t$. If $W(awn) \neq 0$ then at least one of the summands does not vanish and we can conclude that $|r_i| \leq A$ for $i = 1, ..., t-1$. Now taking $C = \max\{A, C_t\}$ we get our result. □

6. Bessel functions for supercuspidal representations

In this section we attach Bessel functions to irreducible supercuspidal representations of $G = GL_n(F)$. This section is not needed in the sequel since we will later attach Bessel function to every irreducible generic representation of $GL_n(F)$. The reason we include this section is that the situation for supercuspidal representations is nicer than the general situation and both the formulas and proofs are simpler.

Lemma 6.1. *Let (π, V) be a supercuspidal representation of G and let $\mathcal{W}(\pi, \psi)$ be the Whittaker model of π (see (2)). Then $\mathcal{W}(\pi, \psi) \subset \mathcal{W}^0(G, \psi)$.*

Proof. Let $W \in \mathcal{W}(\pi, \psi)$. Then by **[11]**, W is compactly supported mod NZ. It follows that for every $\alpha \in \Delta$ and every $w \in \mathbf{W}$ the support of W in BwB has bounded image under α. (Since the support of W in G already has bounded image under α.) Hence $W \in \mathcal{W}^0$. □

The main result that allows the definition of the Bessel functions is the following: (For the proof see Corollary 1.2).

Corollary 6.2. *Let (π, V) be a supercuspidal representation of G and let $W \in \mathcal{W}(\pi, \psi)$. Let $w \in \mathbf{W}$ and fix $g \in BwB$. Then the function*

$$n \to W(gn)$$

from N_w^- to $\mathbf{C}$ is compactly supported in N_w^-.

Let $w \in \mathbf{W}$. We define the subtorus A_w to be

$$A_w = \{a \in A : \ \psi(n) = \psi(n^{aw}), \ \text{for all } n \in N_w^+\}. \tag{42}$$

Here $n^g = gng^{-1}$. It is easy to see that $A_e = Z(G)$ and that $A_{w_0} = A$.

Definition 6.3. We say that w is a relevant Weyl element if $A_w \neq \emptyset$.

It is well known (see **[12]**) that w is relevant if and only if $w = w_0 w_S$ where $S \subset \Delta$ and w_S is the longest Weyl element in the standard parabolic subgroup given by S. The set of relevant Weyl elements is the set of Weyl element of the form

$$\begin{pmatrix} & & & I_{m_1} \\ & & I_{m_2} & \\ & \cdot & & \\ \cdot & & & \\ I_{m_l} & & & \end{pmatrix}$$

where I_m is identity matrix of order m and $m_1 + m_2 + \ldots + m_l = n$.

Fix a relevant Weyl element w and fix $g \in NA_w wN = NA_w wN_w^-$. Let (π, V) be an irreducible supercuspidal representation of G. Let $W \in \mathcal{W}(\pi, \psi)$. Define

$$L_g(W) = \int_{N_w^-} W(gn)\psi^{-1}(n)dn.$$

By Corollary 6.2 this integral is absolutely convergent. Let G act on $\mathcal{W}(\pi, \psi)$ by right translations as in (3).

Lemma 6.4.

$$L_g(\rho(n)W) = \psi(n)L_g(W), \quad n \in N.$$

Proof. This is obvious if $n \in N_w^-$. Assume $n_1 \in N_w^+$. Then for $n \in N_w^-$ and $g = n_2 awn_3$ with $n_2 \in N$, $a \in A_w$ and $n_3 \in N_w^-$ we have

$$\begin{aligned} W(gnn_1) &= W(n_2 awn_3 nn_1) \\ &= \psi(n_2)W(awn_1^{n_3 n} n_3 n_1) \\ &= \psi(n_2)\psi(n_1^{n_3 n})W(awn_3 n_1) \\ &= \psi(n_1)W(n_2 awn_3 n) \\ &= \psi(n_1)W(gn) \end{aligned}$$

Here we have used that N_w^- normalizes N_w^+ and that $\psi(n^{-1}n^+n) = \psi(n^+)$ for every $n \in N_w^-$ and $n^+ \in N_w^+$. Writing $(\rho(n_1)W)(gn) = W(gnn_1)$ and computing the integral defining $L_g(\rho(n_1)W)$ we get our result for $n \in N_w^+$. Since every $n \in N$ can be written in the form $n = n^+ n_-$ for some $n^+ \in N_w^+$ and $n_- \in N_w^-$ we get our result for a general $n \in N$. □

It follows that L_g is a Whittaker functional on π. Hence by the uniqueness of the Whittaker functional we get that there exists a scalar $j_{\pi,\psi,w}(g) \in \mathbf{C}$ such that

$$L_g(W) = j_{\pi,\psi,w}(g)W(e), \quad W \in \mathcal{W}(\pi,\psi). \tag{43}$$

$j_{\pi,\psi,w}$ is a function on $NA_w wN$ which we call the Bessel function associated to π and w. We shall show in Section 8 that $j_{\pi,\psi,w}$ is locally constant on $NA_w wN$. When $w = w_0$ we set $j_\pi = j_{\pi,\psi} = j_{\pi,\psi,w_0}$. To get a formula for $j_{\pi,\psi,w}$ we notice that since π is supercuspidal there exists a function $W \in \mathcal{W}(\pi,\psi)$ such that $W(e) = 1$. (This follows from the existence of a nontrivial Whittaker functional on π.) Hence we get from (43) that

Corollary 6.5. *Let π be a supercuspidal representation of G and let w be a relevant Weyl element. Then there exists $W \in \mathcal{W}(\pi,\psi)$ such that*

$$j_{\pi,\psi,w}(g) = \int_{N_w^-} W(gn)\psi^{-1}(n)dn, \quad g \in NA_w wN.$$

7. Projection into $\mathcal{W}^0(G,\psi)$

In this section we shall show that every $W \in \mathcal{W}(G,\psi)$ can be projected into $\mathcal{W}^0(G,\psi)$ by integrating it on a compact unipotent group versus a character of that group. We start with some preliminary results about Howe vectors. The proofs can be found in ([4], Section 5).

7.1. Howe vectors For a positive integer m we denote by K_m the congruence subgroup of K given by $K_m = e + M_n(P^m)$. We let $A_m = A \cap K_m$. Let

$$d = \begin{pmatrix} 1 & & & & \\ & \varpi^2 & & & \\ & & \varpi^4 & & \\ & & & \ddots & \\ & & & & \varpi^{2n-2} \end{pmatrix}.$$

Let $J_m = d^m K_m d^{-m}$. Notice that J_m is expanding above the main diagonal and shrinking on and below the main diagonal. Let

$$N_m = N \cap J_m. \tag{44}$$

Let $N_m = \bar{N} \cap J_m$ and $\bar{B}_m = \bar{B} \bigcap J_m$. Using similar properties of K_m, it is easy to see that

$$J_m = \bar{N}_m A_m N_m = \bar{B}_m N_m.$$

Moreover, for $\alpha \in \Phi^+$ let

$$J_\alpha = N_\alpha \cap J_m = \{x_\alpha(r) : |r| \leq q^{(2\mathrm{height}(\alpha)-1)m}\} \tag{45}$$

and for $\alpha \in \Phi^-$ let

$$J_\alpha = N_\alpha \cap J_m = \{x_\alpha(r) : |r| \leq q^{-(2\mathrm{height}(\alpha)+1)m}\}. \tag{46}$$

Then

$$N_m = \prod_{\alpha\in\Phi^+} J_\alpha, \quad \bar{N}_m = \prod_{\alpha\in\Phi^-} J_\alpha. \tag{47}$$

We fix a character ψ_F on F as in Section 3. In particular $\psi_F = 1$ identically on the ring of integers O and $\psi_F(P^{-1}) \neq 1$. Let ψ be a character of N obtained from ψ_F as in (34). For $m \geq 1$ we define a character ψ_m on J_m by

$$\psi_m(j) = \psi(n_j)$$

where $j = \bar{b}_j n_j$, $\bar{b}_j \in \bar{B}_m$, $n_j \in N_m$ is the unique decomposition of j. It is easy to see that ψ_m is a character on J_m. For each $W \in \mathcal{W}(G,\psi)$ we define $W_m = W_{N_m,\psi}$ by

$$W_m(g) = W_{N_m,\psi}(g) = \int_{N_m} W(gn)\psi^{-1}(n)dn. \tag{48}$$

Since $N_{m+1} \supset N_m$ it is a simple application of Fubini to show that if $m \geq k$ then

$$W_m(g) = \mathrm{vol}(N_k)^{-1} \int_{N_m} W_k(gn)\psi^{-1}(n)dn. \tag{49}$$

For $g_1 \in G$ we let $(\rho(g_1)W)(g) = W(gg_1)$. The proof of the following Lemma is the same as the proof of Lemma 5.1 in [4].

Lemma 7.1. *Let M be such that $\rho(K_M)W = W$ and let m be an integer such that $m > 3M$. Then*

$$\rho(j)W_m = \psi_m(j)W_m, \quad j \in J_m. \tag{50}$$

Formulating Lemma 7.1 for functions we get that for $m > 3M$

$$W_m(gj) = \psi_m(j)W_m(g) \quad \text{for all } g \in G, j \in J_m. \tag{51}$$

We call a vector W in a representation space of G satisfying (50) (or (51)) a Howe vector. The above Lemma shows that if the representation space affords a nontrivial Whittaker functional then non-zero Howe vectors exist. This property and some uniqueness properties of Howe vectors for irreducible admissible representations of $GL_n(F)$ were established in [8]. We now continue to study the behavior of Whittaker functions satisfying (51).

Lemma 7.2. *Let $w \in \mathbb{W}, a \in A$ and $\alpha \in S^0(w)$. Assume $W \in \mathcal{W}$ satisfies (51) for some $m \geq 1$. Then*

$$W(aw) \neq 0 \implies \alpha(a) \in 1 + P^m.$$

Proof. We divide into two cases. First assume that w is not relevant (see Definition 6.3) , that is, w is not of the form $w = w_S w_0$ for some subset S of simple roots. (Notice that $\{w_S w_0 : S \subset \Delta\} = \{w_0 w_S : S \subset \Delta\}$.) Then by **[14]** Lemma 89, there exists a simple root β such that $\alpha = w(\beta) > 0$ but $w(\beta)$ is not a simple root. Let $r \in P^{-m}$. Then

$$\psi_F(r)W(aw) = W(awx_\beta(r)) = W(x_\alpha(\alpha(a)r)aw) = W(aw). \qquad (52)$$

By our assumptions on the conductor of ψ there exists $r \in P^{-m}$, such that $\psi_F(r) \neq 0$. Hence $W(aw) = 0$ and our statement is trivially true.

Assume $w = w_S w_0$. Then $S = S^0(w)$. (See **[10]**, Section 1.8, ex.2). Let $\alpha \in S$ and let $\beta = w_0 w_S(\alpha)$. β is a positive simple root. Arguing as in (52) we get

$$\psi_F(r)W(aw) = \psi_F(\alpha(a)r)W(aw)$$

for all $|r| \leq q^{-m}$. Hence $W(aw) \neq 0$ implies that $\alpha(a) - 1 \in P^m$ which is the required conclusion. □

Our main theorem of this section is the following. It implies (and in fact is equivalent to) Theorem 1.3 in the introduction.

Theorem 7.3. *Let $W \in \mathcal{W}(G, \psi)$. Then there exists a positive integer M such that $W_m = W_{N_m,\psi} \in \mathcal{W}^0(G, \psi)$ for every $m \geq M$.*

Proof. We need to show that there exists M such that for every fixed $m \geq M$ and every $w \in \mathbb{W}$, the support of W_m in BwB has bounded image under every $|\alpha| \in S^0(w)$. In other words, the statement of the theorem is equivalent to the following statement:

(A) Fix $w \in \mathbb{W}$ and $\alpha \in S^0(w)$. Then there exists an integer $M > 0$ and constants $C < D$ (depending on m) such that if $g \in BwB$ and $W_m(g) \neq 0$ then $C < |\alpha(g)| < D$.

We shall prove statement (A) by induction on $l(w)$.

$l(w) = 0$: That is, $w = e$.

In this case $BwB = B = NA$. By Lemma 7.1 there exists a positive integer M such that W_m satisfies (51) for every $m \geq M$. Let $m \geq M$ and assume that $g = na$ is in the support of W_m. Then $W_m(g) = W_m(na) = \psi(n)W_m(a) \neq 0$. Hence $W_m(a) \neq 0$ and by Lemma 7.2, $\alpha(a) \in 1 + P^m$ for every $\alpha \in S^0(e) = \Delta$. Since $\alpha(g) = \alpha(a)$ we get statement (A) for $w = e$.

For the general case, fix $w \in \mathbf{W}$, $w \neq e$. Let

$$S^-(w) = \{\alpha_1, \ldots, \alpha_l\}$$

and assume that $\text{height}(\alpha_i) \geq \text{height}(\alpha_{i+1})$ for $i = 1, \ldots, l-1$. We can write every $g \in BwB$ uniquely in the form

$$g = nawn_1 = nawx_{\alpha_1}(r_1)x_{\alpha_2}(r_2)\cdots x_{\alpha_l}(r_l) \tag{53}$$

with $n \in N$, $a \in A$, $r_1, \ldots, r_l \in F$. Here

$$n_1 = x_{\alpha_1}(r_1)x_{\alpha_2}(r_2)\cdots x_{\alpha_t}(r_t). \tag{54}$$

First case: Fix $\alpha \in S^0(w)$. Assume $W_m(g) \neq 0$ and assume that $n_1 \in N_m$. Then $W_m(g) = \psi(n_1)W_m(gn_1^{-1})$ hence if we let $g_1 = gn_1^{-1}$ we get that $W_m(g_1) \neq 0$. Now $W_m(g_1) = W_m(naw) = \psi(n)W_m(aw)$. Hence $W_m(aw) \neq 0$ and by Lemma 7.2, $\alpha(a)$ is in a compact set. Since $w \in K$ we have that $\alpha(g_1) = \alpha(a)$. Hence we proved that if g is of the form (53) with $n_1 \in N_m$ and $W(g) \neq 0$ then there exists $r_g \in R = N_m$ such that $\alpha(gr_g)$ is in a fixed compact set. By Lemma 5.4 (b), $\alpha(g)$ is in a fixed compact set.

We shall now consider the second case where $n_1 \notin N_m$. If $n_1 \notin N_m$ then there exists i such that $x_{\alpha_i}(r_i) \notin N_m$. This is equivalent to $r_i = r_i(g)$ satisfying $|r_i(g)| > q^{(2j+1)m}$ where $j = \text{height}(\alpha_i)$. (See (45)).

Lemma 7.4. *Let m be a positive integer and assume that $g \in BwB$ is of the form*

$$g = nawn_1 = nawx_{\alpha_1}(r_1)x_{\alpha_2}(r_2)\cdots x_{\alpha_i}(r_i)$$

with $i \leq l$ and $x_{\alpha_i}(r_i) \notin N_m$. Let $n_2 \in N_m$ and let $g_1 = gn_2$. Then in the decomposition of g_1 into (53) we have $|r_i(g_1)| = |r_i(g)| = |r_i|$.

Proof. We can write

$$n_2 = n_3x_{\alpha_i}(b_i)x_{\alpha_{i+1}}(b_{i+1})\cdots x_{\alpha_l}(r_l)$$

with n_3 a product over the positive root subgroups that are different than $\alpha_i, \ldots, \alpha_t$. This decomposition of n_2 is unique and $x_{\alpha_j}(b_j) \in N_m$ for $j = i, \ldots, l$. It follows from (14) that $r_j(g_1) = b_j$ for $j = i+1, \ldots, l$ and that $r_i(g_1) = r_i + b_i$. Since $|r_i| > q^{(2\text{height}(\alpha_i)+1)m}$ and since $|b_i| \leq q^{(2\text{height}(\alpha_i)+1)m}$ we get that $|r_i(g_1)| = |r_i(g)|$. □

Fix $\alpha \in S^0(w)$. To finish the proof we need to show that there exists a positive integer M such that if $m \geq M$ and if $W_m(g) \neq 0$ for g of the form (53) with $n_1 \notin N_m$ then $\alpha(g)$ is in a fixed compact set. Since $n_1 \notin N_m$ there exists a maximal i in the decomposition of n_1 in (54) such that $x_{\alpha_i}(r_i) \notin N_m$. We shall prove our Theorem by downward induction on this maximal i. That is, our second induction statement is the following:

(B) Fix i, $1 \le i \le l$. There exists an integer $M > 0$ such that if $m \ge M$ and if $W_m(g) \ne 0$ with g of the form (53) with $x_{\alpha_i}(r_i) \notin N_m$ and i is the maximal such index than $\alpha(g)$ is in a fixed bounded set (depending on m, w and W but not on such g).

We consider the case $i = l$. Let M_1 be such that for $m \ge M_1$, W_m satisfies (51) and such that W_m satisfies the induction assumption (A) for every $w_1 \in \mathbb{W}$ such that $l(w_1) < l(w)$ and for every $\alpha_1 \in S^0(w_1)$. That is, we assume that if $g \in Bw_1B$ and if $W_m(g) \ne 0$ then $\alpha_1(g)$ is in a fixed compact set. We can enlarge this fixed compact set to be good for every such w_1 and every such α_1. Let $M = 3M_1$. Assume $m \ge M$ and assume that $W_m(g) \ne 0$ where g is of the form (53) with $x_{\alpha_l}(r_l) \notin N_m$. By our assumption that $\text{height}(\alpha_{j+1}) \le \text{height}(\alpha_j)$ we have that α_l is a simple root and $|r_l| > q^m$. By (49) we have

$$W_m(g) = \frac{1}{\text{vol}(N_{M_1})} \int_{N_m} W_{M_1}(gn)\psi^{-1}(n)dn.$$

Since $W_m(g) \ne 0$ there exists $n_2 \in N_m$ such that $W_{M_1}(gn_2) \ne 0$. Let $g_1 = gn_2$. By Lemma 7.4 we have that

$$|r_l(g_1)| = |r_l(g)| > q^m \ge q^{3M_1}.$$

It follows that $x_{-\alpha_l}(-1/r_l(g_1)) \in J_{M_1}$, hence by (51)

$$W_{M_1}(g_1x_{-\alpha_l}(-1/r_l(g_1))) = W_{M_1}(g_1) \ne 0.$$

Let $g_2 = g_1x_{-\alpha_l}(-1/r_l(g_1))$. By Lemma 2.6, $g_2 \in Bw_1B$ with $l(w_1) < l(w)$. Moreover, by (25), $\alpha \in S^0(w_1)$. Hence by our assumptions on M_1 above, $W_{M_1}(g_2) \ne 0$ implies that $\alpha(g_2)$ is in a fixed compact set. Hence we proved that for every g satisfying the conditions above such that $W_m(g) \ne 0$ there exists $r_g \in R = N_mJ_{M_1}$ such that $\alpha(gr)$ is in a fixed compact set. By Lemma 5.4 (b), $\alpha(g)$ is in a fixed compact set.

We now prove the general case. Fix $1 \ge i < l$. Let M_1 be as in the case $i = l$. By our induction assumption (B) we can also assume (by enlarging M_1) that if $m \ge M_1$ and if $W_m(g) \ne 0$ and if g is of the form (53) with $x_{\alpha_j}(r_j) \notin N_m$ for some $j > i$ then $\alpha(g)$ is in a fixed compact set.

Let $M = 3M_1$ and let $m \ge M$. Assume that $W_m(g) \ne 0$ where g is in the form (53) with $x_{\alpha_i}(r_i) \notin N_m$ and $x_{\alpha_j}(r_j) \in N_m$ for $j > i$. Then

$$W_m(g) = \psi_F(r_{i+1}) \cdots \psi_F(r_l)W_m(nawx_{\alpha_1}(r_1)) \cdots x_{\alpha_i}(r_i)) \ne 0.$$

Let $n_2 = x_{\alpha_l}(-r_1)) \cdots x_{\alpha_{i+1}}(-r_i))$ Then $n_2 \in N_m$ and the above equation implies that for $g_1 = gn_2 = nawx_{\alpha_1}(r_1) \cdots x_{\alpha_i}(r_i)$ we have $W_m(g_1) = W_m(gn_2) \ne 0$. We also have

$$W_m(g_1) = \frac{1}{\text{vol}(N_{M_1})} \int_{N_m} W_{M_1}(g_1n)\psi^{-1}(n)dn.$$

Since $W_m(g_1) \neq 0$ it follows that there exists $n_3 \in N_m$ such that $W_{M_1}(g_1 n_3) \neq 0$. Let $g_2 = g_1 n_3$ By Lemma 7.4 we have that $|r_i(g_2)| = |r_i(g)| = |r_i|$. We divide into two cases. First assume that there exists $j > i$ such that $x_{\alpha_j}(r_j(g_2)) \notin N_{M_1}$. Then it follows by our assumptions on M_1 that $\alpha(g_2)$ is in a fixed compact set. Since $g_2 = gr_g$ for $r \in R = N_m$ we get that $\alpha(g)$ is in a fixed compact set.

Next assume that $x_{\alpha_j}(r_j(g_2)) \in N_{M_1}$ for every $j > i$. Using (51) as above we get that $W_{M_1}(g_3) \neq 0$ where

$$g_3 = nawx_{\alpha_1}(r_1(g_2)) \cdots x_{\alpha_i}(r_i(g_2))$$

and $g_3 = g_2 n_4$ with $n_4 \in N_{M_1}$ Since

$$|r_i(g_2)| = |r_i(g)| > q^{(2\text{height}(\alpha_i)+1)m} \geq q^{(2\text{height}(\alpha_i)+1)3M_1}$$

it follows from (46) that $x_{-\alpha_l}(-1/r_i(g_2)) \in J_{M_1}$ hence by the same arguments as in the case $i = l$ we get that $\alpha(g)$ is in a fixed compact set. □

8. Bessel functions

In this section we attach Bessel functions to irreducible generic representation of $G = GL_n(F)$. The definition of these functions depend on Theorem 5.7 and Theorem 7.3 and is identical to the definition of the Bessel functions in [4]. Since the proofs are the same as in ([4], Section 6) we shall omit them. Given an irreducible generic representation of G we will attach a Bessel function for each relevant Weyl element w. This Bessel function will be defined on a subset of BwB and will be locally constant there. If the representation is supercuspidal then our definition here will coincide with the definition in Section 6 making Section 6 redundant. We are primarily interested in the Bessel function which is attached to the longest Weyl element w_0 which we call the main (or principal) Bessel function. We shall provide full proofs in this case for the sake of completeness.

Let $w \in \mathbb{W}$ be a relevant Weyl element. That is, there exists $S \subset \Delta$ such that $w = w_S w_0$. Let N_w^+ and N_w^- be the subgroups of N as defined in (27). We define the subtorus A_w as in Section 6 to be

$$A_w = \{a \in A : \psi(n) = \psi(n^{aw}), \text{ for all } n \in N_w^+\}.$$

Here $n^g = gng^{-1}$. Let (π, V) be an irreducible generic representation of G and let $W \in \mathcal{W}(\pi, \psi)$. By Theorem 7.3 there exists a positive integer M such that if $m \geq M$ then $W_m \in \mathcal{W}^0$ (See (48) for the definition of W_m.) Fix $m \geq M$

and let $g \in NA_w wN_w^-$. We define

$$L_{g,w}(W) = \frac{1}{\mathrm{vol}(N_m)} \int_{N_w^-} W_m(gn)\psi^{-1}(n)dn. \tag{55}$$

By writing $g = n_1 awn_2$ and using Theorem 5.7 it follows that the integral above converges. (See also Corollary 6.2). The main result of this section is the following:

Proposition 8.1.

(a) $L_{g,w}(W)$ is independent of $m \geq M$.
(b) $L_{g,w}$ is a Whittaker functional on $\mathcal{W}(\pi, \psi)$, that is, for every $n \in N$, $L_{g,w}(\pi(n)W) = \psi(n)L_{g,w}(W)$.
(c) If $W \in \mathcal{W}^0$ then

$$L_{g,w}(W) = \int_{N_w^-} W(gn)\psi^{-1}(n)dn.$$

The proof is the same as in ([4] Proposition 6.1.) We will prove the Proposition for the case $w = w_0$. (see also the introduction for the case $w = w_0$.)

Proof. In that case $N_w^- = N$, $N_w^+ = \{e\}$ and $A_w = A$. Using of the Fubini theorem, it is easy to see that if $m_1 \geq m \geq M$ then

$$\frac{1}{\mathrm{vol}(N_m)} \int_{N_{m_1}} W_m(gn)\psi^{-1}(n)dn = \int_{N_{m_1}} W(gn)\psi^{-1}(n)dn.$$

Since N_{m_1} cover N when $m_1 \to \infty$ it follows that

$$\begin{aligned} L_{g,w_0}(W) &= \frac{1}{\mathrm{vol}(N_m)} \int_N W_m(gn)\psi^{-1}(n)dn \\ &= \lim_{m_1\to\infty} \frac{1}{\mathrm{vol}(N_m)} \int_{N_{m_1}} W_m(gn)\psi^{-1}(n)dn \\ &= \lim_{m_1\to\infty} \int_{N_{m_1}} W(gn)\psi^{-1}(n)dn. \end{aligned}$$

Now (a) and (c) follow from the last line of the above equation. For part (b) we fix $n_1 \in N$ and consider the above limit for $\rho(n_1)W$. Since N_m cover N we have that there exists M_1 such that $n_1 \in N_m$ for all $m > M_1$. Now a simple change of variable in the integral above will give the result. □

By (b), $L_{g,w}$ is a Whittaker functional, hence by the uniqueness of Whittaker functionals it follows that there exists a scalar $j_{\pi,\psi,w}(g)$ such that

$$L_{g,w}(W) = j_{\pi,\psi,w}(g)W(e) \quad g \in NA_w wN, W \in \mathcal{W}(\pi, \psi). \tag{56}$$

We call $j_{\pi,\psi,w}(g)$ the Bessel function attached to w and denote by $j_{\pi,\psi} = j_{\pi,\psi,w_0}(g)$ the Bessel function attached to π. It is easy to see that

$$j_{\pi,\psi,w}(n_1 g n_2) = \psi(n_1)\psi(n_2) j_{\pi,\psi,w}(g), \quad g \in NA_w wN, n_1; n_2 \in N. \tag{57}$$

Lemma 8.2. *There exists $W \in \mathcal{W}^0(\pi,\psi)$ such that*

$$j_{\pi,\psi,w}(g) = \int_{N_w^-} W(gn)\psi^{-1}(n)dn, \quad g \in NA_w wN.$$

Proof. It follows from Theorem 7.3 that there exists $W \in \mathcal{W}^0(\pi,\psi)$ (where $\mathcal{W}^0(\pi,\psi) = \mathcal{W}(\pi,\psi) \cap \mathcal{W}^0(G,\psi)$) such that $W(e) = 1$. The result now follows from Proposition 8.1 (c). □

Corollary 8.3. *$j_{\pi,\psi,w}(g)$ is locally constant on $NA_w wN$.*

Proof. By (57) it is enough to prove that $j_{\pi,\psi,w}(g)$ is locally constant on $A_w w$. Let W be as in Lemma 8.2. By Theorem 5.7 $n \mapsto W(awn)$ is compactly supported on the set $(A^M(w) \cap A_w) \times N_w^-$. It follows from Lemma 8.2 that $j_{\pi,\psi,w}$ is locally constant on $A^M(w) \cap A_w)$. Since $A^M(w)$ cover $A_w)$ when $M \to \infty$ we get our result. □

We end this section by describing the Bessel functions attached to the contragredient representation.

Lemma 8.4. *Let π be a generic representation of G and $\hat{\pi}$ the representation contragredient to π. Then*

$$j_{\hat{\pi},\psi^{-1},w}(g) = j_{\pi,\psi,w_0ww_0}(g^{-1}) \quad g \in NA_w wN.$$

Proof. Let τ be the involution of G defined by $\tau(g) = w_0^t g^{-1} w_0$ For each $W \in \mathcal{W}(\pi,\psi)$ we define $W^\tau(g) = W(\tau(g))$. By **[11]** the mapping $W \mapsto W^\tau$ is a bijection between $\mathcal{W}(\pi,\psi)$ and $\mathcal{W}(\hat{\pi},\psi^{-1})$. If $g \in G$ is written in the Iwasawa decomposition in the form $g = nak$ where n is upper triangular a is diagonal and $k \in GL_n(O)$ then

$$\tau(g) = (w_0^t n^{-1} w_0)(w_0 a^{-1} w_0)(w_0^t k^{-1} w_0)$$

is an Iwasawa decomposition for $\tau(g)$. Hence, if $W \in \mathcal{W}^0(\pi,\psi)$ then we get that $W^\tau \in \mathcal{W}^0(\hat{\pi},\psi^{-1})$. Using Lemma 8.2 we get that

$$j_{\hat{\pi},\psi^{-1},w}(g) = j_{\pi,\psi,\tau(w)}(\tau(g)) \quad g \in NA_w wN.$$

Now $\tau(w) = w_0 w w_0$ and we claim that $j_{\pi,\psi,w_0ww_0}(\tau(g)) = j_{\pi,\psi,w_0ww_0}(g^{-1})$ for all $g \in NA_w wN$.. Since both functions satisfy (57) with ψ^{-1} replacing ψ it is enough to show that they coincide on the set $A_w w$. Since $\tau(g) = g^{-1}$ for all $g \in A_w w$ we get our result. □

Corollary 8.5. *Let $j_{\pi,\psi} = j_{\pi,\psi,w_0}$ be the Bessel function attached to π. Then*

$$j_{\tilde{\pi},\psi^{-1}}(g) = j_{\pi,\psi}(g^{-1}), \quad g \in Bw_0B.$$

9. Orbital integrals

In this section we show that the Bessel functions for the longest Weyl element (the main Bessel function) defined in Section 8 are given locally by orbital integrals. These integrals were studied in [12]. We will do this in two steps. We will show that the Bessel function restricted to a compact set in G is given by an integral of a Whittaker function which is compactly supported mod N. That is, if we restrict ourselves to this small neighborhood, we can replace a Whittaker function in the representation space with a different Whittaker function (not necessarily in the representation space) which is compactly supported mod NZ. Then we use the fact that each Whittaker function which is compactly supported mod NZ comes from an integral of a function in $C_c^\infty(G)$. We will start from the second part. Let ω be a character of Z and let $\mathcal{W}_\omega(G,\psi) \subseteq \mathcal{W}(G,\psi)$ be the subspace of functions $W \in \mathcal{W}(G,\psi)$ satisfying

$$W(gz) = \omega(z)W(g) \quad g \in G, z \in Z. \tag{58}$$

Let $C_c^\infty(G)$ be the space of locally constant functions on G with compact support. For each $f \in C_c^\infty(G)$ we let

$$W_f(g) = W_f^\psi(g) = \int_N f(ng)\psi^{-1}(n)dn.$$

It is clear that $W_f \in \mathcal{W}(G,\psi)$. We also define

$$W_{f,\omega}(g) = \int_Z \int_N f(nzg)\psi^{-1}(n)\omega^{-1}(z)dndz, \quad f \in C_c^\infty(G). \tag{59}$$

It is clear that $W_{f,\omega} \in \mathcal{W}_\omega(G,\psi)$. The image of these maps is well known. (See for example [4], Lemma 7.1). It is given in the following Lemmas:

Lemma 9.1. *Let $f \in C_c^\infty(G)$. Then W_f is compactly supported mod N and the map $f \mapsto W_f$ is a linear map onto the space of compactly supported functions mod N in $\mathcal{W}(G,\psi)$.*

Lemma 9.2. *Let $f \in S(G)$. Then $W_{f,\omega}$ is compactly supported mod NZ and the map $f \mapsto W_{f,\omega}$ is a linear map onto the space of compactly supported functions mod N in $\mathcal{W}_\omega(G,\psi)$.*

Let $|V|$ be the subspace of $|X|$ given by $|V| = \{|\alpha|_{r_1,r_2,\dots,r_n} : r_1+r_2+\ldots+r_n = 0\}$. (see (15)). Let $Q = \{\beta_1, \dots, \beta_{n-1}\}$ be a basis for $|V|$. Let $C_1 < C_2$ be positive constants and define

$$A_Q(C_1, C_2) = \{a \in A : \; C_1 < \beta_i(a) < C_2, \; i = 1, \dots, n-1\}.$$

Lemma 9.3. *A function W on G is compactly supported mod NZ if and only if there exist constants C_1, C_2 such that W is supported on $NA_Q(C_1, C_2)K$.*

Proof. We can write $A_Q(C_1, C_2) = ZA'$ where $A' = \{d(a_1, a_2, \dots, a_{n-1}, 1) \in A_Q(C_1, C_2)\}$. Since Q is a basis, it is clear that A' is compact. Hence if W is supported on $NA_Q(C_1, C_2)K$ then it is compactly supported mod NZ. Now assume W is compactly supported mod NZ. Then W is supported on a set of the form NZR for some compact set R. Since the sets $NA_Q(C_1, C_2)K$ for different choices of C_1 and C_2 are open sets that cover G we get that the sets of the form $NA_Q(C_1, C_2)K$ cover R. Since R is compact there exist constants C_1', C_2' so that $R \subset NA_Q(C_1', C_2')K$. Hence $NZR \subset NA_Q(C_1', C_2')K$. □

For each $w \in \mathbf{W}$ we define the set $M(w) \subset \Delta^*$ as follows:

$$M(w) = \{\alpha^* | \alpha \in \Delta, \; \alpha \notin S^0(w)\}.$$

Remark 9.4. If $w_1 < w$ then by (25), $S^0(w_1) \supseteq S^0(w)$, hence $M(w_1) \subseteq M(w)$.

Let E be a positive constant. We let

$$A_w(E) = \{a \in A : \; |\lambda|(a) > E \;\text{ for every } \lambda \in M(w)\}.$$

Theorem 9.5. *Let $W \in \mathcal{W}^0$, $w \in \mathbf{W}$ and $E > 0$. There exists a function $W_1 \in \mathcal{W}(G, \psi)$ compactly supported mod NZ such that*

$$W_1(n_1 a w n_2) = W(n_1 a w n_2) \tag{60}$$

for all $a \in A_w(E)$ and $n_1, n_2 \in N$.

Remark 9.6. Let $C_1 < C_2$ be positive constants and let $A_{C_1,C_2} = A_\Delta(C_1, C_2)$. By Lemma 9.3 we have that W_1 being compactly supported mod NZ is equivalent to W_1 being supported on a set of the form $NA_{C_1,C_2}K$ for some C_1, C_2. Hence we can find $W_1 \in \mathcal{W}$ compactly supported mod NZ such that (60) holds if and only if we can find constants C_1, C_2 such that the function

$$W_1(g) = \begin{cases} W(g), & \text{if } g \in NA_{C_1,C_2}K; \\ 0, & \text{otherwise.} \end{cases} \tag{61}$$

satisfies (60). Hence, we shall use (61) to define the desired W_1. Notice that if we define W_1 by (61) then $W(g) = 0 \Rightarrow W_1(g) = 0$ hence we only need to prove (60) for $g = n_1 a w n_2$ such that $a \in A_w(E)$ and $W(g) \neq 0$.

Proof. We shall prove this theorem by an induction on $l(w)$ as in the proof of Theorem 4.1, Theorem 5.7 and Theorem 7.3.

$l(w) = 0$

In this case $w = e$, $S^0(w) = \Delta$, $M(w) = \emptyset$ and $A_w(E) = A$. We need to show the existence of a Whittaker function W_1, compactly supported mod NZ such that $W_1 = W$ on B. Since $W \in \mathcal{W}^0$ it follows that the support of W on B is contained in a set of the form NA_{C_1,C_2}. Define W_1 as in (61). Then W_1 satisfies the requirements of the Theorem.

We turn to the general case: $l(w) \geq 1$. Fix $w \in \mathbf{W}$ such that $l(w) \geq 1$.

Remark 9.7. By the induction assumption, and by Remark 9.6, if we are given a set of positive constants $\{E_{w_1} : l(w_1) < l(w)\}$ then there exists a Whittaker function W_1 compactly supported mod NZ such that (60) holds for every w_1 such that $l(w_1) < l(w)$ and every $a \in A^{w_1}(E_{w_1})$.

Fix $E > 0$ and let $a \in A_w(E)$. We need to show the existence of a function W_1 as above such that

$$W_1(awn) = W(awn) \tag{62}$$

for all $a \in A_w(E)$ and all $n \in N_w^-$. Let $S_w^- = \{\alpha_1, \dots, \alpha_l\}$. We can assume that $\text{height}(\alpha_i) \geq \text{height}(\alpha_{i+1})$, $i = 1, \dots, l-1$. Every $n \in N_w^-$ can be written (not uniquely) in the form

$$n = x_{\alpha_1}(r_1) \cdots x_{\alpha_j}(r_j) \tag{63}$$

for $0 \leq j \leq l$. We shall prove by an induction on j that there exists a Whittaker function W_1 as above such that (62) holds for every $a \in A_w(E)$ and every n of the form (63).

$j = 0$.

For $j = 0$ we need to show the existence of W_1 as above such that

$$W_1(aw) = W(aw)$$

for all $a \in A_w(E)$. By the remark above it is enough to consider the case where $a \in A_w(E)$ and $W(aw) \neq 0$. Since every $\beta \in \Delta^*$ is a positive linear combination of positive simple roots (see Remark 3.2), it follows from Remark 5.2 that for every $\beta \in M(w) \subset \Delta^*$ there exists a constant D_β such that

$$|\beta(a)| < D_\beta. \tag{64}$$

Since $a \in A_w(E)$ it follows that for every $\beta \in M(w)$ we have

$$E < |\beta(a)|. \tag{65}$$

It is possible that the set of such a is empty in which case we take $W_1 = 0$ (or W_1 given by (61) for any constants $C_1 < C_2$.) By Lemma 7.2, we have that for every $\alpha \in S^0(W)$ there exist constants $C_\alpha < D_\alpha$ such that

$$C_\alpha < |\alpha(a)| < D_\alpha. \tag{66}$$

Putting together (64), (65), (66) and using that $M(w) \cup S^0(w)$ is a basis for $|V|$ (see Lemma 3.4 (a)) we get by Lemma 9.3 that a satisfying the conditions above is in a set of the form A_{C_1,C_2} for some constants $C_1 < C_2$. Hence we can use (61) to define W_1.

The general case: Assume $j \geq 1$ and let $n \in N_w^+$ be in the form (63). Since W is smooth on the right, there exists a positive constant D such that if $|r| \geq D$ then

$$W(gx_{-\alpha_j}(-r^{-1})) = W(g), \quad g \in G.$$

Assume that n is of the form (63) with $a \in A_w(E)$ and $|r_j| \geq D$. We have

$$W(awn) = W(awnx_{-\alpha_j}(-r_j^{-1})). \tag{67}$$

By Lemma 2.6, $g = awnx_{-\alpha_j}(-r_j^{-1}) \in Bw_1B$ with $w_1 < w$. Moreover, if we write $g = n_1a_1w_1n_2$ for $n_1 \in N$, $a_1 \in A$ and $n_2 \in N_{w_1}^-$ then we have $a_1 = ah_{w(\alpha_j)}(r_j)$. Let $\beta \in M(w_1)$. Since $|r_j| \geq D$ it follows from Remark 2.1 that there exists $C_{\beta,D} > 0$ such that $|\beta(h_{w(\alpha_j)}(r_j))| > C_{\beta,D}$. By Remark 9.4 we have that $\beta \in M(w)$ hence $|\beta(a)| > E$. Hence we get that for every $|r_j| \geq D$ and every $a \in A_w(E)$

$$|\beta(a_1)| = |\beta(ah_{w(\alpha_j)}(r_j))| > EC_{\beta,D}.$$

It follows that if we take $E_1 = \min\{EC_{\beta,D} : \beta \in M(w)\}$ then $a_1 \in A_{w_1}(E_1)$.

Remark 9.8. If $M(w) = \emptyset$ then $M(w_1) = \emptyset$ and $A_{w_1}(E_1) = A$ for every $E_1 > 0$. Hence, in that case, we can take any E_1 that we like and a_1 will be in $A_{w_1}(E_1)$.

It follows from our first induction assumption that there exists a function W_1 given by (61) so that

$$W_1(n_1a_1w_1n_2) = W(n_1a_1w_1n_2) \tag{68}$$

for every $n_1 \in N$, $a_1 \in A_{w_1}(E_1)$ and $n_2 \in N_{w_1}^-$. Since W_1 is also smooth on the right it follows that there exists a constant $D_1 \geq D$ such that

$$W_1(awnx_{-\alpha_j}(-r_j^{-1})) = W_1(awn) \tag{69}$$

when $|r_j| \geq D_1$. Combining (67), (68) and (69) we get that

$$W_1(awn) = W(awn)$$

for every $a \in A_w(E)$ and every n of the form (63) with $|r_j| \geq D_1$. We now consider the case $|r_j| < D_1$. Fix r such that $|r| < D_1$ and let $W' \in \mathcal{W}^0$ be

defined by

$$W' = \rho((x_{\alpha_j}(r))W. \tag{70}$$

By our induction hypothesis on j, there exists a Whittaker function W_1 compactly supported mod NZ such that

$$W_1(awn) = W'(awn) \tag{71}$$

for all $a \in A_w(E)$ and all n of the form

$$n = x_{\alpha_1}(r_1) \cdots x_{\alpha_{j-1}}(r_{j-1}). \tag{72}$$

Let $W_2 = \rho((x_{\alpha_j}(-r))W_1$. Then W_2 is compactly supported mod NZ and from (70) and (71) we have

$$W_2(awnx_{\alpha_j}(r)) = W(awnx_{\alpha_j}(r))$$

for every $a \in A_w(E)$ and every n of the form (72). Since both W and W_2 are smooth on the right it follows that the above equality holds in a neighborhood of r. Since the set $\{r : |r| < D_1\}$ is compact it follows that we can cover it by a finite number of such neighborhoods. Since we can assume that each $W_2 = W_2(r)$ is given by (61) it follows that we can take the constants in (61) such that the equality will hold for all r_j such that $|r_j| < D_1$. □

Remark 9.9. Let ω be a quasi-character of Z. If $W \in \mathcal{W}^0$ satisfies $W(zg) = \omega(z)W(g)$ for every $z \in Z$ and $g \in G$ then W_1 given by (61) satisfies $W_1(zg) = \omega(z)W_1(g)$ for every $z \in Z$ and $g \in G$.

For $W \in \mathcal{W}^0$ and $g \in Bw_0B$ we let

$$J(W, g) = J_{g,w_0}(W) = \int_N W(gn)\psi^{-1}(n)dn.$$

Corollary 9.10. *Let $W \in \mathcal{W}^0$. Let U be a compact set in G. Then there exists a function $W_1 \in \mathcal{W}$ compactly supported mod NZ such that*

$$J(W_1, g) = J(W, g)$$

for every $g \in U \cap Bw_0B$.

Proof. Let $\Delta^* = \{\lambda_1, ..., \lambda_{n-1}\}$. Since U is compact it follows that λ_i is bounded on U for $i = 1, \ldots, n-1$. Hence there exists a constant E such that $\lambda(g) > E$ for every $\lambda \subset \Delta^*$ and every $g \in U$. Let $g \in U \bigcap Bw_0B$. Then $g = n_1aw_0n_2$ for some $n_1, n_2 \in N$ and $a \in A$. Let $\lambda \in \Delta^*$. By Theorem 4.1 we have

$$|\lambda(g)| = |\lambda(n_1aw_0n_2)| = |\lambda(a)||\lambda(w_0n_2)| \leq |\lambda(a)|.$$

Since $\lambda(g) > E$ it follows that $\lambda(a) > E$, hence $a \in A_w^E$. By Theorem 9.5 we can find W_1 in $\mathcal{W}$ compactly supported mod NZ such that

$$W_1(aw_0n) = W(aw_0n)$$

for every $n \in N$ and $a \in A_w(E)$. In particular we have $W_1(gn) = W(gn)$ for every $g \in U \cap Bw_0B$ and $n \in N$. Hence $J(W_1, g) = J(W, g)$ for such g which is what we wanted to prove. □

Let $\phi \in C_c^\infty(G)$ and ω a quasi-character of Z. Let $g \in Bw_0B$. We define

$$J_{\psi,\omega}(\phi, g) = \int_N \int_N \int_Z f(n_1 z g n_2)\psi^{-1}(n_1)\psi^{-1}(n_2)\omega^{-1}(z)dn_1 dn_2 dz.$$

It follows from **[12]** that this integral converges absolutely. It is easy to see that

$$J_{\psi,\omega}(\phi, g) = \int_N W_{\phi,\omega}(gn)\psi^{-1}(n)dn = J(W_{\phi,\omega}, g)$$

where $W_{\phi,\omega}$ is defined by (59).

The following Corollary implies Theorem 1.7 in the introduction.

Corollary 9.11. *Let π be an irreducible admissible generic representation of G with central character ω_π. Let $j_{\pi,\psi}$ be the Bessel function of π as defined in Section 8. Let U be a compact open set in G. Then there exists a function $\phi \in C_c^\infty(G)$ such that*

$$J_{\psi,\omega_\pi}(\phi, g) = j_{\pi,\psi}(g) \tag{73}$$

for every $g \in U \cap Bw_0B$.

Proof. Recall that $j_{\pi,\psi}(g) = j_{\pi,\psi,w_0}(g)$ is the Bessel function associated to the longest Weyl element w_0. By (8.2) there exists $W \in \mathcal{W}^0(\pi, \psi)$ such that $j_{\pi,\psi}(g) = J(W, g)$ for every $g \in Bw_0B$. Since the central character of π is ω_π it follows that $W \in \mathcal{W}^0_{\omega_\pi}(G, \psi) = \mathcal{W}_{\omega_\pi}(G, \psi) \cap \mathcal{W}^0(G, \psi)$. By Corollary 9.10 and by remark 9.6 and Remark 9.9 there exists $W_1 \in \mathcal{W}_\omega(G, \psi)$ which is compactly supported mod NZ such that $J(W_1, g) = J(W, g)$ for every $g \in U \cap Bw_0B$. By Lemma 9.2 there exists $\phi \in C_c^\infty(G)$ such that $W_{\phi,\omega} = W_1$. Hence

$$J_{\psi,\omega}(\phi, g) = J(W_{\phi,\omega}, g) = J(W_1, g) = J(W, g) = j_{\pi,\psi}(g)$$

for every $g \in U \cap Bw_0B$. □

Remark 9.12. If π is supercuspidal then there exits $\phi \in C_c^\infty(G)$ such that (73) holds for every $g \in Bw_0B$. That is, in this case the Bessel function is globally given by an orbital integral. This follows from Remark 5.2, from Lemma 9.2 and (8.2).

10. Concluding remarks

10.1. Simply laced groups Our results in this paper remain valid for split, simply laced algebraic groups over a local non-archimedean field. The main facts that we need is that Lemma 2.6 remains valid in that setting and that the root system of such groups is obtuse (see Section 3). The proofs will be the same as in the $GL(n)$ case. Since Lemma 2.6 is not valid in the setting of a split reductive group we can not yet generalize the results for that case. However, we believe that the results are still true in the general setting of a quasi-split group over a non-archimedean local field.

10.2. Distributions In [3], Bessel functions were defined using a distribution approach for every generic representation of a quasi-split group over a local field. In [2],[4] it is proved that in the case of $GL_2(F)$ and $GL_3(F)$ these are the same as the functions we defined here. Moreover, these functions are locally integrable and give the Bessel distribution everywhere. We believe that this should be true for the Bessel functions defined here. In Corollary 1.9 we have reduced the question of local integrability of the Bessel functions to the Local integrability of the orbital integrals. By [12], these orbital integrals are given asymptotically by certain germs. It remains to find good bounds for these germs.

References

[1] Zeev Averbuch, *Bessel functions of algebraic groups' representations over a local field*, Ph.D. thesis, Tel Aviv University, 1992.

[2] Ehud Moshe Baruch, *On Bessel distributions for gl_2 over a p-adic field*, J. Number Theory **67** (1997), no. 2, 190–202.

[3] ______, *On Bessel distributions for quasi-split groups*, Trans. Amer. Math. Soc. **353** (2001), no. 7, 2601–2614 (electronic).

[4] ______, *Bessel functions for $GL(3)$ over a p-adic field*, Pacific J. of Math. **211** (2003), no. 1, 1–34.

[5] J. W. Cogdell and I. I. Piatetski-Shapiro, *Stability of gamma factors for* SO$(2n+1)$, Manuscripta Math. **95** (1998), no. 4, 437–461.

[6] Stephen Gelbart and I. I. Piatetski-Shapiro, *Distinguished representations and modular forms of half-integral weight*, Invent. Math. **59** (1980), no. 2, 145–188.

[7] Harish-Chandra, *Admissible invariant distributions on reductive p-adic groups*, University Lecture Series, vol. 16, American Mathematical Society, Providence, RI, 1999, Preface and notes by Stephen DeBacker and Paul J. Sally, Jr.

[8] Roger Howe, *Classification of irreducible representations of $GL_2(F)$, F a local field*, preprint, 1978.

[9] James E. Humphreys, *Introduction to Lie algebras and representation theory*, Graduate Texts in Mathematics, vol. 9, Springer-Verlag, New York, 1978.

[10] ———, *Reflection groups and Coxeter groups*, Cambridge Studies in Advanced Mathematics, vol. 29, Cambridge University Press, Cambridge, 1990.

[11] H. Jacquet, I. I. Piatetski-Shapiro, and J. A. Shalika, *Rankin-Selberg convolutions*, Amer. J. Math. **105** (1983), no. 2, 367–464.

[12] Hervé Jacquet and Yangbo Ye, *Distinguished representations and quadratic base change for $GL(3)$*, Trans. Amer. Math. Soc. **348** (1996), no. 3, 913–939.

[13] David Soudry, *The L and γ factors for generic representations of $GSp(4, k) \times GL(2, k)$ over a local non-Archimedean field k*, Duke Math. J. **51** (1984), no. 2, 355–394.

[14] Robert Steinberg, *Lectures on Chevalley groups*, Yale University, New Haven, Conn., 1968, Notes prepared by John Faulkner and Robert Wilson.

Ehud Moshe Baruch
Department of Mathematics
Technion Institute of Technology
Haifa 32000, Israel
embaruch@math.technion.ac.il

The Rankin-Selberg method: an introduction and survey

Daniel Bump

It is a great pleasure to dedicate this paper to Steve Rallis, from whom we have all learned so very much.

This paper has three aims: to explain some ideas, to survey some of the work of the last 15 years in the Rankin-Selberg method, and particularly to indicate the great contributions of Rallis in this field. The scope of this paper is limited to constructions of L-functions, omitting noneulerian integrals, special value results, p-adic theory, and applications to number theory. A previous survey of this topic was given in [**5**]. Due to the existence of this older report, we will not be systematic about describing older work.

I would like to thank David Ginzburg for advice he gave as I prepared the lecture on which this paper is based, and Masaaki Furusawa and the referee for helpful comments on an earlier draft. This work was supported in part by NSF grant DMS-9970841.

1. Langlands L-functions

We assume the reader to be at least somewhat familiar with Langlands L-functions and the functoriality conjecture. See Borel [**4**] for precise statements. Fairly recent expository discussions of the Langlands conjectures may be found in Bump [**6**], Cogdell [**23**] and Knapp [**59**], as well as other papers in the same two volumes as the last two mentioned. We summarize Langlands L-functions only to fix notations.

If G is an algebraic group defined over a field F we will denote by G_F or $G(F)$ the group of F-rational points of G. If F is a global field and v is a place of F we will denote $G(F_v)$ as G_v.

Let G be a quasisplit reductive affine algebraic group over a nonarchimedean local field F with residue cardinality q. There is associated with G a complex analytic group ${}^LG$. It is often taken to be the product of a complex analytic group with a Galois group, but we will mostly ignore the Galois group part.

If π is a spherical representation of G_F there is associated with π a conjugacy class $A_\pi \in {}^LG$. This follows from Langlands' interpretation of the Satake isomorphism, identifying the spherical Hecke algebra on G_F with respect to a special maximal compact subgroup with a ring of class functions on ${}^LG$. If $\rho : {}^LG \longrightarrow \mathrm{GL}(d,\mathbb{C})$ is an analytic representation, then

$$L(s,\pi,\rho) = \det\left(I - \rho(A_\pi)q^{-s}\right)^{-1} \tag{1}$$

is the local Langlands L-function.

If G is a reductive affine algebraic group over a global field F with center Z, and π is an automorphic cuspidal representation with unitary central character $\omega : Z_F\backslash Z_{\mathbb{A}} \longrightarrow \mathbb{C}$, then π acts by right translation on an irreducible invariant subspace of the space $L_0^2(Z_{\mathbb{A}}G_F\backslash G_{\mathbb{A}},\omega)$ of square integrable cusp forms with central character ω. Replacing F with the completion F_v of a global field F at a nonarchimedean place v and π by the constituent π_v of an automorphic cuspidal representation we will denote this local L-function (1) as $L_v(s,\pi_v,\rho)$. There exists a finite set of places S of F such that if v is a place of F not in S then v is nonarchimedean and the localization π_v is spherical. Let $L^S(s,\pi,\rho) = \prod_{v\notin S} L_v(s,\pi_v,\rho)$. Langlands conjectured that $L^S(s,\pi,\rho)$ has analytic continuation and a functional equation under $s \longrightarrow 1-s$.

2. The $\mathrm{GL}(n) \times \mathrm{GL}(m)$ convolution

If π_1 and π_2 are automorphic representations of $\mathrm{GL}(n)$ and $\mathrm{GL}(m)$, respectively, the L-function $L(s,\pi_1\times\pi_2)$ is well understood. If $n = m = 2$, this example was the original discovery of Rankin and Selberg. It is holomorphic, except that when $n = m$ and π_1 and π_2 are contragredient representations, it has a simple pole at $s = 1$. This important fact has many applications.

Cogdell **[22]** is an excellent reference for the $\mathrm{GL}(n)\times\mathrm{GL}(m)$ convolution. It was discussed in Bump **[6]** but in classical language which has not experienced any great revival. We will use this convolution to illustrate some points later so we now review it, referring to the Cogdell article for the reader who wants more detail, and references to the original papers, particularly those of Jacquet, Piatetski-Shapiro and Shalika.

There is a noted difference between the cases $n = m$ and $n \neq m$. If $n = m$, the construction involves an Eisenstein series. Let F be a global field, $\mathbb{A}$ its adele ring. Let P be the standard maximal parabolic subgroup of $\mathrm{GL}(n)$ with Levi factor $\mathrm{GL}(n-1)\times\mathrm{GL}(1)$, and let $\delta : P_{\mathbb{A}} \longrightarrow \mathbb{C}^\times$ be the modular quasicharacter:

$$\delta\begin{pmatrix} h & * \\ & a \end{pmatrix} = |\det(h)|a^{-(n-1)}, \qquad h \in \mathrm{GL}(n-1,\mathbb{A}), a \in \mathbb{A}^\times.$$

If $s \in \mathbb{C}$ let $f_s \in \mathrm{Ind}_{P(\mathbb{A})}^{\mathrm{GL}(n,\mathbb{A})}(\delta_P^s)$. This means that f_s is a smooth function and

$$f_s(pg) = \delta_P(p)^s f_s(g).$$

We assume that the restriction of f_s to a standard maximal compact subgroup of $\mathrm{GL}(n, \mathbb{A})$ is independent of s. There is no real loss of generality in assuming that f_s is a "pure tensor" in $\mathrm{Ind}_{P(\mathbb{A})}^{\mathrm{GL}(n,\mathbb{A})}(\delta_P^s) = \otimes_v \mathrm{Ind}_{P_v}^{\mathrm{GL}(n,F_v)}(\delta_P^s)$. This means that $f_s(g) = \prod_v f_{s,v}(g_v)$ where $f_v \in \mathrm{Ind}_{P_v}^{\mathrm{GL}(n,F_v)}(\delta_P^s)$. We have an Eisenstein series

$$E(g,s) = \zeta(ns) \sum_{P_F \backslash \mathrm{GL}(n,F)} f_s(\gamma g)^s. \tag{2}$$

This series is convergent if re(s) is sufficiently large, and has meromorphic continuation to all s.

Now if ϕ_1 and ϕ_2 are $\mathrm{GL}(n)$ cusp forms in automorphic representations π_1, π_2 we consider

$$\int_{\mathrm{GL}(n,F)Z_{\mathbb{A}} \backslash \mathrm{GL}(n,\mathbb{A})} \phi_1(g)\, \phi_2(g)\, E(g,s)\, dg, \tag{3}$$

where Z is the center of $\mathrm{GL}(n)$. (We will assume for simplicity that the product of the central characters of π_1 and π_2 is trivial.) After "unfolding" this equals

$$\zeta(ns) \int_{N_{\mathbb{A}} Z_{\mathbb{A}} \backslash \mathrm{GL}(n,\mathbb{A})} W_1(g)\, W_2(g)\, f_s(g)\, dg, \tag{4}$$

where W_1 and W_2 are "Whittaker functions" to be described as follows. Let $\psi : \mathbb{A}/F \longrightarrow \mathbb{C}$ be a nontrivial additive character (a notation we will use without comment throughout this paper). Let N be the algebraic subgroup of upper triangular unipotent matrices in $\mathrm{GL}(n)$, and define a character $\psi_N : N_{\mathbb{A}} \longrightarrow \mathbb{C}$ by

$$\psi_N(n) = \psi\left(\sum_{i=1}^{n-1} n_{i,i+1}\right).$$

Then

$$W_1(g) = \int_{N_F \backslash N_{\mathbb{A}}} \phi_1(ng)\, \psi(n)\, dn, \qquad W_2(g) = \int_{N_F \backslash N_{\mathbb{A}}} \phi_2(ng)\, \psi(n)^{-1}\, dn.$$

Note that we use inverse characters ψ and ψ^{-1} in the two Whittaker functions. Due to the well-known uniqueness of Whittaker models, these functions are Euler products. That is (assuming without much loss of generality that $\phi_i = \otimes \phi_{i,v}$ is a "pure tensor" in $\pi_i = \otimes_v \pi_{i,v}$) we have

$$W_i(g) = \prod_v W_{i,v}(g_v), \tag{5}$$

where $W_{i,v}$ is a function on $\mathrm{GL}(n, F_v)$ about which we will have more to say later. Immediately however (5) and (4) show that the integral is an Euler product; in fact (4) equals

$$\prod_v \zeta_v(ns) \int_{N_v Z_v \backslash \mathrm{GL}(n,F_v)} W_{1,v}(g_v)\, W_{2,v}(g_v)\, f_{s,v}(g_v)\, dg_v, \tag{6}$$

where ζ_v is the local Dedekind zeta function. Thus if v is a nonarchimedean place with residue cardinality q_v we have $\zeta_v(ns) = (1 - q_v^{-ns})^{-1}$.

If $m = n - 1$, then the global integral is of *Hecke type*. This means that it does not involve an Eisenstein series. Instead, it looks like:

$$\int_{\mathrm{GL}(n-1,F)\backslash \mathrm{GL}(n-1,\mathbb{A})} \phi_1 \begin{pmatrix} g & \\ & 1 \end{pmatrix} \phi_2(g)\, |\det(g)|^{s-1/2}\, dg.$$

Rankin-Selberg integrals of Hecke type are fairly rare: most Rankin-Selberg integrals which have been discovered involve Eisenstein series. (Many authors do not use the term "Rankin-Selberg" unless there is an Eisenstein series, but this seems to me a questionable distinction since the the Hecke integrals which are employed to represent L-functions are more like Rankin-Selberg integrals than they are unlike them.)

If $m < n - 1$, then it is necessary to "pad" the gap between the two GL's with a unipotent integration. For example if $m = n - 2$ the integration looks like this:

$$\int_{\mathrm{GL}(n-2,F)\backslash \mathrm{GL}(n-2,\mathbb{A})} \int_{(\mathbb{A}/F)^{n-1}} \int_{\mathbb{A}/F} \phi_1 \begin{pmatrix} \boxed{g} & & \xi \\ & 1 & x \\ & & 1 \end{pmatrix} \phi_2(g)$$
$$|\det(g)|^{s-1}\, \psi(x)\, dx\, d\xi\, dg. \tag{7}$$

3. The Casselman-Shalika formula

As we mentioned before, the fact that the integral (5) is an Euler product follows from the well-known uniqueness of Whittaker functions. If (π, V) is an irreducible representation of $\mathrm{GL}(n, F)$, with F local, there can be at most one (up to scalar multiple) linear functional $\Lambda : V \longrightarrow \mathbb{C}$ such that

$$\Lambda(\pi(n)\xi) = \psi_N(n)\Lambda(\xi), \qquad n \in N(F), \xi \in V. \tag{8}$$

If F is archimedean we require that Λ extends to a functional on the space of smooth vectors. The space of functions $W_\xi(g) = \Lambda(\pi(g)\xi)$ with $\xi \in V$ is called the *Whittaker model*. Uniqueness of Whittaker models was proved by Gelfand and Graev first over finite fields, later by Gelfand and Kazhdan, Piatetski-Shapiro, and Shalika over local fields.

The notion of a Whittaker model can be extended to arbitrary split reductive groups. The notion also extends naturally to representations of adele groups, particularly automorphic representations. Not all automorphic representations have Whittaker models. An automorphic representation with a Whittaker model is called *generic*. If $G = \mathrm{GL}(n)$, however, every automorphic cuspidal representation does have a Whittaker model.

Returning to the local situation, assume that π is unramified. This means that it has a *spherical vector* ξ° invariant under the maximal compact subgroup $K = \mathrm{GL}(n, \mathfrak{o})$, where $\mathfrak{o}$ is the maximal order in F. The vector ξ, like the Whittaker functional, is unique up to scalar multiple, and we choose Λ so that $W^\circ = W_{\xi^\circ}$ takes value 1 at the identity. Then W° is the *spherical Whittaker function* for which a remarkable formula exists. This formula was conjectured by Langlands, proved by Shintani (for $\mathrm{GL}(n)$) and Kato (for more general reductive groups). A particularly beautiful proof was found by Casselman and Shalika **[21]**, and due to the importance of this proof it is habitually referred to as the Casselman-Shalika formula, though they did not discover or first prove it.

We begin by stating the formula precisely for $\mathrm{GL}(n)$, then describing it more impressionistically for an arbitrary split reductive group. Since W is left quasi-invariant under N_F and right invariant under K, its values are determined by $W(t)$ as t runs through a set of representatives of the double cosets $N_F\backslash \mathrm{GL}(n, F)/K$. We may take these to be the matrices

$$t_\lambda = \begin{pmatrix} \varpi^{\lambda_1} & & \\ & \ddots & \\ & & \varpi^{\lambda_n} \end{pmatrix},$$

where ϖ is a generator of the maximal ideal $\mathfrak{p}$ of $\mathfrak{o}$, and $\lambda = (\lambda_1, \cdots, \lambda_n)$ are a set of integers. $W(t_\lambda)$ vanishes unless λ is *dominant*, that is

$$\lambda_1 \geqslant \lambda_2 \geqslant \cdots \geqslant \lambda_n. \tag{9}$$

Let $\alpha_1, \cdots, \alpha_n$ be the eigenvalues of the semisimple conjugacy class A_π parametrizing π, as in Section 1. The Casselman-Shalika formula asserts that

$$W^\circ(t_\lambda) = \delta_B(t_\lambda)^{1/2} s_\lambda(\alpha_1, \cdots, \alpha_n),$$

where $\delta_B(t_\lambda) = \prod_i |\varpi|^{\lambda_i(n+1-2i)}$ is the modular quasicharacter of the standard Borel subgroup B_F of $\mathrm{GL}(n, F)$, consisting of upper triangular matrices, and the *Schur polynomial*

$$s_\lambda(x_1,\cdots,x_n)=\frac{\begin{vmatrix} x_1^{\lambda_1+n-1} & x_2^{\lambda_1+n-1} & \cdots & x_n^{\lambda_1+n-1} \\ x_1^{\lambda_2+n-2} & x_2^{\lambda_2+n-2} & \cdots & x_n^{\lambda_2+n-2} \\ \vdots & & & \vdots \\ x_1^{\lambda_n} & x_2^{\lambda_n} & \cdots & x_n^{\lambda_n} \end{vmatrix}}{\begin{vmatrix} x_1^{n-1} & x_2^{n-1} & \cdots & x_n^{n-1} \\ x_1^{n-2} & x_2^{n-2} & \cdots & x_n^{n-2} \\ \vdots & & & \vdots \\ x_1 & x_2 & \cdots & x_n \\ 1 & 1 & \cdots & 1 \end{vmatrix}} \tag{10}$$

The irreducible analytic representations of $\mathrm{GL}(n,\mathbb{C})$ are parametrized by integer sequences λ satisfying (9). In fact, the Weyl character formula shows that the value of $s_\lambda(\alpha_1,\cdots,\alpha_n)$ is precisely equal to $\chi_\lambda(A_\pi)$, where χ_λ is the character of the irreducible representation π_λ of $\mathrm{GL}(n,\mathbb{C})$ with highest weight vector λ, where A_π is the semisimple conjugacy class in the L-group parametrizing π, as explained in Section 1.

Thus we may rewrite the Casselman-Shalika formula

$$W^\circ(t_\lambda)=\begin{cases} \delta_B(t_\lambda)^{1/2}\chi_\lambda(A_\pi) & \text{if } \lambda \text{ is dominant;} \\ 0 & \text{otherwise}. \end{cases} \tag{11}$$

Now suppose that G is an arbitrary split reductive group over a nonarchimedean local field. Let T be a split maximal torus of G. As in the case just considered, there exists a parametrization of the irreducible representations of ${}^LG$ by the dominant elements of $T_F/T_\circ$, and if $t \in T_F$ corresponds in this parametrization to the representation whose character χ_λ has highest weight vector λ determined by t, then (11) is true. This follows from combining the Casselman-Shalika formula (for G) with the Weyl character formula (for ${}^LG$).

This seems like magic, since the initial problem had nothing to do with the representation theory of ${}^LG$. Moreover it is highly applicable magic, as we will see in the next section. In an extraordinary way, the Casselman-Shalika is a window by which the representation theory of the L-group enters into the Rankin-Selberg method.

We can generalize the Casselman-Shalika formula as follows. If H is a subgroup of G, and if ψ is a character of H such that the induced representation $\mathrm{Ind}_H^G(\psi)$ is multiplicity free, and if π is an irreducible representation of G which occurs in $\mathrm{Ind}_H^G(\psi)$, then π can be realized in a unique space $\mathcal{W}_\pi$ of functions on G, invariant under right translation, satisfying $f(hg)=\psi(h)f(g)$ for $h \in H$. Now if K is another subgroup such that $\mathrm{Ind}_K^G(1)$ is multiplicity free, then $\mathcal{W}_\pi$ can contain at most one linearly independent K-fixed element W. In the examples we have in mind, G and H will be p-adic groups and K will be a maximal compact subgroup. The problem is then to parametrize the double cosets in $H\backslash G/K$ and to compute the values of W on a set of representatives. Any such formula potentially has applications to the Rankin-Selberg method. We will not attempt to catalog these, but we will encounter a number of examples of non-Whittaker models in our survey of the known constructions.

4. Symmetric algebra decompositions

Suppose that a Rankin-Selberg integral on a group G unfolds to an integral of the Whittaker function on G. Using the Casselman-Shalika formula, each unramified integral can be expressed in terms of the character values $\chi_\lambda(A_{\pi,v})$.

On the other hand, we are hoping that this is equal to a local Langlands L-function $L_v(s,\pi_v,\rho)$, where $\rho : {}^LG \longrightarrow \mathrm{GL}(d,\mathbb{C})$ is an analytic representation. Let us therefore reconsider (1).

If $\mu_1,\cdots,\mu_d$ are the eigenvalues of $\rho(A_\pi)$ the coefficient of q^{-ks} in (1) is the k-th elementary symmetric polynomial $h_k(\mu_1,\cdots,\mu_d)$, that is, the sum of all monomials of degree k in the μ_i. This is the trace of $\rho(A_\pi)$ on the k-th symmetric power $\vee^k\rho$ of ρ. Thus

$$L_v(s,\pi_v,\rho) = \sum_{k=0}^{\infty} \mathrm{tr}(A_\pi | \vee^k \rho)\, q^{-ks}.$$

Verifying that the local Whittaker integral equals $L_v(s,\pi_v,\rho)$ amounts to computing the decomposition of the symmetric algebra $\bigvee \rho = \bigoplus_k \vee^k\rho$ into irreducible representations of ${}^LG$.

Let us illustrate this with the $\mathrm{GL}(n)\times\mathrm{GL}(n)$ convolution. The integrand is K_v invariant, where $K_v = \mathrm{GL}(n,\mathfrak{o}_v)$, and T_v contains a complete set of coset representatives of $N_v\backslash\mathrm{GL}(n,F_v)/K$. The integral (6) therefore reduces to an integral over the torus:

$$(1-q^{-ns})^{-1}\int_{Z_v\backslash T_v} W_1(t)\,W_2(t)\,f_s(t)\,\delta_B(t)^{-1}\,dt.$$

Applying the Casselman-Shalika formula this equals

$$(1-q^{-ns})^{-1}\sum_{\lambda_1\geqslant\cdots\geqslant\lambda_n=0}\chi_\lambda(A_{\pi_1})\chi_\lambda(A_{\pi_2})q^{-|\lambda|s}. \tag{12}$$

where $|\lambda|$ denotes$\sum_i\lambda_i$. Since we are dividing by Z_v we may integrate over the subtorus of diagonal matrices whose last diagonal entry is trivial; thus $\lambda = (\lambda_1,\cdots,\lambda_{n-1},0)$. Let $\alpha_1,\cdots,\alpha_n$ and $\beta_1,\cdots,\beta_n$ be the eigenvalues of A_{π_1} and A_{π_2}, respectively. Recall that we are assuming that the product of the central characters of π_1 and π_2 is trivial. This means that $\alpha_1\cdots\alpha_n\beta_1\cdots\beta_n = 1$. The

Schur polynomial satisfies

$$\chi_{(\lambda_1,\cdots,\lambda_n)}(\alpha_1,\cdots,\alpha_n) = \left(\prod \alpha_i\right)^{\lambda_n} \chi_{(\lambda_1-\lambda_n,\cdots,\lambda_{n-1}-\lambda_n,0)}(\alpha_1,\cdots,\alpha_n).$$

As a consequence of this we can rewrite (12) as

$$\sum_{\lambda_1 \geqslant \cdots \geqslant \lambda_n \geqslant 0} \chi_\lambda(A_{\pi_1})\,\chi_\lambda(A_{\pi_2})\,q^{-|\lambda|s},$$

where now the summation is over all λ such that $\lambda_1 \geqslant \cdots \geqslant \lambda_n \geqslant 0$. Indeed, the summation over λ_n produces a geometric series whose sum is $(1-q^{-ns})^{-1}$. Now we have a well-known formula of Cauchy, which was missattributed to a later writer in [5].

Proposition 1. **(Cauchy)** *We have*

$$\sum_{\lambda_1 \geqslant \cdots \geqslant \lambda_n \geqslant 0} s_\lambda(\alpha_1,\cdots,\alpha_n)\,s_\lambda(\beta_1,\cdots,\beta_n)q^{-|\lambda|s} = \prod_{i,j}(1-\alpha_i\beta_j q^{-s})^{-1}. \quad (13)$$

Cauchy's identity identifies (6) as the local L-function of the $\mathrm{GL}(n) \times \mathrm{GL}(n)$ convolution.

This is a typical example of how a symmetric algebra decomposition, combined with the Casselman-Shalika formula, underlies a local computation in the Rankin-Selberg method. Some irreducible representations have simple symmetric algebra decompositions, some have complicated ones. Both cases can appear in the Rankin-Selberg method.

Various proofs of Cauchy's identity may be found in the literature. I have not seen the following proof in print; it was told to me some years ago by Rallis.

Since s_λ is a homogeneous polynomial of degree $|\lambda|$ it is sufficient to prove the Cauchy identity with $s=0$. We omit q^{-s} from the following considerations, and assume that $|\alpha_i|, |\beta_j| < 1$.

The group $\mathrm{GL}(n,\mathbb{C}) \times \mathrm{GL}(n,\mathbb{C})$ acts on $\mathrm{Mat}_n(\mathbb{C})$ by left and right multiplication. There is a corresponding action on the ring $\mathcal{P}$ of all polynomial functions on $\mathrm{Mat}_n(\mathbb{C})$, thus:

$$(g_1,g_2)f(x) = f({}^t g_1 x g_2), \qquad g_i \in \mathrm{GL}(n,\mathbb{C}), \quad f \in \mathcal{P}, \quad x \in \mathrm{Mat}_n(\mathbb{C}).$$

This infinite-dimensional representation of $\mathrm{GL}(n,\mathbb{C}) \times \mathrm{GL}(n,\mathbb{C})$ is the symmetric algebra on $\mathbb{C}^n \otimes \mathbb{C}^n \cong \mathrm{Mat}_n(\mathbb{C})$, where $\mathbb{C}^n$ is the standard module of $\mathrm{GL}(n,\mathbb{C})$.

To decompose it into irreducibles, let us first consider the corresponding action on the affine ring $\mathcal{A}$ of $\mathrm{GL}(n,\mathbb{C})$, which is generated by the subring $\mathcal{P}$ together with the function $\det^{-1}$ which is polar on the determinant locus. We will show that as a $\mathrm{GL}(n,\mathbb{C}) \times \mathrm{GL}(n,\mathbb{C})$ module

$$\mathcal{A} \cong \bigoplus_{\lambda_1 \geqslant \cdots \geqslant \lambda_n} \pi_\lambda \otimes \pi_\lambda. \tag{14}$$

Note that in this sum, unlike (13) where $\lambda_n \geqslant 0$, we allow λ_n to be negative. This is a sum over the characters χ_λ of *all* irreducible analytic representations of $\mathrm{GL}(n,\mathbb{C})$.

To prove (14) we begin by noting that the affine ring $\mathcal{A}$ can be identified with the ring of matrix coefficients of the maximal compact subgroup $U(n)$ of $\mathrm{GL}(n,\mathbb{C})$; a *matrix coefficient* of a compact group is a function whose left and right translates span a finite-dimensional vector space. Indeed, it is not hard to see that a polynomial function on $\mathrm{GL}(n,\mathbb{C})$ (which is allowed to have factors of $\det^{-1}$) is determined by its restriction to $U(n)$, and the restriction is a matrix coefficient. Using the Peter-Weyl theorem, the ring of matrix coefficients of the compact group $U(n)$ decomposes into irreducibles as

$$\mathcal{A} \cong \bigoplus \hat{\pi}_\lambda \otimes \pi_\lambda,$$

where the sum is over all irreducible representations, and the action of $U(n) \times U(n)$ is now

$$(g_1, g_2)f(x) = f(g_1^{-1} x g_2), \qquad g_i \in U(n), \quad f \in \mathcal{A}, \quad x \in U(n).$$

Such a decomposition of the ring of matrix coefficients is valid for any compact group. Now (14) follows as a decomposition over $U(n)$ on noting that the outer automorphism $g \longrightarrow {}^t g^{-1}$ interchanges a representation and its contragredient. As a decomposition over $\mathrm{GL}(n,\mathbb{C})$ it follows from the fact that restriction to $U(n)$ gives a bijection between the irreducible analytic representations of $\mathrm{GL}(n,\mathbb{C})$ and the irreducible representations of $U(n)$.

Not every element of $\mathcal{A}$ can be extended to an element of $\mathcal{P}$. The representations which occur in $\mathcal{P}$ are those whose matrix coefficients are polynomials (not involving $\det^{-1}$). In terms of the parametrization by highest weight vectors, this means that $\lambda_n \geqslant 0$, and the Cauchy identity is now proved.

5. Doubling variables

Typically a Rankin-Selberg integral involves a pair of groups, $G \subset H$, and a pair of automorphic forms on G and H; one is a cusp form, the other an Eisenstein series. The Rankin Selberg method can involve:

I. a cusp form on H, an Eisenstein series on G;

II. a cusp form on G, Eisenstein series on H;

III. more complicated unipotent integrations, etc.

Case I is the easiest to analyze.

Case II leads to a double coset decomposition. Suppose that ϕ is a cusp form on G and $E(g,s) = \sum_{P_F \backslash H_F} f_s(\gamma g)$ is an Eisenstein series on H. We have

$$\int_{G(F)\backslash G(\mathbb{A})} \phi(g)\, E(g,s)\, dg =$$

$$\sum_{\gamma \in P_F \backslash H_F / G_F} \int_{{}^{\gamma}G(\mathbb{A})\backslash G(\mathbb{A})} f_s(\gamma k) \quad \left[\int_{{}^{\gamma}G(F)\backslash {}^{\gamma}G(\mathbb{A})} \phi(gk)\, dg\right] dk, \tag{15}$$

where ${}^{\gamma}G = G \cap \gamma^{-1} H \gamma$. If ${}^{\gamma}G$ contains the unipotent radical of a parabolic subgroup of G then the inner integral vanishes by cuspidality. If all goes well, only one particular γ will contribute, and that integral can be reduced to a Whittaker one, or to another unique model.

A type II example is worked out in Bump **[6]** Section 3.10.

As an example of this type of integration, Piatetski-Shapiro and Rallis **[70]** gave a construction which included the standard L-functions of all classical groups. The symplectic case of the same construction was found independently by Böcherer **[3]**. This construction is known as the "doubling variables" method.

For definiteness let us assume that $G = O(n) \times O(n)$. Let $H = O(2n)$. Take an Eisenstein series on H for the parabolic P whose Levi factor is $\mathrm{GL}(1) \times O(2n-2)$, and consider the integral

$$\int_{G(F)\backslash G(\mathbb{A})} \phi_1(g_1)\, \overline{\phi_2(g_2)}\, E\big((g_1, g_2), s\big)\, dg_1 \times dg_2.$$

To use (15) one must compute the double cosets $P\backslash H/G$, and show that only one double coset contributes to the integral. The integral unfolds to

$$\int_{O(n,F)\backslash O(n,\mathbb{A})} \langle \phi_1, \pi(g)\, \phi_2 \rangle\, f_s\big(\gamma(g,g)\big)\, dg,$$

for suitable $\gamma \in H_F$. Here

$$\langle \phi_1, \pi(g)\phi_2 \rangle = \int_{O(n,F)\backslash O(n,\mathbb{A})} \phi_1(x) \overline{\phi_2(xg)}\, dx$$

is the usual L^2 inner product, with $\pi(g)\, \phi_2(x) = \phi_2(xg)$ since the group acts by right translation. We see that we must take ϕ_1 and ϕ_2 in the same automorphic representation of $O(n)$, if the integral is nonzero. The matrix coefficient $\langle \phi_1, \pi(g)\phi_2 \rangle$ is Eulerian, and its local constituent is the spherical function at almost all places.

This construction was quite different from anything seen before. The appearance of matrix coefficients is not quite unprecedented since they also appear in Godement and Jacquet **[49]**. It works equally well with symplectic or unitary groups.

6. The Rankin-Selberg and Langlands-Shahidi methods contrasted

The Rankin-Selberg method seeks to construct Langlands L-functions as integrals of automorphic forms. An alternative approach to this problem is the Langlands-Shahidi method, in which the Langlands L-function is realized as the Whittaker coefficient of an Eisenstein series on a larger group. See Shahidi [78] for a survey of the recent spectacular successes of the Langlands-Shahidi method.

Over the years, there has been a change in how these two methods are viewed, as we will now explain.

It should be born in mind that the Langlands-Shahidi method constructs a fixed and known list of L-functions. To construct a Langlands L-function of an automorphic form on G by this method, one needs to find a maximal parabolic subgroup of a larger group H whose Levi subgroup is isomorphic to G. One may read off from the Dynkin diagrams of the various semisimple Lie groups the Levi factors of their parabolic subgroups, and then a bit of additional work is needed to determine what Langlands L-function is obtained. Those with connected Levi subgroup were listed in Langlands **[62]**, and those with disconnected factors are listed in [77].

By contrast, the Rankin-Selberg method does not construct a fixed and known list of L-functions. There is no classification theorem for L-functions constructible by the Rankin-Selberg method, only a set of examples and heuristics. By now, Rankin-Selberg constructions are becoming harder to find, but there remains the tantalizing possibility that new ideas will bring new examples.

Both the Rankin-Selberg and Langlands-Shahidi methods fall far short of the goal of constructing all Langlands L-functions. Both construct enough L-functions for nontrivial applications.

Until around 1985, comparatively few Rankin-Selberg constructions were known. The $\mathrm{GL}(n)\times\mathrm{GL}(m)$ construction and the Jacquet-Godement construction of the standard L-function were known. The exterior square construction of Jacquet and Shalika **[55]** was known but not yet published. There were also some constructions for classical groups due to Novodvorsky, Piatetski-Shapiro, Gelbart and Piatetski-Shapiro, and others, though the most important constructions for the classical groups were not yet available. There were some important symplectic constructions of Andrianov, Shimura's integrals using theta functions, and Asai's constructions. The survey of Bump **[6]**, written shortly after this time, gives an indication of this state of affairs.

In 1985 it was suspected by many that all Rankin-Selberg L-functions were also constructible by the Langlands-Shahidi method, and it was possible to believe that the Rankin-Selberg list was a proper subset of the Langlands-Shahidi list. Nevertheless it was thought that when the Rankin-Selberg method did work it gave more information.

These opinions seemed reasonable in 1985 but they did not stand the test of time. In the spring of 1988 Piatetski-Shapiro, Rallis and Schiffman **[73]** found a Rankin-Selberg construction for the standard L-function of $G_2 \times \mathrm{GL}(2)$. (Surprisingly, the standard L-function of $G_2 \times \mathrm{GL}(2)$ proved a little easier than the standard L-function of G_2 by itself, which was constructed much later by Ginzburg **[39]**.) Since $G_2 \times \mathrm{GL}(2)$ is not the Levi factor of a parabolic subgroup, this L-function cannot be constructed as a Langlands-Shahidi L-function. After this example, the Rankin-Selberg list was no longer a subset of the Shahidi list. Moreover, since this time, the Rankin-Selberg list has expanded so that there are now only a few Langlands-Shahidi L-functions for which Rankin-Selberg constructions are not known.

The second belief, that the Rankin-Selberg gives more precise information than the Langlands-Shahidi method, has also proved unfounded. A strong motivation for developing both methods was found in the converse theorem, developed by Piatetski-Shapiro and Cogdell, and applied by them and other collaborators **[25]**. If G is a classical group, its (connected) L-group is also a classical group and has a "standard" representation ${}^LG \longrightarrow \mathrm{GL}(r)$, so there is a standard L-function. We may tensor this standard representation with the standard representation of the (connected) L-group $\mathrm{GL}(n, \mathbb{C})$ of $\mathrm{GL}(n)$, and obtain the standard L-functions of $G \times \mathrm{GL}(n)$, an Euler product of degree rn associated with a cusp form on G and a cusp form on $\mathrm{GL}(n)$. The converse theorem shows that if functional equations for such L-functions can be proved, together with analyticity and boundedness in vertical strips, then the cusp form on G can be lifted to a cusp form on $\mathrm{GL}(r)$, in accordance with the functoriality conjecture.

Happily, these $G \times \mathrm{GL}(n)$ L-functions are quite approachable by both the Rankin-Selberg and Langlands-Shahidi methods, though the Rankin-Selberg constructions were only known in special cases in 1985. The $G \times \mathrm{GL}(n)$ constructions are much easier than, say, $G \times G$. I believe that the only direct Rankin-Selberg construction for $G \times G$ which is known where G is a classical group of rank > 1 is the case of the spin $\times$ spin L-function for $G \times G$ when $G = \mathrm{GSp}(4)$, in Jiang **[56]**. (Of course if the lifting from classical groups is known, these L-functions can be constructed by first lifting to $\mathrm{GL}(r)$, so *indirect* constructions are possible.) The same is true for the Langlands-Shahidi method—if G is a classical group, the $G \times \mathrm{GL}(n)$ L-functions can be constructed, but the $G \times G$ constructions cannot be.

Once Rankin-Selberg constructions were found for the $G \times \mathrm{GL}(n)$ L-functions it was believed for a time that these would be better suited for obtaining control of the poles needed for the converse theorem. During the early 1990's, much work was devoted toward developing the Rankin-Selberg method for $G \times \mathrm{GL}(n)$.

But it was shown by Moeglin and Waldspurger **[64]** for the $\mathrm{GL}(n) \times \mathrm{GL}(n)$ convolution that Shahidi's method could precisely determine the poles, filling a gap in the literature, since although the Rankin-Selberg construction was well known, but the details in print were quite sketchy. After this, the Langlands-Shahidi method was developed by Kim and Shahidi, culminating in the work of Cogdell, Kim, Piatetski-Shapiro and Shahidi **[24]** proving liftings from classical groups to $\mathrm{GL}(r)$. Around this time and in the same current Ramakrishnan **[75]** proved the automorphicity of the Rankin-Selberg L-function on $\mathrm{GL}(2) \times \mathrm{GL}(2)$.

Either the Rankin-Selberg method or the Langlands-Shahidi method would probably have proved sufficient for the lifting program, though ultimately it was the Langlands-Shahidi method that was used.

Although the above history presents the Rankin-Selberg method and the Langlands-Shahidi method as competing programs, they are related to each other. Certain definitions, ultimately applied in the Langlands-Shahidi method were inspired by corresponding definitions first found in the Rankin-Selberg method.

7. $G \times \mathrm{GL}(n)$, G orthogonal

The importance of the $G \times \mathrm{GL}(n)$ constructions, with G a classical group, was explained in the previous section. The completion of the case where G is orthogonal and π generic, begun by Piatetski-Shapiro, Novodvorsky and Gelbart, was completed by Ginzburg and Soudry.

After the early work of Novodvorsky and Piatetski-Shapiro there came the work of Gelbart and Piatetski-Shapiro **[36]** which treated $G \times \mathrm{GL}(n)$ in the special case where the rank of G is n. In a nutshell, this construction is as follows. If $G = \mathrm{SO}(2n+1)$, one constructs with the given cusp form on $\mathrm{GL}(n)$ an Eisenstein series on $\mathrm{SO}(2n)$, restricts the cusp form from $\mathrm{SO}(2n+1)$ to $\mathrm{SO}(2n)$ and integrates against this Eisenstein series. Or if $G = \mathrm{SO}(2n-1)$, one can still integrate against the same Eisenstein series on $\mathrm{SO}(2n)$, restricting it to $\mathrm{SO}(2n-1)$.

Alternatively, if $G = \mathrm{SO}(2n)$, one forms an Eisenstein series on $\mathrm{SO}(2n+1)$ with the $\mathrm{GL}(n)$ cusp form, restricts to $\mathrm{SO}(2n)$, and integrates against the cusp form there. One can do the same thing if $G = \mathrm{SO}(2n+2)$.

This is a special case of a more general construction for $\mathrm{SO}(2r+1) \times \mathrm{GL}(n)$ and $\mathrm{SO}(2r) \times \mathrm{GL}(n)$. We see that in 1985 the construction was known when r and n are equal or nearly so.

In the general case one proceeds in the same way—for $\mathrm{SO}(2r+1) \times \mathrm{GL}(n)$ one makes an Eisenstein series with the $\mathrm{GL}(n)$ cusp form on $\mathrm{SO}(2n)$, and for $\mathrm{SO}(2r) \times \mathrm{GL}(n)$ one makes an Eisenstein series on $\mathrm{SO}(2n+1)$. In either case, one has an Eisenstein series and a cusp form, one on an odd orthogonal group and one on an even orthogonal group. If the ranks differ, however, one must "pad" the difference between the two orthogonal groups with a unipotent integration. We saw this already with the $\mathrm{GL}(n) \times \mathrm{GL}(m)$ convolutions in (7). This is not a serious complication but may have delayed the discovery of these integrals by a few years. If $r > n$ this construction was found by Ginzburg [37].

Let us illustrate the need for a unipotent integration with an example. We describe the Rankin-Selberg convolution for $\mathrm{SO}(7) \times \mathrm{GL}(2)$. Make an Eisenstein series on $\mathrm{SO}(4)$ with the $\mathrm{GL}(2)$ cusp form τ. One may embed $\mathrm{SO}(4)$ into $\mathrm{SO}(7)$ and integrate against the $\mathrm{SO}(7)$ cusp for ϕ as follows:

$$\int_{O(4,F)\backslash O(4,\mathbb{A})} \int_{\mathbb{A}/F} \int_{(\mathbb{A}/F)^4} E_\tau(h,s)\,\psi(r)$$
$$\phi\left(\begin{pmatrix} 1 & & & & x_1 & & \\ & 1 & & & x_2 & & \\ x_4 & x_3 & 1 & r & * & * & * \\ & & & 1 & -r & & \\ & & & & 1 & & \\ & & & & * & 1 & \\ & & & & * & & 1 \end{pmatrix} \begin{pmatrix} \boxed{h} & & \boxed{h} \\ & I_3 & \\ \boxed{h} & & \boxed{h} \end{pmatrix}\right) dx_i\, dr\, dh$$

here $\psi : \mathbb{A}/F \longrightarrow \mathbb{C}$ is a nontrivial additive character. The need to pad the integration with a unipotent group explains the fact that the equal rank case was discovered historically earlier. (Although the case $n = 1$ was also discovered early, by Novodvorsky.)

The $\mathrm{SO}(4) \times \mathrm{GL}(3)$ construction is outwardly similar: one makes an Eisenstein series on $\mathrm{SO}(7)$ with the $\mathrm{GL}(3)$ cusp form and considers the same integral, with the same unipotent subgroup, except that the roles of the cusp form and Eisenstein series are interchanged.

However there is an important difference between the cases n small and n large. If $n \leqslant r$ (where $G = \mathrm{SO}(2r)$ or $\mathrm{SO}(2r+1)$) then the decomposition of the symmetric algebra of the L-group is simple and easily comprehensible. In the classification of Section 5, this is a Type I integral, which unfolds to a straightforward integral of the Whittaker function. The unramified computation is therefore simple.

If $n > r$, then the decomposition of the symmetric algebra is chaotic. The integration is of Type II, and the unramified computation seems intractable.

Luckily Soudry [82] p. 97 was able to solve this problem very ingeniously. We have already observed a similarity between the constructions for $\mathrm{SO}(7) \times \mathrm{GL}(2)$ and $\mathrm{SO}(4) \times \mathrm{GL}(3)$. More generally, we consider the constructions for $\mathrm{SO}(2r+1) \times \mathrm{GL}(n)$ and $\mathrm{SO}(2n) \times \mathrm{GL}(r)$ to be in duality. Note that not only are the global integrals similar, with the roles of the cusp form and Eisenstein series interchanged, but the local integrals are Euler products of the same degree $2rn$. (The L-groups are $\mathrm{Sp}(2r) \times \mathrm{GL}(n)$ and $\mathrm{SO}(2n) \times \mathrm{GL}(r)$.) Soudry was able to find a method of transforming the (hard) unramified computation for $SO(2r+1) \times GL(n)$ with n large into the (easy) unramified computation for $SO(2n) \times GL(r)$.

For automorphic cuspidal representations which are not generic, including those over orthogonal groups which may not be split or quasisplit, a complete and extensive theory was given by Ginzburg, Piatetski-Shapiro and Rallis **[44]**. The Whittaker model is replaced by more general unique models (Bessel model). Modifying (8) for the Bessel model the unipotent group N is replaced by another which may involve a smaller orthogonal group as well as a normal unipotent subgroup. An earlier such construction, which is a special case of the general one of Ginzburg, Piatetski-Shapiro and Rallis, but which may still be useful as an example is Furusawa **[32]**. In this paper Furusawa constructs the L-function for $SO(4) \times GL(2)$

Yet another construction was found by Murase and Sugano **[66]**, who consider the restriction of an Eisenstein series induced from the parabolic subgroup with Levi factor $GL(1) \times O(n-1)$ to $O(n+1)$, restricted to $O(n)$ and integrated against a cusp form there. This construction thus involves cusp forms on $O(n-1)$ and $O(n)$.

This example illustrates an observation, made by Bernstein, that there is a strong tendency for any integral of an $O(n)$ automorphic form against an $O(n-1)$ automorphic form, if nonzero, to be an L-function. One may take a cusp form on the larger group and an Eisenstein series on the smaller one or conversely. The explanation for this is contained in a uniqueness principle, to be discussed in the next section. Example of this type are Murase and Sugano **[67]** and the Appendix by Furusawa to Bump, Friedberg and Furusawa **[8]**. This phenomenon means that orthogonal constructions may be fairly abundant and difficult to classify.

8. Uniqueness principles

We have already seen one uniqueness principle underlying the $GL(n) \times GL(n)$ convolution, namely that the global Rankin-Selberg integral (3) unfolds to a unique model, the Whittaker model. Yet the global integral (3) itself reflects a uniqueness theorem: this integral may be regarded as an invariant trilinear form on the product of three representations spaces, π_1, π_2 and (corresponding to the Eisenstein series) $\mathrm{Ind}_P^{GL(n)}(\delta_P^s)$. There is a uniqueness for such trilinear forms, which we may state locally, though it is true also globally.

Theorem 1. *Let F be a local field, let (π_1, V_1) and (π_2, V_2) be irreducible admissible representations of $GL(n, F)$, and let $V_3(s) = \mathrm{Ind}_{P(F)}^{GL(n,F)}(\delta_P^s)$. If s is* *in general position, the space of invariant trilinear forms* $T : V_1 \times V_2 \times V_3(s) \longrightarrow \mathbb{C}$ *is at most one dimensional.*

See Jacquet, Piatetski-Shapiro and Shalika **[53]** for this result, and Gelbart and Piatetski-Shapiro **[36]** for a discussion of how this sort of result is proved, using the Bernstein-Zelevinsky theory of derivatives. Although this is a strictly local statement, the proof is parallel to the unfolding of the global integral (3).

An important paradigm is that uniqueness theorems in representation theory correspond to constructions of L-functions, or sometimes special values of L-functions.

To explain what this means suppose that G is a reductive algebraic group over a global field F and H a subgroup, and suppose that (τ, V_τ) is an automorphic representation of H such that the induction of τ to G is multiplicity free. Let (π, V_π) be an automorphic representation of G. We can consider the integral

$$\int_{H(F)\backslash H(\mathbb{A})} f(h)\phi(h)\, dh, \qquad f \in V_\tau, \phi \in V_\pi.$$

This integral is an H-invariant bilinear form $V_\tau \times V_\pi \longrightarrow \mathbb{C}$, and by Frobenius reciprocity, the fact that $\mathrm{Ind}(\tau)$ is multiplicity free means that this functional is unique. Therefore we expect that unless the integral vanishes identically (which often happens) it may represent a value of an L-function. If f is an Eisenstein series, we may get an L-function of π, or if ϕ is an Eisenstein series, we may get an L-function of τ. If neither automorphic form is an Eisenstein series, we may get a special value or residue of an L-function.

As an example, we may take $G = \mathrm{GL}(n) \times \mathrm{GL}(n)$ and $H = \mathrm{GL}(n)$ embedded diagonally in G, and take $\tau = \mathrm{Ind}_P^G(\delta_P^s)$. The uniqueness theorem quoted above may be paraphrased by saying that inducing τ to G gives a multiplicity free representation. Therefore we expect the Rankin-Selberg integral (3) to be Eulerian.

We may now explain the remark at the end of the last section. In an influential lecture of May, 1989 at the Piatetski-Shapiro symposium Bernstein pointed out, among other things, that $O(n-1)$ inside $O(n)$ is a full multiplicity one subgroup, namely if τ is any irreducible representation of $O(n-1)$ its induction to $O(n)$ is multiplicity free. As a consequence, orthogonal integrals of the type described at the end of the last section have a strong tendency to be Eulerian.

9. The symmetric and exterior squares

It is expected that a self-dual automorphic representation π of $\mathrm{GL}(n)$ is a lift from a classical group. To see this, note that $L(s, \pi \times \pi) = L(s, \pi \times \hat{\pi})$ has a simple pole from the general properties of the Rankin-Selberg convolution. We can factor

$$L(s,\pi\times\pi)=L(s,\pi,\vee^2)L(s,\pi,\wedge^2),$$

into the symmetric and exterior square L-functions, corresponding to the representations

$$\vee^2:\mathrm{GL}(n,\mathbb{C})\longrightarrow \mathrm{GL}\left(\tfrac{n(n+1)}{2},\mathbb{C}\right),\qquad \wedge^2:\mathrm{GL}(n,\mathbb{C})\longrightarrow \mathrm{GL}\left(\tfrac{n(n-1)}{2},\mathbb{C}\right).$$

Thus one of these L-functions has a pole, the other doesn't. It follows from a Rankin-Selberg representation that $L(s,\pi,\wedge^2)$ can have a pole only of n is even. A heuristic based on Galois representations (explained in the introduction to **[15]**) shows that the possibilities should be:

	Polar L-function	Lift From
$n=2m$	$L(s,\pi,\vee^2)$	SO($2m$)
$n=2m$	$L(s,\pi,\wedge^2)$	SO($2m+1$)
$n=2m+1$	$L(s,\pi,\vee^2)$	Sp($2m$)

In Ginzburg, Rallis and Soudry **[48]** it is shown that indeed, the existence of a pole of $L(s,\pi,\wedge^2)$ implies that π is a lift from SO($2m+1$), and the other predicted liftings can be similarly realized.

Shimura **[79]** and **[80]** gave two Rankin-Selberg constructions involving the Jacobi theta function

$$\theta(z)=\sum_{n=-\infty}^{\infty} e^{\pi i n^2 z},\quad z=x+iy, y>0. \tag{16}$$

Both constructions have many important generalizations, and they should be mentioned together because they are in some sense dual to each other.

This θ is a modular form of weight 1/2, and in our present view is actually an automorphic form on the metaplectic double cover $\widetilde{\mathrm{GL}}(2)$ of GL(2). This is a central extension:

$$1\longrightarrow\{\pm1\}\longrightarrow\widetilde{\mathrm{GL}}(2)\longrightarrow\mathrm{GL}(2)\longrightarrow 1. \tag{17}$$

Given an automorphic representation of $\widetilde{\mathrm{GL}}(2)$, either $\{\pm1\}$ acts trivially, in which case the automorphic representation is actually a representation of GL(2), or it acts nontrivially, in which case we say the representation is *genuine*. The automorphic form θ, like all automorphic forms of half integral weight, belongs to a genuine automorphic representation of $\widetilde{\mathrm{GL}}(2)$.

One may consider a convolution varying (3) of the form:

$$\int_{Z_{\mathbb{A}}\,\mathrm{GL}(2,F)\backslash\,\mathrm{GL}(2,\mathbb{A})}\theta(g)\,\phi(g)\,E(g,s)\,dg. \tag{18}$$

The integrand must be invariant under $\{\pm 1\}$ (see (17)), so two of these forms must be genuine, and one must be nongenuine. We can, if we prefer, just integrate over SL(2) since the theta function is essentially supported there modulo the center.

If ϕ is a cusp form of half integral weight $k+\frac{1}{2}$, this is an Euler product of degree two, and the functional equation is identical to that of a modular form of weight $2k$. This leads to the Shimura correspondence.

If ϕ is an ordinary automorphic form (not genuine) this is an Euler product of degree 3, corresponding to the symmetric square representation ${}^L\mathrm{GL}(2) = \mathrm{GL}(2,\mathbb{C}) \longrightarrow \mathrm{GL}(3,\mathbb{C})$. This construction was generalized to GL(3) by Patterson and Piatetski-Shapiro **[69]** and to GL(n) by Bump and Ginzburg **[15]**.

For the exterior square L-function, there are two constructions known, related to each other and both important. We will describe these constructions for GL($2n$). Let $\phi \in V$ where (π, V) is an automorphic cuspidal representation of GL($2n$).

The first construction, due to Jacquet and Shalika **[55]** involves the following integral:

$$\int_{Z_{\mathbb{A}}\,\mathrm{GL}(n,F)\backslash\,\mathrm{GL}(n,\mathbb{A})} \int_{\mathrm{Mat}_n(F)\backslash\,\mathrm{Mat}_n(\mathbb{A})} \phi\left(\begin{pmatrix} I & X \\ & I \end{pmatrix}\begin{pmatrix} g & \\ & g \end{pmatrix}\right) \overline{\psi(\mathrm{tr}\ X)} E(g,s)\, dX\, dg,$$

where the Eisenstein series $E(g,s)$ is the same one which appeared in (3). From the fact that this represents the exterior square L-function, and from the fact that the residue at $s=1$ of $E(g,s)$ is constant, we see that the period:

$$\int_{Z_{\mathbb{A}}\,\mathrm{GL}(n,F)\backslash\,\mathrm{GL}(n,\mathbb{A})} \int_{\mathrm{Mat}_n(F)\backslash\,\mathrm{Mat}_n(\mathbb{A})} \phi\left(\begin{pmatrix} I & X \\ & I \end{pmatrix}\begin{pmatrix} g & \\ & g \end{pmatrix}\right) \overline{\psi(\mathrm{tr}\ X)}\, dX\, dg$$

is nonzero if and only if $L(s,\pi,\wedge^2)$ has a pole at $s=1$. Conjecturally, we have already observed, this is true if and only if π is a functorial lift from SO($2n+1$). This is the *Shalika period*. It corresponds to a functional proved unique by Jacquet and Rallis **[54]**, namely the *Shalika functional*. The corresponding model is the *Shalika model*.

Specifically, let (π, V) be an irreducible representation of GL(n, F) over a local field F. A Shalika functional Λ is one which satisfies

$$\Lambda\left(\pi\begin{pmatrix} I & X \\ & I \end{pmatrix}\pi\begin{pmatrix} g & \\ & g \end{pmatrix}v\right) = \psi(\mathrm{tr}\, X)\Lambda(v).$$

By the result of Jacquet and Rallis, if it exists, it is unique; and our expectation is that it exists if π is self-dual, and a local lifting from SO($2n+1$).

The second Rankin-Selberg integral was found by Bump and Friedberg [7]. It involves two parameters and represents a product of two L-functions. It is:

$$\int_{Z_{\mathbb{A}}(\mathrm{GL}(n,F)\times\mathrm{GL}(n,F))\backslash(\mathrm{GL}(n,\mathbb{A})\times\mathrm{GL}(n,\mathbb{A}))} \phi\begin{pmatrix} g_1 & \\ & g_2\end{pmatrix}$$

$$\left|\frac{\det(g_1)}{\det(g_2)}\right|^{s_1-1/2} E(g_1,s_2)\,d(g_1\times g_2).$$

It represents $L(s_1,\pi)L(s_2,\pi,\wedge^2)$. From this (taking the residue at $s=1$) we learn that the integral

$$\int_{Z_{\mathbb{A}}\left(\mathrm{GL}(n,F)\times\mathrm{GL}(n,F)\right)\backslash\left(\mathrm{GL}(n,\mathbb{A})\times\mathrm{GL}(n,\mathbb{A})\right)} \phi\begin{pmatrix} g_1 & \\ & g_2\end{pmatrix}\left|\frac{\det(g_1)}{\det(g_2)}\right|^{s-1/2} d(g_1\times g_2)$$

is nonvanishing only if $L(s,\pi,\wedge^2)$ has a pole, in which case it represents the standard L-function. Friedberg and Jacquet [30] studied this integral directly without making use of the Bump-Friedberg construction. They showed that it unfolds to the Shalika model and represents the standard L-function.

Another construction unfolding to the Shalika model may be found in Bump, Friedberg and Ginzburg [9]. This is a construction on $\mathrm{SO}(2n+2)$. However if $n=2$, making use of the relationship between SO(6) and GL(4), it may be interpreted as a Rankin-Selberg integral on GL(4) which vanishes unless the cuspidal automorphic form π on GL(4) is such that $L(s,\pi,\wedge^2)$ has a pole at $s=1$. It represents a degree 5 L-function. This is not a Langlands L-function attached to π. However assuming that $L(s,\pi,\wedge^2)$ has a pole, it is expected that π is the lift of an automorphic represenation π' of SO(5), and this is the degree 5 L-function of π'.

10. $G\times\mathrm{GL}(n)$, G symplectic

Rankin-Selberg constructions for symplectic groups tend to look a little different from constructions for orthogonal groups. Typically they are Shimura integals like (18) involving a theta function. It should be remembered that $\mathrm{GL}(2)=\mathrm{GSp}(2)$ and $\mathrm{SL}(2)=\mathrm{Sp}(2)$, so (18) can be regarded as an archetypal symplectic integral. The phenomenon of a "See-Saw," described in Kudla [61] and developed in Rallis [74] should be borne in mind in approaching symplectic integrals, and their relationship with orthogonal ones. A typical see-saw looks like this:

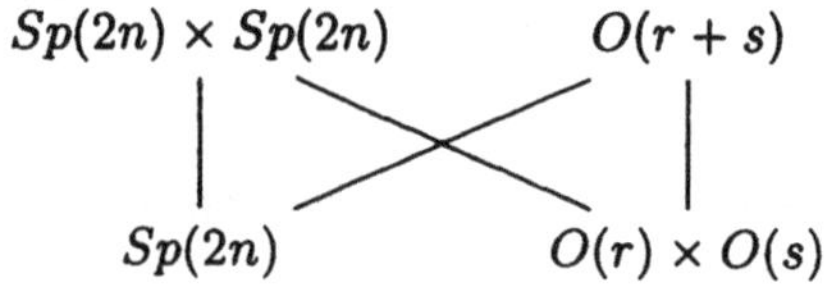

All groups in the picture fit into one large metaplectic group $\widetilde{\mathrm{Sp}}(2n(r+s))$ in such a way that the centralizer of $\mathrm{Sp}(2n)$ is $O(r+s)$, and the centralizer of $\mathrm{Sp}(2n) \times \mathrm{Sp}(2n)$ is $O(r) \times O(s)$. If an $\mathrm{Sp}(2n)$ in this diagram is paired with $O(k)$, k odd, we should write it as $\widetilde{\mathrm{Sp}}(2n)$ since the metaplectic cover does not split over it in this case. Two out of three of the numbers r, s and $r+s$ will be odd, the other even (unless they are all even). This means that two out of three of the symplectic groups on the left will be metaplectic groups $\widetilde{\mathrm{Sp}}(2n)$, the third a symplectic group $\mathrm{Sp}(2n)$.

The vertical lines are embeddings; on the left we have the diagonal embedding of $\mathrm{Sp}(2n)$ in $\mathrm{Sp}(2n) \times \mathrm{Sp}(2n)$. The diagonal lines are theta lifts.

By the general theory of dual reductive pairs and Howe duality, restricting the automorphic Weil representation of $\widetilde{\mathrm{Sp}}(2n(r+s))$ produces automorphic theta kernels for theta lifts of automorphic forms from $\mathrm{Sp}(2n)$ to $O(r+s)$, and from $O(r) \times O(s)$. This allows us to transfer Rankin-Selberg constructions from orthogonal groups to symplectic groups. The symplectic integrals appear as trilinear expressions like (18), involving three automorphic forms, typically one of them (say the lift from $O(s)$) a recognizable theta series, another an Eisenstein series, and yet another a cusp form.

We note that depending on how it is embedded in $\widetilde{\mathrm{Sp}}(2n(r+s))$, the metaplectic cover may or may not split over $\mathrm{Sp}(2n)$. In the see-saw, one of the three $\mathrm{Sp}(2n)$'s is paired with $O(k)$ with k even, the cover splits; otherwise it does not split.

As an example, take $r = 2n$ and $s = 1$. The see-saw looks as follows:

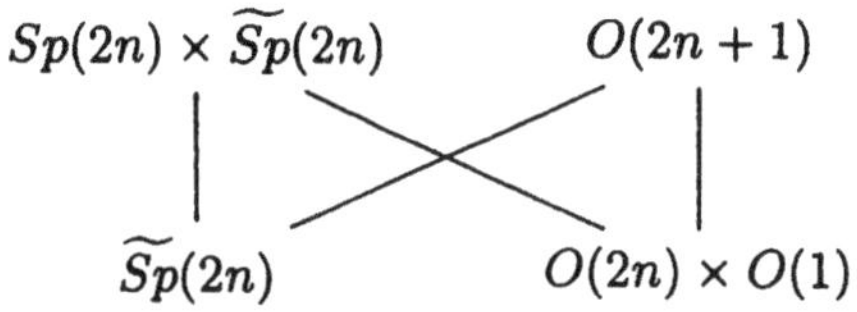

On the top $\widetilde{\mathrm{Sp}}(2n)$ we have the theta lift from $O(1)$, which is a theta function generalizing Jacobi's theta function (16). We will take a cusp form ϕ on $\mathrm{Sp}(2n)$ and an Eisenstein series on the bottom $\widetilde{\mathrm{Sp}}(2n)$. For the Eisenstein series, we can take a cusp form τ on $\mathrm{GL}(n)$ (Levi factor of the Siegel parabolic) and make an Eisenstein series on $\widetilde{\mathrm{Sp}}(2n)$. This is possible because the metaplectic double cover splits over GL(n). The integral on the left side is:

$$\int_{\mathrm{Sp}(2n,F)\backslash\,\mathrm{Sp}(2n,\mathbb{A})} \phi(g)\, E_\tau(g,s)\, \theta(g)\, dg. \tag{19}$$

The corresponding integral on the orthogonal side is the $O(2n) \times \mathrm{GL}(n)$ convolution of Gelbart and Piatetski-Shapiro **[36]**, already explained in Section 7. The symplectic integral just explained is also in Gelbart and Piatetski-Shapiro **[36]**.

If the lift from Sp($2n$) to $O(2n)$ in this diagram is zero, we enlarge the orthogonal groups on the right and lift to $O(2n+2)$ instead.

For generic cusp forms, a complete theory of Rankin-Selberg integrals for $\mathrm{Sp}(2r) \times \mathrm{GL}(n)$ was given by Ginzburg, Rallis and Soudry **[47]**. This has many features in common with the orthogonal theory. First, the "equal rank" case, just discussed, is simpler, in avoiding any unipotent integration. In the unequal rank case, the cases $r < n$ and $r > n$ are different, but in duality. The case where $r < n$ is the hardest, and requires Soudry's idea of transforming the hard case into the easy case. All these features are perfectly parallel with the orthogonal case.

As we explained at the end of Section 7, orthogonal examples may be abundant and difficult to classify. By duality, the same may be true with symplectic constructions. One fairly unique construction is that of Murase and Sugano **[65]**, involving a Heisenberg model that does not fit into the above description.

11. Metaplectic groups

Many symplectic integrals will have variants for genuine cusp forms on the metaplectic group. Just as with Shimura's integral, we may interchange the roles in (19) of the Eisenstein series and the cusp form. This means that the cusp form will be genuine and the Eisenstein series will not. As a consequence we obtain an Eulerian Rankin-Selberg associated with a genuine cusp form on $\widetilde{\mathrm{Sp}}(2n)$ and a GL(n) cusp form. This L-function has degree $2n \times n$ instead of degree $(2n+1)\times n$, a phenomenon already noted with the Shimura integral (18) when $n = 1$. Similarly, the metaplectic integrals of Zhuravlev **[88]** resemble those of Andrianov **[1]** but give an L-function of degree $2n$ instead of $2n+1$. This shows that the standard L-function on Sp($2n$) has degree $2n+1$, but the standard L-function on the double cover $\widetilde{\mathrm{Sp}}(2n)$ has degree $2n$.

Matsumoto **[63]** defined a metaplectic cover of degree r generalizing (17) over G_F (F local) or $G_{\mathbb{A}}$ (F global) if F contains the r-th roots of unity and G is simply connected. If G is not simply connected you may still define a cover but you may require more roots of unity. For example the double cover of $O(n)$ is defined if F contains the fourth roots of unity.

Although Langlands' definition of the L-group does not include higher metaplectic covers, the Shimura correspondence and its generalizations show that an L-group can be defined. Savin **[76]** computed Iwahori Hecke algebras for covering groups and the answer can be inferred from his results. In general, Savin's computations show that the L-group of the metaplectic r-fold cover of $\mathrm{Sp}(2n)$ should be defined as $O(2n+1,\mathbb{C})$ if r is odd, and $\mathrm{Sp}(2n,\mathbb{C})$ if r is even. This is why the standard L-function on $\mathrm{Sp}(2n)$ has degree $2n+1$, but the standard L-function on the double cover $\widetilde{\mathrm{Sp}}(2n)$ has degree $2n$.

Generally Rankin-Selberg integrals on covering groups resemble (as global integrals) known Rankin-Selberg integrals on non-metaplectic groups. The Kazhdan-Patterson generalized theta series on the r-fold cover **[58]** of $\mathrm{GL}(r)$ often appears as a *deus ex machina*, convolution with which will produce an Euler product. Examples of this type are described in the next section.

12. Nonunique models

Most known examples of Rankin-Selberg integrals unfold to unique models such as the Whittaker model. Among the more exotic constructions, the Whittaker model is predominant. In a few known cases, the integral unfolds to a model which is not unique, but for which the Rankin-Selberg integral nevertheless represents an Euler product. Andrianov **[1]** considered a Rankin-Selberg integral which superficially resembles (19), but in which there is no cusp form in the Eisenstein series, and in which the theta function is a lift from $O(n)$ rather than $O(1)$. In order to prove that the integral represented the standard L-function on $\mathrm{Sp}(2n)$, of degree $2n+1$, a difficult procedure involving finding recursion relations between Fourier coefficients was required. Zhuravlev **[88]** considered the corresponding problem on $\widetilde{\mathrm{Sp}}(2n)$, obtaining the standard L-function (of degree $2n$).

Meanwhile Bump and Hoffstein **[20]**, **[19]** found Euler products for cusp forms on the metaplectic n-fold cover of $\mathrm{GL}(r)$ analogous to Shimura **[80]** making use of generalized theta series on the n-fold of $\mathrm{GL}(n)$, constructed by Kazhdan and Patterson. This construction, later extended by T. Suzuki **[84]**, was proved by a difficult procedure analogous to that used by Andrianov. (Alas, no corresponding analog of Shimura **[79]** is known, though see Bump, Ginzburg and Hoffstein **[18]** for a construction of the symmetric cube L-functions on $\mathrm{GL}(2)$ using the generalized theta series on the 3-fold cover of $\mathrm{GL}(3)$.)

Goetze **[50]** showed that the Rankin-Selberg convolution $O(5)\times\mathrm{GL}(3)$, using the global integral of Gelbart and Piatetski-Shapiro **[36]**, but working on the 3-fold metaplectic cover, taking a cusp form on $O(5) \cong \mathrm{PGSp}(4)$ and the Kazhdan-Patterson theta function on the 3-fold cover of GL(3), gives an Euler product. In general it is expected that the Ginzburg-Soudry construction for $O(n) \times \mathrm{GL}(r)$ described in Section 7 can be used to construct Euler products associated with automorphic forms on the metaplectic covers of all orthogonal groups.

Piatetski-Shapiro and Rallis [**72**] reconsidered this situation, and reproved Andrianov's integrals by representation theoretic considerations. They found that the integral unfolded to a nonunique model, but that the local integral, applied to a spherical vector in any model of the representation, produced the same local factor.

Bump, Furusawa and Ginzburg [**12**] gave further examples of constructions of L-functions involving nonunique models, and recently I was shown another very good example which is apparently still unpublished and which I will therefore not describe here. It is an important question whether these examples are common or rare. It is conceivable that they are common but hard to discover because adequate heuristics are not yet known.

13. Spin L-functions

If $G = \mathrm{PGSp}(2n)$, the L-group is the topological double cover $\mathrm{spin}(2n+1)$ of $\mathrm{SO}(2n+1)$, which has an irreducible representation of degree 2^n, the spin representation. In a few cases if (π, V) is an automorphic cuspidal representation of G a construction of $L(s, \pi, \mathrm{spin})$ is known. We include the possibility that $G = \mathrm{PGSp}(2n_1) \times \mathrm{PGSp}(2n_2)$, which has a spin L-function of degree $2^{n_1+n_2}$. Note that $\mathrm{GSp}(2) = \mathrm{GL}(2)$.

For GSp(4), there is a Hecke construction, due to Novodvorsky [**68**]. A useful current reference is Takloo-Bighash [**85**]. Novodvorsky gave a Rankin-Selberg construction for $\mathrm{GSp}(4) \times \mathrm{GL}(2)$ in the same article. Furusawa [**31**] gave a Rankin-Selberg construction for this same L-function which (unlike Novodvorsky's) unfolds not to a Whittaker model but to a Bessel model, and is therefore applicable to holomorphic Siegel modular forms (which are not generic). He used this construction to prove a special value result.

For GSp(6), $\mathrm{GSp}(6) \times \mathrm{GL}(2)$, GSp(8) and GSp(10), Rankin-Selberg constructions were found by Bump and Ginzburg [**17**], [**14**]. The case of GSp(6) is the subject of Vo [**87**]. Finally the spin L-function for $\mathrm{GSp}(4) \times \mathrm{GSp}(4)$ is constructed by Jiang [**56**].

Similarly if $G = \mathrm{PGO}(2n)$, the L-group is $\mathrm{spin}(2n)$. This group admits two spin representations both of degree 2^{n-1}. Rankin-Selberg constructions were found in a few cases by Ginzburg [**40**].

Bump, Friedberg and Ginzburg **[10]** gave examples of constructions producing the product of two L-functions in separate variables, one of which is the spin L-function. These are thus reminiscent of the construction of Bump and Friedberg of the exterior square L-function.

Except for Furusawa's construction, all integrals discussed above apply only to generic cusp form, hence do not apply to Siegel modular forms. Other constructions of the (degree 4) spin L-function for GSp(4) are known, but we omit mention of the older ones surveyed in **[5]**. One novel construction of the degree four L-function which was found later is that of Kohnen and Skoruppa **[60]**, which involves two cusp forms F and G. If G is in the Maass space the Dirichlet series produces the degree 4 L-function of F. Extensions of this work can be found in Gritsenko **[51]** and Heim **[52]**.

14. The towers and the magic square

Ginzburg and Rallis **[45]** found four Rankin-Selberg integrals which involve the same G_2 Eisenstein series, and which can be given a uniform construction. There is an underlying heuristic behind their paper from Kac **[57]**, which contains a list (Tables II and III, with imporant corrections at the end of **[26]**) that allows one to read off the normalizing factors of the Eisenstein series appearing in many Rankin-Selberg convolutions. Kac' table can be regarded as a list of L-group representations for which one seeks Rankin-Selberg constructions. It is remarkable how many important constructions can be found in this list. The invariant degrees $d_1, \cdots, d_l$ which appear in the table will be the degrees of the L-functions which appear as normalizing factors of the Eisenstein series, giving a clue which will allow some constructions to be guessed. For example, with $\mathrm{SL}_n \otimes \mathrm{SL}_n$ Kac gives a single degree $d = n$, from which we infer the normalizing factor $\zeta(ns)$ in (2).

Ginzburg and Rallis **[45]** began with the observation that three familiar Rankin-Selberg convolutions can be regarded as forming a tower. These are:

G	${}^L G$	representation
$\mathrm{GL}(n)$	$\mathrm{GL}(n,\mathbb{C})$	symmetric square
$\mathrm{GL}(n) \times \mathrm{GL}(n)$	$\mathrm{GL}(n,\mathbb{C}) \times \mathrm{GL}(n,\mathbb{C})$	Rankin-Selberg
$\mathrm{GL}(2n)$	$\mathrm{GL}(2n,\mathbb{C})$	exterior square

These three constructions form a "tower" meaning that they should be viewed together. From their heuristic based on Kac' paper this is true. An aspect of this is that the invariant degree d in Kac' Table II is the same for these three cases, just n, reflecting the fact that the Rankin-Selberg constructions all involve the same "mirabolic" Eisenstein series on $\mathrm{GL}(n)$, with normalizing factor $\zeta(ns)$. (It is true that in the case of the symmetric square, the Eisenstein series is on the double cover but in their view this is not an important distinction.)

Next they give another tower:

G	${}^LG$	representation
GL(2)	GL(2, $\mathbb{C}$)	symmetric fourth power
GL(3)	GL(3, $\mathbb{C}$)	adjoint (degree 15)
SO(7)	Sp(4, $\mathbb{C}$)	second fundamental
F_4	F_4	standard

The first construction, of the fourth symmetric power of GL(2), is described in Ginzburg [**42**]. The second of these constructions was previously found by Ginzburg [**38**]. The remaining two are new in this paper. They remark that the degree 14 L-function on SO(7) is dual to the construction of Piatetski-Shapiro, Rallis and Schiffmann [**73**] of a degree 14 L-function on $G_2 \times \mathrm{GL}(2)$. The duality phenomenon was explained in Section 7.

Ginzburg [**41**] gave two more towers. First, a tower of constructions involving a GL(3) mirabolic tower, which is an extension of the GL(n) mirabolic tower already described, adding a fourth entry when $n = 3$:

G	${}^LG$	representation
GL(3)	GL(3, $\mathbb{C}$)	symmetric square
GL(3) × GL(3)	GL(3, $\mathbb{C}$) × GL(3, $\mathbb{C}$)	Rankin-Selberg
GL(6)	GL(6, $\mathbb{C}$)	exterior square
GE_6	GE_6	standard

Ginzburg gives another tower involving an Eisenstein series on the Siegel parabolic subgroup of GSp(6). This tower, as he decribes it, has five levels. I will assume, for reasons which will presently become clear, that the first two levels in this tower as he presents it are red herrings. The correct first level of the tower, I think, must be the construction of Bump, Friedberg and Ginzburg [**11**] of the degree 14 L-function of SO(7), corresponding to the *third* fundamental representation of the L-group Sp(6, $\mathbb{C}$). This appears in Kac' table II as $\wedge_0^3\mathfrak{sp}(6)$ and as with the other cases for this tower, $d = 4$ in Kac' list. *Caution*: both the second and third fundamental representations of Sp(6, $\mathbb{C}$) have degree 14, and both appear in this discussion. With this modification the third tower is:

G	${}^LG$	representation
SO(7)	Sp(6)	third fundamental
GL(6)	GL(6, $\mathbb{C}$)	exterior cube
GSO(12)	spin(12)	spin
GE_7	GE_7	standard

The exterior cube L-function on GL(6) was constructed by Ginzburg and Rallis [**46**], the spin L-function on GSO(12) was constructed by Ginzburg [**40**], and the E_7 construction is in Ginzburg [**41**].

In the exceptional groups one encounters the *Magic Square* of Freudenthal [**29**].

Geometry	$\mathbb{R}$	$\mathbb{C}$	$\mathbb{H}$	$\mathbb{O}$
2-dimensional elliptic	B_1	A_2	C_3	F_4
2-dimensional projective	A_2	$A_2 \times A_2$	A_5	E_6
5-dimensional symplectic	C_3	A_5	D_6	E_7
hypersymplectic	F_4	E_6	E_7	E_8

The magic square has appeared in many different ways. From Freudenthal's viewpoint, four different geometries, one of them (hypersymplectic) being an exotic one which does not fit into an infinite family, were considered over four different fields. It will be noted that the L-groups in the first tower of Ginzburg and Rallis, and the second and third towers found by Ginzburg (with the third tower modified as explained above) coincide exactly with the first three columns of the magic square.

Does this mean that we should expect Rankin-Selberg constructions to be found for the fourth tower? We do not know the answer to this question, but there is some reason to be pessimistic. For one thing, the heuristic based on Kac' list does not predict any such tower. In the next section we will explain another "numerological" heuristic and investigate what it predicts for the fourth tower.

15. Numerology

Suppose that some Rankin-Selberg integral, involving a cusp form and an Eisenstein series, unfolds to a Whittaker model. Let $\boldsymbol{W}$ be the dimension of the unipotent subgroup in the Whittaker model, let $\boldsymbol{E}$ be the dimension of the unipotent radical in the Eisenstein series, and let $\boldsymbol{D}$ be the dimension of the entire integration in the Rankin-Selberg integral. Then

$$\boldsymbol{D} = \boldsymbol{W} + \boldsymbol{E}.$$

This numerical indentity can be quite handy in searching for Rankin-Selberg integrals. We will apply it in a case where it fails, but it was used with success to find some of the integrals in Section 12. A different dimensional heuristic is found in Garrett [**34**].

As an example, let us consider the $\mathrm{GL}(n) \times \mathrm{GL}(n)$ Rankin-Selberg integral (3). The integration is over $Z\backslash \mathrm{GL}(n)$, or $n^2 - 1$ dimensions. There are two Whittaker functions, each contributing $\frac{1}{2}n(n-1)$ to $\boldsymbol{W}$, so $\boldsymbol{W} = n^2 - n$. Finally, the Eisenstein series is over the parabolic subgroup with Levi factor $\mathrm{GL}(n-1) \times \mathrm{GL}(1)$, and its unipotent radical has dimension $\boldsymbol{E} = n - 1$.

Now let us consider second, third and fourth examples from the third tower. In each case, there is a parabolic subgroup of G with Levi decomposition MU_k, and a character ψ_U of U_k whose stabilizer in M is Sp(6). The dimension of integration $\boldsymbol{D}$ is therefore known, as is the dimension of the Whittaker model. Subtracting, $\boldsymbol{E} = 6$ so we are looking in each case for an integration over a parabolic subgroup of Sp(6) with a unipotent radical of dimension 6. This is the Siegel parabolic.

	ρ of L-gp	Integration	D	W	$D-W$
A_5	ω_3 of GL(6)	Sp(6)	21	15	6
D_6	spin_{12}	$\mathrm{Sp}(6) \times U_{15}$	36	30	6
E_7	standard	$\mathrm{Sp}(6) \times U_{48}$	69	63	6

This heuristic is quite handy in searching for Rankin-Selberg integrals. Can a similar scenario pertain for a hypothetical fourth tower?

	Integration	D	W	$D-W$
E_6	F_4	52	36	16
E_7	$F_4 \times U_{27}$	79	63	16
E_8	$F_4 \times U_{84}$	136	120	16

Note that $\boldsymbol{D} - \boldsymbol{W}$ is constant. So far so good, but if the scenario is to be identical to that of the third tower we would be looking for a parabolic subgroup of F_4 with a unipotent radical of dimension 16. Such a parabolic does not exist.

16. Miscellany

An early Rankin-Selberg construction unfolding to a non-Whittaker model is Sugano [**83**].

A construction on SU(2, 1) was found independently by Shintani [**81**] and Gelbart and Piatetski-Shapiro [**35**]. Further unitary constructions were found by Tamir [**86**].

Bump and Ginzburg [16] and [13] have two related constructions of the adjoint L-function on GL(4), of degree 15, and a degree 27 L-function on split spin(9), corresponding to the second fundamental representation of the L-group PGSp(8). Both these constructions use the same Eisenstein series on the exceptional group F_4.

Ginzburg [43] has a construction for the adjoint L-function of Sp(4), an Euler product of degree 10.

In Section 9 we saw that the symmetric and exterior square L-functions could be used to distinguish the images of two different lifts from $O(2n)$ and $O(2n+1)$ to $GL(2n)$. The Asai L-function [2], which is a Galois twisted version of the $GL(2) \times GL(2)$ convolution plays a role similar in distinguishing the images of two different types of base change lift, and showing that their images are disjoint. Flicker [27], [28] gave a similar Galois twisted analog of the $GL(n) \times GL(n)$ and applied it to quadratic base change. Piatetski-Shapiro and Rallis [71] gave an analog of the Asai L-function for GL(2) and *cubic* Galois extensions. This is a twisted version of the important construction Garrett [33] of a $GL(2) \times GL(2) \times GL(2)$ Rankin-Selberg construction based on the restriction of an Eisenstein series of Sp(6). We note that a $GL(2) \times GL(2) \times GL(2)$ can also be obtained from the Gelbart-Piatetski-Shapiro $O(4) \times GL(2)$ construction since $O(4)$ is isogenous to $GL(2) \times GL(2)$, but this alternative method does not give the twisted L-function.

References

[1] A. Andrianov. Euler expansions of the theta-transform of Siegel modular forms of genus n. *Mat. Sb. (N.S.)*, 105(147)(291–341), 1978.

[2] T. Asai. On certain dirichlet series associated with hilbert modular forms and rankin's method. *Math. Ann.*, 226:81–94, 1977.

[3] S. Böcherer. Über die Funktionalgleichung automorpher L-Funktionen zur Siegelschen Modulgruppe. *J. Reine Angew. Math.*, 362:146–168, 1985.

[4] A. Borel. Automorphic L-functions. In *Automorphic Forms, Representations and L-Functions, Part 2*, volume 33 of *Proc. Sympos. Pure Math.*, pages 27–61. AMS, 1979.

[5] D. Bump. The Rankin-Selberg Method: A Survey. In K. Aubert, E. Bombieri, and D. Goldfeld, editors, *Number Theory, Trace Formulas and Discrete Groups, A Symposium in Honor Atle Selberg*. Academic Press, 1989.

[6] D. Bump. *Automorphic forms and representations*, volume 55 of *Cambridge Studies in Advanced Mathematics*. Cambridge University Press, Cambridge, 1997.

[7] D. Bump and S. Friedberg. The exterior square automorphic L-functions on $GL(n)$. In *Festschrift in honor of I. I. Piatetski-Shapiro on the occasion of his*

sixtieth birthday, Part II (Ramat Aviv, 1989), volume 3 of *Israel Math. Conf. Proc.*, pages 47–65, 1990.

[8] D. Bump, S. Friedberg, and M. Furusawa. Explicit formulas for the Waldspurger and Bessel models. *Israel J. Math.*, 102:125–177, 1997.

[9] D. Bump, S. Friedberg, and D. Ginzburg. Whittaker-orthogonal models, functoriality, and the rankin-selberg method. *Invent. Math.*, 109:55–96, 1992.

[10] D. Bump, S. Friedberg, and D. Ginzburg. Rankin-Selberg integrals in two complex variables. *Math. Ann.*, 313:731–761, 1999.

[11] D. Bump, S. Friedberg, and D. Ginzburg. A Rankin-Selberg integral using the automorphic minimal representation of $SO(7)$. *J. Ramanujan Math. Soc.*, 15:81–124, 2000.

[12] D. Bump, M. Furusawa, and D. Ginzburg. Non-unique models in the Rankin-Selberg method. *J. Reine Angew. Math.*, 468:77–111, 1995.

[13] D. Bump and D. Ginzburg. An L-function of degree 27 for Spin(9). The Ramanujan Journal, to appear.

[14] D. Bump and D. Ginzburg. Spin L-functions on symplectic groups. *Internat. Math. Res. Notices*, 8:153–160, 1992.

[15] D. Bump and D. Ginzburg. Symmetric square L-functions on $\mathrm{GL}(r)$. *Ann. of Math.*, 136:137–205, 1992.

[16] D. Bump and D. Ginzburg. The adjoint l-function of $\mathrm{GL}(4)$. *J. Reine Angew. Math.*, 505:119–172, 1998.

[17] D. Bump and D. Ginzburg. Spin L-functions on GSp_8 and GSp_{10}. *Trans. Amer. Math. Soc.*, pages 875–899, 2000.

[18] D. Bump, D. Ginzburg, and J. Hoffstein. The symmetric cube. *Invent. Math.*, 125:413–449, 1996.

[19] D. Bump and J. Hoffstein. Some Euler products associated with cubic metaplectic forms on $\mathrm{GL}(3)$. *Duke Math. J.*, 53:1047–1072, 1986.

[20] D. Bump and J. Hoffstein. On Shimura's correspondence. *Duke Math. J.*, 55:661–691, 1987.

[21] W. Casselman and J. Shalika. The unramified principal series of p-adic groups II, The Whittaker function. *Compositio Math.*, 41:207–231, 1980.

[22] J. Cogdell. Analytic Theory of L-functions for $GL(n)$. In J. Bernstein and S. Gelbart, editors, *An Introduction to the Langland Program*, pages 197–228. Birkhäuser, 2003.

[23] J. Cogdell. Dual Groups and Langlands Functoriality. In J. Bernstein and S. Gelbart, editors, *An Introduction to the Langland Program*, pages 251–268. Birkhäuser, 2003.

[24] J. Cogdell, H. Kim, I. Piatetski-Shapiro, and F. Shahidi. On lifting from classical groups to GL_N. *Publ. Math. Inst. Hautes Études Sci.*, 93:5–30, 2001.

[25] J. Cogdell and I. Piatetski-Shapiro. Converse theorems, functoriality, and applications to number theory. In *Proceedings of the International Congress of Mathematicians, Vol. II (Beijing, 2002)*, pages 119–128. Higher Ed. Press, Beijing, 2002.

[26] J. Dadok and V. Kac. Polar representations. *J. Algebra*, 92:504–524, 1985.

[27] Y. Flicker. Twisted tensors and Euler products. *Bull. Soc. Math. France*, 116(3):295–313, 1988.

[28] Y. Flicker. On zeroes of the twisted tensor L-function. *Math. Ann.*, 297:199–219, 1993.

[29] H. Freudenthal. Lie groups in the foundations of geometry. *Advances in Math.*, 1:145–190, 1964.

[30] S. Friedberg and H. Jacquet. Linear periods. *J. Reine Angew. Math.*, 443:91–139, 1993.

[31] M. Furusawa. On L-functions for $GSp(4) \times GL(2)$ and their special values. *J. Reine Angew. Math.*, 438:187–218, 1993.

[32] Masaaki Furusawa. On Fourier coefficients of Eisenstein series on $\mathrm{SO}(5, 2)$. *Amer. J. Math.*, 115(4):823–860, 1993.

[33] P. Garrett. Decomposition of Eisenstein series: Rankin triple products. *Ann. of Math.*, 125:209–235, 1987.

[34] P. Garrett. Euler factorization of global integrals. In *Automorphic forms, automorphic representations, and arithmetic (Fort Worth, TX, 1996), Part 2*, volume 66 of *Proc. Sympos. Pure Math.*, pages 35–101. Amer. Math. Soc., 1999.

[35] S. Gelbart and I. Piatetski-Shapiro. Automorphic forms and L-functions for the unitary group. In *Lie group representations, II (College Park, Md., 1982/1983)*, volume 1041 of *Lecture Notes in Math.*, pages 141–184. Springer, Berlin, 1984.

[36] S. Gelbart and I. Piatetski-Shapiro. *Explicit Constructions of L-functions*, volume 1254 of *Lecture Notes in Mathematics*, chapter L-Functions for $G \times GL(n)$, pages 53–146. Springer Verlag, 1987.

[37] D. Ginzburg. L-functions for $\mathrm{SO}_n \times \mathrm{GL}_k$. *J. Reine Angew. Math.*, 405:156–180, 1990.

[38] D. Ginzburg. A Rankin-Selberg integral for the adjoint representation of GL_3. *Invent. Math.*, 105:571–588, 1991.

[39] D. Ginzburg. On the standard L-function for G_2. *Duke Math. J.*, 69:315–333, 1993.

[40] D. Ginzburg. On spin L-functions for orthogonal groups. *Duke Math. J.*, 77:753–798, 1995.

[41] D. Ginzburg. On standard L-functions for E_6 and E_7. *J. Reine Angew. Math.*, 465:101–131, 1995.

[42] D. Ginzburg. On the symmetric fourth power L-function of GL_2. *Israel J. Math.*, 92:157–184, 1995.

[43] D. Ginzburg. A Rankin-Selberg integral for the adjoint L-function of Sp_4. *Israel J. Math.*, 95:301–339, 1996.

[44] D. Ginzburg, I. Piatetski-Shapiro, and S. Rallis. *L-functions for the orthogonal group*, volume 611 of *Mem. Amer. Math. Soc.* AMS, 1997.

[45] D. Ginzburg and S. Rallis. A tower of Rankin-Selberg integrals. *Internat. Math. Res. Notices*, 5:201–208, 1994.

[46] D. Ginzburg and S. Rallis. The exterior cube L-function for $GL(6)$. *Compositio Math.*, 123:243–272, 2000.

[47] D. Ginzburg, S. Rallis, and D. Soudry. L-functions for symplectic groups. *Bull. Soc. Math. France*, 126:181–244, 1998.

[48] David Ginzburg, Stephen Rallis, and David Soudry. On explicit lifts of cusp forms from GL_m to classical groups. *Ann. of Math. (2)*, 150(3):807–866, 1999.

[49] R. Godement and H. Jacquet. *Zeta Functions of Simple Algebras*, volume 260 of *Lecture Notes in Mathematics*. Springer Verlag, 1972.

[50] T. Goetze. Euler products associated to metaplectic automorphic forms on the 3-fold cover of GSp(4). *Trans. Amer. Math. Soc*, 350:975–1011, 1998.

[51] Valeri Gritsenko. Arithmetical lifting and its applications. In *Number theory (Paris, 1992–1993)*, volume 215 of *London Math. Soc. Lecture Note Ser.*, pages 103–126. Cambridge Univ. Press, Cambridge, 1995.

[52] Bernhard E. Heim. Pullbacks of Eisenstein series, Hecke-Jacobi theory and automorphic l-functions. In *Automorphic forms, automorphic representations, and arithmetic (Fort Worth, TX, 1996)*, volume 66 of *Proc. Sympos. Pure Math.*, pages 201–238. Amer. Math. Soc., Providence, RI, 1999.

[53] H. Jacquet, I. Piatetskii-Shapiro, and J. Shalika. Rankin-Selberg convolutions. *Amer. J. Math.*, 105(2):367–464, 1983.

[54] H. Jacquet and S. Rallis. Uniqueness of linear periods. *Compositio Math.*, 102:65–123, 1996.

[55] H. Jacquet and J. Shalika. Exterior square L-functions. In *Automorphic forms, Shimura varieties, and L-functions, Vol. II (Ann Arbor, MI, 1988)*, number 11 in Perspect. Math., pages 143–226. Academic Press, 1990.

[56] D. Jiang. *Degree* 16 *standard L-function of* $GSp(2) \times GSp(2)$, volume 588 of *Mem. Amer. Math. Soc.* AMS, 1996.

[57] V. Kac. Some remarks on nilpotent orbits. *J. Algebra*, 64:190–213, 1980.

[58] D. Kazhdan and S. Patterson. Metaplectic forms. *Inst. Hautes Études Sci. Publ. Math.*, 59:35–142, 1984.

[59] A. Knapp. Introduction to the Langlands program. In *Representation theory and automorphic forms (Edinburgh, 1996)*, volume 61 of *Proc. Sympos. Pure Math.*, pages 245–302. AMS, 1997.

[60] W. Kohnen and N. Skoruppa. A certain Dirichlet series attached to Siegel modular forms of degree two. *Invent. Math.*, 95:541–558, 1989.

[61] S. Kudla. Seesaw dual reductive pairs. In *Automorphic forms of several variables (Katata, 1983)*, volume 46 of *Progr. Math*, pages 244–268. Birkhäuser, 1984.

[62] R. Langlands. *Euler Products*. Yale University Press, 1971.

[63] H. Matsumoto. Sur les sous-groupes arithmétiques des groupes semi-simples déployés. *Ann. Sci. École Norm. Sup. (4)*, 2:1–62, 1969.

[64] C. Moeglin and J.-L. Waldspurger. Le spectre résiduel de $GL(n)$. *Ann. Sci. École Norm. Sup.*, 22:605–674, 1989.

[65] A. Murase and T. Sugano. Whittaker-Shintani functions on the symplectic group of Fourier-Jacobi type. *Compositio Math.*, 79:321–349, 1991.

[66] A. Murase and T. Sugano. Shintani function and its application to automorphic L-functions for classical groups. I. The case of orthogonal groups. *Math. Ann.*, 299:17–56, 1994.

[67] A. Murase and T. Sugano. Shintani functions and Rankin-Selberg convolution, I and II (Japanese). *Sūrikaisekikenkyūsho Kōkyūroku*, 909(154–159,160–168), 1995.

[68] M. Novodvorsky. Automorphic l-functions for symplectic group GSp(4). In *Automorphic forms, representations and L-functions (Proc. Sympos. Pure Math., Oregon State Univ., Corvallis, Ore., 1977), Part 2*, Proc. Sympos. Pure Math., XXXIII, pages 87–95, Providence, R.I., 1979. Amer. Math. Soc.

[69] S. Patterson and I. Piatetski-Shapiro. The symmetric-square L-function attached to a cuspidal automorphic representation of GL_3. *Math. Ann.*, 283:551–572, 1989.

[70] I. Piatetski-Shapiro and S. Rallis. *Explicit Constructions of L-functions*, volume 1254 of *Lecture Notes in Mathematics*, chapter L-Functions for the classical groups, pages 1–52. Springer Verlag, 1987.

[71] I. Piatetski-Shapiro and S. Rallis. Rankin triple l-functions. *Compositio Math.*, 64(1):31–115, 1987.

[72] I. Piatetski-Shapiro and S. Rallis. A new way to get Euler products. *J. Reine Angew. Math.*, 392:110–124, 1988.

[73] I. Piatetski-Shapiro, S. Rallis, and G. Schiffmann. Rankin-Selberg integrals for the group G_2. *Amer. J. Math.*, 114:1269–1315, 1992.

[74] S. Rallis. *L-functions and the oscillator representation*, volume 1245 of *Lecture Notes in Mathematics*. Springer Verlag, 1987.

[75] D. Ramakrishnan. Modularity of the Rankin-Selberg L-series, and multiplicity one for $SL(2)$. *Ann. of Math.*, 152:45–111, 2000.

[76] G. Savin. Local Shimura correspondence. *Math. Ann.*, 280:185–190, 1988.

[77] F. Shahidi. On the Ramanujan conjecture and finiteness of poles for certain L-functions. *Ann. of Math. (2)*, 127(3):547–584, 1988.

[78] F. Shahidi. Automorphic L-functions and functoriality. In *Proceedings of the International Congress of Mathematicians, Vol. II (Beijing, 2002)*, pages 655–666. Higher Ed. Press, Beijing, 2002.

[79] G. Shimura. On modular forms of half integral weight. *Ann. of Math. (2)*, 97:440–481, 1973.

[80] G. Shimura. On the holomorphy of certain Dirichlet series. *Proc. London Math. Soc*, 31:79–98, 1975.

[81] T. Shintani. On automorphic forms on unitary groups of order 3. Unpublished, 1979.

[82] D. Soudry. *Rankin-Selberg convolutions for* $SO_{2l+1} \times GL_n$*: local theory*, volume 500 of *Mem. Amer. Math. Soc.* AMS, 1993.

[83] T. Sugano. On holomorphic cusp forms on quaternion unitary groups of degree 2. *J. Fac. Sci. Univ. Tokyo Sect. IA Math.*, 31(3):521–568, 1985.

[84] T. Suzuki. Distinguished representations of metaplectic groups. *Amer. J. Math.*, 120:723–755, 1998.

[85] R. Takloo-Bighash. L-functions for the p-adic group GSp(4). *Amer. J. Math.*, 122:1085–1120, 2000.

[86] Boaz Tamir. On L-functions and intertwining operators for unitary groups. *Israel J. Math.*, 73(2):161–188, 1991.

[87] S. Vo. The spin L-function on the symplectic group GSp(6). *Israel J. Math.*, 101:1–71, 1997.

[88] V. Zhuravlev. Euler expansions of theta-transforms of Siegel modular forms of half-integral weight and their analytic properties. *Mat. Sb. (N.S.)*, 123(165):174–194, 1984.

Daniel Bump
Department of Mathematics
Stanford University
Stanford, CA 94035-2125. USA

Asymptotics for sums of twisted L-functions and applications

Gautam Chinta, Solomon Friedberg, and Jeffrey Hoffstein

Abstract. Let $n \geq 2$, let F be a global field containing a full set of n-th roots of unity, and let π be an isobaric automorphic representation of $GL_r(\mathbb{A}_F)$. We establish asymptotic estimates for the sum of the n-th order twisted L-functions of π, $L(s, \pi \otimes \chi)$, for s such that $\mathrm{Re}(s) > \max(1-1/r, 1/2)$ if $n = 2$ and $\mathrm{Re}(s) > 1 - 1/(r+1)$ if $n > 2$. As an application we establish new nonvanishing theorems for twists of given order, including a simultaneous nonvanishing result. When $n = 2$ we use this information on asymptotics to prove that the twisted L-values at $s = 1$ give rise to a distribution function.

1991 Mathematics Subject Classification:Primary 11F70, Secondary 11F66, 11F67, 11M41, 11M45, 11N64, 22E55

1. Introduction

Let F be a global field and let π be a unitary isobaric automorphic representation of $GL_r(\mathbb{A}_F)$ with L-function $L(s, \pi)$. In this paper we study the family of twisted L-functions $L(s, \pi \otimes \chi)$, where χ ranges over the idèle class characters of *fixed* finite order. In particular we establish the existence of nonvanishing twists for given s inside the critical strip but sufficiently close to the edge.

The existence of nonvanishing twists of some finite order for π on $GL(2)$ was established by Rohrlich **[22]**. Building on this work, Barthel and Ramakrishnan **[3]** show that for π a cuspidal automorphic representation of $GL_r(\mathbb{A}_F)$, $r \geq 3$, and given s with $\mathrm{Re}(s) > 1 - 1/r$ there exist infinitely many primitive ray class characters χ such that $L(s, \pi \otimes \chi) \neq 0$. Under the stronger assumption of temperedness, Barthel and Ramakrishnan find infinitely many nonvanishing twists for s satisfying $\mathrm{Re}(s) > 1 - 2/(r+1)$. Luo, Rudnick and Sarnak **[20]** show a similar result in the case of a degree r Rankin-Selberg convolution $L(s, \pi_1 \times \tilde{\pi}_1)$ (so r is a square), replacing the assumption of temperedness by the observation that the Rankin-Selberg L-series has positive coefficients. By Langlands functoriality, this L-series is conjecturally of the form $L(s, \pi)$ for some isobaric automorphic representation π.

We sharpen these results by showing that, in fact, for any global field F and any isobaric automorphic representation π of $GL_r(\mathbb{A}_F)$, $r \geq 2$, for given s with $\mathrm{Re}(s) > 1 - 1/r$ there exist infinitely many *quadratic* characters χ such that $L(s, \pi \otimes \chi) \neq 0$. In addition, if the field F contains the n-th roots of unity for some $n > 2$, then for s with $\mathrm{Re}(s) > 1 - 1/(r+1)$ there exist infinitely many characters χ *of order exactly* n such that $L(s, \pi \otimes \chi) \neq 0$. In both these results the characters may be specified at a finite number of places.

The methods of Barthel-Ramakrishnan combined with the large sieve can also be used to give nonvanishing results for twists by characters of fixed order near $\mathrm{Re}(s) = 1$. For example, when $n = 2$ and the base field is $\mathbb{Q}$, the real character sum estimate of Heath-Brown **[16]** can be used to obtain a result as strong as ours. However, the results of this paper—in particular, the uniformity of the interval of nonvanishing over all number fields—seem to be stronger than can be obtained by the large sieve in its present form. Our methods are based on properties of automorphic L-functions and double Dirichlet series.

Our results are formulated for a general isobaric automorphic representation. Let π_j, $1 \leq j \leq k$, be cuspidal automorphic representations of $GL_{r_j}(\mathbb{A}_F)$ with $r = \sum_{j=1}^{m} r_j$. Recall that Langlands's theory of Eisenstein series allows one to construct an isobaric automorphic representation $\pi = \boxplus_{j=1}^{m} \pi_j$ of $GL_r(\mathbb{A}_F)$ such that the standard L-function of π is given by

$$L(s, \pi) = \prod_{j=1}^{m} L(s, \pi_j). \tag{1.1}$$

Since $L(s, \pi \otimes \chi) = \prod_{j=1}^{m} L(s, \pi_j \otimes \chi)$, our nonvanishing result establishes a simultaneous nonvanishing theorem for twists.

These results on nonvanishing are consequences of our main theorem, which we now state.

Let $n \geq 2$ be a fixed integer, and let F be a global field with n n-th roots of unity. Let S be a finite set of places containing all archimedean places, all places dividing n, and such that the class number of the ring of S-integers is 1. To each square-free ideal d prime to S one may attach (as in Fisher-Friedberg **[11]**) an idèle class character χ_d, of order n for $d \neq (1)$, such that if $d = (d_0)$ is principal with d_0 sufficiently congruent to 1 then χ_d is the character attached by class field theory to the extension $F(\sqrt[n]{d_0})/F$. Let C be a sufficiently large ideal supported on S (see Section 2 for the precise condition) and let H_C be the (narrow) ray class group modulo C. For π an automorphic representation let $L_S(s, \pi)$ denote its standard L-function with the factors at $v \in S$ removed. Then we have:

Theorem 1.1. *Let π be a unitary isobaric automorphic representation of the group $GL_r(\mathbb{A}_F)$ (not necessarily cuspidal) which is unramified outside S. Let*

A be a class in the ray class group H_C. Fix $s \in \mathbb{C}$ with $\mathrm{Re}(s) > \max(1-1/r, \frac{1}{2})$ *if* $n = 2$ *and* $\mathrm{Re}(s) > 1 - 1/(r+1)$ *if* $n > 2$.

1. *Let $k > 0$ be a sufficiently large integer, depending on F, n, r. Then for any $\epsilon > 0$*

$$\sum_{\substack{d \in A \\ d \text{ sq-free} \\ 1<|d|<X}} L_S(s, \pi \otimes \chi_d)\left(1 - \frac{|d|}{X}\right)^k = \frac{c_S(s)X}{k+1} + O_\epsilon(X^{\frac{1}{2}+\epsilon})$$

   as $X \to \infty$, where the sum is over square-free integral ideals d in the ray class A of absolute norm between 1 *and X, and where $c_S(s)$ is a constant given explicitly below (see equation (2.3)). If S is sufficiently large then $c_S(s)$ is nonzero.*

2. *Let $n = 2$, let π be self-contragredient, and let s be real. If $L_S(s, \pi \otimes \chi_d) \geq 0$ for all d then*

$$\sum_{\substack{d \in A \\ d \text{ sq-free} \\ 1<|d|<X}} L_S(s, \pi \otimes \chi_d) \sim c_S(s)X$$

   as $X \to \infty$, where the sum is over square-free integral ideals d in the ray class A of absolute norm between 1 *and X.*

Similar results hold for the full L-function $L(s, \pi \otimes \chi_d)$ in place of $L_S(\pi \otimes \chi_d)$. See Section 3.

Corollary 1.2. *Let π be a unitary isobaric automorphic representation of $GL_r(\mathbb{A}_F)$, $r \geq 2$. Let A be a class in the ray class group H_C. Let* $\mathrm{Re}(s) > 1 - 1/(r+1)$. *Then there exist infinitely many characters χ_d, $d \in A$, of order n such that*

$$L(s, \pi \otimes \chi_d) \neq 0.$$

If $n = 2$, the conclusion is true for $\mathrm{Re}(s) > 1 - 1/r$.

Since Theorem 1.1 applies to all isobaric automorphic representations it also gives information about moments; see (1.1). This information is sufficient to establish the existence of a distribution function $F(y)$ giving the limiting proportion of the quadratically twisted L-values $L(1, \pi \otimes \chi)$ of magnitude at most y. For the simplest case, that is, for $L(s, \chi_d)$ the family of Dirichlet L-series associated to the Kronecker symbols χ_d, such a distribution result was obtained (in fact, for $d > 0$ and $\mathrm{Re}(s) > 3/4$) by Chowla and Erdös [8]. The analogue of Theorem 1.1 was established by Barban [2], who showed that for

$k \in \mathbb{N}$,

$$\sum_{1<-d<X} L(1,\chi_d)^k \sim r_k X,$$

where the r_k are explicitly given non-zero constants. This asymptotic estimate for the k-th moments together with a theorem of Fréchet and Shohat [13] allows one to establish a Chowla-Erdös type value distribution result for $d<0$ and $s=1$; see [2] and Section 3 below.

In Section 3 we shall use a similar method to show

Corollary 1.3. *Suppose that $\pi = \boxplus_{j=1}^m \pi_j$ is a unitary isobaric automorphic representation. Let $n=2$. Fix S, A as above. Then there exists a non-decreasing right-continuous function $F_{\pi,A}:\mathbb{R}\to[0,1]$ such that*

$$\lim_{X\to\infty} \frac{|\{d\in A, d \textit{ sq-free}, |d|<X : |L(1,\pi\otimes\chi_d)|<y\}|}{|\{d\in A, d \textit{ sq-free}, |d|<X\}|} \to F_{\pi,A}(y)$$

at all points of continuity y of $F_{\pi,A}(y)$.

In an earlier version of this paper, Corollary 1.3 was stated only in the case where each factor π_j was tempered. We thank the referee for prompting us to realize that the temperedness assumption was unnecessary.

Notice that if L is a motivic L-function of odd degree then 1 is a critical value in the sense of Deligne and these L-values are of arithmetic interest. For example,

Corollary 1.4. *Let k be a totally real number field. Let A be the trivial ray class in H_C. Then there exists a non-decreasing right-continuous function $F_k:\mathbb{R}\to[0,1]$ such that*

$$\lim_{X\to\infty} \frac{|\{(d)\in A, (d) \textit{ sq-free}, d\gg 0, |d|<X : \frac{h(k(\sqrt{-d}))}{w_d Q_d \sqrt{f_d}} < y\}|}{|\{d\in A, d \textit{ sq-free}, |d|<X\}|} \to F_k(y)$$

for all points of continuity y of $F_k(y)$. Here $h(k(\sqrt{-d}))$ is the relative class number of $k(\sqrt{-d})/k$, w_d is the number of roots of unity in $k(\sqrt{-d})$, f_d is the norm of the conductor of χ_{-d} and $Q_d = 1$ or 2 as in [**24**], *Theorem 4.12.*

This follows from the Dirichlet class number formula for $k(\sqrt{-d})/k$ (see [**24**], for example). Similarly, one may adjoin $\sqrt{d}$ where $d\in k$ has exactly one positive embedding. A moment result in such a case was proved with certain restrictions on k by Peter in [**21**].

Distribution results similar to Cors. 1.3 and 1.4 can also be obtained by the use of the large sieve as noted above; we thank Peter Sarnak for pointing this out to us. Also, in the case of the Dirichlet L-series $L(s,\chi_d)$, much more is known about the distribution function, see for example [**15**] and the references there.

The proofs of our results are based on a double Dirichlet series obtained by summing the twisted L-functions. Such series first arose from the study of certain Hecke-Rankin-Selberg type integrals of metaplectic Eisenstein series (see **[6]**), but it is not apparent that the series used here may be so-obtained. Instead, our work is based on the convexity principle for holomorphic functions of two complex variables, whose use to study such series was first observed by Bump, Friedberg and Hoffstein **[7]**. In **[10, 11, 12]** a similar point of view was taken. In particular, in **[10]** the approach of searching for a multiple Dirichlet series which is a weighted sum of twisted L-functions with a full meromorphic continuation in all variables was formalized. Unfortunately the desired weighting factors are not easy to obtain (see **[7, 11, 12]** for some low rank instances when $n = 2$). In this paper we take an opposite point of view – studying a multiple Dirichlet series which is an unweighted sum of twisted L-functions. This object has more limited meromorphic continuation but we show that this continuation still has consequences. We analyze the multiple Dirichlet series in several ways (one related to metaplectic Eisenstein series), and use this analysis to establish its continuation to different overlapping tube domains. By the convexity principle the double Dirichlet series then continues to the convex hull of the union of these regions. This information allows us to establish Theorem 1.1 by Tauberian methods.

2. The double Dirichlet series

Fix $n \geq 2$ and let F be a global field with n n-th roots of unity. In the number field case, let $\mathcal{O}$ denote the ring of integers of F, and let S_f be a finite set of non-archimedean places containing all places dividing n and such that the ring of S_f-integers $\mathcal{O}_{S_f}$ has class number 1. Let S_∞ denote the set of archimedean places and let $S = S_f \cup S_\infty$. Similarly, in the function field case, let $\mathcal{C}$ denote the smooth projective curve with function field F and choose a finite set of places S such that the divisors supported on S represent all classes in Pic($\mathcal{C}$). Let $\left(\frac{a}{*}\right)$ be the power residue symbol attached to the extension $F(\sqrt[n]{a})$ of F. To properly formulate the double Dirichlet series, let us extend the n-th power residue symbol as in Fisher-Friedberg **[11]**. We review the definition.

For each place v, let F_v denote the completion of F at v. Suppose first that F is a number field. For v nonarchimedean, let P_v denote the corresponding ideal of $\mathcal{O}$. Let $C = \prod_{v \in S_f} P_v^{n_v}$ with $n_v \geq 1$ sufficiently large that $a \in F_v$, $\mathrm{ord}_v(a - 1) \geq n_v$ implies that $a \in (F_v^\times)^n$. Let H_C be the ray class group modulo C (the narrow ray class group if $n = 2$ and F has real embeddings), and let $R_C = H_C \otimes \mathbb{Z}/n\mathbb{Z}$. Write the finite group R_C as a direct product of cyclic groups, choose a generator for each, and let $\mathcal{E}_0$ be a set of ideals of $\mathcal{O}$ prime to S which represent these generators. For each $E_0 \in \mathcal{E}_0$ choose

$m_{E_0} \in F^\times$ such that $E_0 \mathcal{O}_{S_f} = m_{E_0} \mathcal{O}_{S_f}$. Let $\mathcal{E}$ be a full set of representatives for R_C of the form $\prod_{E_0 \in \mathcal{E}_0} E_0^{n_{E_0}}$, $n_{E_0} \in \mathbb{Z}$. If $E = \prod_{E_0 \in \mathcal{E}_0} E_0^{n_{E_0}}$ is such a representative, then let $m_E = \prod_{E_0 \in \mathcal{E}_0} m_{E_0}^{n_{E_0}}$. Note that $E\mathcal{O}_{S_f} = m_E \mathcal{O}_{S_f}$ for all $E \in \mathcal{E}$. For convenience only we suppose that $\mathcal{O} \in \mathcal{E}$ and $m_{\mathcal{O}} = 1$.

Let $J(S)$ denote the group of fractional ideals of $\mathcal{O}$ prime to S_f. Let $I, I_1 \in J(S)$ be coprime. Write $I = (m)EG^n$ with $E \in \mathcal{E}$, $m \in F^\times$, $m \equiv 1 \bmod C$, m positive under all real embeddings of F (we write $m \gg 0$; this condition is trivial unless $n = 2$) and $G \in J(S)$, $(G, I_1) = 1$. Then as in **[11]**, the n-th power residue symbol $\left(\frac{m m_E}{I_1}\right)$ is defined, and if $I = (m')E'G'^n$ is another such decomposition, then $E' = E$ and $\left(\frac{m' m_E}{I_1}\right) = \left(\frac{m m_E}{I_1}\right)$.

In view of this define the n-th power residue symbol $\left(\frac{I}{I_1}\right)$ by $\left(\frac{I}{I_1}\right) = \left(\frac{m m_E}{I_1}\right)$ and the character χ_I by $\chi_I(I_1) = \left(\frac{I}{I_1}\right)$. This depends on the choices above, but we suppress this from the notation. Let S_I denote the support of the conductor of χ_I. One may check that if $I = I'G^n$, then $\chi_I(I_1) = \chi_{I'}(I_1)$ whenever both are defined. This allows one to extend χ_I to a character of all ideals of $J(S \cup S_I)$. One has, as in **[11]**

Proposition 2.1. Reciprocity – *Let I, $J \in J(S)$ be coprime, and $\alpha(I, J) = \chi_I(J)\chi_J(I)^{-1}$. Then $\alpha(I, J)$ depends only on the images of I and J in R_C.*

The definitions in the function field case are similar, but it is move traditional to work with divisors and write them additively. For more details see **[11, 12]**, where the case $n = 2$ is discussed in detail. Throughout this section we keep the number field notation, but the argument for function fields is the same.

For d an integral ideal, let $|d|$ denote its absolute norm. Let $\mathcal{I}(S)$ denote the set of integral ideals prime to S_f. Let π be an isobaric automorphic representation of $GL_r(\mathbb{A}_F)$, not necessarily cuspidal, which is unramified outside S and let $L_S(s, \pi \otimes \chi_d)$ be the L-function for π twisted by the character χ_d, with the primes in S removed. (Note that the Euler factor is also 1 at the primes dividing d.)

Before proceeding to give the proof of Theorem 1.1, let us record a necessary observation about the functional equation of such a π. So let π be as above. Then there is a partition $r = \sum_{j=1}^m r_j$ and a set of unitary cuspidal automorphic representations π_j such that $\pi \cong \boxplus_{j=1}^m \pi_j$. If ξ is an idèle class character then the twisted L-function $L(s, \pi \otimes \xi)$ satisfies a functional equation

$$L(s, \pi \otimes \xi) = \epsilon(s, \pi \otimes \xi)\, L(1 - s, \tilde{\pi} \otimes \xi^{-1}), \tag{2.1}$$

where $\epsilon(s, \pi \otimes \xi)$ is the epsilon factor of $\pi \otimes \xi$.

Lemma 2.2. *Let $d, e \in \mathcal{O}(S)$ be square-free, and suppose that $\chi_d = \chi_e \chi_m$ with $m \in F^\times$, $m \equiv 1 \bmod C$ and $m \gg 0$ if $n = 2$ (this equality holds if and only if d, e project to the same class in R_C). Then*

$$\epsilon(s, \pi \otimes \chi_d) = \epsilon(1/2, \chi_m)^r \chi_\pi(d/e) |d/e|^{r(1/2-s)} \epsilon(s, \pi \otimes \chi_e).$$

Here $\epsilon(1/2, \chi_m)$ is given by a (normalized) n-th order Gauss sum, as in Tate's thesis. If $n = 2$ then it is identically 1.

Proof. We have $\epsilon(s, \pi) = |D^r f(\pi)|^{1/2-s} \epsilon(1/2, \pi)$ where D is the different of F, $f(\pi)$ is the conductor of π, and $\epsilon(1/2, \pi)$ is the central value of the epsilon-factor of π. Since π is unramified outside S, this central value is a product of local factors for $v \in S$ (each depending also on the choice of additive character). Now $\chi_d = \chi_e \chi_m$ where $m \equiv 1 \bmod C$. By the hypotheses on C, this implies that the local character $\chi_{m,v} = 1$ for all $v \in S$, and the local factors for $v \in S$ are equal. Since we are working with the twisted representations $\pi \otimes \chi_d$, $\pi \otimes \chi_e$, we must also compute the contributions from the places v dividing d or e. At these places each $\pi_{j,v}$ is unramified principal series and the comparison is obtained from the comparison of GL_1-epsilon-factors. These are computed as in Tate's thesis. (For more details see Lemma 2.2 in **[11]** and the proof of Theorem 3.1 in **[12]**.) □

Proof of Theorem 1.1. We turn to the double Dirichlet series construction. The L-function for π is of the form

$$L(s, \pi) = L_S(s, \pi)\, L(s, \pi; S)$$

where $L(s, \pi; S)$ is the contribution from places $v \in S$ and where the contribution from the places prime to S may be written as a Dirichlet series

$$L_S(s, \pi) = \sum_{m \in \mathcal{I}(S)} a(m)\, |m|^{-s}$$

(m ranging over the ideals of $\mathcal{O}$ prime to S). Let A be a class in H_C. Let

$$Z_A(s, w; \pi) = \sideset{}{'}\sum_{d \in A} \frac{L_S(s, \pi \otimes \chi_d)}{|d|^w},$$

where the sum is over integral ideals d in A, and the prime on summation indicates d square-free. Since $L(s, \pi)$ converges absolutely for $\mathrm{Re}(s) > 1$, the double Dirichlet series $Z_A(s, w; \pi)$ is absolutely convergent for $\mathrm{Re}(s) > 1, \mathrm{Re}(w) > 1$. We are interested in obtaining the analytic continuation of $Z_A(s, w; \pi)$ to a region containing $\mathrm{Re}(s) = 1, \mathrm{Re}(w) = 1$.

On the one hand, we have the convexity bound for $L(s, \pi \otimes \chi_d)$ which follows from the functional equation (2.1). If $L(s, \pi \otimes \chi_d)$ is entire then

$$L_S(s, \pi \otimes \chi_d) \ll_\epsilon \max\{1, |d|^{r(1-\sigma)/2+\epsilon}, |d|^{r(1/2-\sigma)+\epsilon}\},$$

for $\epsilon > 0$, where $\sigma = \mathrm{Re}(s)$. If ℓ is the maximum order pole of $L(s, \pi \otimes \chi_d)$ for $d \in A$ then we conclude that $(s-1)^\ell Z_A(s, w; \pi)$ is holomorphic for

$$\mathrm{Re}(w) > \max\{1, \frac{r}{2}(1 - \mathrm{Re}(s)) + 1, r(\frac{1}{2} - \mathrm{Re}(s)) + 1\}.$$

To extend past this domain let us interchange the order of summation. We obtain

$$\begin{aligned} Z_A(s, w; \pi) &= \sum_{d \in A}{}' \frac{L_S(s, \pi \otimes \chi_d)}{|d|^w} \\ &= \sum_{m \in \mathcal{I}(S)} \frac{a_m}{|m|^s} \sum_{d \in A,\ (d,m)=1}{}' \frac{\chi_d(m)}{|d|^w}. \end{aligned}$$

For $m \in \mathcal{I}(S)$, let $[m]$ denote the class of m in H_C. Also let h_C be the order of the ray class group H_C and $\hat{H}_C$ be its group of characters. Upon using the power reciprocity law as formulated in Proposition 2.1 and the orthogonality relations, for $m \in \mathcal{I}(S)$ we obtain

$$\begin{aligned} \sum_{d \in A, (d,m)=1}{}' \frac{\chi_d(m)}{|d|^w} &= \alpha(A, [m]) \sum_{d \in A, (d,m)=1}{}' \frac{\chi_m(d)}{|d|^w} \\ &= \frac{\alpha(A, [m])}{h_C} \sum_{\eta \in \hat{H}_C} \eta^{-1}(A)\, D_m(w, \eta\chi_m), \end{aligned}$$

where

$$D_m(w, \eta\chi_m) = \sum_{d \in \mathcal{I}(S), (d,m)=1}{}' \frac{\eta(d)\chi_m(d)}{|d|^w}.$$

If μ now denotes the Möbius function of ideals of $\mathcal{O}$, then

$$\begin{aligned} D_m(w, \eta\chi_m) &= \sum_{d \in \mathcal{I}(S), (d,m)=1} \sum_{e^2 | d} \mu(e) \frac{\eta(d)\chi_m(d)}{|d|^w} \\ &= \sum_{e \in \mathcal{I}(S), (e,m)=1} \frac{\mu(e)\, \eta^2(e)\, \chi_m^2(e)}{|e|^{2w}} \sum_{d \in \mathcal{I}(S), (d,m)=1} \frac{\eta(d)\chi_m(d)}{|d|^w} \\ &= \frac{L_{m \cup S}(w, \eta\chi_m)}{L_{m \cup S}(2w, \eta^2\chi_m^2)}, \end{aligned} \tag{2.2}$$

where the subscripts indicate that the primes in S and those dividing m are removed from the last quotient of GL_1 L-functions. Thus $D_m(w, \eta\chi_m)$ is holomorphic for $\mathrm{Re}(w) \geq 1/2$ if $\eta\chi_m$ is not the trivial character; otherwise, it is meromorphic in this region with only a simple pole at $w = 1$. Note that this occurs only when $\eta = 1$ and m is a perfect n-th power.

It is the residue at this pole which gives the main term in Theorem 1.1. Note that for $\mathrm{Re}(s) > 1$,

$$\begin{aligned} c_S(s) &= \mathrm{Res}_{w=1}\, Z_A(s,w;\pi) \\ &= \frac{\kappa}{\zeta_F(2)\, h_C} \left(\prod_{v \in S} \frac{\zeta_{F,v}(2)}{\zeta_{F,v}(1)} \right) \sum_{m \in \mathcal{I}(S)} \frac{a_{m^n}}{|m|^{ns}} \prod_{v|m} \frac{\zeta_{F,v}(2)}{\zeta_{F,v}(1)}, \end{aligned} \tag{2.3}$$

where $\zeta_F = \prod \zeta_{F,v}$ is the Dedekind zeta function of F and κ is its residue at $w = 1$. We show later in this proof that this Dirichlet series representation, initially valid for $\mathrm{Re}(s) > 1$, remains true for $\mathrm{Re}(s) > \max(1 - 1/r, \frac{1}{2})$ if $n = 2$ and for $\mathrm{Re}(s) > 1 - 1/(r+1)$ if $n > 2$. Moreover, we will show that for S sufficiently large, the residue is non-zero for s in this range.

Next let us note that for $\epsilon > 0, \mathrm{Re}(w) \geq 1/2, w \neq 1$ and $\mathrm{Im}\, w = t$,

$$\sum_{m \in \mathcal{I}(S), |m| < X} |D_m(w, \eta\chi_m)|^2 \ll_{C,\epsilon} X^{1+\epsilon}(1 + |t|)^K. \tag{2.4}$$

Here K is a constant depending on the ground field F, and on n, r. Indeed, this follows from the evaluation (2.2) since the $L(w, \eta\chi_m)$ occur as Fourier coefficients of the Eisenstein series on the n-fold cover of GL_n induced from the theta function on the n-fold cover of GL_{n-1} (see Suzuki **[23]** and Banks, Bump, and Lieman **[1]**); alternatively it follows by using the associated double Dirichlet series as in Diaconu **[9]**. See also **[14]**.

Moreover, from the properties of the Rankin-Selberg L-function $L(s, \pi \otimes \tilde{\pi})$, it follows that

$$\sum_{|m| < X} |a_m|^2 \ll X(\log X)^\lambda$$

for some integer λ. Therefore by the Cauchy-Schwarz inequality,

$$\sum_{|m| < X} |a_m D_m(w, \eta\chi_m)|^2 \ll_{A,\epsilon} X^{1+\epsilon}(1 + |t|)^K. \tag{2.5}$$

By partial summation, we conclude that $(w-1)Z_A(s,w;\pi)$ is holomorphic for $\mathrm{Re}(s) > 1$, $\mathrm{Re}(w) > 1/2$ and that

$$(w-1)Z_A(s,w;\pi) \ll_{A,\epsilon} (1 + |t|)^K.$$

Combining this with the region of absolute convergence deduced from convexity estimates on $L_S(s, \pi \otimes \chi_d)$, we have established the holomorphicity of $(w-1)(s-1)^\ell Z_A(s,w;\pi)$ on the union of the tube domains

$$T_1 = \{(s,w) : \mathrm{Re}(w) > 1/2, \mathrm{Re}(s) > 1\}$$

and

$$T_2 = \{(s,w) : \mathrm{Re}(w) > \max(1, \tfrac{r}{2}(1 - \mathrm{Re}(s)) + 1, r(\tfrac{1}{2} - \mathrm{Re}(s)) + 1)\}.$$

In the case $n = 2$ we can further enlarge the region of continuation as follows. For convenience let us suppose that $r \geq 2$; the case $r = 1$ is similar but simpler. We apply the functional equation (2.1). (See also Remark 2.5 below.) Note that for $d \in A$ the product of local L-factors $L(s, \pi \otimes \chi_d; S)$ depends only on A; so we write it $L(s, \pi \otimes \chi_A; S)$. Also, by Lemma 2.2 we may write $\epsilon(s, \pi \otimes \chi_d) = A(s)\, \chi_\pi(d) |d|^{r(s-1/2)}$ where $A(s)$ is a monomial function depending only on π and A. Thus we have

$$\begin{aligned} Z_A(s, w; \pi) &= \sum_{d \in A}{}' \frac{L_S(s, \pi \otimes \chi_d)}{|d|^w} \\ &= \frac{L(1-s, \tilde{\pi} \otimes \chi_A; S)}{L(s, \pi \otimes \chi_A; S)} \sum_{d \in A}{}' \frac{\chi_\pi(d)\, L_S(1-s, \tilde{\pi} \otimes \chi_d)}{|d|^{w+r(s-1/2)}}. \end{aligned}$$

By interchanging summation and using the Cauchy-Schwarz inequality as above, one thus sees that $Z_A(s, w; \pi)$ has holomorphic continuation to the tube domain

$$T_3 := \{(s, w) : \mathrm{Re}(s) < 0, \mathrm{Re}(w) > -r\,\mathrm{Re}(s) + r/2 + 1/2\}.$$

By the analytic continuation of holomorphic functions of two complex variables on tube domains to the convex hull **[17]**, the function $(s-1)^\ell (w-1) Z_A(s, w; \pi)$ thus has holomorphic continuation to the convex hull of the tube domain $T_1 \cup T_2$ for $n \geq 3$ (resp. $T_1 \cup T_2 \cup T_3$ for $n = 2$).

Note that $L_S(s, \pi \otimes \chi_d)$ is entire for $d \in A$ unless $d = (1)$. Hence $Z_A(s, w; \pi)$ can have a pole at $s = 1$ only if $(1) \in A$. If this is the case, we will find it more convenient to work with $Z_A(s, w; \pi) - L_S(s, \pi)$. Let us define

$$Z_A^0(s, w; \pi) = \begin{cases} Z_A(s, w; \pi) - L_S(s, \pi) & \text{if } (1) \in A \\ Z_A(s, w; \pi) & \text{otherwise.} \end{cases}$$

Observe that

$$\mathrm{Res}_{w=1}\, Z_A^0(s, w; \pi) = \mathrm{Res}_{w=1}\, Z_A(s, w; \pi).$$

Theorem 1.1 will now be a consequence of a standard Tauberian argument, as follows. First observe that as the convex hull of the tube domains is a domain of holomorphy it follows from the Phragmen-Lindelöf principle that the function $(w-1) Z_A^0(s, w; \pi) \ll_{A,\epsilon} (1 + |t|)^K$ in the entire domain. Choose k to be a positive integer with $k > K + 1$, with K as in (2.4). Applying the integral transform

$$\frac{1}{2\pi i} \int_{2-i\infty}^{2+i\infty} \frac{x^w\, dw}{w(w+1)\dots(w+k)} = \begin{cases} \frac{1}{k!}(1 - 1/x)^k & \text{if } x > 1 \\ 0 & \text{if } 0 < x \leq 1 \end{cases}$$

we obtain first

$$\frac{1}{2\pi i}\int_{2-i\infty}^{2+i\infty}\frac{Z_A^0(s,w;\pi)\,x^w dw}{w(w+1)\ldots(w+k)}=\frac{1}{k!}\sum_{\substack{d\in A, d\neq(1)\\ d \text{ sq-free}}} L_S(s,\pi\otimes\chi_d)\left(1-\frac{|d|}{x}\right)^k.$$

Moving the line of integration to $\operatorname{Re} w = 1/2+\epsilon$, for $\epsilon > 0$, the above equals

$$\frac{1}{(k+1)!}\operatorname{Res}_{w=1} Z_A^0(s,w;\pi)x+\frac{1}{2\pi i}\int_{1/2+\epsilon-i\infty}^{1/2+\epsilon+i\infty}\frac{Z_A^0(s,w;\pi)\,x^w dw}{w(w+1)\ldots(w+k)},$$

as $Z_A^0(s,w;\pi)$ is an analytic function of w in this region except for a simple pole at $w = 1$. The integral converges absolutely and is bounded above by $x^{1/2+\epsilon}$, so letting $c_S(s) = \operatorname{Res}_{w=1} Z_A(s,w;\pi)$ we obtain part (1) of Theorem 1.1. In the case of part (2) the values being averaged are non-negative, so the Wiener–Ikehara Tauberian theorem applies to give an unweighted asymptotic result.

To complete the proof of Theorem 1.1 we must now analyze the residue of $Z_A(s,w;\pi)$ at $w = 1$. This is given for $\operatorname{Re}(s) > 1$ by the expression (2.3). Thus this residue is a nonzero constant times the Euler product

$$\prod_{(v,S)=1}\left(1+\frac{q_v}{q_v+1}\sum_{g=1}^{\infty}\frac{a_{P_v^{ng}}}{q_v^{ngs}}\right) \tag{2.6}$$

where we have set $|P_v| = q_v$ for convenience. We analyze the convergence of this product separately in the cases $n > 2$ and $n = 2$.

Suppose that the Satake parameters of π_v are given by $\alpha_{v,i}$, $1 \le i \le r$. Then by the result of Luo, Rudnick and Sarnak [20], $|\alpha_{v,i}| \le q_v^{1/2-1/(r^2+1)}$. Thus

$$|a_{P_v^{ng}}| = \sum_{k_1+\cdots+k_r=ng}\prod_i \alpha_{v,i}^{k_i} \le p_r(ng)q_v^{ng(1/2-1/(r^2+1))},$$

where $p_r(ng)$ is the number of ordered partitions of ng into r non-negative integral pieces. If $r = 2$ one may do better using the estimate of Kim and Shahidi [18], namely $|\alpha_{v,i}| < q_v^{1/9}$. For later use, define

$$\delta=\delta_{r,n}=\begin{cases}1/2-1/(r+1)+1/(r^2+1) & \text{if } r>2, n>2\\ 1/2-1/r+1/(r^2+1) & \text{if } r>2, n=2\\ 7/18 & \text{if } r=2.\end{cases}$$

Also, note the trivial bound

$$p_r(ng) \le (ng+1)^r.$$

First, suppose $n \ge 3$. Recall that an infinite product $\prod_v(1+b_v)$ converges absolutely if and only if the sum $\sum_v |b_v|$ converges. For $\operatorname{Re}(s) > 1-1/(r+1)$

we have

$$\sum_{(v,S)=1}\sum_{g=1}^{\infty}\frac{|a_{P_v^{ng}}|}{q_v^{ng\,\mathrm{Re}(s)}} \le \sum_{(v,S)=1}\sum_{g=1}^{\infty}\frac{p_r(ng)}{q_v^{ng\delta}}$$
$$\le \sum_{(v,S)=1}\sum_{g=1}^{\infty}\frac{(ng+1)^r}{q_v^{ng\delta}}.$$

This series converges absolutely because $n\delta > 1$.

Now suppose instead that $n = 2$. To prove the absolute convergence of the Euler product (2.6), we make the observation that the residue is well approximated by the symmetric square L-function, which converges absolutely for $\mathrm{Re}(s) > 1$ by Rankin-Selberg theory. Note that if $\pi = \boxplus_{j=1}^{m}\pi_j$ then

$$L(s,\pi,\mathrm{Sym}^2) = \prod_{j=1}^{m} L(s,\pi_j,\mathrm{Sym}^2) \prod_{1\le j<j'\le m} L(s,\pi_j\times\pi_{j'})$$

so we make use of the analytic properties of Rankin-Selberg convolutions as well. Let

$$L(s,\pi,\mathrm{Sym}^2) = \prod_v T_v(s),$$

where

$$T_v(s) = \prod_{1\le i\le j\le r}(1-\alpha_{v,i}\alpha_{v,j}q_v^{-s})^{-1} = \prod_{1\le i\le r'}(1-\beta_{v,i}q_v^{-s})^{-1},$$

say, with $r' = r(r+1)/2$ and $|\beta_{v,i}| \le q_v^{1-2/(r^2+1)}$. Then for $\mathrm{Re}(s) > 1-1/r$,

$$T_v(2s) = 1 + \frac{a(P_v^2)}{q_v^{2s}} + \sum_{g\ge 2}\frac{1}{q_v^{2gs}}\sum_{k_1+\cdots k_{r'}=g}\beta_{v,1}^{k_1}\cdots\beta_{v,r'}^{k_{r'}}$$
$$= 1 + \frac{a(P_v^2)}{q_v^{2s}} + O_r(q_v^{-4\delta}).$$

Similarly,

$$1+\frac{q_v}{q_v+1}\sum_{g=1}^{\infty}\frac{a_{P_v^{2g}}}{q_v^{2gs}} = 1+\left(1+O\left(\frac{1}{q_v}\right)\right)\left(\frac{a(P_v^2)}{q_v^{2s}}+\sum_{g\ge 2}\frac{a_{P_v^{2g}}}{q_v^{2gs}}\right)$$
$$= 1+\frac{a(P_v^2)}{q_v^{2s}} + O_r(q_v^{-1-2\delta}).$$

Hence the quotient

$$\frac{1+\frac{q_v}{q_v+1}\sum_{g=1}^{\infty}\frac{a_{P_v^{2g}}}{q_v^{2gs}}}{T_v(2s)} = 1 + O_r(\max(q_v^{-4\delta}, q_v^{-1-2\delta})),$$

and the product over all places

$$\prod_v (1 + O_r(\max(q_v^{-4\delta}, q_v^{-1-2\delta})))$$

converges absolutely. From the absolute convergence of the symmetric square L-function for $\mathrm{Re}(s) > 1$ the convergence of (2.6) follows.

In all cases the Euler product (2.6) is absolutely convergent for $\mathrm{Re}(s)$ as in Theorem 1.1. It is hence non-zero provided none of the terms vanishes. We may guarantee that none of the terms vanishes by throwing out a finite number of primes, i.e. by enlarging the set S. Explicitly, we desire q_v sufficiently large that

$$\frac{q_v}{q_v+1}\sum_{g=1}^{\infty}\frac{a_{P_v^{ng}}}{q_v^{ngs}}$$

is less than 1 in absolute value, for $\mathrm{Re}(s) > 1 - 1/r$ if $n = 2$ or for $\mathrm{Re}(s) > 1 - 1/(r+1)$ if $n \geq 3$. Using the absolute bound for the Fourier coefficients, for such s one has

$$\left|\frac{q_v}{q_v+1}\sum_{g=1}^{\infty}\frac{a_{P_v^{ng}}}{q_v^{ngs}}\right| \leq \sum_{g=1}^{\infty}\frac{p_r(ng)}{q_v^{ng\delta}}.$$

For the partition function

$$p_r(ng) = \binom{ng+r-1}{r-1} = \frac{(ng+r-1)\cdots(ng+1)}{(r-1)!}$$

we use the bounds

$$p_r(ng) \leq \frac{(2r)^{r-1}}{(r-1)!},$$

for $ng \leq r$, and by Stirling's formula

$$p_r(ng) \leq A^{ng}$$

if $ng > r$, for some absolute constant A. Therefore, if q_v is sufficiently large,

$$\begin{aligned}\sum_{g=1}^{\infty}\frac{p_r(ng)}{q_v^{ng\delta}} &\leq \sum_{g=1}^{[r/n]} + \sum_{g\geq r/n}\\ &\leq \frac{(2r)^{r-1}}{(r-1)!}\sum_{g=1}^{[r/n]} q_v^{-ng\delta} + \sum_{g\geq r/n}(A/q_v^{\delta})^{ng}\\ &\leq C_1\frac{(2r)^{r-1}}{(r-1)!}q_v^{-n\delta} + C_2(A/q_v^{\delta})^{r}\end{aligned}$$

for some absolute constants C_1, C_2. This quantity will be less than 1 provided

$$q_v^{n\delta} >> \max\left\{\frac{(2r)^{r-1}}{(r-1)!}, A^n\right\}.$$

This concludes the proof of Theorem 1.1. □

Remark 2.3. Changing the choices made in defining the characters χ_d above would amount to changing π to $\pi \otimes \chi$ where χ is an idèle class character which is obtained from a character of R_C.

Remark 2.4. In fact, the proof of Theorem 1.1 does not make use of the full hypothesis that π is an automorphic representation of $GL_r(\mathbb{A}_F)$. If $D(s)$ is any Eulerian Dirichlet series, absolutely convergent for $\mathrm{Re}(s) > 1$, with suitable functional equations under GL_1 twists, satisfying an estimate for the growth of the coefficients as in (2.5), and such that the residue (2.6) is well-behaved, then the proof of Theorem 1.1 goes through and the growth estimates of Theorem 1.1 hold. For example, this holds for any Dirichlet series in the Selberg class which has suitable functional equations under GL_1 twists.

Remark 2.5. One may also make use of the functional equation to increase the region of continuation of $Z_A(s, w; \pi)$ in the cases $r = 1$, n arbitrary and $r = 2$, $n = 3$. We shall not pursue these cases here since in fact further work allows one to continue an appropriate modified double Dirichlet series to all of $\mathbb{C}^2$. (See Friedberg-Hoffstein-Lieman **[14]** for the first case and Brubaker-Friedberg-Hoffstein **[5]** for the second.)

3. Value distribution of $L(1, \pi \otimes \chi_d)$

Let $\{\mu_n\}$ be a sequence of Borel probability measures on $\mathbb{R}$. One says that this sequence converges weakly to a measure μ if

$$\int f d\mu_n \to \int f d\mu, \quad \text{as } n \to \infty$$

for all continuous, compactly supported functions f on $\mathbb{R}$. Suppose F (resp. F_n) is the distribution function of μ (resp. μ_n), i.e., F is the non-decreasing right-continuous function defined by

$$F(x) = \mu((-\infty, x]).$$

Then $\mu_n \to \mu$ weakly if and only if

$$\lim_{n\to\infty} F_n(x) = F(x)$$

for all points of continuity x of F. The k-th moment of a measure μ on $\mathbb{R}$ is defined to be

$$M_k[\mu] := \int_{\mathbb{R}} x^k d\mu(x) = \int x^k dF(x)$$

if this limit exists.

A theorem of Fréchet and Shohat [13],[4] allows us to deduce weak convergence from convergence of moments. Precisely: let $\{\mu_n\}$ be a sequence of Borel probability measures on $\mathbb{R}$ such that for every positive integer k, the limit

$$M_k := \lim_{n\to\infty} M_k[\mu_n]$$

exists and the power series

$$\Psi(w) := \sum_{k\geq 0} \frac{M_k w^k}{k!}$$

has positive radius of convergence. Then there exists a measure μ such that

1. μ_n converges weakly to μ,
2. for all positive integers $k, M_k = M_k[\mu]$, and
3. for $t \in \mathbb{R}$ sufficiently small, the characteristic function ψ of μ,

   $$\psi(t) := \int_{\mathbb{R}} e^{itz} d\mu z,$$

   is given by

   $$\Psi(it) = \psi(t).$$

We now turn to the proof of Corollary 1.3. For a positive integer X, we define the probability measure μ_X (depending on A) by

$$\mu_X([0,y)) := \frac{|\{d \in A, d \text{ sq-free}, |d| < X : |L(1, \pi\otimes\chi_d)|^2 < y\}|}{|\{d \in A, d \text{ sq-free}, |d| < X\}|}$$

for all $y > 0$. Fix S as in Theorem 1.1. Let $n = 2$. Since for $d \in A$ the local character $\chi_{d,v}$ is independent of d for $v \in S$, it follows that the quotient $L(1, \pi\otimes\chi_d)/L_S(1, \pi\otimes\chi_d) = g_{S,A}$ is independent of d. (This constant gives the dependence of the distribution function on the ray class A.) Let $L_S(s, \pi\boxplus\tilde{\pi})^k = \sum b_k(m)|m|^{-s}$. By Theorem 1.1 applied to $\pi^{\boxplus k} \boxplus \tilde{\pi}^{\boxplus k}$, for each $k \in \mathbb{N}$ there exists a constant c_k given by the Euler product

$$c_k = \frac{\kappa |g_{S,A}|^{2k}}{\zeta_F(2)\, h_C} \left(\prod_{v\in S} \frac{\zeta_{F,v}(2)}{\zeta_{F,v}(1)}\right) \prod_{(v,S)=1} \left(1 + \frac{q_v}{q_v+1} \sum_{g=1}^{\infty} \frac{b_k(P_v^{2g})}{q_v^{2g}}\right)$$

such that

$$\lim_{X\to\infty} M_k[\mu_X] = c_k.$$

Let β_v denote the absolute value of the largest Satake parameter of π at v. By the trivial bound $\beta_v \le q_v^{1/2}$, we know that $\beta_v/q_v < 3/4$, say. Now

$$b_k(P_v^m) = \sum_{i_1+\cdots+i_k=m} b_1(P_v^{i_1})\cdots b_1(P_v^{i_k})$$

and

$$|b_1(P_v^m)| \le \beta_v^m p_{2r}(m),$$

where $p_{2r}(m)$ is the partition function in Section 2. Hence

$$|b_k(P_v^m)| \le \beta_v^m \sum_{i_1+\cdots+i_k=m} p_{2r}(i_1)\cdots p_{2r}(i_k) = \beta_v^m p_{2rk}(m).$$

In order to bound c_k we need to bound the Euler product

$$X_k = \prod_{(v,S)=1}\left(\sum_{g=0}^{\infty}\frac{b_k(P_v^{2g})}{q_v^{2g}}\right).$$

Each term of the above Euler product is bounded by

$$\sum_{g=0}^{\infty}\frac{\beta_v^{2g}}{q_v^{2g}}p_{2rk}(2g) = \frac{1}{2}\left(\frac{1}{(1-\beta_v/q_v)^{2rk}} + \frac{1}{(1+\beta_v/q_v)^{2rk}}\right).$$

Write

$$X_k = \prod_{(v,S)=1} Q_k(\beta_v/q_v).$$

Taking two derivatives and using Taylor's theorem,

$$Q_k(x) = 1 + O(x^2k^2(1+C_r|x|)^{k-2})$$

in the interval $|x| < 3/4$, for some constant C_r depending only on r.

Then

$$\log X_k = S_0 + S_1 + S_2 + S_3$$

where

$$S_0 = \sum_{q_v<k} \log Q_k(\beta_v/q_v),$$
$$S_1 = \sum_{k\le q_v<k^2, 1\le\beta_v\le q_v/k} \log Q_k(\beta_v/q_v),$$
$$S_2 = \sum_{k\le q_v<k^2, \beta_v>q_v/k} \log Q_k(\beta_v/q_v),$$
$$S_3 = \sum_{q_v>k^2} \log Q_k(\beta_v/q_v).$$

In treating these sums below, we make repeated use of the estimates

$$\sum_{q_v<x} \beta_v \ll x/\log x \quad \text{and} \quad \sum_{q_v<x} \beta_v^2 \ll x/\log x.$$

In these two estimates and in those below, the implicit constants will depend on π.

First,

$$S_0 \ll \sum_{q_v<k} \log(k^2(\beta_v^2/q_v^2)(1+C_r\beta_v/q_v)^{k-2}),$$

so

$$S_0 \ll \sum_{q_v<k} \log k + k\sum_{q_v<k} \log(1+C_r\beta_v/q_v) \ll k + k\sum_{q_v<k} \beta_v/q_v \ll k\log\log k.$$

For $k \le q_v < k^2, 1 \le \beta_v \le q_v/k$, we make the estimate

$$\begin{aligned} Q_k(\beta_v/q_v) &= 1 + O((\beta_v/q_v)^2k^2(1+C_r(\beta_v/q_v))^{k-2}) \\ &= 1 + O((\beta_v/q_v)^2k^2(1+C_r/k)^{k-2}) \\ &= 1 + O((\beta_v/q_v)^2k^2). \end{aligned}$$

Thus

$$S_1 \ll \sum_{k\le q_v<k^2, 1\le\beta_v\le q_v/k} (k\beta_v/q_v)^2 \ll k^2 \sum_{k\le q_v<k^2} \beta_v^2/q_v^2 \ll k/\log k.$$

For $k \le q_v < k^2, \beta_v > q_v/k$, we make the estimate

$$\log Q_k(\beta_v/q_v) \ll \log((k\beta_v/q_v)^2(1+\beta_v/q_v)^k) \ll k\beta_v/q_v,$$

from which we get

$$S_2 \ll \sum_{k\le q_v<k^2, \beta_v>q_v/k} k\beta_v/q_v \ll k\log\log k.$$

Finally, for the last case $q_v > k^2$, the trivial bound $\beta_v \leq \sqrt{q_v}$ implies

$$(1 + C_r\beta_v/q_v)^{k-2} \ll 1$$

from which we get

$$S_3 \ll \sum_{q_v > k^2} k^2\beta_v^2/q_v^2 \ll k/\log k.$$

Thus $\log X_k \ll k \log\log k$.

Hence

$$c_k \ll \exp(Ck \log\log(k)),$$

where both the implicit constant and the constant C depend on π. By the theorem of Fréchet and Shohat described above, Corollary 1.3 holds. □

We remark that this last estimate shows that the moment generating function Ψ is entire.

References

[1] W. Banks, D. Bump and D. Lieman, Whittaker-Fourier coefficients of metaplectic Eisenstein series. Compositio Math. 135 (2003), no. 2, 153–178.

[2] M. B. Barban, The "large sieve" method and its application to number theory. (Russian) Uspehi Mat. Nauk 21 (1966), no. 1, 51–102.

[3] L. Barthel and D. Ramakrishnan, A nonvanishing result for twists of L-functions of GL(n). Duke Math. J. 74 (1994), no. 3, 681–700.

[4] P. Billingsley, Probability and Measure. Second edition. Wiley Series in Probability and Mathematical Statistics: Probability and Mathematical Statistics. John Wiley & Sons, Inc., New York, 1986. xiv+622 pp.

[5] B. Brubaker, S. Friedberg and J. Hoffstein, A nonvanishing theorem for cubic twists of $GL(2)$, Invent. Math. 160 (2005), no. 1, 31-58.

[6] D. Bump, S. Friedberg and J. Hoffstein, On some applications of automorphic forms to number theory. Bull. Amer. Math. Soc. (N.S.) 33 (1996), no. 2, 157–175.

[7] D. Bump, S. Friedberg and J. Hoffstein, Sums of twisted $GL(3)$ automorphic L-functions, in: Contributions to Automorphic Forms, Geometry and Arithmetic, eds. H. Hida, D. Ramakrishnan and F. Shahidi. Johns Hopkins Univ. Press, 2004, pp. 131-162.

[8] S. Chowla and P. Erdös, A theorem on the distribution of the values of L-functions. J. Indian Math. Soc. (N.S.) 15, (1951), 11–18.

[9] A. Diaconu, Mean square values of Hecke L-series formed with n^{th} order characters. Invent. Math. 157 (2004), no. 3, 635-684.

[10] A. Diaconu, D. Goldfeld, J. Hoffstein, Multiple Dirichlet series and moments of zeta and L-functions. Compositio Math. 139 (2003), no. 3, 297–360.

[11] B. Fisher and S. Friedberg, Double Dirichlet series over function fields. Compositio Math. 140 (2004), no. 3, 613–630.

[12] B. Fisher and S. Friedberg, Sums of twisted GL(2) L-functions over function fields. Duke Math. J. 117 (2003), no. 3, 543–570.

[13] M. Fréchet and J. Shohat, A proof of the generalized second-limit theorem in the theory of probability. Trans. Amer. Math. Soc. 33 (1931), no. 2, 533–543.

[14] S. Friedberg, J. Hoffstein, and D. Lieman, Double Dirichlet series and the n-th order twists of Hecke L-series. Math. Ann. 327 (2003), no. 2, 315–338.

[15] A. Granville and K. Soundararajan, The distribution of values of $L(1, \chi_d)$. Geom. Funct. Anal. 13 (2003), no. 5, 992–1028.

[16] D. R. Heath-Brown, A mean value estimate for real character sums. Acta Arith. 72 (1995), no. 3, 235–275.

[17] L. Hörmander, An Introduction to Complex Analysis in Several Variables. Third edition. North-Holland Mathematical Library, 7. North-Holland Publishing Co., Amsterdam, 1990. xii+254 pp.

[18] H. Kim and F. Shahidi, Cuspidality of symmetric powers with applications. Duke Math. J. 112 (2002), no. 1, 177–197.

[19] W. Luo, Values of symmetric square L-functions at 1. J. Reine Angew. Math. 506 (1999), 215–235.

[20] W. Luo, Z. Rudnick, and P. Sarnak, On the generalized Ramanujan conjecture for GL(n), in: Automorphic Forms, Automorphic Representations, and Arithmetic (Fort Worth, TX, 1996), 301–310, Proc. Sympos. Pure Math., 66, Part 2, Amer. Math. Soc., Providence, RI, 1999.

[21] M. Peter, Momente der Klassenzahlen binärer quadratischer Formen mit ganzalgebraischen Koeffizienten. (German) [Moments of class numbers of binary quadratic forms with algebraic integer coefficients] Acta Arith. 70 (1995), no. 1, 43–77.

[22] D. Rohrlich, Nonvanishing of L-functions for GL(2). Invent. Math. 97 (1989), no. 2, 381–403.

[23] T. Suzuki, Metaplectic Eisenstein series and the Bump-Hoffstein conjecture. Duke Math. J. 90 (1997), no. 3, 577–630.

[24] L. Washington, Introduction to Cyclotomic Fields. Second edition. Graduate Texts in Mathematics, 83. Springer-Verlag, New York, 1997. xiv+487 pp.

Gautam Chinta
Brown University
Dept. of Mathematics
Providence, RI 02912. USA
chinta@math.brown.edu

Solomon Friedberg
Boston College
Dept. of Mathematics
Chestnut Hill. MA 02467-3806, USA
friedber@bc.edu

Jeffrey Hoffstein
Brown University
Dept. of Mathematics
Providence. RI 02912. USA
jhoff@math.brown.edu

Partial Bessel functions for quasi-split groups

*J.W. Cogdell, I.I. Piatetski-Shapiro, and F. Shahidi**

To Steve Rallis, with admiration, on the occasion of his sixtieth birthday

The question of functoriality is a central one in the theory of automorphic forms and representations. There has been recent progress on the proof of this conjecture for globally generic representations of the split classical groups by combining the converse theorem with the Langlands-Shahidi method of controlling automorphic L-functions **[6, 7]**. We refer the reader to those papers for an exposition of the general process.

One crucial local result needed in these proofs is the *stability of the local γ-factors* for the split classical groups. This seemingly technical result allows us to finesse the lack of the local Langlands conjecture at the finite number of finite places where the generic cuspidal representation is ramified. For the split odd orthogonal results this was established in **[8]** and then used in establishing functoriality from SO_{2n+1} to GL_{2n} in **[6]**. For the functoriality for even orthogonal groups and symplectic groups, the results of **[8]** were extended to these cases in Section 4 of **[7]**. In these later cases, the stability was established by expressing the local γ-factor as a Mellin transform of a certain partial Bessel function **[16]** and then using results on the asymptotics of these partial Bessel functions established in the split case in **[8]**.

In order to facilitate the establishment of functoriality for general quasi-split groups, we turn to the problem of stability of local γ-factors for generic representations of these groups. In this paper, we extend the results of **[8]** and **[7]** on the asymptotics of certain partial Bessel functions of representations to the quasi-split case. The definition of the partial Bessel functions of interest can be found in Section 3 and the statement of our main result about them can be found in Section 7 of this paper. The arguments are essentially those found in **[8]** and **[7]** generalized to the quasi-split setting. Even though the modifications are minor in places, we have chosen to reproduce the arguments in full both to make this paper self contained and to be sure there is no question as to their validity in this increased generality.

*During the work on this project J.C. was partially supported by the NSA through grant MDA 904-03-1-0043 and by the Vaughn Foundation, I.I.P-S. by the NSF through grant DMS-0072450, and F.S. by the NSF through grant DMS-0200325.

In a subsequent paper **[9]** we will turn to the combination of these asymptotics with the expression of the local γ-factor as a Mellin transform of related Bessel functions in **[16]** to obtain the general stability result for these γ-factors. Then progress on functoriality for quasi-split groups can proceed.

We conclude the paper with an Appendix on the existence of the (full) Bessel function of a generic representation associated to certain minimal Weyl elements (see Section 8). While this existence is not needed for the applications to functoriality, such Bessel functions are analogs of (twisted) orbital integrals (cf. Remark 3.2 of **[16]**) and are expected to play a fundamental role in the relative trace formula. Their existence was assumed in **[16]** (cf. Section 3 of that paper) for heuristic purposes, even though it was not used for the results established there; those results also require only the partial Bessel functions. We prove the existence of the Bessel function for quasi-split groups here, albeit in the limited context of certain minimal Weyl elements.

We would like to thank the referee for bringing to our attention a false proof in an earlier version of this paper and for the suggestions made on streamlining the presentation of the paper. We would also like to once again acknowledge that the genesis of many of the ideas for analyzing Bessel functions are to be found in the thesis of Averbuch **[1]**.

1. Preliminaries

Let k be a non-archimedean local field of characteristic zero, with ring of integers $\mathcal{O}$ and maximal ideal $\mathcal{P}$. Let $q = q_k$ be the order of the residue field $\mathcal{O}/\mathcal{P}$. Let $\Gamma = Gal(\overline{k}/k)$ denote the absolute Galois group of k.

Let G be a connected reductive algebraic group which is defined and quasi-split over k. We shall assume in addition that the center $\mathrm{Z} = \mathcal{Z}(\mathrm{G})$ is connected and cohomologically trivial to first order, i.e., $H^1(\Gamma, \mathrm{Z}) = \{0\}$. Since the ultimate goal of this paper is an application to the stability of the local γ-factor of a generic representation of $\mathrm{G}(k)$ under highly ramified twists, we know by Proposition 5.4 of **[16]** and the Appendix to **[7]** that these restrictions have no effect on the applicability of our results.

Fix a Borel subgroup $\mathrm{B} = \mathrm{TU}$ over k with unipotent radical U and maximal torus T. Let A be the maximal k-split subtorus of T; then $\mathrm{T} = \mathcal{Z}(\mathrm{A})$.

Let $\widetilde{\Phi} = \Phi(\mathrm{T}, \mathrm{G})$ be the set of (non-restricted) roots of T in G **[4, 13]**. The choice of U then defines a set of positive roots $\widetilde{\Phi}^+$ and simple roots $\widetilde{\Delta}$ of T in U. Let K/k be Galois splitting field of G and let $\Gamma_K = Gal(K/k)$. Since G is quasi-split over k, and split over K, both $\widetilde{\Phi}$ and $\widetilde{\Delta}$ decompose into a finite number of Γ or Γ_K orbits **[13]**.

Let $\Phi = \Phi(\mathrm{A}, \mathrm{G})$ be the set of (restricted) roots of G with respect to A. Again, the choice of U then defines a set of positive roots Φ^+ and simple roots Δ for A. The root system Φ may have multiple roots. Let Φ_{nd} be the non-divisible roots of Φ. For each $\alpha \in \Phi_{nd}$ we will let (α) denote the set of roots which are positive multiples of α. The possibilities are either $(\alpha) = \{\alpha\}$ or $(\alpha) = \{\alpha, 2\alpha\}$ **[2]**. The Γ or Γ_K orbits in $\widetilde{\Delta}$ are in one-to-one correspondence with the restricted simple roots Δ **[3]**. Given $\alpha \in \Delta$ we will let $\widetilde{\Delta}_\alpha$ denote the Γ-orbit of roots $\tilde{\alpha} \in \widetilde{\Delta}$ which restrict to α.

Let W denote the (relative) Weyl group of A in G, i.e., $\mathcal{N}(\mathrm{A})/\mathcal{Z}(\mathrm{A})$.

1.1. Splittings. For each $\tilde{\alpha} \in \widetilde{\Delta}$ let $K_{\tilde{\alpha}}$ be the splitting field of the simple root $\tilde{\alpha}$. So $k \subset \mathrm{K}_{\tilde{\alpha}} \subset K$. Let $\{x_{\tilde{\alpha}} \mid \tilde{\alpha} \in \widetilde{\Delta}\}$ be a Steinberg splitting of G (cf. Section 4.1.3 of **[5]**). So if $\mathrm{U}_{\tilde{\alpha}} \subset \mathrm{G}_K$ is the root subgroup corresponding to $\tilde{\alpha}$ over K, we have

(i) $x_{\tilde{\alpha}} : \mathbb{G}_a \to \mathrm{U}_{\tilde{\alpha}}$ is an isomorphism over $K_{\tilde{\alpha}}$;

(ii) $x_{\gamma(\tilde{\alpha})} = \gamma \circ x_{\tilde{\alpha}} \circ \gamma^{-1}$ for $\gamma \in \Gamma_K$.

This splitting determines an associated Chevalley system $\{x_{\tilde{\alpha}} \mid \tilde{\alpha} \in \widetilde{\Phi}\}$ for G over K called a Chevalley-Steinberg system for G (see Section 4.1.3 of **[5]**). Recall that such a splitting always satisfies $Ad(t)x_{\tilde{\alpha}}(u) = x_{\tilde{\alpha}}(\tilde{\alpha}(t)u)$ for $t \in \mathrm{T}(K)$. A Chevalley-Steinberg system then defines compatible root datum $(\mathrm{T}, (\mathrm{U}_{\tilde{\alpha}})_{\tilde{\alpha}\in\widetilde{\Phi}})$ for $\mathrm{G}(K)$ and $(\mathrm{T}, (\mathrm{U}_\alpha)_{\alpha\in\Phi})$ for $\mathrm{G}(k)$ **[5]**. The choice of a splitting gives representatives for the (absolute) Weyl group elements $w_{\tilde{\alpha}} \in \widetilde{W} = \mathcal{N}(T)/\mathrm{T}$ associated to the simple reflections for $\tilde{\alpha} \in \widetilde{\Delta}$ via

$$w_{\tilde{\alpha}} = x_{\tilde{\alpha}}(1)x_{-\tilde{\alpha}}(-1)x_{\tilde{\alpha}}(1)$$

(see Section 3.2.1 of **[5]**).

The choice of a splitting fixes the natural homomorphisms from the usual simply connected rank one groups into G. If $\alpha \in \Delta$ is such that $(\alpha) = \{\alpha\}$ and we let $\tilde{\alpha} \in \widetilde{\Delta}$ be a root of T restricting to α, then the associated rank one group G^α is isomorphic to $\mathrm{R}_{K_{\tilde{\alpha}}/k}SL_2$. In this case the associated k-splitting gives $x_\alpha = \mathrm{R}_{K_{\tilde{\alpha}}/k}x_{\tilde{\alpha}}$ and $x_\alpha : \mathrm{R}_{K_\alpha/k}\mathbb{G}_a \to \mathrm{U}_\alpha = \mathrm{U}_{(\alpha)}$. If $u \in K_{\tilde{\alpha}}$ then

$$x_\alpha(u) = \prod_{\tilde{\beta}\in\widetilde{\Delta}_\alpha} x_{\tilde{\beta}}(u_{\tilde{\beta}}) \quad \text{with} \quad u_{\gamma(\tilde{\alpha})} = \gamma(u).$$

If $\alpha \in \Delta$ is such that $(\alpha) = \{\alpha, 2\alpha\}$ and we let $\tilde{\alpha}, \tilde{\alpha}' \in \tilde{\Delta}_\alpha$ be two roots of T restricting to α such that $\tilde{\alpha} + \tilde{\alpha}'$ is again a root. Then $\tilde{\alpha}$ and $\tilde{\alpha}'$ have the same splitting field, $K_{\tilde{\alpha}}$ which is a quadratic extension of the splitting field $K_{\tilde{\alpha}+\tilde{\alpha}'}$ of $\tilde{\alpha} + \tilde{\alpha}'$. For simplicity, let us denote $K'_{\tilde{\alpha}} = K_{\tilde{\alpha}+\tilde{\alpha}'}$. Then the associated rank one group G^α is isomorphic to $\mathrm{R}_{K'_{\tilde{\alpha}}/k} SU_3$. Let $\mathrm{H}_{\tilde{\alpha}}$ denote the subvariety of $K_{\tilde{\alpha}} \times K_{\tilde{\alpha}}$, considered as a vector space of dimension 4 over $K'_{\tilde{\alpha}}$ defined by $v + v^\sigma = u^\sigma u$, where σ is the non-trivial Galois automorphism in $Gal(K_{\tilde{\alpha}}/K'_{\tilde{\alpha}})$, equipped with the group law

$$(u, v)(u', v') = (u + u', v + v' + u^\sigma u').$$

Then $\mathrm{U}_{(\alpha)} \simeq \mathrm{R}_{K_{\tilde{\alpha}}/K'_{\tilde{\alpha}}} \mathrm{H}_{\tilde{\alpha}}$, with the pair (u, v) corresponding to the unipotent matrix

$$\mu(u, v) = \begin{pmatrix} 1 & -u^\sigma & -v \\ 0 & 1 & u \\ 0 & 0 & 1 \end{pmatrix} \in SU_3.$$

Then the splitting is given as follows. Choose a splitting $x'_{\tilde{\alpha}} : \mathbb{G}_a \to \mathrm{U}_{\tilde{\alpha}}$. Then $x_{\tilde{\alpha}'} = \sigma \circ x_{\tilde{\alpha}} \circ \sigma^{-1}$ and $x_{\tilde{\alpha}+\tilde{\alpha}'} = \mathrm{int}\, w_{\tilde{\alpha}'}^{-1} \circ x_{\tilde{\alpha}}$ (see [5], Sction 4.1.9). Then

$$x_\alpha(u, v) = \prod_{\{\tilde{\beta}, \tilde{\beta}'\}} x_{\tilde{\beta}}(u_{\tilde{\beta}}) x_{\tilde{\beta}+\tilde{\beta}'}(-v_{\tilde{\beta}}) x_{\tilde{\beta}'}(u^\sigma_{\tilde{\beta}})$$

for $(u, v) \in \mathrm{H}_{\tilde{\alpha}} \subset K_{\tilde{\alpha}} \times K_{\tilde{\alpha}}$, where the product is over distinct pairs $\{\tilde{\beta}, \tilde{\beta}'\} \in \tilde{\Delta}_\alpha$ with $\tilde{\beta} + \tilde{\beta}'$ a root. Here, for each $\tilde{\beta}$ we choose $\gamma \in Gal(K/k)$ such that $\tilde{\beta} = \gamma(\tilde{\alpha})$; then $\tilde{\beta}' = \gamma(\tilde{\alpha}')$, $x_{\tilde{\beta}} = \gamma \circ x_{\tilde{\alpha}} \circ \gamma^{-1}$, $x_{\tilde{\beta}'} = \gamma \circ x_{\tilde{\alpha}'} \circ \gamma^{-1}$, $x_{\tilde{\beta}+\tilde{\beta}'} = \gamma \circ x_{\tilde{\alpha}+\tilde{\alpha}'} \circ \gamma^{-1}$, $u_{\tilde{\beta}} = \gamma(u)$, and $v_{\tilde{\beta}} = \gamma(v)$. Note that the image of $x_{\tilde{\alpha}}(u, v)$ in $\mathrm{U}_\alpha/\mathrm{U}_{2\alpha}$ only depends on u and will be denoted $\overline{x}_\alpha(u)$. The map $u \mapsto \overline{x}_\alpha(u)$ gives an isomorphism of k-vector spaces of $K_{\tilde{\alpha}}$ onto $\mathrm{U}_{(\alpha)}/\mathrm{U}_{2\alpha}$. We shall use this notation in the case of $(\alpha) = \{\alpha\}$ as well, taking $\mathrm{U}_{2\alpha}$ to be trivial.

The splitting, through the isomorphisms with the simply connected rank one groups, gives representatives for the (relative) Weyl group elements $w_\alpha \in W$ associated to the simple reflections for $\alpha \in \Delta$ (see Sections 4.1.5 and 4.1.9 of [5]). We can then choose representatives for each $w \in W$ by means of a reduced decomposition and this choice of the w_α. This is independent of the choice of decomposition.

Let $\omega : k^\times \to \mathbb{Z}$ be the valuation on k, with associated normalized absolute value $|u| = q_k^{-\omega(u)}$ where q_k is the order of the residue class field of k. This extends uniquely to give a compatible valuation on K and each $K_{\tilde{\alpha}}$, which we will also denote by ω since it is unique, with associated absolute values. The root datum $(\mathrm{T}, (\mathrm{U}_{\tilde{\alpha}})_{\tilde{\alpha} \in \tilde{\Phi}})$ is naturally valued by the maps $\varphi_{\tilde{\alpha}} : \mathrm{U}_{\tilde{\alpha}} \to \mathbb{R}$

defined by

$$\varphi_{\tilde{\alpha}}(x_{\tilde{\alpha}}(u)) = \omega(u) \quad \text{for } u \in K^{\times}.$$

This valuation on the root datum descends to a valuation $\varphi_\alpha : \mathrm{U}_\alpha \to \mathbb{R}$ of the associated root datum $(\mathrm{T}, (\mathrm{U}_\alpha)_{\alpha\in\Phi})$ for $\mathrm{G}(k)$ (see Section 4.2.2 and Theorem 4.2.3 of [5]). If $\alpha \in \Delta$ is such that $(\alpha) = \{\alpha\}$ then we have

$$\varphi_\alpha(x_\alpha(u)) = \omega(u) \quad \text{for } u \in K_{\tilde{\alpha}}^{\times}.$$

If the root $\alpha \in \Delta$ is multiple, so $(\alpha) = \{\alpha, 2\alpha\}$, then

$$\varphi_\alpha(x_\alpha(u,v)) = \frac{1}{2}\omega(v) \leq \omega(u) \quad \text{for} \quad (u,v) \in \mathrm{H}_{\tilde{\alpha}},\ (u,v) \neq (0,0)$$

while

$$\varphi_{2\alpha}(x_\alpha(0,v)) = \omega(v) \quad \text{for} \quad v \in K_{\tilde{\alpha}}^{\times},\ Tr_{K_{\tilde{\alpha}}/K'_{\tilde{\alpha}}}(v) = 0.$$

The valued root datum allows us to define a natural exhaustive family of compact open subgroups of $\mathrm{U}(K)$ and hence $\mathrm{U}(k)$. Enumerate the simple roots $\tilde{\Delta} = \{\tilde{\alpha}_1, \ldots, \tilde{\alpha}_r\}$. If $\tilde{\beta} \in \tilde{\Phi}^+$ is a positive root occurring in U then $\tilde{\beta}$ has a unique expression of the form $\tilde{\beta} = n_1\tilde{\alpha}_1 + \cdots + n_r\tilde{\alpha}_r$ with n_j a non-negative integer. Then as usual we set $\mathrm{ht}(\tilde{\beta}) = n_1 + \cdots + n_r$, the height of the positive root $\tilde{\beta}$. For each positive integer M we define a concave function $f_M : \tilde{\Phi} \to \mathbb{R} \cup \{\infty\}$ by $f_M(\tilde{\beta}) = -\mathrm{ht}(\tilde{\beta})M$ if $\tilde{\beta} \in \tilde{\Phi}^+$ and $f_M(\tilde{\beta}) = \infty$ if $\tilde{\beta} \in \tilde{\Phi}^-$. This then allows us to define an open compact subgroup $\mathrm{U}_{\tilde{\beta},f_M} \subset \mathrm{U}_{\tilde{\alpha}}(K)$ by $\mathrm{U}_{\tilde{\beta},f_M} = \varphi_{\tilde{\beta}}^{-1}([f_M(\tilde{\beta}), \infty])$ (see Section 6.2 of [4]). If $\tilde{\beta} \in \tilde{\Phi}^+$ then $\mathrm{U}_{\tilde{\beta},f_M} = \varphi_{\tilde{\beta}}^{-1}([M\mathrm{ht}(\tilde{\beta}), \infty])$ while if $\tilde{\beta} \in \tilde{\Phi}^-$ then $\mathrm{U}_{\tilde{\beta},f_M} = \{1\}$. Then, following Bruhat and Tits, we define a corresponding subgroup $\mathrm{U}_{f_M} \subset \mathrm{G}(K)$ as the subgroup generated the $\mathrm{U}_{\tilde{\beta},f_M}$ (see Section 6.4 of [4]). In our case, this will be a compact open subgroup of $\mathrm{U}(K)$ and as $M \to \infty$ these will exhaust $\mathrm{U}(K)$. As the standard commutation relations show (see also Section 6.1 of [4]) in our case we can describe U_{f_M} simply as those elements of $\mathrm{U}(K)$ of the form

$$u = \prod_{\tilde{\beta}\in\tilde{\Phi}^+} x_{\tilde{\beta}}(u_{\tilde{\beta}}) \quad \text{with } |u_{\tilde{\beta}}|_K \leq q_K^{M\mathrm{ht}(\tilde{\beta})}.$$

Since the valued root datum descends to $(\mathrm{T}, (\mathrm{U}_\beta)_{\beta\in\Phi})$ and the function f_M is Galois invariant, the subgroups U_{f_M} will also descend to subgroups $\mathrm{U}_{f_M} \subset \mathrm{U}(k)$. These will play a role in what follows. This family of open compact subgroups will also satisfy the conditions needed in [16] (see particularly Section 6 therein).

1.2. Generic characters of U. The notion of splitting is also necessary to define the concept of a generic character of U(k) and generic representations. Let ψ be a non-trivial additive character of k. If $u \in \mathrm{U}(k)$ then we can write

$$u = \prod_{\tilde{\beta} \in \tilde{\Phi}^+} x_{\tilde{\beta}}(u_{\tilde{\beta}})$$

with the $u_{\tilde{\beta}} \in K$ satisfying $\gamma(u_{\tilde{\beta}}) = u_{\gamma(\tilde{\beta})}$ for all $\gamma \in \Gamma_K$. Then we can extend ψ to a non-degenerate character of U(k) relative to this splitting by setting

$$\psi(u) = \psi\left(\sum_{\tilde{\alpha} \in \tilde{\Delta}} u_{\tilde{\alpha}}\right).$$

Note that the Galois invariance of the $\tilde{\Delta}$ ensures that $\sum u_{\tilde{\alpha}} \in k$.

The abelianization U^{ab} of U is isomorphic to the direct sum of the abelianization of the root groups $\mathrm{U}_{(\alpha)}$ for $\alpha \in \Delta$ and we have $\mathrm{U}_{(\alpha)}^{ab} \simeq \mathrm{U}_{(\alpha)}/\mathrm{U}_{2\alpha} \simeq \mathrm{R}_{K_{\tilde{\alpha}}/k}\mathbb{G}_a$, with the last isomorphism being given by $u_{\tilde{\alpha}} \in K_{\tilde{\alpha}} \mapsto \overline{x}_\alpha(u_{\tilde{\alpha}})$. Thus we have

$$\psi(u) = \psi\left(\sum_{\alpha \in \Delta} \overline{x}_\alpha(u_{\tilde{\alpha}})\right) = \prod_{\alpha \in \Delta} \psi\left(\mathrm{Tr}_{K_{\tilde{\alpha}}/k}(u_{\tilde{\alpha}})\right).$$

If we let $\psi_\alpha = \psi \circ \mathrm{Tr}_{K_{\tilde{\alpha}}/k}$ then we see that under the isomorphism $\mathrm{U}^{ab} \simeq \oplus \mathrm{R}_{K_{\tilde{\alpha}}/k}\mathbb{G}_a$ we have $\psi = \prod \psi_\alpha$.

The representatives of $w \in W$ fixed above will be compatible with ψ in the following sense. For every subset $\theta \subset \Delta$ and $w \in W$ such that $w(\theta) \subset \Delta$ we have

$$\psi(u) = \psi(Ad(w)u)$$

for all $u \in \mathrm{U}_\theta$, the unipotent radical of the Levi subgroup M_θ of the parabolic subgroup P_θ associated to θ (see Section 1.4 below and Section 3 of [**15**]). When we speak of generic representations of G(k) we will always mean generic with respect to this character ψ of U(k).

1.3. A splitting of the torus. Recall that we have assumed that the center $\mathrm{Z} = \mathcal{Z}(\mathrm{G})$ of G is connected and cohomologically trivial to first order, i.e., $H^1(\Gamma, \mathrm{Z}) = \{0\}$.

Enumerate the (non-restricted) simple roots of T as $\tilde{\Delta} = \{\tilde{\alpha}_1, \ldots, \tilde{\alpha}_r\}$. Let $K_{\tilde{\alpha}_i} = K_i \supset \mathcal{O}_i \supset \mathcal{P}_i$ denote the ring of integers and maximal ideal for the field of definition of $\tilde{\alpha}_i$. Then for every $t \in \mathrm{T}(k)$ we have $\tilde{\alpha}_i(t) \subset K_i$. For every n-tuple $\underline{M} = (M_1, \dots, M_r)$ of positive integers let us set

$$\mathrm{T}_{\underline{M}} = \{t \in \mathrm{T}(k) \mid \tilde{\alpha}_i(t) \in 1 + \mathcal{P}_i^{M_i},\ 1 \le i \le r\}.$$

Note that $\mathrm{Z}(k) \subset \mathrm{T}_{\underline{M}}$ for all $\underline{M}$. For later purposes (see Section 6) we would like to be able to split the center off of $\mathrm{T}_{\underline{M}}$ for $\underline{M}$ sufficiently large.

We have the short exact sequence

$$0 \longrightarrow \mathrm{Z} \longrightarrow \mathrm{T} \xrightarrow{\rho} \mathrm{T}_{ad} \longrightarrow 0$$

where T_{ad} is a Cartan in the adjoint group of G. We can take ρ to be

$$\rho(t) = (\tilde{\alpha}_1(t), \dots, \tilde{\alpha}_r(t)) \in (\overline{k})^r.$$

Then

$$\mathrm{T}_{ad} \simeq \{(\tilde{\alpha}_1(t), \dots, \tilde{\alpha}_r(t)) \mid t \in \mathrm{T}\}.$$

Any exact sequence of tori splits. Let j be a splitting, i.e., an injection $j : \mathrm{T}_{ad} \to \mathrm{T}$ such that $\rho \circ j = id$. Applying cohomology we find

$$0 \longrightarrow \mathrm{Z}^\Gamma \longrightarrow \mathrm{T}^\Gamma \longrightarrow \mathrm{T}_{ad}^\Gamma \longrightarrow H^1(\Gamma, \mathrm{Z}) = 0.$$

Thus

$$0 \longrightarrow \mathrm{Z}(k) \longrightarrow \mathrm{T}(k) \xrightarrow{\rho} \mathrm{T}_{ad}(k) \longrightarrow 0$$

is exact and split and consequently

$$\mathrm{T}(k) \simeq \mathrm{Z}(k)\mathrm{T}_{ad}(k).$$

Observe that $\mathrm{Z}(k)$ is the center of $\mathrm{G}(k)$. Moreover

$$\mathrm{T}_{ad}(k) \simeq \{(\tilde{\alpha}_1(t), \dots, \tilde{\alpha}_r(t)) \mid t \in \mathrm{T}(k)\}.$$

Thus

$$\mathrm{T}_{ad}(k) \subset K_1^\times \times \cdots \times K_r^\times$$

where K_i/k are as above. Let

$$O_i = \{\tilde{\alpha}_i(t) \mid t \in \mathrm{T}(k)\} \subset K_i^\times.$$

Observe that each O_i is open in $K_i^\times$. Take the M_i large enough that

$$(1 + \mathcal{P}_1^{M_1}) \times \cdots \times (1 + \mathcal{P}_r^{M_r}) \subset \mathrm{T}_{ad}(k)$$

and let

$$\mathrm{T}_{\underline{M}}^1 = j((1 + \mathcal{P}_1^{M_1}) \times \cdots \times (1 + \mathcal{P}_r^{M_r})).$$

Observe that

$$\rho(\mathrm{T}_{\underline{M}}) = (1 + \mathcal{P}_1^{M_1}) \times \cdots \times (1 + \mathcal{P}_r^{M_r})$$

and then

$$T_{\underline{M}} = Z(k)\mathrm{T}^1_{\underline{M}}.$$

Thus we have proved the following lemma.

Lemma 1.1. *With the notation above, for* $\underline{M} = (M_1, \ldots, M_n)$ *sufficiently large we have the splitting*

$$T_{\underline{M}} = Z(k)\mathrm{T}^1_{\underline{M}}.$$

We finally note that if G is as above and its center Z has $H^1(\Gamma, \mathrm{Z}) = \{0\}$, then the same will be true for the center Z(M) of any Levi subgroup M of G. This follows by using induced tori to which one can apply Shapiro's lemma. Thus we have similar splittings for the maximal tori of any Levi subgroup of G. We will present a detailed proof of this in [**9**].

1.4. Bruhat decomposition. Let W denote the (relative) Weyl group of A in G, i.e., $\mathcal{N}(\mathrm{A})/\mathcal{Z}(\mathrm{A})$ and let S denote the set of simple reflections in W corresponding to the choice of simple roots Δ as above. Then $(\mathrm{G}(k), \mathrm{B}(k), \mathrm{N}(k), S)$ is a Tits system, where we have let $\mathrm{N} = \mathcal{N}(\mathrm{A})$, and the pair (W, S) is a Coxeter system [**2**]. Hence the basic results on the Bruhat decomposition remain valid in this case.

We recall some standard results, all of which can be found in Section 21 of [**2**]. Let $\mathfrak{g}$ be the Lie algebra of G. For $\alpha \in \Phi$, let $\mathfrak{g}_\alpha$ be the corresponding eigenspace in $\mathfrak{g}$ and $\mathfrak{g}_{(\alpha)} = \oplus_{\beta \in (\alpha)} \mathfrak{g}_\beta$. For each α there is a unique closed unipotent k-subgroup $\mathrm{U}_{(\alpha)}$ normalized by T with Lie algebra $\mathfrak{g}_{(\alpha)}$. (These were described in Section 1.1 for $\alpha \in \Delta$.) U is then directly spanned by the $\mathrm{U}_{(\alpha)}$ with $\alpha \in \Phi^+_{nd}$.

For $w \in W$ let

$$\Phi^+_w = \{\alpha \in \Phi^+_{nd} \mid w(\alpha) > 0\} \quad \text{and} \quad \Phi^-_w = \{\alpha \in \Phi^+_{nd} \mid w(\alpha) < 0\}$$

and set

$$\mathrm{U}^+_w = \mathrm{U}_{\Phi^+_w} = \mathrm{U} \cap w^{-1}\mathrm{U}w \quad \text{and} \quad \mathrm{U}^-_w = \mathrm{U}_{\Phi^-_w} = \mathrm{U} \cap w^{-1}\mathrm{U}^-w.$$

Then

$$\mathrm{U} = \mathrm{U}^+_w \mathrm{U}^-_w$$

and

$$w\mathrm{U}^+_w w^{-1} \subset \mathrm{U} \quad \text{and} \quad w\mathrm{U}^-_w w^{-1} \subset \mathrm{U}^-.$$

For each $w \in W$ we let $C(w) = \mathrm{B}w\mathrm{B} = \mathrm{B}w\mathrm{U}_w^-$ be the corresponding Bruhat cell. The (k-rational) Bruhat decomposition for G is then

$$G = \bigcup_{w \in W} C(w) = \bigcup_{w \in W} \mathrm{B}w\mathrm{U}_w^-.$$

The relative closure of the Bruhat cells are described as follows (see Theorem 21.26 and Proposition 21.27 of [**2**]). For $w \in W$, let $w = w_{\alpha_1} \cdots w_{\alpha_n}$, $\alpha_i \in \Delta$, be a reduced decomposition of w. Let

$$S(w) = \{w_{\alpha_{i_1}} \cdots w_{\alpha_{i_m}} \mid 1 \leq i_1 < \cdots < i_m \leq n\}.$$

Then

$$\overline{C(w)} = \bigcup_{w' \in S(w)} C(w').$$

Since the Bruhat order on W can be characterized by $w' \leq w$ iff $w' \in S(w)$ [**12**], we may also write this as

$$\overline{C(w)} = \bigcup_{w' \leq w} C(w').$$

Let $\theta \subset \Delta$. Let $[\theta]$ denote the set of k-roots which are linear combinations of the elements of θ. Set $\Psi(\theta) = \Phi^+ - [\theta]$. Let

$$\mathrm{A}_\theta = \left(\bigcap_{\alpha \in [\theta]} \ker \alpha \right)^\circ.$$

Then $[\theta] = \Phi(\mathrm{A}, \mathcal{Z}(\mathrm{A}_\theta))$ and $\mathcal{Z}(\mathrm{A}_\theta) = \langle \mathrm{T}, \mathrm{U}_{(\alpha)} \mid \alpha \in [\theta] \rangle$. Then

$$\mathrm{P}_\theta = \mathcal{Z}(\mathrm{A}_\theta) \ltimes \mathrm{U}_{\Psi(\theta)}$$

is a standard parabolic subgroup. $\mathrm{P}_\emptyset = \mathrm{B}$. If we set $W_\theta = \langle w_\alpha \mid \alpha \in \theta \rangle$ then we also have

$$\mathrm{P}_\theta = \mathrm{B}W_\theta\mathrm{B}.$$

If α is a simple root and we let $\mathrm{P}_\alpha = \mathrm{P}_{\{\alpha\}}$ then

$$\mathrm{P}_\alpha = C(w_\alpha) \cup \mathrm{B}.$$

2. Partial Bessel functions

2.1. Finite field heuristics. In order to motivate what follows, let us take $\mathbb{F}$ to be a finite field and G a quasi-split reductive algebraic group over $\mathbb{F}$. The basics of the theory of Bessel functions for representations of algebraic groups over finite fields can be found in **[10, 11]**.

We may retain all the notation and concepts of Section 1. Suppose that (π, V_π) is a generic representation of G($\mathbb{F}$). Over a finite field, generic representations will tend to have both Whittaker functionals and Whittaker vectors. So suppose that $v_W \in V_\pi$ satisfies $\pi(u)v_W = \psi(u)v_W$ for $u \in \mathrm{U}(\mathbb{F})$ and $\Lambda = \Lambda_W$ satisfies $\Lambda(\pi(u)v) = \psi(u)\Lambda(v)$ for all $u \in \mathrm{U}(\mathbb{F})$ and all $v \in V_\pi$. Moreover, assume these choices are normalized so that $\Lambda(v_W) = 1$. Then $J_\pi(g) = \Lambda(\pi(g)v_W)$ is a function in the Whittaker model of π and it satisfies

$$J_\pi(ugu') = \psi(u)J_\pi(g)\psi(u') \qquad u, u' \in \mathrm{U}(\mathbb{F}).$$

This is the Bessel function of the representation.

If we restrict this function to the various Bruhat cells, then the restriction of J_π to $C(w)$ is not identically zero iff for every $\alpha \in \Delta$ we have that $w\alpha > 0$ implies $w\alpha \in \Delta$. Let $J_{\pi,w}$ denote the restriction of J_π to $C(w)$. Write $C(w) = \mathrm{B}w\mathrm{B} = \mathrm{UT}w\mathrm{U}_w^-$. Then $J_{\pi,w}(utwu') = \psi(u)J_{\pi,w}(tw)\psi(u')$ is essentially a function on T($\mathbb{F}$) and one can check that its restriction to the split A vanishes unless $a \in \mathrm{A}_w = \{a \in \mathrm{A} \mid w\alpha(a) = 1 \text{ for all simple } \alpha \text{ with } w\alpha > 0\}$. This function on A_w carries the information of the restriction $J_{\pi,w}$ of the Bessel function J_π to the Bruhat cell associated to w.

Finally, for what follows, note that one has a formula for $J_{\pi,w}(aw)$ given by

$$J_{\pi,w}(aw) = |\mathrm{U}_w^-(\mathbb{F})|^{-1} \int_{\mathrm{U}_w^-(\mathbb{F})} W_{v_W}(awu)\psi^{-1}(u)\, du$$

where, as usual, the Whittaker model is defined by $W_v(g) = \Lambda(\pi(g)v)$ for $v \in V_\pi$.

2.2. Weyl group elements that support Bessel functions. Now let k once again be a non-archimedean field of characteristic 0.

We say that an element $w \in W$ supports a Bessel function if for every $\alpha \in \Delta$ we have that $w\alpha > 0$ implies $w\alpha \in \Delta$. So every simple root which remains positive under the action of w must remain simple. Note that $w = e$ and $w = w_\ell$, the long element of W, always support Bessel functions. By Lemma 89 of **[17]** (page 257), which is valid for quasi-split groups, we have that w supports a Bessel function iff $w = w_\ell w_\ell^\theta$ for $\theta = \theta_w = \{\alpha \in \Delta \mid w\alpha > 0\} \subset \Delta$ and $w_\ell^\theta \in W_\theta$ the corresponding long Weyl element. This implies that there are actually $2^{|\Delta|}$ Weyl elements which support Bessel functions and to each $\theta \subset \Delta$ we have associated a Weyl element $w_\theta = w_\ell w_\ell^\theta$ which supports a Bessel function.

To the set θ is associated the standard parabolic subgroup P_θ as above. Then, in this case, we have $\Phi_w^+ = [\theta_w]$ so that $\mathrm{U}_w^+ = \mathrm{U}_\theta$ and $\mathrm{U}_w^- = \mathrm{U}_{\Psi(\theta)}$. Furthermore, if we set

$$\begin{aligned}\mathrm{A}_w &= \{a \in \mathrm{A} \mid w\alpha(a) = 1 \text{ for all simple } \alpha \text{ with } w\alpha > 0\} \\ &= \{a \in \mathrm{A} \mid w\alpha(a) = 1 \text{ for all } \alpha \in \theta\}\end{aligned}$$

then $\mathrm{A}_w^0 = \mathrm{A}_{w(\theta)}$ is the split component of the center of the Levi component of the associate parabolic $\mathrm{P}_{w(\theta)}$. (Note that in [8] we implicitly assumed that $w(\theta) = \theta$, which was indeed true in the case of interest to us.)

We recall from [8] the following proposition.

Proposition 2.1. *Let $w, w' \in W$ support Bessel functions. Write $w = w_\ell w_\ell^\theta$ and $w' = w_\ell w_\ell^{\theta'}$ for subsets $\theta, \theta' \subset \Delta$. Then $w' \leq w$ iff $\theta \subset \theta'$. In particular, if θ is of the form $\theta = \Delta - \{\alpha\}$, i.e., if P_θ is a maximal parabolic subgroup, then the only Weyl elements $w' \leq w$ which support Bessel functions are $w' = w$ and $w' = e$.*

Proof: Let $\theta = \theta_w$ and $\theta' = \theta_{w'}$. By Example 3, Section 5.9 of [12], we know that $w' \leq w$ iff $w_\ell w \leq w_\ell w'$ and this last is equivalent to $w_\ell^\theta \leq w_\ell^{\theta'}$. Since w_ℓ^θ is the long element of W_θ, then by the compatibility of the Bruhat ordering (Section 5.5 and 5.10 of [12]) any reduced expression for w_ℓ^θ contains only the basic reflections r_α for $\alpha \in \theta$ and by Section 1.8 of [12], we know that each simple reflection r_α with $\alpha \in \theta$ occurs. The same is true of $w_\ell^{\theta'}$ with respect to θ'. However, $w_\ell^\theta \leq w_\ell^{\theta'}$ can be characterized by w_ℓ^θ occurring as a sub-expression of a reduced expression for $w_\ell^{\theta'}$. Thus we must have $\theta \subset \theta'$. □

We will say that $w \in W$ is a minimal Weyl element supporting a Bessel function if w supports a Bessel function and the only $w' \in W$ with $w' \leq w$ which support Bessel functions are $w' = w$ and $w' = e$, i.e., the associated parabolic subgroup is maximal.

2.3. Bessel functions. Let (π, V_π) be an irreducible admissible generic representation of $\mathrm{G}(k)$. We fix a splitting of U and a non-degenerate character ψ of $\mathrm{U}(k)$ as in Section 1.2. Let $\mathcal{W}(\pi,\psi)$ be the associated Whittaker model of π. The functions $W \in \mathcal{W}(\pi,\psi)$ satisfy $W(ug) = \psi(u)W(g)$ for $u \in \mathrm{U}(k)$.

If π is a generic representation with Whittaker model $\mathcal{W}(\pi,\psi)$ then to each $w \in W$ which supports a Bessel function we may associate a formal Bessel function $J_{\pi,w}(a)$ for $a \in A_w$ by

$$J_{\pi,w}(a) = \int_{\mathrm{U}_w^-(k)} W_v(awu)\psi^{-1}(u)\, du$$

for any choice of $W_v \in \mathcal{W}(\pi, \psi)$ for which $W_v(e) = 1$.

Assuming that the integral exists, this is independent of the choice of $v \in V_\pi$, since the map

$$v \mapsto \int_{U_w^-(k)} W_v(awu)\psi^{-1}(u)\, du$$

defines a ψ-Whittaker functional on V_π. To see this, consider $\pi(u')v$. Under the decomposition $U = U_w^+ U_w^-$ we write $u' = u'^+ u'^-$. Then

$$\begin{aligned}\int_{U_w^-(k)} W_{\pi(u')v}(awu)\psi^{-1}(u)\, du &= \int_{U_w^-(k)} W_v(awuu')\psi^{-1}(u)\, du \\ &= \int_{U_w^-(k)} W_v(awu'^+ uu'^-)\psi^{-1}(u)\, du \\ &= \psi(u'^-) \int_{U_w^-(k)} W_v(awu'^+ u)\psi^{-1}(u)\, du \\ &= \psi(u'^-)\psi(Ad(aw)u'^+) \int_{U_w^-(k)} W_v(awu)\psi^{-1}(u)\, du \\ &= \psi(u'^-)\psi(Ad(aw)u'^+) \int_{U_w^-(k)} W_v(awu)\psi^{-1}(u)\, du.\end{aligned}$$

Now, since $u'^+ \in U_w^+ = U_\theta$, then $Ad(w)u'^+ \in U_{w(\theta)}$. Also, $a \in A_w = \mathcal{Z}(M_{w(\theta)})$. Thus $Ad(aw)u'^+ = Ad(w)u'^+$ and so by compatibility

$$\psi(Ad(aw)u'^+) = \psi(Ad(w)u'^+) = \psi(u'^+).$$

Hence

$$\int_{U_w^-(k)} W_{\pi(u')v}(awu)\psi^{-1}(u)\, du = \psi(u') \int_{U_w^-(k)} W_v(awu)\psi^{-1}(u)\, du$$

and this integral defines a Whittaker functional on V_π. By the uniqueness of the Whittaker model, this must be a non-zero multiple of the standard Whittaker functional $v \mapsto W_v(e)$. So there is a constant $J_{\pi,w}(a)$

$$\int_{U_w^-(k)} W_v(awu)\psi^{-1}(u)\, du = J_{\pi,w}(a)W_v(e).$$

Since v was chosen so that $W_v(e) = 1$ this gives

$$\int_{U_w^-(k)} W_v(awu)\psi^{-1}(u)\, du = J_{\pi,w}(a)$$

independent of v with this property.

The convergence of the integrals defining the Bessel functions is subtle and in general we do not have a proof. For Bessel functions attached to a minimal Weyl element which supports a Bessel function we gave an argument for convergence in [8] for the case of split groups (see Lemmas 4.2 and 4.3 as well as Proposition 4.2 of [8]). The convergence for the quasi-split case can be proven by the same argument, utilizing the compact open subgroups U_{f_M} of Section 1.1 in place of the corresponding $X(M)$ of [8]. Since we will not use this full Bessel function in the main theorem or its applications, we will forgo presenting the details of the proof at this point. However, since we believe these Bessel functions to be very significant, we do present the proof of convergence in this most simple case in an appendix at the end of this paper (see Section 8).

2.4. Partial Bessel functions For our purposes, an equally important notion is that of a *partial Bessel function*. Let Y be any open compact subgroup of U_w^- and let $v \in V_\pi$ with $W_v(e) = 1$. Then we define

$$j_{v,w,\mathrm{Y}}(a) = \int_{\mathrm{Y}} W_v(awy)\psi^{-1}(y)\, dy$$

again for $a \in \mathrm{A}_w$. Note that now there are no convergence problems since Y is open and compact and W_v is smooth (so the integral reduces to a finite sum). Note that if $\{\mathrm{Y}_i\}$ is an increasing exhaustive family of open compact subgroups, then at least formally $J_{\pi,w} = \lim_i j_{v,w,\mathrm{Y}_i}$.

3. Approximate Whittaker Vectors

In our finite field heuristics, the vector $v \in V_\pi$ which we used to form the Bessel function was a Whittaker vector. Over local fields we do not have Whittaker vectors. However, we can form a family of approximate Whittaker vectors, which we will need, as follows.

Let (π, V_π) be a generic representation with Whittaker model $\mathcal{W}(\pi)$. Let $v \in V_\pi$ be any vector such that $W_v(e) = 1$. Let $\mathrm{U}_0 \subset \mathrm{U}(k)$ be an open compact subgroup. Then we set

$$v_{\mathrm{U}_0} = \frac{1}{\mathrm{Vol}(\mathrm{U}_0)} \int_{\mathrm{U}_0} \psi^{-1}(u)\pi(u)v\, du.$$

Since the representation is smooth, so that v has a compact open stabilizer, the integration is in fact a finite sum and each $v_{\mathrm{U}_0} \in V_\pi$. If one considers the corresponding Whittaker function, it will satisfy

$$W_{v_{U_0}}(ugu_0) = \psi(u)W_{v_{U_0}}(g)\psi(u_0)$$

for $u \in U$ and $u_0 \in U_0$. Note that

$$W_{v_{U_0}}(g) = \frac{1}{\mathrm{Vol}(U_0)} \int_{U_0} \psi^{-1}(u)W_v(gu)\, du$$

so that

$$W_{v_{U_0}}(e) = W_v(e) = 1$$

so that $v_{U_0} \neq 0$.

We will call v_{U_0} an approximate Whittaker vector for π with respect to U_0.

For any $v \in V_\pi$ let $\mathrm{Stab}(v) = \{g \in G(k) \mid \pi(g)v = v\}$. This is a compact open subgroup of $G(k)$. For U_0 a compact open subgroup of $U(k)$, let

$$T_{U_0} = \{t \in T(k) \mid \psi(u_0) = \psi(Ad(t)u_0) \quad \text{for all } u_0 \in U_0\}.$$

Proposition 3.1. *Let $v \in V_\pi$ such that $W_v(e) = 1$. Then for any compact open subgroup $U_0 \subset U(k)$ we have*

$$\mathrm{Supp}(W_{v_{U_0}}) \cap B(k)\mathrm{Stab}(v_{U_0}) \subset U(k)T_{U_0}\mathrm{Stab}(v_{U_0}).$$

Proof: Let $W_0 = W_{v_{U_0}}$ and let $K_0 = \mathrm{Stab}(v_{U_0})$. Let $g = utk \in B(k)K_0 = U(k)T(k)K_0$. Then $W_0(utk) = \psi(u)W_0(t)$. Now for any $u_0 \in U_0$ we have

$$\psi(u_0)W_0(t) = W_0(tu_0) = W_0(tu_0t^{-1}t) = \psi(tu_0t^{-1})W_0(t).$$

Therefore $0 = (\psi(u_0) - \psi(Ad(t)u_0))W_0(t)$. Thus $W_0(t) \neq 0$ implies $\psi(u_0) = \psi(Ad(t)u_0)$ for all $u_0 \in U_0$. □

4. Partial Bessel functions and approximate Whittaker vectors

In our finite field heuristics in Section 2, the Bessel function was obtained as the Whittaker function associated to a Whittaker vector. In this section we see that we have retained such a relation, at least for the partial Bessel functions and the approximate Whittaker vectors.

Let $w \in W$ be a Weyl group element that supports a Bessel function. For U_* an open compact subgroup of $U(k)$ we will uniformly assume that U_* satisfies the property

$$U_* = U^+_{*,w}U^-_{*,w} \quad \text{where} \quad U^+_{*,w} = U_* \cap U^+_w, \quad U^-_{*,w} = U_* \cap U^-_w.$$

Accordingly, we will let $\mathrm{Y}_* = \mathrm{U}^-_{*,w}$, the associated open compact subgroup of U^-_w. We begin with a proposition which will be useful in the next section as well.

Proposition 4.1. *Let* $\mathrm{U}_0 \subset \mathrm{U}_1$ *be two open compact subsets of* $\mathrm{U}(k)$ *as above. Let* $F(g)$ *be a smooth function on* G *satisfying* $F(ug) = \psi(u)F(g)$ *for all* $u \in \mathrm{U}(k)$, *i.e.,* F *is any smooth Whittaker function but not necessarily in the Whittaker model of* π. *Let* χ *be a smooth function on* G *satisfying* $\chi(ugu_0) = \chi(u)$ *for* $u \in \mathrm{U}(k)$ *and* $u_0 \in \mathrm{U}_0$. *Let*

$$\tilde{F}(g) = \frac{1}{\mathrm{Vol}(\mathrm{U}_0)} \int_{\mathrm{U}_0} F(gu_0)\psi^{-1}(u_0)\, du_0.$$

Assume

$$I(a) = \int_{\mathrm{Y}_1} F(awy)\chi(awy)\psi^{-1}(y)\, dy$$

is convergent for all $a \in \mathrm{A}_w$. *Then so is*

$$\tilde{I}(a) = \int_{\mathrm{Y}_1} \tilde{F}(awy)\chi(awy)\psi^{-1}(y)\, dy$$

and $I(a) = \tilde{I}(a)$.

Proof: Let $c_0 = \mathrm{Vol}(\mathrm{U}_0)$. Inserting the definition of $\tilde{F}$ into $\tilde{I}(a)$ we have

$$\tilde{I}(a) = \frac{1}{c_0} \int_{\mathrm{Y}_1} \int_{\mathrm{U}_0} F(awyu_0)\psi^{-1}(u_0)\, du_0\ \chi(awy)\psi^{-1}(y)\, dy.$$

Interchanging the order of the compact integrations (which are actually finite sums since the functions involved are smooth) we have

$$\tilde{I}(a) = \frac{1}{c_0} \int_{\mathrm{U}_0} \int_{\mathrm{Y}_1} F(awyu_0)\chi(awy)\psi^{-1}(yu_0)\, dy\, du_0.$$

In the inner integral

$$I_{u_0}(a) = \int_{\mathrm{Y}_1} F(awyu_0)\chi(awy)\psi^{-1}(yu_0)\, dy$$

write $u_0 = u_0^- u_0^+$ with $u_0^\pm \in \mathrm{U}^\pm_{0,w}$. Then we have

$$I_{u_0}(a) = \psi^{-1}(u_0^+) \int_{\mathrm{Y}_1} F(awyu_0^- u_0^+)\chi(awy)\psi^{-1}(yu_0^-)\, dy.$$

As χ is left invariant under $U_{0,w}^-$ we may perform a change of variables and absorb u_0^- into y, leaving

$$I_{u_0}(a) = \psi^{-1}(u_0^+) \int_{Y_1} F(awyu_0^+)\chi(awy)\psi^{-1}(y)\, dy.$$

Now, as $U_{0,w}^+ \subset U_{1,w}^+$, this normalizes $U_{1,w}^- = Y_1$. Furthermore, as conjugation does not effect the value of the character $\psi(y)$, we may perform a new change of variable in y by conjugating by u_0^+ to arrive at

$$\begin{aligned} I_{u_0}(a) &= \psi^{-1}(u_0^+) \int_{Y_1} F(awu_0^+ y)\chi(awy)\psi^{-1}(y)\, dy \\ &= \psi^{-1}(u_0^+)\psi(Ad(aw)u_0^+) \int_{Y_1} F(awy)\chi(awy)\psi^{-1}(y)\, dy. \end{aligned}$$

But since $a \in A_w$ and ψ was chosen to be compatible with the splitting, we have

$$\psi(Ad(aw)u_0^+) = \psi(u_0^+).$$

Hence, for each u_0 we have $I_{u_0}(a) = I(a)$. and hence $\tilde{F}(a) = F(a)$. □

If we now let $v \in V_\pi$ such that $W_v(e) = 1$ and take $F(g) = W_v(g)$ and $\chi \equiv 1$, then we see that $\tilde{F}(g) = W_{v_0}(g)$ where W_{v_0} is the approximate Whittaker vector associated to v and the open compact subgroup U_0. The conclusion of the proposition is then an equality of partial Bessel functions $j_{v,w,Y_0}(a) = j_{v_0,w,Y_0}(a)$. On the other hand, since $Y_0 \subset U_0$, we see that in this case

$$\begin{aligned} j_{v_0,w,Y_0}(a) &= \int_{Y_0} W_{v_0}(awy)\psi^{-1}(y)\, dy = \int_{Y_0} W_{v_0}(aw)\psi(y)\psi^{-1}(y)\, dy \\ &= \mathrm{Vol}(Y_0) W_{v_0}(aw). \end{aligned}$$

Thus we have the following corollary.

Corollary 4.2. *For this class of* Y_0 *we always have*

$$j_{v,w,Y_0}(a) = \mathrm{Vol}(Y_0) W_{v_0}(aw).$$

So the partial Bessel function is indeed given by the Whittaker function of an approximate Whittaker vector.

As a second corollary, we can investigate how these partial Bessel functions behave as we increase the compact open subgroup. Suppose that U_0 and U_1 are compact open subgroups of $U(k)$ as above with $U_0 \subset U_1$. First, taking $F = W_{v_0}$ and $\chi \equiv 1$ in Proposition 4.1 and using that since $U_0 \subset U_1$ we have $\tilde{F} = W_{v_1}$, we have $j_{v_0,w,Y_1} = j_{v,w,Y_1}$. On the other hand, if we begin with v_1

and then compute the partial Bessel function with respect to U_0 then automatically $W_{v_1}(gu_0) = \psi(u_0)W_{v_1}(g)$ and we see $j_{v_1,w,Y_0}(a) = \mathrm{Vol}(Y_0)W_{v_1}(aw) = \mathrm{Vol}(Y_0)\mathrm{Vol}(Y_1)^{-1}j_{v,w,Y_1}(a)$.

Corollary 4.3. *If* $U_0 \subset U_1 \subset U(k)$ *with* U_i *compact open as above then we have the relations*

$$j_{v_0,w,Y_1} = j_{v,w,Y_1} \quad \textit{and} \quad j_{v_1,w,Y_0}(a) = (Y_1 : Y_0)^{-1}j_{v,w,Y_1}(a).$$

5. Asymptotics of Bessel functions I

We now turn to an investigation of the asymptotics of the Bessel functions attached to a minimal Weyl elements w which supports a Bessel function. From our finite field heuristics, we expect these Bessel function to be supported on the cell $C(w)$, vanish as we approach bounding cells which do not support Bessel functions, and have non-zero asymptotics as we approach the cell $C(e)$ associated to the identity. To investigate the contribution from the other boundary cells, let us number the Weyl elements w' such that $C(w')$ lies on the boundary of $C(w)$ in a convenient fashion. More precisely, let $S_w = \{w' \in W \mid w' \le w\}$, let $s = s_w = |S_w|$, and enumerate the elements of S_w so that $w'_1 = e$, $w'_s = w$ and if $w'_i \le w'_j$ then $i \le j$.

We will be considering various open compact subgroups of $U(k)$. If we denote one such by U_i, we will always work under our standard assumption

$$U_i = U^+_{i,w}U^-_{i,w} \quad \text{where} \quad U^+_{i,w} = U_i \cap U^+_w, \quad U^-_{i,w} = U_i \cap U^-_w$$

and let $Y_i = U^-_{i,w}$ be the associated open compact subgroup of U^-_w with which we form our partial Bessel functions. Similarly, if $v \in V_\pi$ with $W_v(e) = 1$ we will let v_i be the associated approximate Whittaker vector

$$v_i = \mathrm{Vol}(U_i)^{-1} \int_{U_i} \psi^{-1}(u)\pi(u)v \, du.$$

We will also set $j_i = j_{v,w,Y_i} = j_{v_j,w,Y_i}$ whenever $U_j \subset U_i$.

We begin with two (large) open compact subgroups $U_1 \subset U_s \subset U(k)$ as above. We will fix U_1, but U_s will be pushed to be large enough to contain all other U_i constructed. Take $v \in V_\pi$ with $W_v(e) = 1$. Let $K_0 \subset \mathrm{Stab}(v)$ be an open compact subgroup of $G(k)$ fixing v. Let

$$K_1 = \bigcap_{u \in U_1} u^{-1}K_0u.$$

Since $K_0 \cap U_1$ is compact and open in U_1 we see that this intersection is in fact finite and K_1 is another compact open subgroup of $G(k)$ which fixes v. $B(k)K_1$ is then a neighborhood of $B(k)e$ in $B(k)\backslash G(k)$.

Let χ_1 be the characteristic function of $B(k)K_1$. Let $H_1 = W_{v_1}\chi_1$ and $H_1' = W_{v_1}(1-\chi_1)$. We may accordingly decompose $j_s(a) = I_1(a) + I_1'(a)$ for $a \in A_w$, where

$$I_1(a) = \int_{Y_1} H_1(awy)\psi^{-1}(y)\, dy \quad \text{and} \quad I_1'(a) = \int_{Y_1} H_1'(awy)\psi^{-1}(y)\, dy.$$

Proposition 5.1. $I_1(a) = \int_{Y_s} W_v(awy)\chi_1(awy)\psi^{-1}(y)\, dy.$

Proof: This is simply Proposition 4.1 applied to $F = W_v$, $\chi = \chi_1$, and $U_1 \subset U_s$. We need only check that χ_1 satisfies the hypotheses of that proposition. Now χ_1 is the characteristic function of $B(k)K_1$, so the left invariance under $U(k)$ is clear. Now let $u_1 \in U_1$. Then $\chi_1(gu_1) = 1$ iff $gu_1 \in B(k)K_1$ iff $g \in B(k)K_1u_1^{-1}$. Since $u_1 \in U_1 \subset B(k)$, we have $B(k)K_1u^{-1} = B(k)u_1K_1u_1^{-1} = B(k)K_1$. Hence $\chi_1(gu_1) = \chi_1(g)$. We can now apply Proposition 4.1. □

Note that for K_1 sufficiently small, this integral depends only on the behavior of the approximate Whittaker vector near the cell $C(e)$ on the boundary of $C(w)$.

Note first that the function H_1' satisfies the following properties:

(i) $H_1'(ug) = \psi(u)H_1'(g)$ for all $u \in U(k)$;

(ii) $H_1'(gu_1) = \psi(u_1)H_1'(g)$ for all $u_1 \in U_1$;

(iii) $H_1'(gk_1) = H_1'(g)$ for all $k_1 \in K_1$;

(iv) $\mathrm{Supp}^\circ(H_1') \cap B(k)K_1 = \emptyset$.

Here, and throughout, we will let Supp° denote the "open support" of a function, i.e., $\mathrm{Supp}^\circ(H) = \{g \mid H(g) \neq 0\}$ and so the usual support is given by $\mathrm{Supp}(H) = \overline{\mathrm{Supp}^\circ(H)}$.

We would now like to inductively define an increasing sequence of compact open subgroups $U_1 \subset U_2 \subset \cdots \subset U_{s-1} \subset U_s \subset U(k)$, enlarging U_s if necessary, which all satisfy the decomposition properties above, a decreasing sequence compact open subgroups $K_1 \supset K_2 \supset \cdots \supset K_{s-1}$, and functions $H_i'(g)$ for $1 \leq i < s$ such that:

(i) $H_i'(ug) = \psi(u)H_i'(g)$ for all $u \in U(k)$;

(ii) $H_i'(gu_i) = \psi(u_i)H_i'(g)$ for all $u_i \in \mathrm{U}_i$;

(iii) $H_i'(gk_i) = H_i'(g)$ for all $k_i \in \mathrm{K}_i$;

(iv) $\mathrm{Supp}(H_i') \cap (\cup_{j\leq i} C(w_j')) = \emptyset$.

Suppose we have constructed U_i, K_i, and H_i' for $i < j$.

Let $S_{j-1}' = \mathrm{Supp}^\circ(H_{j-1}')$. Let $p : \mathrm{G}(k) \to \mathrm{B}(k)\backslash\mathrm{G}(k)$. Then $p(S_{j-1}' \cap C(w_j'))$ is compact in $p(C(w_j'))$. Since $p(C(w_j'))$ is a single $\mathrm{U}(k)$ orbit in $\mathrm{B}(k)\backslash\mathrm{G}(k)$, and in fact homeomorphic to $\mathrm{U}^-_{w_j'}$, there exists a compact open subgroup $\mathrm{U}_{j-1}' \subset \mathrm{U}(k)$ such that $p(w_j'\mathrm{U}_{j-1}') \supset p(S_{j-1}') \cap p(C(w_j'))$. Take U_j large enough so that $\mathrm{U}_j \supset \langle \mathrm{U}_{j-1}', \mathrm{U}_{j-1}\rangle$ and set

$$\mathrm{K}_j = \bigcap_{u_j\in \mathrm{U}_j} u_j^{-1}\mathrm{K}_{j-1}u_j.$$

As before, this is really a finite intersection.

Consider

$$\widetilde{H}_j(g) = \frac{1}{\mathrm{Vol}(\mathrm{U}_j)}\int_{\mathrm{U}_j} H_{j-1}'(gu_j)\psi^{-1}(u_j)\, du_j.$$

Then $\widetilde{H}_j(gk_j) = \widetilde{H}_j(g)$ for all $k_j \in \mathrm{K}_j$.

Let $\widetilde{S}_j = \mathrm{Supp}^\circ(\widetilde{H}_j)$. Then $p(\widetilde{S}_j) \subset p(S_j'\mathrm{U}_j)$ and $\overline{p(\widetilde{S}_j)} \cap \overline{p(C(w_j'))} \subset p(w_j'\mathrm{U}_j)$. Let χ_j be the characteristic function of $\mathrm{B}(k)w_j'\mathrm{U}_j\mathrm{K}_j$. This set gives a compact neighborhood of $w_j'\mathrm{U}_j$ in $\mathrm{B}(k)\backslash\mathrm{G}(k)$. Then set $H_j = \widetilde{H}_j\chi_j$ and $H_j' = \widetilde{H}_j(1-\chi_j)$. The function H_j' then satisfies the required properties (i)–(iv) above.

As for the functions H_i, they satisfy properties similar to those of H_1, namely:

(a) $H_i(ug) = \psi(u)H_i(g)$ for all $u \in \mathrm{U}(k)$;

(b) $H_i(gu_i) = \psi(u_i)H_i(g)$ for all $u_i \in \mathrm{U}_i$;

(c) $H_i(gk_i) = H_i(g)$ for all $k_i \in \mathrm{K}_i$;

(d) $\mathrm{Supp}^\circ(H_i) \subset \mathrm{B}(k)w_i'\mathrm{U}_i\mathrm{K}_i$.

However, for $1 < i < s$ we have an even stronger property.

Lemma 5.2. *If $1 < i < s$ and U_i is sufficiently large then $H_i \equiv 0$.*

Proof: Since $1 < i < s$ the Weyl element w_i' does not support a Bessel function. By (d) above, $H_i(g) \neq 0$ implies that $g \in \mathrm{B}(k)w_i'\mathrm{U}_i\mathrm{K}_i$. Writing $g = utw_i'u_ik_i$ with the implied notation we see $H_i(utw_j'u_ik_i) = \psi(uu_i)H_i(tw_i')$. Since w_i' does not support a Bessel function, there exists a simple root $\alpha \in \Delta$ such that $w_i'\alpha$ is positive but but not simple. Then for U_i sufficiently large, there exists $u_i \in \mathrm{U}_i \cap \mathrm{U}_{(\alpha)}$ such that $\psi(u_i) \neq 1$. Then

$$\psi(u_i)H_i(tw_i') = H_i(tw_i'u_i) = H_i((Ad(tw_i')u_i)tw_i') = H_i(tw_i')$$

since $Ad(tw_i')u_i \in \mathrm{U}_{(w_i'\alpha)}$. Hence $H_i'(g) = 0$ for $g \in \mathrm{B}(k)w_i'\mathrm{U}_i\mathrm{K}_i$ as well. Hence $H_i \equiv 0$. □

Now return to the second part of our incomplete Bessel function, namely $I_1'(a)$. We can similarly define

$$I_i'(a) = \int_{\mathrm{Y}_s} H_i'(awy)\psi^{-1}(y)\, dy.$$

Lemma 5.3. *For every $1 \leq i < s$ we have $I_1'(a) = I_i'(a)$.*

Proof: We proceed by finite induction, with the $i = 1$ case being true by definition. Suppose we know the lemma for all j with $1 \leq j < i$. Recall that

$$\widetilde{H}_i(g) = \frac{1}{\mathrm{Vol}(\mathrm{U}_i)} \int_{\mathrm{U}_i} H_{i-1}'(gu_i)\psi^{-1}(u_i)\, du_i$$

and $H_i'(g) = \widetilde{H}_i(g)\chi_i(g)$. Let us set

$$\widetilde{I}_i(a) = \int_{\mathrm{Y}_s} \widetilde{H}_i(awy)\psi^{-1}(y)\, dy.$$

If we now apply Proposition 4.1 with $\mathrm{U}_i \subset \mathrm{U}_s$, $F = H_{i-1}'$ and $\chi \equiv 1$ we have $\widetilde{I}_i(a) = I_{i-1}'(a)$. On the other hand, since $\widetilde{H}_i = H_i + H_i' = H_i'$ we have $\widetilde{I}_i(a) = I_i'(a)$. Thus, by induction, $I_i'(a) = I_{i-1}'(a) = I_1'(a)$. □

Lemma 5.4. *For each i with $1 \leq i < s$ let us set $\chi_i' = \prod_{1\leq j \leq i}(1 - \chi_j)$. Let $\widetilde{\chi}$ be a function on $\mathrm{G}(k)$ such that $\widetilde{\chi}(ugu_i) = \widetilde{\chi}(g)$ for all $u \in \mathrm{U}(k)$ and $u_i \in \mathrm{U}_i$. Then*

$$I_i'(a) = \int_{\mathrm{Y}_s} W_v(awy)\chi_i'(awy)\psi^{-1}(y)\, dy.$$

Proof: For each i, let $\widetilde{\chi}_i$ be a function on $\mathrm{G}(k)$ such that $\widetilde{\chi}(ugu_i) = \widetilde{\chi}(g)$ for all $u \in \mathrm{U}(k)$ and $u_i \in \mathrm{U}_i$. Let

$$I_i'(a, \widetilde{\chi}_i) = \int_{\mathrm{Y}_s} H_i'(awy)\widetilde{\chi}_i(awy)\psi^{-1}(y)\, dy$$

so that we obtain $I_i'(a)$ itself by taking $\widetilde{\chi}_i \equiv 1$.

By induction, we will show that

$$I_i'(a, \widetilde{\chi}_i) = \int_{\mathrm{Y}_s} W_v(awy)\chi_i'(awy)\widetilde{\chi}_i(awy)\psi^{-1}(y)\, dy.$$

In the case $i = 1$, then by definition

$$\begin{aligned} I_1'(a, \widetilde{\chi}_1) &= \int_{\mathrm{Y}_s} H_1'(awy)\widetilde{\chi}_1(awy)\psi^{-1}(y)\, dy \\ &= \int_{\mathrm{Y}_s} W_{v_1}(awy)(1 - \chi_1(awy))\widetilde{\chi}_1(awy)\psi^{-1}(y)\, dy \end{aligned}$$

and the result follows from Proposition 4.1.

Now suppose the statement is true for all j with $1 \le j < i$. Then

$$\begin{aligned} I_i'(a, \widetilde{\chi}_i) &= \int_{\mathrm{Y}_s} H_i'(awy)\widetilde{\chi}_i(awy)\psi^{-1}(y)\, dy \\ &= \int_{\mathrm{Y}_s} \widetilde{H}_i(awy)(1 - \chi_i(awy))\widetilde{\chi}_i(awy)\psi^{-1}(y)\, dy. \end{aligned}$$

Now apply Proposition 4.1 with $\mathrm{U}_i \subset \mathrm{U}_l$, $F = H_{i-1}'$ and $\chi = (1 - \chi_i)\widetilde{\chi}_i$. We then obtain

$$I_i'(a, \widetilde{\chi}_i) = \int_{\mathrm{Y}_s} H_{i-1}'(awy)(1 - \chi_i(awy))\widetilde{\chi}_i(awy)\psi^{-1}(y)\, dy.$$

Now we apply our induction hypothesis with $\widetilde{\chi}_{i-1} = (1 - \chi_i)\widetilde{\chi}_i$ to obtain

$$\begin{aligned} I_i'(a, \widetilde{\chi}_i) &= \int_{\mathrm{Y}_s} W_v(awy)\chi_{i-1}'(awy)(1 - \chi_i(awy))\widetilde{\chi}_i(awy)\psi^{-1}(y)\, dy \\ &= \int_{\mathrm{Y}_s} W_v(awy)\chi_i'(awy)\widetilde{\chi}_i(awy)\psi^{-1}(y)\, dy. \end{aligned}$$

Now, applying this with $\widetilde{\chi}_i \equiv 1$ gives the desired statement. □

Recall that we had decomposed $j_s(a)$ as $j_s(a) = I_1(a) + I_1'(a)$ As a consequence of these lemmas we may write

$$I_1'(a) = I_{s-1}'(a) = \int_{\mathrm{Y}_s} W_v(awy)\chi_{s-1}'(awy)\psi^{-1}(y)\, dy$$

where $\chi_{s-1}' = \prod_{1 \le j \le s-1}(1 - \chi_j)$. Since χ_j was the characteristic function of $\mathrm{B}(k)w_j'\mathrm{U}_j\mathrm{K}_j$ we see that $\mathrm{Supp}(\chi_{s-1}')$ is the characteristic function of the compliment of

$$\bigcup_{1 \le j < s} \mathrm{B}(k)w_j'\mathrm{U}_j\mathrm{K}_j$$

and by construction this is a neighborhood of

$$\partial C(w) = \bigcup_{1\le j<s} C(w'_j).$$

Hence there will be an open compact neighborhood Ω of e in U_w^- such that $C(w) \cap \mathrm{Supp}(\chi'_{s-1}) = \mathrm{B}(k)w\Omega$. If we then let χ_Ω denote the characteristic function of Ω we have

$$I'_1(a) = \int_{\mathrm{Y}_s} W_v(awy)\chi_\Omega(y)\psi^{-1}(y)\,dy.$$

Note that Ω does not depend on our choice of U_s. Hence choosing U_s sufficiently large, we can assume that $\Omega \subset \mathrm{U}_{s,w}^- = \mathrm{Y}_s$.

Thus we have established the following proposition.

Proposition 5.5. *Let w be a minimal Weyl element supporting a Bessel function and let $v \in V_\pi$ with $W_v(e) = 1$. Then there exist a compact open subgroup K_1 of $\mathrm{G}(k)$ and a compact open neighborhood Ω of e in U_w^- such that for every sufficiently large compact open subgroup U_s of $U(k)$*

$$j_{v,w,\mathrm{Y}_s}(a) = \int_{\mathrm{Y}_s} W_v(awy)\chi_1(awy)\psi^{-1}(y)\,dy + \int_\Omega W_v(awy)\psi^{-1}(y)\,dy$$

where χ_1 is the characteristic function of $\mathrm{B}(k)\mathrm{K}_1$.

As expected, this gives that there are two contributions to the Bessel function associated to w – one from the cell $C(w)$ associated to w itself and one from $C(e)$, the only cell on the boundary of $C(w)$ that supports a Bessel function.

As a corollary, note that since v is a smooth vector of V_π and Ω is compact then

$$v_\Omega = \int_\Omega \psi^{-1}(y)\pi(y)v\,dy$$

is a finite sum, so that $v_\Omega \in V_\pi$ and

$$\int_\Omega W_v(awy)\psi^{-1}(y)\,dy = W_{v_\Omega}(aw).$$

This then gives the following corollary.

Corollary 5.6. *Let w be a minimal Weyl element supporting a Bessel function and let $v \in V_\pi$ with $W_v(e) = 1$. Then there exist a compact open subgroup K_1 of $G(k)$ and a vector $v_\Omega \in V_\pi$ such that for every sufficiently large compact* *open subgroup* U_s *of* $U(k)$

$$j_{v,w,Y_s}(a) = \int_{Y_s} W_v(awy)\chi_1(awy)\psi^{-1}(y)\, dy + W_{v_\Omega}(aw)$$

where χ_1 *is the characteristic function of* $B(k)K_1$.

6. The contribution from near $C(e)$

For applications functoriality as in **[8, 6, 7]** it is essential that we be able to show that the first integral occurring in the expression of the Bessel function in Corollary 5.6 – the contribution from the cell $C(e)$ – is only mildly dependent on the representation π. In fact, it will depend only on the central character ω_π of π. For this purpose it is easier if we take U_1 to be one of the U_{f_M} of Section 1.1. This is clearly permissible since this family of compact open subgroups is cofinal and satisfy the decomposition properties of Section 4. (In fact, we could have taken each U_j as $U_{f_{M_j}}$ with $M = M_1 \leq M_2 \leq \cdots \leq M_s$.)

We are interested in the first term

$$I_1(a) = \int_{Y_s} W_v(awy)\chi_1(awy)\psi^{-1}(y)\, dy \tag{6.1}$$

of Corollary 5.6, which by Proposition 5.1 we know is equal to

$$I_1(a) = \int_{Y_s} W_{v_1}(awy)\chi_1(awy)\psi^{-1}(y)\, dy \tag{6.2}$$

with

$$v_1 = \mathrm{Vol}(U_1)^{-1} \int_{U_1} \psi^{-1}(u)\pi(u)v\, du.$$

Recall that χ_1 is the characteristic function of $B(k)K_1$ with K_1 open and compact. It is easy to see that, by construction, $K_1 \subset \mathrm{Stab}(v_1)$. Then applying Proposition 3.1 we see that the support of $W_{v_1}\chi_1$ is contained in $U(k)T_{U_1}\mathrm{Stab}(v_1)$.

Consider now T_{U_1}. By definition,

$$T_{U_1} = \{t \in T(k) \mid \psi(u_1) = \psi(Ad(t)u_1) \quad \text{for all } u_1 \in U_1\}.$$

Now take $U_1 = U_{f_M}$. We have seen in Section 1.2 that we can write our generic character as

$$\psi(u) = \psi\left(\sum_{\alpha\in\Delta} \overline{x}_\alpha(u_{\tilde{\alpha}})\right) = \prod_{\alpha\in\Delta} \psi\left(\mathrm{Tr}_{k_{\tilde{\alpha}}/k}(u_{\tilde{\alpha}})\right),$$

or, letting $\psi_\alpha = \psi \circ \mathrm{Tr}_{K_{\tilde\alpha}/k}$, we have $\psi = \prod \psi_\alpha$ under the isomorphism $\mathrm{U}^{ab} \simeq \oplus \mathrm{R}_{K_{\tilde\alpha}/k}\mathrm{G}_a$. Thus, the condition that t lies in $\mathrm{T}_{\mathrm{U}_1}$ becomes

$$\psi_\alpha(u_{\tilde\alpha} - \tilde\alpha(t)u_{\tilde\alpha}) = 1$$

for all $u_{\tilde\alpha} \in K_{\tilde\alpha}$ with $\omega(u_{\tilde\alpha}) \geq -M$. If we normalize our additive character of k in such a way that it is trivial on $\mathcal{O}$ but not on $\mathcal{P}^{-1}$, i.e., $\psi(u) = 1$ for $\omega(u) \geq 0$ but there exists u with $\omega(u) = -1$ such that $\psi(u) \neq 1$, then $\psi_\alpha(u_{\tilde\alpha}) = 1$ for $\omega(u_{\tilde\alpha}) \geq -d_{\tilde\alpha}$ where $\mathcal{P}_{\tilde\alpha}^{-d_{\tilde\alpha}}$ is the inverse different of $K_{\tilde\alpha}$. Thus our condition that t lie in $\mathrm{T}_{\mathrm{U}_1}$ becomes

$$\omega((1 - \tilde\alpha(t))u_{\tilde\alpha}) \geq -d_{\tilde\alpha}$$

for all $u_{\tilde\alpha}$ with $\omega(u_{\tilde\alpha}) \geq -M$, i.e.,

$$\tilde\alpha(t) \in 1 + \mathcal{P}_{\tilde\alpha}^{M-d_{\tilde\alpha}} \quad \text{for all} \quad \tilde\alpha \in \tilde\Delta.$$

Recalling the notation of Section 1.3, and writing $d_i = d_{\tilde\alpha_i}$, we have established the following lemma.

Lemma 6.1. *Suppose* $\mathrm{U}_1 = \mathrm{U}_{f_M}$ *with* M *sufficiently large. Then for* $\underline{M} = (M, \ldots, M) - (d_1, \ldots, d_r)$ *we have*

$$\mathrm{T}_{\mathrm{U}_1} = \mathrm{T}_{\underline{M}} = \mathrm{Z}(k)\mathrm{T}^1_{\underline{M}}.$$

Proof: The first equality follows from the argument preceeding the statement of the lemma. The second equality then follows from Lemma 1.1 for M sufficiently large. □

Return now to our expression (6.2) for the contribution to the Bessel function from the small cell $C(e)$. Then the support of $W_{v_1}\chi_1$ is contained in $\mathrm{U}(k)\mathrm{T}_{\mathrm{U}_1}\mathrm{K}_1$ with $\mathrm{K}_1 \subset \mathrm{Stab}(v_1)$, and we can now write this as $\mathrm{U}(k)\mathrm{Z}(k)\mathrm{T}^1_{\underline{M}}\mathrm{K}_1$, taking $\mathrm{U}_1 = \mathrm{U}_{f_M}$. Let us write this decomposition as

$$g = u(g)z(g)t^1(g)k_1(g) \quad \text{for} \quad g \in \mathrm{U}(k)\mathrm{Z}(k)\mathrm{T}^1_{\underline{M}}\mathrm{K}_1.$$

Then we can write

$$W_{v_1}(awy)\chi_1(awy) = \psi(u(awy))\omega_\pi(z(awy))W_{v_1}(t^1(awy))$$

for $awy \in \mathrm{U}(k)\mathrm{Z}(k)\mathrm{T}^1_{\underline{M}}\mathrm{K}_1$, or $y \in (aw)^{-1}\mathrm{U}(k)\mathrm{Z}(k)\mathrm{T}^1_{\underline{M}}\mathrm{K}_1$.

For $\mathrm{U}_1 = \mathrm{U}_{f_M}$ we have that

$$v_1 = \mathrm{Vol}(\mathrm{U}_{f_M})^{-1} \int_{\mathrm{U}_{f_M}} \psi^{-1}(u)\pi(u)v\, du.$$

Then if $t^1 \in \mathrm{T}^1_{\underline{M}}$ we see that

$$\pi(t^1)v_1 = \mathrm{Vol}(\mathrm{U}_{f_M})^{-1} \int_{\mathrm{U}_{f_M}} \psi^{-1}(Ad(t^1)u)\pi(u)\pi(t^1)v \, du.$$

Since $\mathrm{T}^1_{\underline{M}} \subset \mathrm{T}_{\underline{M}} = \mathrm{T}_{\mathrm{U}_1}$ we see that

$$\psi(Ad(t^1)u) = \psi(u)$$

for all $u \in \mathrm{U}_1 = \mathrm{U}_{f_M}$. Hence we see that if M is sufficiently large that $\mathrm{T}^1_{\underline{M}} \subset \mathrm{Stab}(v)$, then $\mathrm{T}^1_{\underline{M}} \subset \mathrm{Stab}(v_1)$ so that

$$W_{v_1}(t^1(awy)) = W_{v_1}(e) = W_v(e) = 1.$$

Note that the vector v was fixed at the beginning of our construction and precedes any other choices made. Hence we see that if M is taken sufficiently large we have the support of $W_{v_1}\chi_1$ is precisely $\mathrm{U}(k)\mathrm{Z}(k)\mathrm{T}^1_{\underline{M}}\mathrm{K}_1$.

Combining these analyses, we arrive at the following proposition.

Proposition 6.2. *Let $v \in V_\pi$ be such that $W_v(e) = 1$. Choose M sufficiently large so that $\mathrm{T}^1_{\underline{M}} \subset \mathrm{Stab}(v)$ and that Lemma 1.1 holds. Take $\mathrm{U}_1 = \mathrm{U}_{f_M}$. Then for $a \in \mathrm{A}_w$ we have*

$$\begin{aligned} I_1(a) &= \int_{\mathrm{Y}_s} W_v(awy)\chi_1(awy)\psi^{-1}(y) \, dy \\ &= \int_{\mathrm{Y}_s \cap (aw)^{-1}\mathrm{U}(k)\mathrm{Z}(k)\mathrm{T}^1_{\underline{M}}\mathrm{K}_1} \psi(u(awy))\omega_\pi(z(awy))\psi^{-1}(y) \, dy \end{aligned}$$

which depends on the representation π only through its central character ω_π.

7. Asymptotics of Bessel functions II

Let us now return to our partial Bessel function $j_{v,w,\mathrm{Y}_s}(a)$ from Corollary 5.6

$$j_{v,w,\mathrm{Y}_s}(a) = \int_{\mathrm{Y}_s} W_v(awy)\chi_1(awy)\psi^{-1}(y) \, dy + W_{v_\Omega}(aw).$$

Recall that the first term, namely $I_1(a)$ from (6.1), is the contribution to the Bessel function from near the small cell $C(e)$ while the second term $W_{v_\Omega}(aw)$ is the contribution from the "interior" of the cell $C(w)$.

Consider first the contribution from $C(w)$. We may rewrite this as

$$j^{C(w)}_{v,w,\mathrm{Y}_s}(a) = W_{v_\Omega}(aw) = W_{\pi(w)v_\Omega}(a). \tag{7.1}$$

In this form it is seen to be given by the value of a Whittaker function for a vector $\pi(w)v_\Omega \in V_\pi$. As a function of $a \in \mathrm{A}_w$ this then is compactly supported as $\alpha(a) \to \infty$ for simple roots $\alpha \in \Delta$, as all smooth Whittaker functions are, and satisfies asymptotics determined by the representation π as $\alpha(a) \to 0$. Hence its asymptotics are well understood and are determined by the representation π. Also, since this contribution is given by a fixed Whittaker function, it is smooth as a function of $a \in \mathrm{A}_w$.

The contribution from the cell $C(e)$, which we will now denote by $j^{C(e)}_{v,w,\mathrm{Y}_s}(a)$, is quite interesting and was analyzed in the last section. From Proposition 6.2 we know that $j^{C(e)}_{v,w,\mathrm{Y}_s}(a) = I_1(a)$ is actually given by a quite complicated exponential sum, namely

$$j^{C(e)}_{v,w,\mathrm{Y}_s}(a) = \int_{\mathrm{Y}_s \cap (aw)^{-1}\mathrm{U}(k)\mathrm{Z}(k)\mathrm{T}^1_M \mathrm{K}_1} \psi(u(awy))\omega_\pi(z(awy))\psi^{-1}(y)\, dy. \quad (7.2)$$

Since the contribution from $C(w)$ vanishes as $\alpha(a) \to \infty$ we see that the asymptotics of the Bessel function $j_{v,w,\mathrm{Y}_s}(a)$ as $\alpha(a) \to \infty$ are completely given by this exponential sum. Fortunately for our application, even though this sum is complicated, it depends on the representation π only through its central character ω_π.

Combining the above, we finally arrive at the main theorem of this paper.

Theorem 7.1. *Let (π, V_π) be a generic representation of $\mathrm{G}(k)$ and let $v \in V_\pi$ satisfy $W_v(e) = 1$. Let $w \in W$ be a Weyl group element which supports a Bessel function and is a minimal non-trivial such. Then for every sufficiently large compact open subgroup $\mathrm{Y} \subset \mathrm{U}^-_w(k)$ we have that the partial Bessel function $j_{v,w,\mathrm{Y}}(a)$ can be decomposed as*

$$j_{v,w,\mathrm{Y}}(a) = j^{C(e)}_{v,w,\mathrm{Y}}(a) + j^{C(w)}_{v,w,\mathrm{Y}}(a)$$

where $j^{C(w)}_{v,w,\mathrm{Y}}(a)$ is given by (7.1) and is a smooth function of $a \in \mathrm{A}_w$ which vanishes as $\alpha(a) \to \infty$ and $j^{C(e)}_{v,w,\mathrm{Y}}(a)$ is given by (7.2) and is dependent only on the central character ω_π of π.

We will return to the application of this result to the stability of local γ-functions for generic representations of quasi-split groups in a subsequent paper [9].

8. Appendix: On the existence of $J_{\pi,w}(a)$.

In this Appendix we will present the arguments for the convergence of the Bessel function $J_{\pi,w}(a)$ attached to a minimal Weyl element w which supports a Bessel function from Section 2.3. We begin with some preliminaries which generalize a construction of Steinberg [17] to the quasi-split setting.

8.1. Coefficients of Steinberg Let $\mathfrak{g}$ denote the Lie algebra of G. Recall [2] that we have the decomposition of the Lie algebra of G given by

$$\mathfrak{g} = \mathfrak{g}_0 \oplus \bigoplus_{\alpha \in \Phi_{nd}} \mathfrak{g}_{(\alpha)}.$$

For each non-divisible root $\alpha \in \Phi_{nd}$ chose a k-rational basis $\{X_{\alpha,i_\alpha} \mid 1 \leq i_\alpha \leq \dim(\mathfrak{g}_{(\alpha)})\}$ of $\mathfrak{g}_{(\alpha)}$. Similarly choose a k-rational basis $\{H_1, \ldots, H_r\}$ of $\mathfrak{g}_0 = \mathfrak{t}$, the Lie algebra of the maximal k-torus T. We may assume this basis is an extension of a basis of $\mathfrak{a}$, the Lie algebra of A.

Let $N = \dim_k \mathfrak{u}$ be the dimension of the maximal unipotent subgroup U of G, or its Lie algebra $\mathfrak{u}$. Consider the Nth exterior product of $\mathfrak{g}$, $\wedge^N \mathfrak{g}$ with the basis consisting of wedge products of the basis $\{X_{\alpha,i_\alpha}, H_j\}$ of $\mathfrak{g}$. G acts k-rationally on $\mathfrak{g}$, and hence on $\wedge^N \mathfrak{g}$, by the adjoint action.

Let

$$Y_e = \wedge_{\alpha>0} \wedge_{i_\alpha} X_{\alpha,i_\alpha}.$$

This is one of our canonical basis vectors. Similarly, for any $w \in W$, let

$$Y_w = \wedge_{\alpha>0} \wedge_{i_{w\alpha}} X_{w\alpha,i_{w\alpha}}.$$

For any $g \in G$ define $c_w(g)$ to be the coefficient of Y_w in the expansion of $Ad(g)Y_e$ in our chosen basis for $\wedge^N \mathfrak{g}$. Since the adjoint action is k-rational, this is a well defined rational function of g. Note that $Ad(w)Y_e = c_w(w)Y_w$ with $c_w(w) \neq 0$.

From the relation $[\mathfrak{g}_\alpha, \mathfrak{g}_\beta] \subset \mathfrak{g}_{\alpha+\beta}$ it follows that for $u \in$ U we have $Ad(u)Y_e = c_e(u)Y_e$ with $c_e(u) \neq 0$. Then for $t \in$ T we also have $Ad(t)Y_e = c_e(t)Y_e$ with $c_e(t) \neq 0$ being essentially the modulus character of t. As noted above, for $w \in W$ we have $Ad(w)Y_e = c_w(w)Y_w$ by definition. Finally, if $u \in$ U we have $Ad(u)Y_w = cY_w +$ "higher order terms", with respect to the ordering given by sums of positive roots, with $c \neq 0$. Thus $Ad(\mathrm{B}w\mathrm{B})Y_e \subset k^\times Y_w +$ "higher order terms", i.e., c_w is non-vanishing on the Bruhat cell BwB. Then the proof of Theorem 23 of Steinberg [17] (p.127) gives the following result.

Proposition 8.1. *Let $w, w' \in W$. Then the following three conditions are equivalent:*

(i) $\mathrm{B}w'\mathrm{B} \subset \overline{\mathrm{B}w\mathrm{B}}$*;*

(ii) $w' \leq w$*;*

(iii) $c_{w'}(g)$ *is not identically zero on* $\mathrm{B}w\mathrm{B}$.

Note that as a consequence of this, we see that for $w = w_\ell$ the long Weyl element we have that c_{w_ℓ} is non-vanishing precisely on the big Bruhat cell $C(w_\ell)$. Moreover, Lemma 52 of Steinberg **[17]** (p.123), generalized to the quasi-split situation, gives that c_e is non-vanishing precisely on the translated large cell $\mathrm{U}^-\mathrm{T}\mathrm{U} = w_\ell^{-1}C(w_\ell)$. To compute the support of the other coefficient functions, we have the following lemma.

Lemma 8.2. *For any* $w \in W$,

$$c_w(wg) = c_e(g)c_w(w).$$

Proof: This is an elementary calculation. On the one hand, by definition we have $Ad(wg)Y_e = c_w(wg)Y_w + \cdots$. On the other hand,

$$\begin{aligned} Ad(wg)Y_e = Ad(w)Ad(g)Y_e &= Ad(w)(c_e(g)Y_e + \cdots) \\ &= c_e(g)c_w(w)Y_w + \cdots \end{aligned}$$

Thus $c_w(wg) = c_e(g)c_w(w)$. □

An equivalent formulation is that $c_w(g) = c_e(w^{-1}g)c_w(w)$. Thus we have $c_w(g) \neq 0$ iff $c_e(w^{-1}g) \neq 0$ iff $w^{-1}g \in \mathrm{U}^-\mathrm{B}$ iff $g \in w\mathrm{U}^-\mathrm{B} = ww_\ell^{-1}\mathrm{U}w_\ell\mathrm{B}$.Thus we have the following generalization of the above statements.

Proposition 8.3. $c_w(g)$ *is non-vanishing precisely on* $w\mathrm{U}^-\mathrm{B}$.

The above lemma also lets us explicitly evaluate the Steinberg coefficients. First recall that for $t \in \mathrm{T}$ we have the modulus character $\delta(t) = \delta_{\mathrm{B}}(t) = \det(Ad(t)|\mathrm{U})$. Then we see that by definition $c_e(t) = \delta(t)$. Hence on the Bruhat cell $C(e) = \mathrm{B}$ we have $c_e(tu) = \delta(t) \neq 0$, as claimed. Next consider c_w as a function on $C(w)$ If we write $g \in C(w)$ as $g = u_1wtu_2$ we see that

$$\begin{aligned} Ad(u_1wtu_2)Y_e &= Ad(u_1wt)Y_e = Ad(u_1w)c_e(t)Y_e \\ &= c_e(t)Ad(u_1)c_w(w)Y_w = c_e(t)c_w(w)Y_w + \text{"higher order terms"} \end{aligned}$$

since $u_1 \in \mathrm{U}(k)$. Thus $c_w(u_1wtu_2) = \delta(t)c_w(w)$. For our purposes, a more useful formula will be $c_w(u_1twu_2) = c_w(w)\delta(w^{-1}tw)$. Note that the function appearing, $\delta_w(t) = \delta(w^{-1}tw) = \det(Ad(t)|w\mathrm{U}w^{-1})$, is the modulus character associated to the conjugate Borel $\mathrm{B}^w = w\mathrm{B}w^{-1}$.

Proposition 8.4. *Let $w \in W$ and let $S(wt)$ denote the slice $\mathrm{U}(k)wt\mathrm{U}(k)$ of the Bruhat cell $C(w)$. Then c_w is constant on this slice and $c_w(S(tw)) = \delta(t)c_w(w)$. Moreover, c_w is constant on the slice $S(tw)$ of $C(w)$ and on this slice $c_w(S(tw)) = \delta_w(t)c_w(w)$.*

8.2. The convergence of $J_{\pi,w}(a)$ We now return to the situation of Sections 2 and 3.

Proposition 8.5. *Let $v \in V_\pi$ such that $W_v(e) = 1$. For any compact open subgroup $\mathrm{U}_0 \subset \mathrm{U}(k)$ let $v_0 = v_{\mathrm{U}_0}$ be the associated approximate Whittaker vector in V_π. Let $w \in W$, $w \neq e$. For $t \in \mathrm{T}(k)$ consider the slice $S(tw) = \mathrm{U}(k)tw\mathrm{U}(k) = \mathrm{U}(k)tw\mathrm{U}_w^-$ of the Bruhat cell $C(w) = \mathrm{B}w\mathrm{U}_w^-$. Then there is an open compact subgroup K_0' such that the restriction of W_{v_0} to $S(tw)$ vanishes on $S(tw) \cap \mathrm{BK}_0'$.*

Proof: Let $W_0 = W_{v_0}$ and let $\mathrm{K}_0 = \mathrm{Stab}(v_0)$. If K' is any subgroup of K_0 then as in the proof of the Proposition 3.1 we have that $\mathrm{Supp}(W_0) \cap \mathrm{B}(k)\mathrm{K}' \subset \mathrm{U}(k)\mathrm{T}_{\mathrm{U}_0}\mathrm{K}'$. Hence the proposition will follow if, given $w \neq e$ and $t \in \mathrm{T}(k)$, we can find an open neighborhood K_0' of e such that $S(tw) \cap \mathrm{U}(k)\mathrm{T}_{\mathrm{U}_0}\mathrm{K}_0' = \emptyset$. Taking inverses, it suffices to find a K_0' such that $S(w^{-1}t^{-1}) \cap \mathrm{K}_0'\mathrm{T}_{\mathrm{U}_0}\mathrm{U}(k) = \emptyset$.

Recall from Section 8.1 that to any $w \in W$ there is associated a rational (hence continuous) function $c_w(g)$ such that $C(w) \subset \overline{C(w')}$ iff $c_w(g)$ is not identically zero on $C(w')$. In particular, for $w \neq e$ we have $c_w(b) = 0$ for all $b \in \mathrm{B}(k) = C(e)$. On the other hand, we have seen that $c_w(g)$ is a non-zero constant on either of the slices $S(wt) = \mathrm{U}(k)wt\mathrm{U}(k)$ or $S(tw)$.

Now consider the restriction of $c_{w^{-1}}$ to the open set $\mathrm{K}'\mathrm{T}_{\mathrm{U}_0}\mathrm{U}(k)$. Then for $k't'u \in \mathrm{K}'\mathrm{T}_{\mathrm{U}_0}\mathrm{U}(k)$ we have $Ad(k't'u)Y_e = c_e(t')Ad(k')Y_e$ so that $c_{w^{-1}}(k't'u) = c_e(t')c_{w^{-1}}(k')$. Since $t' \in \mathrm{T}_{\mathrm{U}_0}$ satisfies $\psi(u_0) = \psi(Ad(t')u_0)$ for all u_0 in the compact open subgroup U_0 and $c_e(t')$ is defined by the adjoint action of t' on Y_e, we see that $|c_e(t')|$ is bounded above and below on $\mathrm{T}_{\mathrm{U}_0}$. On the other hand, $c_{w^{-1}}$ is continuous and $c_{w^{-1}}(e) = 0$. Hence given any $L > 0$ we can find a compact open neighborhood K_0' of e such that for $k' \in \mathrm{K}_0'$ we have $|c_{w^{-1}}(k')| \leq q^{-L}$. In particular, we can choose K_0' such that for all $t' \in \mathrm{T}_{\mathrm{U}_0}$ and $k' \in \mathrm{K}_0'$ we have $|c_e(t')c_{w^{-1}}(k')| < |c_{w^{-1}}(t^{-1})|$. Hence, for this K_0' we have that $\mathrm{K}_0'\mathrm{T}_{\mathrm{U}_0}\mathrm{U}(k)$ and $S(w^{-1}t^{-1})$ are separated by the values taken by the function $c_{w^{-1}}$. □

Proposition 8.6. *Suppose that $w \in W$ does not support a Bessel function. Let $\mathrm{Y} \subset \mathrm{U}_w^-$ be open and compact. Then for every sufficiently large open compact subset $\mathrm{U}_0 \subset \mathrm{U}(k)$ there exists a compact open subgroup $\mathrm{K}_0 \subset \mathrm{G}(k)$ such that if $v_0 = v_{\mathrm{U}_0}$ we have the restriction of W_{v_0} to $\mathrm{B}(k)w\mathrm{Y}\mathrm{K}_0$ vanishes identically.*

Proof: For U_0 any compact open subgroup of $\mathrm{U}(k)$ let $\mathrm{U}_{0,w}^+ = \mathrm{U}_0 \cap \mathrm{U}_w^+$ and $\mathrm{U}_{0,w}^- = \mathrm{U}_0 \cap \mathrm{U}_w^-$. By enlarging U_0 if necessary, we may assume that

(i) $\mathrm{U}_0 = \mathrm{U}_{0,w}^+ \mathrm{U}_{0,w}^-$,

(ii) $Y \subset \mathrm{U}_{0,w}^-$

Let $\mathrm{K}_0 = \mathrm{Stab}(v_0)$ and $W_0 = W_{v_0}$. Then if $bwyk \in \mathrm{B}(k)w\mathrm{Y}\mathrm{K}_0$ with $b = ut \in \mathrm{U}(k)\mathrm{T}(k)$ then $W_0(utwyk) = \psi(u)W_{v_0}(tw)\psi(y)$.

For any $u_0^+ \in \mathrm{U}_0^+$ we have

$$\psi(u_0^+)W_0(tw) = W_0(twu_0^+) = \psi(Ad(t)wu_0^+w^{-1})W_0(tw).$$

Since w does not support a Bessel function, there is a simple root α such that $w\alpha$ is positive but not simple. Hence for U_0 sufficiently large we can find $u_0^+ \in \mathrm{U}_{0,w}^+$ lying in the root subgroup corresponding to α such that $\psi(u_0^+) \neq 1$ but $\psi(Ad(t)wu_0^+w^{-1}) \equiv 1$ independent of t. Hence $W_0(tw) \equiv 0$. □

We now consider $w \in W$ a minimal Weyl element supporting a Bessel function. Let $S_w = \{w' \in W \mid w' \leq w\}$ and let $s = s_w = |S_w|$. Enumerate the elements of S_w so that $w_1' = e$, $w_s' = w$ and if $w_i' \leq w_j'$ then $i \leq j$. Recall that we have defined a canonical exhaustive sequence of open compact subgroup $\mathrm{U}_{f_M} \subset \mathrm{U}(k)$ in Section 1.1 indexed by positive integers M. In order to simplify the notation, we will now denote this subgroup by $\mathrm{U}(M) = \mathrm{U}_{f_M}$. Similarly, we will set $v_M = v_{\mathrm{U}(M)}$ and $W_M = W_{v_M}$.

Proposition 8.7. *Let w be a minimal Weyl element which supports a Bessel function as above. Fix a slice $S(tw) = \mathrm{U}(k)tw\mathrm{U}(k)$ of the Bruhat cell $C(w)$. For every sufficiently large open compact subgroup $\mathrm{U}(M) \subset \mathrm{U}(k)$ there is a compact open subgroup $\mathrm{K}(M) \subset \mathrm{G}(k)$ such that the restriction of W_M to the slice $S(tw)$ vanishes on $S(tw) \cap (\cup_{i=1}^{s-1} C(w_i')\mathrm{K}(M))$, where $C(w_i')$ is the Bruhat cell $\mathrm{B}(k)w_i'\mathrm{U}_{w_i'}^-$.*

Proof: We will prove this by induction on j. More precisely, we prove that for each $j < s$ there exists a compact open subgroup of the form $\mathrm{U}(M_j)$ such that for every compact open subgroup $\mathrm{U}(M) \supset \mathrm{U}(M_j)$ there is an open compact subgroup $K_j(M) \subset G(k)$ such that W_M vanishes on $S(tw) \cap (\cup_{i=1}^{j} C(w_i')K_j(M))$. Then the $U(M)$ and $K(M)$ of the theorem will be those associated to $j = s-1$.

If $j = 1$ then $w_j' = e$ and we may take $M_1 = 1$ and $U(M)$ and $K_1(M)$ to be those from Proposition 8.5.

We now assume the statement for j, that is, there exists M_j such that for all $U(M) \supset U(M_j)$ there exists $K_j(M)$ such that we have that W_M vanishes on $S(tw) \cap (\cup_{i=1}^{j} C(w_i')K_j(M))$.

For simplicity, let us write $V_j(M) = \cup_{i=1}^{j} C(w_i')K_j(M)$ and let $V_j = V_j(M_j)$.

Consider $C(w_{j+1}')$. Then V_j is a neighborhood of $\partial C(w_{j+1}') = \overline{C(w_{j+1}')} - C(w_{j+1}')$. Then there exists an open compact $Y_{j+1} \subset U_{w_{j+1}'}^-$ such that

$$B(k)w_{j+1}'Y_{j+1} \supset C(w_{j+1}') - V_j.$$

Now take N sufficiently large so that $U(N)_{w_{j+1}'}^- \supset Y_{j+1}$ and let $M_{j+1} = \max(N, M_j)$.

Assume that $M' \geq M_{j+1}$, i.e., $U(M') \supset U(M_{j+1})$. Let

$$V_j'(M') = \bigcap_{u \in U(M')} V_j u.$$

Since $(U(M') \cap K_j(M_j))\backslash U(M')$ is finite, this is actually a finite intersection and $V_j'(M')$ is open and still contains $\partial C(w_{j+1}') = \cup_{i=1}^{j} C(w_j')$. In fact, if we let

$$K_j'(M') = \bigcap_{u \in U(M')} u^{-1}K_j(M_j)u$$

then

$$V_j'(M') = \bigcup_{i=1}^{j} C(w_i')K_j'(M').$$

Now $C(w_{j+1}') - V_j \subset B(k)w_{j+1}'U(N)_{w_{j+1}'}^-$ so that for all $u \in U(M')$ we have

$$C(w_{j+1}') - V_j u = (C(w_{j+1}') - V_j)u \subset B(k)w_{j+1}'U(N)_{w_{j+1}'}^- u$$

and hence

$$\begin{aligned} C(w_{j+1}') - V_j'(M') &\subset \bigcup_{u \in U(M')} B(k)w_{j+1}'U(N)_{w_{j+1}'}^- u \\ &= B(k)w_{j+1}'U(M') \\ &= B(k)w_{j+1}'U(M')_{w_{j+1}'}^-. \end{aligned}$$

Take for $K_{j+1}(M')$ any compact open subgroup of $K'_j(M') \cap \mathrm{Stab}(v_{M'})$ and let $V_{j+1}(M') = \cup_{i=1}^{j+1} C(w'_i) K_{j+1}(M')$. Note that

$$V_{j+1}(M') \subset (V'_j(M') \cup B(k) w'_{j+1} U(M')^-_{w'_{j-1}} \mathrm{Stab}(v_{M'})).$$

Now consider the restriction of $W_{M'}$ to $S(tw) \cap V_{j+1}(M')$. Let $g \in S(tw) \cap V_{j+1}(M')$. If $g \in V'_j(M')$ then we write

$$W_{M'}(g) = \frac{1}{\mathrm{Vol}(U(M'))} \int_{U(M')} W_{M_j}(gu) \psi^{-1}(u)\, du.$$

Since $g \in V'_j(M')$, then $gu \in V'_j(M')u \subset V_j$. Since W_{M_j} vanishes on $S(tw) \cap V_j$ we have $W_{M'}(g) = 0$ in this case. On the other hand, if instead $g \in S(tw) \cap B(k) w'_{j+1} U(M)^-_{w'_{j+1}} \mathrm{Stab}(v_{M'})$, then

$$W_{M'}(g) = W_{M'}(utw'_{j+1} u^-_{j+1} k) = \psi(u) W_0(tw'_{j+1}) \psi(u^-_{j+1}).$$

But since w'_{j+1} does not support a Bessel function, we see that $W_{M'}(tw'_{j+1}) = 0$ as in the proof of Proposition 8.6.

Therefore $W_{M'}$ vanishes on $S(tw) \cap V_{j+1}(M')$. This completes the induction and thereby the proof of the proposition. □

This proposition then leads to the existence of $J_{\pi,w}$ for w a minimal Weyl element that supports a Bessel function. To see this, consider

$$J_{\pi,w}(a) = \int_{U^-_w} W_v(awu) \psi^{-1}(u)\, du$$

for $a \in A_w$ and $v \in V_\pi$ with $W_v(e) = 1$. Note that for any M we have by an elementary change of variables that

$$\int_{U^-_w} W_v(awu) \psi^{-1}(u)\, du = \int_{U^-_w} W_{v_M}(awu) \psi^{-1}(u)\, du$$

in the notation of Proposition 8.7. Fixing $a \in A_w$, by Proposition 8.7 there exists an M' such that $W_{v_{M'}}$ vanishes on $S(aw) \cap (\cup_{i=1}^{s-1} B(k) C(w'_j) K(M')$, that is, $W_{v_{M'}}(awu)$, as a function of $u \in U^-_w$, is compactly supported. Hence this integral converges and $J_{\pi,w}(a)$ exists.

Theorem 8.8. *If π is a generic representation of $G(k)$ and w is a minimal Weyl element that supports a Bessel function, then the associated (full) Bessel function $J_{\pi,w}(a)$ exists as a function on A_w.*

References

[1] Z. Averbuch, Bessel functions of algebraic groups' representations over a local field. Theses, Tel-Aviv University, 1992.

[2] A. Borel, Linear Algebraic Groups. Graduate Texts in Mathematics **126**. Springer-Verlag, New York, 1991.

[3] A. Borel and J. Tits, *Groupes réductifs*. Publ. Math. IHES **27** (1965), 55–150.

[4] F. Bruhat and J. Tits, *Groupes réductifs sur un corps local, Chapitre I*. Publ. Math. IHES **41** (1972), 5–252.

[5] F. Bruhat and J. Tits, *Groupes réductifs sur un corps local, Chapitre II*. Publ. Math. IHES **60** (1984), 5–184.

[6] J.W. Cogdell, H.H. Kim, I.I. Piatetski-Shapiro, and F. Shahidi, *On lifting from classical groups to GL_N*. Publ. Math. IHES **79** (2001), 5–30.

[7] J.W. Cogdell, H.H. Kim, I.I. Piatetski-Shapiro, and F. Shahidi, *Functoriality for the classical groups*, Publ. Math. IHES **99** (2004), 163–233.

[8] J.W. Cogdell and I.I. Piatetski-Shapiro, *Stability of gamma factors for $SO(2n+1)$*, manuscripta math. **95** (1998), 437–461.

[9] J.W. Cogdell, I.I. Piatetski-Shapiro, and F. Shahidi, *Stability of γ-factors for quasi-split groups*. In preparation.

[10] I.M. Gelfand and M.I. Graev, *The construction of irreducible representations of simple algebraic groups over a finite field*. Dokl. Akad. Nauk USSR **147** (1962), 529–532.

[11] S.I. Gelfand, *Representations of the general linear group over a finite field*. Lie Groups and Their Representations (I.M. Gelfand, Ed.), John Wiley and Sons, New York, 1975, 119–132.

[12] J.E. Humphreys, Reflection Groups and Coxeter Groups. Cambridge Studies in Advanced mathematics **29**. Cambridge University Press, Cambridge, 1990.

[13] I. Satake, Classification Theory of Semi-simple Algebraic Groups. Lecture Notes in Pure and Applied Mathematics **3**. Marcel Dekker, New York, 1971.

[14] F. Shahidi, *Local coefficients as Artin factors for real groups*. Duke Math. J. **52** (1985), 973–1007.

[15] F. Shahidi, *A proof of Langlands' conjecture on Plancherel measures; Complementary series for p-adic groups*. Annals of Math. **132** (1990), 273–330.

[16] F. Shahidi, *Local coefficients as Mellin transforms of Bessel functions: Towards a general stability*. IMRN **2002**, No. 39, (2002), 2075–2119.

[17] R. Steinberg, Lectures on Chevalley Groups. Mimeographed notes, Yale University, 1967.

J. W. Cogdell
Department of Mathematics
Oklahoma State University
Stillwater, OK 74078, USA
Current address:
Department of Mathematics
Ohio State University
Columbus. OH 43210. USA
cogdell@math.ohio-state.edu

I. I. Piatetski-Shapiro
Department of Mathematics
Yale University
New Haven, CT 06520. USA
and
School of Mathematics
Tel Aviv University
Tel Aviv 69978, Israel
ilya@math.yale.edu

F. Shahidi
Department of Mathematics
Purdue University
West Lafayette, IN 47907, USA
shahidi@math.purdue.edu

Non-tempered Arthur packets of G_2

Wee Teck Gan and Nadya Gurevich

to Professor S. Rallis
with admiration and best wishes

§1. Introduction

In his famous conjecture, J. Arthur gave a description of the discrete spectrum $L^2_{disc}(G(F)\backslash G(\mathbb{A}))$ of a reductive linear algebraic group G defined over a number field F. According to this conjecture, the irreducible constituents of L^2_{disc} can be partitioned into equivalence classes called A-packets. In this survey article, we describe some recent construction and characterization of the (non-tempered) A-packets of the split exceptional group of type G_2. These results are contained in the papers [GGJ], [G] and [GG]. For the classical groups of small rank, the construction of these packets has been carried out by the work of various people. These include: the work of J. Rogawski [R] for $U(3)$, the work of I. Piatetski-Shapiro [PS] and D. Soudry [So] for $PGSp_4$, and the recent work of K. Konno and T. Konno [KK] for $U(2,2)$. As we shall see in the article, the case of G_2 offers a number of (pleasant) surprises which do not appear for classical groups.

We begin by giving a brief description of Arthur's conjecture for the discrete spectrum, assuming for simplicity that G is split and simply-connected, so that the dual group has no center. The references are of course [A1] and [A2].

(1.1) Decomposition of discrete spectrum. Loosely speaking, Arthur's conjecture is a classification of the constituents of L^2_{disc} into near equivalence classes (this is not entirely true, but for the group G_2, it is expected to be so). Here, we say that two representations π_1 and π_2 of $G(\mathbb{A})$ are nearly equivalent if for almost all places v, $\pi_{1,v}$ and $\pi_{2,v}$ are isomorphic: this is an equivalence relation.

More precisely, Arthur speculated that the discrete spectrum possesses a decomposition

$$L^2_{disc}(G(F)\backslash G(\mathbb{A})) = \widehat{\bigoplus_{\psi}} L^2_{\psi},$$

where the Hilbert space direct sum runs over equivalence classes of A-parameters ψ, i.e. admissible maps

$$\psi : L_F \times SL_2(\mathbb{C}) \longrightarrow \hat{G}$$

where L_F denotes the conjectural Langlands group of F and $\hat{G}$ is the complex dual group of G. For any ψ, the space L^2_ψ is a direct sum of nearly equivalent representations and can be more precisely described as follows.

(1.2) Local A-packets. The global A-parameter ψ gives rise to a local A-parameter

$$\psi_\nu : L_{F_\nu} \times SL_2(\mathbb{C}) \longrightarrow \hat{G}$$

for each place ν of F. We denote by S_{ψ_v} the group of components of $Cent_{\hat{G}}(Im(\psi_\nu))$. To each irreducible representation η_v of S_{ψ_v}, Arthur speculated that one can attach a unitarizable admissible (possibly reducible, possibly zero) representation π_{η_v} of $G(F_v)$. The set

$$A_{\psi_\nu} = \left\{\pi_{\eta_\nu} \ : \ \eta_\nu \in \widehat{S}_{\psi_\nu}\right\}$$

is called a local A-packet. It is required that

- for almost all v, π_{η_v} is irreducible and unramified if η_v is the trivial character 1_v. In fact, for almost all v, π_{η_v} is the unramified representation whose Satake parameter is:

  $$s_{\psi_v} = \psi_v \left(Fr_v \times \begin{pmatrix} q_v^{1/2} & \\ & q_v^{-1/2} \end{pmatrix} \right),$$

  where Fr_v is a Frobenius element at v and q_v is the number of elements of the residue field at v;
- the distribution

  $$\sum_{\eta_v} \eta_v(z_o) Tr(\pi_{\eta_v})$$

  is stable, and certain identities involving transfer of character distributions to endoscopic groups of $G(F_v)$ should hold. Here, z_0 is the image of the non-trivial central element of $SL_2(\mathbb{C})$ in S_{ψ_v}

These requirements may not characterize the set A_{ψ_v} but they come pretty close; for more details, see [A2].

(1.3) Global A-packets. With the local packets A_{ψ_v} at hand, we may define the global A-packet by:

$$A_\psi = \{\pi = \otimes_v \pi_{\eta_v} : \pi_{\eta_v} \in A_{\psi_v} \, \eta_v = 1_v \text{ for almost all } v\}.$$

It is a set of nearly equivalent representations of $G(\mathbb{A})$, indexed by the irreducible representations of the compact group $\mathcal{S}_{\psi,\mathbb{A}} := \prod_v S_{\psi,v}$. If $\eta = \bigotimes_v \eta_v$ is an irreducible character of $\mathcal{S}_{\psi,\mathbb{A}}$, then we may set

$$\pi_\eta = \bigotimes_v \pi_{\eta_v}.$$

This is possible because for almost all v, $\eta_v = \mathbf{1}_v$ and π_{1_v} is required to be unramified by the above.

(1.4) Multiplicity formula. The space L^2_ψ will be the sum of the elements of A_ψ with some multiplicities. More precisely, Arthur attached to ψ a quadratic character ϵ_ψ of S_ψ. Now if η is an irreducible character of $\mathcal{S}_{\psi,\mathbb{A}}$, we set

$$m_\eta = \frac{1}{\# S_\psi} \cdot \left(\sum_{s \in S_\psi} \epsilon_\psi(s) \cdot \eta(s) \right).$$

Then Arthur conjectures that there is a $G(\mathbb{A})$-equivariant embedding

$$\iota_\psi : \bigoplus_\eta m_\eta \pi_\eta \hookrightarrow L^2_{disc}(G(F)\backslash G(\mathbb{A})).$$

The image of this embedding is the subspace L^2_ψ.

(1.5) Tempered and non-tempered parameters. An A-parameter ψ is called *tempered* if ψ is trivial when restricted to $SL_2(\mathbb{C})$. In this case, the representations in A_ψ are conjectured to be tempered. A non-tempered A-parameter tends to factor through a subgroup $\hat{H} \subset \hat{G}$ with H an endoscopic group of G. Of course, this is the case for some tempered parameters too. In any case, the representations corresponding to such a parameter are considered more degenerate. One might thus hope to construct the packets A_{ψ_v} and the embedding $L^2_\psi \hookrightarrow L^2_{disc}$ by lifting from a group smaller than G.

(1.6) CAP representations. Some of the earlier constructions of A-packets for classical groups of low rank were couched in the language of **CAP** representations, rather than in the framework of Arthur's conjectures. Recall that a cuspidal representation π of G is said to be **CAP** *with respect to a parabolic subgroup* P of G if it is nearly equivalent to the irreducible constituents of an induced representation $Ind_P^G \tau$ with τ a cuspidal representation of the Levi factor of P. Arthur's conjecture implies that the CAP representations are precisely the cuspidal representations belonging to the non-tempered A-packets of G.

(1.7) A list of problems. Given an A-parameter ψ, let us formulate a list of problems which describe more precisely what we would like to achieve.

PROBLEM A: Give a natural construction of the local A-packets A_{ψ_v}.

How can we justify that our naturally constructed set A_{ψ_v} is the correct A-packet? Ideally, we would like to be able to verify that the set A_{ψ_v} that was constructed is stable. Unfortunately, we have nothing to offer on this. However, one can give strong global justification for the authenticity of A_{ψ_v} by addressing:

PROBLEM B: With the local packets defined, construct the subspace $L^2_{\psi} \subset L^2_{disc}$ as prescribed by Arthur's conjecture.

After resolving Problem B, one knows that if η is an irreducible character of $S_{\psi,\mathbf{A}}$, then the multiplicity $m_{disc}(\pi_\eta)$ of π_η in L^2_{disc} satisfies

$$m_{disc}(\pi_\eta) \geq m_\eta,$$

at least when π_η is irreducible. Though this is nice, it does not rule out the occurrence of π_η when $m_\eta = 0$. Thus, we would like to resolve:

PROBLEM C: Find the true discrete multiplicity of π_η. For example, decide if $m_{disc}(\pi_\eta) = m_\eta$.

Note that it is not always the case that $m_{disc}(\pi_\eta) = m_\eta$. For example, for the group SL_n ($n \geq 3$), Blasius has shown in [B] that there are cuspidal representations which occur as members of global A-packets corresponding to inequivalent A-parameters. Another example of this phenomenon is the group SO_{2n}, where one has inequivalent A-parameters which are conjugate by the outer automorphism group.

Finally, suppose that we have resolved the three problems above; is there still any doubt about the authenticity of our local packets A_{ψ_v}? Certainly. The reason is that the representations π_η are not required to be irreducible. It may happen that the true π_η is the sum of 2 irreducible representations, but the π_η we defined is irreducible; in other words, we have missed out one of the two constituents of the true π_η. In that case, we will not be able to detect the missing representation by resolving Problems A-C. This leads us to:

PROBLEM D: Decide if the subspace $L^2_{\psi_\tau}$ constructed in Problem B is a full near equivalence class. More precisely, if $\pi \subset L^2_{disc}$ is nearly equivalent to the representations in A_ψ (i.e. $\pi_v \cong \pi_{1_v}$ for almost all v), is $\pi \subset L^2_\psi$?

Again, we do not expect the statement in Problem D to hold all the time, in view of the examples of Blasius [B]. However, a positive solution to Problem D will show beyond any reasonable doubt that our local packets are the right ones. Indeed, if σ is a representation of $G(F_v)$ which should be in A_{ψ_v}, then σ should occur as the local component of a global representation $\pi \subset L^2_{disc}$ which is nearly equivalent to the representations in A_ψ. A positive solution to Problem D will show that we have already captured all such candidates σ.

In this paper, which is purely expository, we discuss our recent results on the above four problems for the non-tempered A-parameters of the split exceptional group G_2.

§2. The group G_2

We begin by introducing some basic notations and facts about the structure of G_2 and its representations. More importantly, we describe the 4 families of non-tempered A-parameters of G_2.

(2.1) The group. The group G_2 can be realized as the automorphism group of the (split) octonion algebra $\mathbb{O}$. Like the quaternion algebra, the octonion algebra carries a quadratic norm form N and a linear trace form Tr and these are preserved by G_2. Let V be the space of trace zero elements in $\mathbb{O}$, equipped with the quadratic form $q = -N$. Then G_2 acts as automorphisms of (V, q), giving an embedding $G_2 \hookrightarrow SO(V, q) \cong SO_7$.

(2.2) Maximal parabolics. The group G_2 has two conjugacy classes of maximal parabolic subgroups. One of them is the Heisenberg parabolic $P = M \cdot N$, with N a 5-dimensional Heisenberg group. Denote the other maximal parabolic by $Q = L \cdot U$. Its unipotent radical U is a 3-step nilpotent group. In both cases, the Levi subgroups are isomorphic to GL_2 and we fix these isomorphisms.

(2.3) The subgroups SU_3^K. Let K be an étale quadratic algebra, which determines a quasi-split special unitary group SU_3^K. There is a natural embedding $SU_3^K \hookrightarrow G_2$, which is unique up to conjugacy. To give one such embedding, we take a trace zero element x of $\mathbb{O}$ such that the subalgebra $F[x]$ generated by x is isomorphic to K. The stabilizer of x in G_2 is then isomorphic to SU_3^K.

(2.4) Representations. Now we introduce some notations for the representations of $G_2(F_v)$. Let τ be a tempered representation of $GL_2(F_v)$ and $s > 0$. In the following, we shall use standard notions for the representations of PGL_2. For example, St denotes the Steinberg representation and $\pi(\mu_1, \mu_2)$ denotes a principal series representation. The induced representations

$$I_P(\tau, s) = Ind_P^{G_2}\delta_P^{1/2} \cdot \tau \cdot |det|^s \quad \text{and} \quad I_Q(\tau, s) = Ind_Q^{G_2}\delta_Q^{1/2} \cdot \tau \cdot |det|^s$$

have unique irreducible quotient $J_P(\tau, s)$ and $J_Q(\tau, s)$.

(2.5) Fourier coefficients. Now we come to the global setting. Let $\mathcal{A}(G_2)$ be the space of automorphic forms on G_2 and let $\mathcal{A}_2 = \mathcal{A} \cap L^2$ be the space of square-integrable automorphic forms.

For $f \in \mathcal{A}(G_2)$, we will consider its generic Fourier coefficients along the unipotent radical N of P. More precisely, if χ is a unitary character of $N(\mathbb{A})$ trivial on $N(F)$, then the χ-th Fourier coefficient of f is the function on $G_2(\mathbb{A})$ given by:

$$f_\chi(g) = \int_{N(F)\backslash N(\mathbb{A})} \overline{\chi(n)} \cdot f(ng)\, dn.$$

We let l_χ be the linear map on $\mathcal{A}(G_2)$ given by $l_\chi(f) = f_\chi(1)$.

Now the Levi factor $M(F)$ acts naturally on the set of χ's which are trivial on $N(F)$. It is known that the generic $M(F)$-orbits are naturally parametrized by étale cubic F-algebras (cf. [GGJ]). If E is an étale cubic F-algebra, we let ψ_E be a character which is in the orbit indexed by E. If $\pi \subset \mathcal{A}(G_2)$ is irreducible, we say that π has non-zero *E-Fourier coefficient* if l_{ψ_E} is non-zero on π. It is known [G, Thm. 3.1] that if π is not the trivial representation, then π has a non-zero E-coefficient for some étale E.

We say that π is *E-distinguished* if the only non-zero (generic) Fourier coefficient of π along N is the one corresponding to E.

(2.6) The stabilizer M_{ψ_E}. The stabilizer in M of ψ_E is a finite group scheme S_E which is a twisted form of the finite constant group scheme S_3. More precisely, if B is any commutative F-algebra,

$$S_E(B) = Aut_B(E \otimes_F B).$$

Another way to define S_E is given in (3.2).

If π_v is a representation of $G_2(F_v)$, then $M_{\psi_E}(F_v)$ acts naturally on $Hom_{N(F_v)}(\pi_v, \mathbb{C}_{\psi_E})$. Similarly, if π is a representation of $G_2(\mathbb{A})$, then $M_{\psi_E}(\mathbb{A})$ acts naturally on $Hom_{N(\mathbb{A})}(\pi, \mathbb{C}_{\psi_E})$.

(2.7) The A-parameters of G_2. We conclude this section by describing the A-parameters of G_2. Recall that the complex dual group of G is $G_2(\mathbb{C})$, and there are 4 conjugacy classes of non-trivial homomorphisms $SL_2(\mathbb{C}) \longrightarrow G_2(\mathbb{C})$ corresponding to the 4 non-trivial unipotent conjugacy classes in $G_2(\mathbb{C})$:

$$\mathcal{O}_{long} < \mathcal{O}_{short} < \mathcal{O}_{subreg} < \mathcal{O}_{reg}.$$

There are thus 4 families of non-tempered A-parameters for G_2. We list them below:

- $\psi|_{SL_2}$ corresponds to the regular orbit. The corresponding local and global A-packets are singletons consisting of the trivial representation of $G(F_\nu)$ and $G(\mathbb{A})$ respectively.

- $\psi|_{SL_2}$ corresponds to the subregular orbit. The corresponding packets were studied in [GGJ] and [G]. The results are described in §3.

- $\psi|_{SL_2}$ gives the short root SL_2 in G_2. The corresponding packets were constructed in [GG], and the results are described in §5.

- $\psi|_{SL_2}$ gives the long root SL_2 in G_2. This is an ongoing project, which we hope to complete in the near future.

§3. Cubic unipotent A-packets

In this section, we describe the construction of the A-packets A_ψ where $\psi|_{SL_2}$ is the subregular SL_2. The subregular map $SL_2(\mathbb{C}) \to G_2(\mathbb{C})$ is not injective; it factors through $SO_3(\mathbb{C})$. The centralizer of this $SO_3(\mathbb{C})$ in $G_2(\mathbb{C})$ is isomorphic to the symmetric group S_3 on 3 letters, so that $SO_3(\mathbb{C}) \times S_3$ is a subgroup of $G_2(\mathbb{C})$. With this information, we can now write down all subregular parameters ψ.

(3.1) Cubic unipotent parameters. Let E be an étale cubic F-algebra. Then E corresponds to a conjugacy class of maps

$$\rho_E : L_F \longrightarrow Gal(\overline{F}/F) \longrightarrow S_3.$$

Using ρ_E and the natural projection map from $SL_2(\mathbb{C})$ to $SO_3(\mathbb{C})$, we have an A-parameter

$$\psi_E : L_F \times SL_2(\mathbb{C}) \longrightarrow S_3 \times SO_3(\mathbb{C}) \hookrightarrow G_2(\mathbb{C}).$$

These maps ψ_E are the cubic unipotent A-parameters.

(3.2) The centralizer of ψ_E. Via ρ_E, the Galois group $Gal(\overline{F}/F)$ acts on S_3 by conjugation. This defines a finite group scheme S_E, which is a twisted form of the constant group scheme S_3. The group S_E was defined earlier in (2.6). It is not difficult to see that

$$S_{\psi_E} = S_E(F) \qquad \text{and} \qquad S_{\psi_{E,v}} = S_E(F_v).$$

Thus, $S_{\psi_E,\mathbb{A}} = S_E(\mathbb{A})$.

(3.3) The split case. As an example, consider the case when $E = E_0$ is the split algebra $F \times F \times F$. In this case, ρ_{E_0} is the trivial map, and so we have:

$$\begin{cases} S_{\psi_{E_0}} = S_{\psi_{E_0},v} = S_3 \\ S_{\psi_{E_0},\mathbb{A}} = S_3(\mathbb{A}). \end{cases}$$

The map $S_\psi \to S_{\psi,\mathbb{A}}$ is simply the natural embedding $S_3(F) \hookrightarrow S_3(\mathbb{A})$. Thus, for each place v, the corresponding local A packet has 3 members indexed by the irreducible characters of S_3:

$$A_{\psi_{E_0}} = \{\pi_{1_v}, \pi_{r_v}, \pi_{\epsilon_v}\}$$

where ϵ_v is the sign character of S_3 and r_v is the 2-dimensional one.

(3.4) Construction of local packets. We now explain the results of [GGJ], which give a construction of the cubic unipotent A-packets.

The étale cubic algebra E determines a simply-connected quasi-split group $Spin_8^E$ of type D_4, whose outer automorphism group is isomorphic to S_E. Let H_E be the disconnected linear algebraic group $Spin_8^E \rtimes S_E$. For each place v of F, the group $H_E(F_v)$ has a distinguished representation Π_{E_v} known as the minimal representation. For example, when $E = E_0$, $\Pi_{E_{0,v}}$ is a particular extension to $H_{E_0}(F_v)$ of the unramified representation of $Spin_8(F_v)$ whose Satake parameter is

$$\iota \begin{pmatrix} q_v^{1/2} & \\ & q_v^{-1/2} \end{pmatrix}$$

where $\iota : SL_2(\mathbb{C}) \longrightarrow PGSO_8(\mathbb{C})$ is the map associated to the subregular unipotent orbit of the dual group $PGSO_8(\mathbb{C})$.

Now the action of S_E on $Spin_8^E$ has pointwise stabilizer isomorphic to G_2. Thus H contains the subgroup $S_E \times G_2$, and one may restrict the representation Π_{E_v} to the subgroup $S_E(F_v) \times G_2(F_v)$ to get:

$$\Pi_{E_v} = \bigoplus_{\eta_v \in \widehat{S_E(F_v)}} \eta_v \otimes \pi_{\eta_v}.$$

In the papers [HMS] and [V], Huang-Magaard-Savin (for non-archimedean v) and Vogan (for archimedean v) showed the following theorem:

(3.5) Theorem *Each π_{η_v} is a non-zero irreducible unitarizable representation and the π_{η_v}'s are mutually distinct. Moreover, the representations π_{η_v} can be completely determined, and π_{1_v} is unramified with Satake parameter $s_{\psi_{E,v}}$.*

For example, when $E = E_0$ is the split algebra, then

$$\pi_{1_v} = J_Q(\pi(1,1),1), \qquad \pi_{r_v} = J_P(St, 1/2)$$

and π_{ϵ_v} is supercuspidal of depth zero if v is finite.

In view of these results, it is natural to take $A_{\psi_{E,v}}$ to be the set of representations π_{η_v} furnished by the theorem. This is our solution to Problem A.

(3.6) Construction of the subspace $L^2_{\psi_E}$. We come now to Problem B. Consider thus the global situation. The quadratic character ϵ_{ψ_E} is equal to the trivial character in this case. Thus, if our definition of the local packets is correct, we expect to construct a subspace of L^2_{disc} isomorphic to $\bigoplus_\eta \eta^{S_E(F)} \otimes \pi_\eta$.

If $\Pi_E = \otimes_v \Pi_{E_v}$, then as an abstract representation of $S_3(\mathbb{A}) \times G_2(\mathbb{A})$, we have:

$$\Pi_E = \bigoplus_\eta \eta \otimes \pi_\eta$$

as $\eta = \otimes_v \eta_v$ ranges over the irreducible representations of $S_E(\mathbb{A})$.

Using residues of Eisenstein series, one can construct a $Spin_8^E(\mathbb{A})$-equivariant embedding

$$\Theta : \Pi_E \hookrightarrow \mathcal{A}_2(Spin_8^E)$$

of Π_E into the space of square-integrable automorphic forms of $Spin_8^E$. We may now define a $G_2(\mathbb{A})$-equivariant map ι_E as follows:

$$\iota_E : \Pi_E = \oplus_\eta \eta \otimes \pi_\eta \xrightarrow{\Theta} \mathcal{A}_2(Spin_8^E) \xrightarrow{\text{restriction}} \{\text{functions on } G_2(F)\backslash G_2(\mathbb{A})\}.$$

Let us write $\eta = \eta^{S_E(F)} \oplus (\eta^{S_E(F)})^\perp$. Then it is easy to see that ι_E is the zero map when restricted to the subspace $\bigoplus_\eta (\eta^{S_E(F)})^\perp \otimes \pi_\eta$. There is thus no loss in restricting ι_E to the subspace $\bigoplus_\eta \eta^{S_E(F)} \otimes \pi_\eta$. Then the following is the main theorem proved in [GGJ]:

(3.7) Theorem *(i) The image of ι_E is contained in $\mathcal{A}_2(G_2)$.*

(ii) The restriction of ι_E to the subspace $\oplus_\eta \eta^{S_E(F)} \otimes \pi_\eta$ is injective.

Thus, the theorem provides a construction of the subspace $L^2_{\psi_E}$, which resolves Problem B and implies:

(3.8) Corollary *We have:*

$$m_{disc}(\pi_\eta) \geq m_\eta.$$

(3.9) Properties. We remark that the representations in $L^2_{\psi_E}$ are E-distinguished, and the cuspidal ones are CAP with respect to the Borel subgroup (resp. the maximal parabolic Q) if E is Galois (resp. non-Galois). Here, we say that E is non-Galois if it is a non-Galois cubic field extension of F, and is Galois otherwise. In the final section of the paper, we discuss whether these properties characterize the A-packets A_{ψ_E}.

(3.10) Some striking consequences. We briefly highlight two striking consequences of the theorem. Suppose for simplicity that $E = E_0$ is the split algebra, so that $S_{\psi_E} = S_{\psi_{E,v}} = S_3$. For S a non-empty finite set of places of F, let

$$\eta_S = (\otimes_{v \in S} r_v) \bigotimes (\otimes_{v \notin S} \mathbf{1}_v).$$

Then the representation

$$\pi_{\eta_S} = (\otimes_{v \in S} \pi_{r_v}) \bigotimes (\otimes_{v \notin S} \pi_{1_v})$$

occurs in $L^2_{\psi_{E_0}}$ with multiplicity equal to the multiplicity of the trivial representation in $r \otimes r \otimes .... \otimes r$ ($\#S$ times). A quick computation gives:

$$m_{disc}(\pi_{\eta_S}) \geq \frac{1}{6} \cdot (2^{\#S} + (-1)^{\#S} 2),$$

On the other hand, the residual spectrum of G_2 has been determined by H. Kim [K] and S. Zampera [Z]. Their results show that L^2_{res} has the multiplicity one property, and further that $m_{res}(\pi_{\eta_S}) = 1$ if $\#S \neq 1$. Thus, we have:

(3.11) Theorem *(i) If $\#S \geq 3$, the representation π_{η_S} occurs in both the residual spectrum and the cuspidal spectrum.*

(ii) As $\#S \to \infty$, the cuspidal and discrete multiplicities of the π_{η_S}'s are unbounded.

(3.12) Exact multiplicity formula. Theorem 3.7 and its Corollary 3.8 already provide some justification for our definition of local packets. However, when $m_\eta = 0$, the inequality in Corollary 3.8 does not rule out the occurrence of π_η in the discrete spectrum. In [G], this inequality is strengthened to an equality, thus resolving Problem C for the cubic unipotent A-packets:

(3.13) Theorem *Fix the étale cubic algebra E, and let η be an irreducible character of $S_E(\mathbb{A})$. Then*

$$m_{disc}(\pi_\eta) = m_\eta.$$

The proof of this theorem is reminiscent of the proof of the multiplicity-one theorem for cusp forms on GL_n. It is a consequence of the following two results:

- Let ψ_E be a character of $N(\mathbb{A})$ trivial on $N(F)$ which lies in the $M(F)$-orbit indexed by E, and let l_{ψ_E} be as defined in (2.5). Then the assignment $\iota \mapsto l_{\psi_E} \circ \iota$ defines an *injective* map

$$\mathrm{Hom}_{G_2(\mathbb{A})}(\pi_\eta, \mathcal{A}(G_2)) \longrightarrow \mathrm{Hom}_{N(\mathbb{A})}(\pi_\eta, \mathbb{C}_{\psi_E})^{M_{\psi_E}(F)}.$$

- For each place v, we have an isomorphism of $M_{\psi_E}(F_v) \cong S_E(F_v)$-modules:

$$Hom_{N(F_v)}(\pi_{\eta_v}, \mathbb{C}_{\psi_E}) \cong \chi_{K_{E_v}} \cdot \eta_v^\vee,$$

  where $\chi_{K_{E_v}}$ is the quadratic character associated to the discriminant (quadratic) algebra K_{E_v} of E_v. Note that $\chi_{K_E} = \prod_v \chi_{K_{E_v}}$ is trivial when restricted to $M_{\psi_E}(F) = S_E(F)$.

We shall postpone the discussion of Problem D to §6.

§4. Tempered representations with unbounded multiplicities

The representations π_{η_S} of Theorem 3.11 are very degenerate: their local components are non-tempered and non-generic, and are the so-called unipotent representations. This may lead one to think that the phenomenon of unbounded cuspidal multiplicities only happens for very degenerate representations. However, as we explain in this section, it should already occur for representations in some tempered L-packets. We shall discuss how we intend to construct these tempered representations of arbitrarily high cuspidal multiplicities.

(4.1) Some tempered parameters. We begin by considering some tempered A-parameters, i.e. those for which ψ is trivial on $SL_2(\mathbb{C})$. Let us start with a cuspidal representation τ of PGL_2 such that

$$\tau_v = \begin{cases} \text{Steinberg representation for } v \in S_\tau; \\ \text{an unramified representation for } v \notin S_\tau \end{cases}$$

for some non-empty finite set S_τ of finite places of F. Conjecturally, τ corresponds to a map $\phi_\tau : L_F \longrightarrow SL_2(\mathbb{C})$. Because of our assumptions, the map ϕ_τ is surjective; in fact, for $v \in S_\tau$, the local parameter ϕ_{τ_v} is already surjective, since it corresponds to the Steinberg representation.

Now we construct an A-parameter for G_2 using ϕ_τ as follows:

$$\psi_\tau : L_F \longrightarrow SL_2(\mathbb{C}) \rightarrow SO_3(\mathbb{C}) \hookrightarrow G_2(\mathbb{C}).$$

Then we have:

$$\begin{cases} S_{\psi_\tau} = S_{\psi_\tau,v} = S_3 & \text{for all } v \in S_\tau. \\ S_{\psi_\tau,v} = \{1\} & \text{for all } v \notin S_\tau. \end{cases}$$

In particular, Arthur's conjecture predicts that the local packets have the following form:

$$A_{\psi_\tau,v} = \begin{cases} \{\pi'_{1_v}, \pi'_{r_v}, \pi'_{\epsilon_v}\} & \text{if } v \in S_\tau; \\ \{\pi'_{1_v}\} & \text{if } v \notin S_\tau. \end{cases}$$

Moreover, the representations in the local packets should be tempered.

In fact, the parameter ψ_τ is an example of Langlands parameter considered by Lusztig. Hence, in this case, the local packet $A_{\psi_\tau,v}$ has already been defined, and it does consist of 3 discrete series representations (see [GrS]) when $v \in S_\tau$. Indeed, for $v \in S_\tau$, $A_{\psi_\tau,v}$ is related to the cubic unipotent packet $A_{\psi_{E_0},v}$ by the Aubert-Iwahori-Matsumoto involution.

Finally, if we set

$$\pi_\tau = \left(\otimes_{v\in S_\tau} \pi'_{r_v}\right) \bigotimes \left(\otimes_{v\notin S_\tau} \pi'_{1_v}\right),$$

then Arthur's conjecture implies that

$$m_{disc}(\pi_\tau) \geq \frac{1}{6} \cdot (2^{\#S_\tau} + (-1)^{\#S_\tau} 2).$$

In fact, since the representation π'_τ is tempered, it cannot occur in the residual spectrum, and so we should have:

$$m_{cusp}(\pi_\tau) \geq \frac{1}{6} \cdot (2^{\#S_\tau} + (-1)^{\#S_\tau} 2).$$

Now one can find cuspidal representations τ of PGL_2 of the above type and with S_τ as big as one wishes (using the trace formula for example). Hence, Arthur's conjecture predicts that one can find a family of tempered representations of $G_2(\mathbb{A})$ whose cuspidal multiplicities are unbounded.

(4.2) Potential construction. Finally, we would like to explain how we expect to demonstrate the lower bound on $m_{cusp}(\pi_\tau)$ predicted by Arthur's conjecture.

The parameter

$$\psi_\tau : L_F \longrightarrow SO_3(\mathbb{C}) \hookrightarrow G_2(\mathbb{C})$$

actually factors as:

$$\psi_\tau : L_F \longrightarrow SO_3(\mathbb{C}) \hookrightarrow SL_3(\mathbb{C}) \hookrightarrow G_2(\mathbb{C}).$$

Hence, instead of lifting the cuspidal representation τ of PGL_2 directly to G_2, one may first lift it to a cuspidal representation of PGL_3. This is precisely the Gelbart-Jacquet lift, and we denote this cuspidal representation of PGL_3 by $GJ(\tau)$. Note that

$$GJ(\tau)_v = \begin{cases} \text{the Steinberg representation } St_v \text{ if } v \in S_\tau; \\ \text{a specific unramified representation if } v \notin S_\tau. \end{cases}$$

Now it turns out that $PGL_3 \times G_2$ is a dual pair in the split (adjoint) exceptional group of type E_6. This suggests that we may use exceptional theta correspondence to lift $GJ(\tau)$ from PGL_3 to G_2: hopefully we will get the representation π_τ. For this to work out, one should first verify that under local theta correspondence, the Steinberg representation St_v of $PGL_3(F_v)$ lifts to the representation π'_τ of $G_2(F_v)$. However, it was shown in [GS] that the theta lift of St_v is equal to $\pi'_1 \oplus \pi'_\epsilon$. So this doesn't work out as planned.

Thankfully, a homomorphism $L_F \longrightarrow SL_3(\mathbb{C})$ is not just a Langlands parameter for PGL_3; it is also a parameter for any inner form of PGL_3. Such an inner form is of the form $PD^\times$ where D is a degree 3 division algebra. Over a p-adic field F_v, there are two such division algebras: D_v and its opposite D_v^{opp}. Being opposite algebras, their groups of invertible elements define isomorphic algebraic groups. Thus, locally, PGL_3 has precisely one inner form $PD^\times$.

Now under the local Jacquet-Langlands correspondence, the Steinberg representation St_v corresponds to the trivial representation $\mathbf{1_v}$ of $PD^\times(F_v)$. Moreover, $PD^\times \times G_2$ is a dual pair in an inner form of E_6. It was shown in [S] that the local theta lift of $\mathbf{1_v}$ is indeed equal to π'_τ.

Hence we are led to the following strategy for embedding π_τ into L^2_{cusp}. Choose a global division algebra D of degree 3 which is ramified precisely at the set S_τ. Then one lifts τ from PGL_2 to G_2 as follows:

$$\begin{array}{ccccccc} PGL_2 & \xrightarrow{\text{Gelbart-Jacquet}} & PGL_3 & \xrightarrow{\text{Jacquet-Langlands}} & PD^\times & \xrightarrow{\text{theta lift}} & G_2 \\ \tau & \longrightarrow & GJ(\tau) & \longrightarrow & JL_D(GJ(\tau)) & \longrightarrow & \Theta(JL_D(GJ(\tau))). \end{array}$$

As an abstract representation, $\Theta(JL_D(GJ(\tau)))$ is either zero or is isomorphic to π_τ.

How does the multiplicity $\frac{1}{6} \cdot (2^{\#S_\tau} + (-1)^{\#S_\tau} 2)$ arise in this case? The answer lies in the following lemma:

(4.3) Lemma *The number of global division algebras of degree 3 ramified precisely at a set S is equal to*

$$\frac{1}{3} \cdot (2^{\#S} + (-1)^{\#S} 2).$$

In particular, the number of inner forms of PGL_3 which are ramified at the set S is half of the above number.

Note that the various inner forms of the lemma are non-isomorphic as algebraic groups, but their groups of adelic points are abstractly isomorphic. Thus the reason for the high multiplicity here is the failure of Hasse principle for the inner forms of PGL_3!

In order for the above strategy to work, it remains to resolve the following problems:

PROBLEMS:

- Show the non-vanishing of the theta lift $\Theta(JL_D(GJ(\tau)))$;
- Show that the various $\Theta(JL_D(GJ(\tau)))$'s generate linearly independent copies of π_τ in L^2_{cusp}.

At the moment, these issues are unresolved.

§5. Short root A-packets

In this section, we shall consider the short root A-packets, which have been constructed in [GG]. We first give a description of the short root A-parameters.

(5.1) Short root A-parameters. Observe that

$$SL_{2,l} \times_{\mu_2} SL_{2,s} \subset G_2$$

where $(SL_{2,l}, SL_{2,s})$ is a pair of commuting SL_2's corresponding to a pair of mutually orthogonal long and short roots. Indeed, the centralizer of one of these SL_2's is the other SL_2.

Now suppose that τ is a cuspidal representation of PGL_2. Conjecturally, τ corresponds to an irreducible representation

$$\rho_\tau : L_F \longrightarrow SL_2(\mathbb{C}).$$

We define an Arthur parameter by

$$\psi_\tau : L_F \times SL_2(\mathbb{C}) \xrightarrow{\rho_\tau \times id} SL_{2,l}(\mathbb{C}) \times SL_{2,s}(\mathbb{C}) \xrightarrow{\quad i \quad} G_2(\mathbb{C}).$$

Thus, the short root A-parameters are indexed by cuspidal representations of PGL_2. In particular, it makes sense to talk about them even though we don't know what L_F is.

The centralizer S_{ψ_τ} of ψ_τ is isomorphic to $\mathbb{Z}/2\mathbb{Z}$. Moreover, the local component groups $S_{\psi_{\tau,v}}$ are given by

$$S_{\psi_\tau,v} = \begin{cases} 1, & \text{if } \rho_{\tau,v} \text{ is reducible} \\ \mathbb{Z}/2\mathbb{Z}, & \text{if } \rho_{\tau,v} \text{ is irreducible.} \end{cases}$$

Note the the condition $\rho_{\tau,v}$ is irreducible is equivalent to τ_v being a discrete series representation of $PGL_2(F_v)$.

(5.2) Local Arthur packets. Now Arthur's conjecture predicts that for each place v, the local A-packets $A_{\tau,v}$ has the form:

$$A_{\tau,v} = \begin{cases} \{\pi_{\tau_v}^+\}, & \text{if } \tau_v \text{ is not discrete series} \\ \{\pi_{\tau_v}^+, \pi_{\tau_v}^-\}, & \text{if } \tau_v \text{ is discrete series.} \end{cases}$$

Here, $\pi_{\tau_v}^+$ is indexed by the trivial character of $S_{\tau,v}$.

Of course, we know what π_v^+ has to be for almost all v. Hence, our task is to define the remaining elements of $A_{\tau,v}$. In fact, it is not difficult to guess what π_v^+ is in general. Indeed, for almost all v where τ_v is unramified, π_v^+ is the spherical representation with Satake parameter $s_{\psi_{\tau,v}}$, and this representation can be alternatively expressed as $J_P(\tau_v, 1/2)$. Thus it is natural to guess that

$$\pi_v^+ \cong J_P(\tau_v, 1/2) \quad \text{for all } v.$$

As we explain later, this is essentially, but not completely, correct.

(5.3) Global A-packets. Let S_τ be the set of places v where τ_v is discrete series, so that the global A-packet has $2^{\# S_\tau}$ elements. To describe the multiplicity of $\pi_\eta \in A_\tau$ in $L^2_{\psi_\tau}$, we need to know the quadratic character ϵ_{ψ_τ} of S_{ψ_τ}. It turns out that ϵ_{ψ_τ} is the non-trivial character of $S_{\psi_\tau} \cong \mathbb{Z}/2\mathbb{Z}$ if and only if $\epsilon(\tau, 1/2) = -1$.

Now if $\pi = \otimes_v \pi_v^{\epsilon_v} \in A_\tau$, then the multiplicity associated to π by Arthur's conjecture is:

$$m(\pi) = \begin{cases} 1, & \text{if } \epsilon_\pi := \prod_v \epsilon_v = \epsilon(\tau, 1/2); \\ 0, & \text{if } \epsilon_\pi = -\epsilon(\tau, 1/2). \end{cases}$$

Thus, we expect to construct a subspace $L^2_{\psi_\tau}$ of L^2_{disc} which is isomorphic to

$$\bigoplus_{\pi \in A_\tau : \epsilon_\pi = \epsilon(\tau,1/2)} \pi.$$

The structure of these A-packets should remind the reader of the Saito-Kurokawa packets for $PGSp_4$ [PS], and more importantly, of the results of Waldspurger on the discrete spectrum of the metaplectic group $\widetilde{SL}_2$. We review the relevant facts below, since they provide a clue for the construction of the short root A-packets.

(5.4) The Weil representations of $\widetilde{SL}_2(F_v)$. Fix a non-trivial unitary character ψ_v of F_v. Then associated to a quadratic character χ_v of $F_v^\times$ (possibly trivial) is a Weil representation ω_{χ_v} of the metaplectic group $\widetilde{SL}_2(F_v)$. The representation can be realized on the space $S(F_v)$ of Schwarz functions on F_v. As a representation of $\widetilde{SL}_2(F_v)$, ω_{χ_v} is reducible. In fact, it is the direct sum of two irreducible representations:

$$\omega_{\chi_v} = \omega^+_{\chi_v} \oplus \omega^-_{\chi_v},$$

where $\omega^+_{\chi_v}$ (resp. $\omega^-_{\chi_v}$) consists of the even (resp. odd) functions in $S(F_v)$. If v is a finite place, then ω^-_χ is supercuspidal and ω^+_χ is not.

(5.5) Waldspurger's lift and packets for $\widetilde{SL}_2(F_v)$. In [W1] and [W2], Waldspurger defined a surjective map Wd_{ψ_v} from the set of irreducible (genuine) representations of $\widetilde{SL}_2(F_v)$ which are not equal to $\omega^+_{\chi_v}$ for any χ_v to the set of infinite dimensional representations of $PGL_2(F_v)$. This leads to a partition of the set of such representations of $\widetilde{SL}_2(F_v)$, indexed by the infinite dimensional representations of $PGL_2(F_v)$. Namely, if τ_v is such a representation of $PGL_2(F_v)$, we set

$$\tilde{A}_{\tau_v} = \text{ inverse image of } \tau_v \text{ under } Wd_{\psi_v}.$$

It turns out that

$$\#\tilde{A}_{\tau_v} = \begin{cases} 2 \text{ if } \tau_v \text{ is discrete series;} \\ 1 \text{ if } \tau_v \text{ is not.} \end{cases}$$

In the first case, the set $\tilde{A}_{\tau_v}$ has a distinguished element $\sigma^+_{\tau_v}$, which is characterized by the fact that $\sigma^+_{\tau_v} \otimes \tau_v$ is a quotient of the Weil representation $\omega^{(3)}_{\psi_v}$ of $\widetilde{SL}_2(F_v) \times PGL_2(F_v)$. The other element of $\tilde{A}_{\tau_v}$ will be denoted by $\sigma^-_{\tau_v}$. In the second case, we shall let $\sigma(\tau_v)^+$ be the unique element in $\tilde{A}_{\tau_v}$ and set $\sigma(\tau_v)^- = 0$. Observe that these "packets" of $\tilde{SL}_2(F_v)$ has the same structure as the local A-packets associated to short root A-parameters. This similarity extends to the global situation as well, as we now explain.

(5.6) Cusp forms of $\widetilde{SL}_2(\mathbb{A})$. Let $\widetilde{SL}_2(\mathbb{A})$ be the two-fold cover of $SL_2(\mathbb{A})$, and fix a non-trivial unitary character $\psi = \prod_v \psi_v$ of $F\backslash\mathbb{A}$. Let $\tilde{\mathcal{A}}_2$ denote the space of square-integrable genuine automorphic forms on $\widetilde{SL}_2(\mathbb{A})$. Then there is an orthogonal decomposition

$$\tilde{\mathcal{A}}_2 = \tilde{\mathcal{A}}_{00} \bigoplus \left(\bigoplus_{\chi} \tilde{\mathcal{A}}_{\chi} \right).$$

Here, χ runs over all quadratic characters (possibly trivial) of $F^\times\backslash\mathbb{A}^\times$.

Let us describe the space $\tilde{\mathcal{A}}_\chi$ more concretely. If $\omega_\chi = \otimes_v \omega_{\chi_v}$ is the global Weil representation attached to χ, then the formation of theta series gives a map $\theta_\chi : \omega_\chi \longrightarrow \tilde{\mathcal{A}}_2$, whose image is the space $\tilde{\mathcal{A}}_\chi$. To describe the decomposition of $\tilde{\mathcal{A}}_\chi$, for a finite set S of places of F, let us set

$$\omega_{\chi,S} = (\otimes_{v\in S}\omega_{\chi_v}^-) \otimes (\otimes_{v\notin S}\omega_{\chi_v}^+)$$

so that

$$\omega_\chi = \bigoplus_S \omega_{\chi,S}.$$

Then we have

$$\tilde{\mathcal{A}}_\chi \cong \bigoplus_{\#S \text{ even}} \omega_{\chi,S}.$$

Moreover, $\omega_{\chi,S}$ is cuspidal if and only if S is non-empty.

(5.7) Near equivalence classes. In the beautiful paper [W2], Waldspurger has described the near equivalence classes of representations in $\tilde{\mathcal{A}}_{00}$. Earlier, in [W1], he has shown that $\tilde{\mathcal{A}}_{00}$ satisfies multiplicity onc. Let us describe his results.

Given a cuspidal automorphic representation $\tau = \otimes_v \tau_v$ of PGL_2, we define a set of irreducible unitary representations of $\widetilde{SL}_2(\mathbb{A})$ as follows. Recall that for each place v, we have a local "packet"

$$\tilde{A}_{\tau_v} = \{\sigma_{\tau_v}^+, \sigma_{\tau_v}^-\}$$

where $\sigma_{\tau_v}^- = 0$ if τ_v is not discrete series. Now set

$$\tilde{A}_\tau = \{\sigma = \otimes_v \sigma_{\tau_v}^{\epsilon_v} : \epsilon_v = \pm \text{ and } \epsilon_v = + \text{ for almost all } v\}.$$

This is the global "packet" of $\widetilde{SL}_2(\mathbb{A})$ associated to τ.

For $\sigma = \otimes_v \sigma_{\tau_v}^{\epsilon_v} \in \tilde{A}_\tau$, let us set

$$\epsilon_\sigma = \prod_v \epsilon_v.$$

Then we have:

$$\tilde{\mathcal{A}}_{00} = \bigoplus_{\text{cuspidal } \tau} \tilde{\mathcal{A}}(\tau)$$

where each $\tilde{\mathcal{A}}(\tau)$ is a near equivalence class of cuspidal representations and is given by:

$$\tilde{\mathcal{A}}(\tau) = \bigoplus_{\sigma \in \tilde{A}_\tau : \epsilon_\sigma = \epsilon(\tau, 1/2)} \sigma.$$

Again, notice the formal similarity between $\tilde{\mathcal{A}}(\tau)$ and $L^2_{\psi_\tau}$.

(5.8) Rallis-Schiffmann lifting. In view of the above results of Waldspurger on the discrete spectrum for $\widetilde{SL}_2(\mathbb{A})$, a reasonable strategy for the construction of the short root A-packets is construct a lifting of the "packets" of $\tilde{SL}_2$ to G_2. That such a lifting is possible was discovered by Rallis and Schiffmann.

In the paper [RS], Rallis and Schiffmann constructed a lifting of cuspidal automorphic forms from $\widetilde{SL}_2$ to G_2. This was achieved by exploiting the fact that $SL_2 \times G_2$ is a subgroup of $SL_2 \times SO(V, q) \cong SL_2 \times SO_7$, which is the classical dual pair in Sp_{14}. The lifting is then defined using the theta kernel furnished by the Weil representation $\omega_\psi^{(7)}$ of $\widetilde{Sp}_{14}$ (associated to ψ and the quadratic space (V, q)).

The surprising discovery of Rallis-Schiffmann is that, despite restricting from SO_7 to the smaller group G_2, one still obtains a correspondence of representations. More precisely, if σ is an irreducible cuspidal representation of $\widetilde{SL}_2(\mathbb{A})$, let $V(\sigma)$ be the theta lift of σ; it is the (non-zero) subspace of $\mathcal{A}(G_2)$ generated by functions:

$$\theta(f\,\varphi)(g) = \int_{SL_2(F)\backslash \widetilde{SL}_2(\mathbb{A})} \theta(\varphi)(gh) \cdot \overline{f(h)}\, dh$$

where $f \in \sigma$, $\varphi \in \omega_\psi^{(7)}$ and $\theta : \omega_\psi^{(7)} \longrightarrow \mathcal{A}_2(\widetilde{Sp}_{14})$ is given by the formation of theta series. Now the main results of Rallis-Schiffmann are:

(5.9) Theorem *(i) $V(\sigma)$ is contained in the space of cusp forms if and only if the theta lift (associated to ψ) of σ to SO_3 (studied by Waldspurger) is zero.*

(ii) The cuspidal representations of G_2 which are not orthogonal to the lifts from $\widetilde{SL}_2$ are characterized as those having a non-zero period with respect to some quasi-split SU_3^K.

(iii) The local correspondence of unramified *representations is precisely determined. In particular, if $\sigma \subset \widetilde{A}(\tau)$, then for each irreducible constituent π of $V(\sigma)$, π_v has Satake parameter $s_{\psi_{\tau,v}}$ for almost all v.*

(iv) The cuspidal representations in $V(\sigma)$ are CAP with respect to P if $\sigma \subset \widetilde{A}_{00}$, and are CAP with respect to the Borel subgroup if $\sigma \subset \widetilde{A}_\chi$ for some χ. This gives the first construction of CAP representations of G_2.

The results of [RS], especially part (ii) of the theorem, certainly suggest that the short root packets can be constructed by the Rallis-Schiffmann lifting. We should also mention that the Rallis-Schiffmann lifting was further investigated by Li-Schwermer [LS].

(5.10) The results of [GG]. In [GG], we completed the study initiated in [RS] by giving a complete analysis of the theta correspondence from $\widetilde{SL}_2$ to G_2, in the style of Waldspurger's analysis of the Shimura correspondence. In particular, we gave a precise determination of the representation $V(\sigma)$.

The first step in this is the complete determination of the *local* theta correspondence. More precisely, if v is a p-adic place of F and σ_v an irreducible representation of $\widetilde{SL}_2(F_v)$, the maximal σ_v-isotypic quotient of $\omega_{\psi_v}^{(7)}$ can be expressed as $\sigma_v \otimes \theta(\sigma_v)$, where $\theta(\sigma_v)$ is a smooth representation of $G_2(F_v)$. Let $\Theta(\sigma_v)$ be the maximal semisimple quotient of $\theta(\sigma_v)$. When v is archimedean, there is an analogous (but slightly different) definition of $\Theta(\sigma_v)$. The main local result of [GG] is:

(5.11) Theorem *$\Theta(\sigma_v)$ can be completely determined for any σ_v (to the extent that classification of representations of $G_2(F_v)$ is known). In particular, $\Theta(\sigma_v)$ is irreducible except when $\sigma_v = \omega_{\psi_v}^{\pm}$ (the even and odd Weil representations of $\tilde{S}L_2(F_v)$ associated to ψ_v). In these two exceptional cases, we have:*

$$\begin{cases} \Theta(\omega_{\psi_v}^{+}) = \pi_{1_v} \oplus \pi_{r_v} \\ \Theta(\omega_{\psi_v}^{-}) = \pi_{r_v} \oplus \pi_{\epsilon_v}. \end{cases}$$

Here, $\{\pi_{1_v}, \pi_{r_v}, \pi_{\epsilon_v}\}$ is the cubic unipotent packet attached to the split cubic algebra E_0.

We should note that for a real place v, the lifting of most discrete series representations was already determined by Li-Schwermer [LS].

(5.12) Global results. We now turn to the global situation. For any cuspidal σ on $\widetilde{SL}_2$, we have defined the theta lift $V(\sigma)$. In fact, one can define the theta lift of any square-integrable automorphic representation of $\widetilde{SL}_2$ by a regularization of the theta integral. Thus, one can speak of the regularized theta lift $V(\sigma)$ of any $\sigma \subset \widetilde{A}_2$.

One can show that $V(\sigma)$ is contained in the space of square-integrable automorphic forms. Thus $V(\sigma)$ is semisimple and is a non-zero summand of $\Theta(\sigma) := \otimes_v \Theta(\sigma_v)$. Theorem 5.11 immediately shows that $V(\sigma) \cong \Theta(\sigma)$, whenever $\sigma_v \neq \omega^{\pm}_{\psi_v}$ for any place v, since $\Theta(\sigma)$ is irreducible in that case. However, when $\Theta(\sigma)$ is reducible, there are more than one possibilities for $V(\sigma)$. The determination of $V(\sigma)$ in this case is easily the trickiest part of [GG]. The result is:

(5.13) Theorem *(i) If $\sigma \subset \mathcal{A}_{00}$, then $V(\sigma) \cong \Theta(\sigma)$.*

(ii) Suppose that K is an étale quadratic F-algebra with associated quadratic character χ_K. Then (the L^2-completion of) the regularized theta lift of $\tilde{\mathcal{A}}_{\chi_K}$ is equal to the space $L^2_{\psi_{F\times K}}$.

Thus, part (ii) of the theorem gives us an alternative construction of the subspaces of L^2_{disc} associated to the cubic unipotent A-packets corresponding to those cubic algebras E which are not fields.

(5.14) Definition of local Arthur packets. Now we can use Theorem 5.11 to give a natural candidate for the packet $A_{\tau,v}$. Recall that τ_v determines a set $\tilde{A}_{\tau_v}$ of representations of $\widetilde{SL}_2(F_v)$, which has 2 or 1 elements $\sigma^{\pm}_{\tau_v}$, depending on whether τ_v is discrete series or not. We set

$$\pi^{\pm}_{\tau_v} = \Theta(\sigma^{\pm}_{\tau_v}).$$

This defines the packet $A_{\tau,v}$ and is our solution to Problem A.

(5.15) A special case. We would like to highlight the case when τ_v is the Steinberg representation St. In this case,

$$\sigma^{+}_{v} = \omega^{-}_{\psi_v}.$$

Thus, according to our definition,

$$\pi^{+}_{\tau_v} = \Theta(\omega^{-}_{\psi_v}) = \pi_{\tau_v} + \pi_{\epsilon_v} = J_P(St, 1/2) + \pi_\epsilon.$$

In particular, $\pi^{+}_{\tau_v}$ is reducible. For the case of split p-adic groups, this is the first instance we know in which the representation in a packet can be reducible, and this is quite surprising at first sight. As we mentioned before, the initial guess would be to take $\pi^{+}_{\tau_v}$ simply as $J_P(St, 1/2)$. For future reference, we set

$$\pi^{++}_{\tau_v} = \begin{cases} J_P(St, 1/2), & \text{if } \tau_v = St; \\ \pi^{+}_{\tau_v}, & \text{otherwise.} \end{cases}$$

Thus $\pi^{++}_{\tau_v}$ is a certain distinguished constituent of $\pi^{+}_{\tau_v}$.

(5.16) The subspace $L^2_{\psi_\tau}$. Why is the above definition of local packets reasonable? For one thing, when τ_v is unramified, $\Theta(\sigma^+_{\tau_v})$ is indeed irreducible and unramified with the required Satake parameter $s_{\psi_\tau,v}$. Secondly, Theorem 5.13(i) says that for a given τ, the global theta correspondence constructs a subspace of L^2_{disc} isomorphic to

$$\bigoplus_{\sigma\in\tilde{A}_\tau:\epsilon_\sigma=\epsilon(\tau,1/2)} \Theta(\sigma).$$

With our definition of the local packets, this gives the desired subspace $L^2_{\psi_\tau}$, thus resolving Problem B. This provides compelling global justification for taking $\pi^+_{\tau_v}$ to be reducible when τ_v is Steinberg.

(5.17) SL_3-distinguished cusp forms. Let us highlight a corollary of Theorem 5.13. It pertains to the question of whether there are cuspidal representations of G_2 with non-zero SL_3-period. Such cuspidal representations should be very scarce. Examples are the cuspidal representations belonging to the cubic unipotent A-packet associated to the split algebra E_0.

It wasn't known previously if other SL_3-distinguished cuspidal representations exist. Now, as a consequence of our global theorem, we know that they do. Indeed, when $\sigma \subset \mathcal{A}_{00}$ is such that its theta lift to SO_3 is non-zero, we know from [RS] that $V(\sigma)$ is not totally contained in $\mathcal{A}_{cusp}(G_2)$. However, this does not exclude the possibility that $V(\sigma)\cap\mathcal{A}_{cusp}(G_2)$ is non-zero. In [RS, Pg. 823], Rallis-Schiffmann remarked that they do not know if such a possibility can actually happen. Theorem 5.13 implies that it does. Indeed, this occurs when σ_v is the odd Weil representation $\omega^-_{\psi_v}$ for some finite place v, because $V(\sigma) \cong \Theta(\sigma)$ and $\Theta(\sigma_v)$ contains the supercuspidal representation π_{ϵ_v} for such a place v. For such a σ, any irreducible constituent of $V(\sigma) \cap \mathcal{A}_{cusp}(G_2)$ will have a non-zero SL_3-period.

§6. Rankin-Selberg integral and near equivalence classes

In this section, we consider Problem D for the cubic unipotent and short root A-packets that we defined in the earlier sections. Recall that what we want is the following analog of the strong multiplicity one theorem:

- If π is an irreducible constituent of L^2_{disc} and π is nearly equivalent to the representations in an A-packet A_ψ, then π is contained in L^2_ψ. In other words, L^2_ψ is a full near equivalence class.

Let's consider the case of the short root A-packets. Hence, suppose that we have an irreducible constituent π of L^2_{disc} which is nearly equivalent to the representations in A_{ψ_τ}, for some cuspidal representation τ of PGL_2. Assume moreover that $\pi \subset \mathcal{A}_{cusp}(G_2)$. In view of Theorem 5.9, if we can show that π has a SU_3^K-period for some quadratic K, we will at least have shown that π is not orthogonal to $L^2_{\psi_\tau}$, and thus π is *isomorphic* to some irreducible constituent of $L^2_{\psi_\tau}$.

Because the cubic unipotent packets associated to $E = F \times K$ can also be constructed by the Rallis-Schiffmann lifting, the same consideration applies to those cuspidal π which are nearly equivalent to representations in $L^2_{\psi_{F\times K}}$.

(6.1) A Rankin-Selberg integral. The discussion above leads us to consider a certain Rankin-Selberg integral. More precisely, let B_K be a Borel subgroup of SU_3^K and consider the family of principal series representations

$$I_K(s) = Ind_{B_K(\mathbb{A})}^{SU_3^K(\mathbb{A})} \delta_{B_K}^s$$

For a standard section $f_s \in I_K(s)$, one has the Eisenstein series $E(g, s, g)$ which is defined for $Re(s) >> 0$ by

$$E(f, s, g) = \sum_{\gamma \in B_K(F) \backslash SU_3^K(F)} f_s(\gamma g)$$

and has a meromorphic continuation to $\mathbb{C}$. At $s = 1$, $E(f, s, g)$ has a pole of order at most 2 (resp. 1) if K is split (resp. non-split). Moreover, this highest possible order of pole is attained for some f, and the leading terms of the Laurent expansion at $s = 1$ of $E(f, s, g)$ (as f varies) span the space of constant functions in L^2_{disc}.

Thus we are led to consider the following integral

$$J_K(f, \varphi, s) = \int_{SU_3^K(F) \backslash SU_3^K(\mathbb{A})} \varphi(g) \cdot E(f, s, g)\, dg$$

where $\varphi \in \pi$. This defines a meromorphic function in s. In [GG, §15], the analytic properties of this Rankin-Selberg integral were studied in some detail. The main result is:

(6.2) Proposition *(i) Suppose that π is nearly equivalent to the representations in $L^2_{\psi_{F\times K}}$. Then*

$$J_K(f, \varphi, s) = \zeta_K^S(2s - 1) \cdot L^S(\chi_K, 4s - 2) \cdot d_S(f, \varphi, s)$$

where S is a large finite set of places of F containing the archimedean ones, and $d_S(f, \varphi, s)$ is a meromorphic function (the bad factor). We can choose f and φ so that $d_S(f, \varphi, s)$ is non-zero at $s = 1$.

(ii) Suppose that π is nearly equivalent to the representations in $L^2_{\psi_\tau}$ and π has non-zero $(F \times K)$-Fourier coefficient with K a quadratic field. Then

$$J_K(f,\varphi,s) = \frac{\zeta_K^S(2s-1)}{L^S(\chi_K,4s-1)} \cdot L(\tau,4s-3/2)\cdot d_S(f,\varphi,s)$$

where S is a large finite set of places of F containing the archimedean ones, and $d_S(f,\varphi,s)$ is a meromorphic function (the bad factor). We can choose f and φ so that $d_S(f,\varphi,s)$ is non-zero at $s=1$.

As a consequence of the proposition, we have the following theorem, which resolves Problem D for some cubic unipotent A-packets.

(6.3) Theorem *Suppose that the étale cubic algebra $E = F \times K$ is not a field. Then $L^2_{\psi_E}$ is a full near equivalence class.*

PROOF. After the exact multiplicity formula of Thm. 3.13, it remains to show that if $\pi \subset L^2_{disc}$ is nearly equivalent to the representations in $L^2_{\psi_E}$, then π is isomorphic to an irreducible constituent of $L^2_{\psi_E}$.

Suppose first that π has non-zero projection $P_{res}(\pi)$ to the residual spectrum. Now the residual spectrum has been determined completely by Kim [K] and Zampera [Z]. From their results, one sees that $P_{res}(\pi)$ is contained in $P_{res}(L^2_{\psi_E})$. Hence, we may assume that $\pi \subset \mathcal{A}_{cusp}(G_2)$. Then Prop. 6.2(i) shows that π has non-zero SU_3^K-period. Thus by Thm. 5.9(ii), π is not orthogonal to the theta lift of some cuspidal σ of $\widetilde{SL}_2$. The local result Thm. 5.11 implies that such a σ must be contained in $\widetilde{A}_{\chi_K}$. Thus, π is not orthogonal to $L^2_{\psi_E}$, as desired. □

To apply Prop. 6.2 to the short root A-packet A_{ψ_τ}, we note:

(6.4) Proposition *If $\pi \subset \mathcal{A}_{cusp}(G_2)$ is nearly equivalent to the representations in A_{ψ_τ}, then π has non-zero $(F \times K)$-coefficient for some quadratic field K.*

The proof of this proposition is decidedly roundabout; we refer the reader to [GG, §16] for the details. In any case, we obtain:

(6.5) Theorem *Suppose that $\pi \subset \mathcal{A}_{cusp}(G_2)$ is nearly equivalent to the representations in $L^2_{\psi_\tau}$. Then*

(i) π is not orthogonal to $L^2_{\psi_\tau}$. In particular, π is isomorphic to a representation in $L^2_{\psi_\tau}$.

(ii) the projection of $L^2_{\psi_\tau}$ onto $L^2_{cusp}(G_2)$ is a full near equivalence class in $\mathcal{A}_{cusp}(G_2)$.

To conclude this section, we note the following corollary, which settles Problem C for most representations in the short root packet A_τ:

(6.6) Corollary *Suppose that π_0 is an irreducible constituent of $\pi \in A_\tau$. Then*

$$m_{disc}(\pi_0) = m(\pi)$$

except possibly in the following case: $\pi_0 \cong \pi_\tau^{++}$ and $L(\tau, 1/2) \neq 0$. *Here $\pi_\tau^{++} = \otimes_v \pi_{\tau_v}^{++}$ and $\pi_{\tau_v}^{++}$ was defined in (5.15). In this exceptional case, $m(\pi) = 1$ but we only know that $m_{disc}(\pi_0) = 1$ or 2.*

PROOF. Away from the exceptional case, this follows from Theorem 6.5(ii) because such a π_0 does not occur in the residual spectrum, by the results of [K] and [Z]. So

$$m_{disc}(\pi_0) = m_{cusp}(\pi_0) = \text{multiplicity of } \pi_0 \text{ in } L^2_{\psi_\tau} = m(\pi).$$

In the exceptional case, $\pi_0 = \pi_\tau^{++}$ does occur in the residual spectrum, and the ambiguity in $m_{disc}(\pi_0)$ arises because we don't know that it cannot also occur in the cuspidal spectrum. By Theorem 6.5(ii), this ambiguity will be resolved if one shows that the copy of π_τ^{++} in $L^2_{\psi_\tau}$ (constructed by theta lifting) is *orthogonal* to the space of cusp forms. Unfortunately, we only know that it is not contained in the space of cusp forms. □

§7. Some conjectures

In this final section, we highlight some unanswered questions regarding the cubic unipotent A-packets. This is an ongoing project of the authors and D.-H. Jiang.

(7.1) Another Rankin-Selberg integral. For the cubic unipotent A-packets associated to $E = F \times K$, Problems A-D have all been resolved. The resolution of Problem D, in particular, relies on the alternative construction of the packets by Rallis-Schiffmann lifting. When E is a field, however, this alternative construction is not available. Thus, it is desirable to have a more uniform treatment of Problem D for the cubic unipotent A-packets.

To this end, we are led to consider another Rankin-Selberg integral. Suppose that E is Galois. Then the minimal representation Π_E of $Spin_8^E(\mathbb{A})$ is the unique irreducible quotient of a certain degenerate principal series

$$I_{P_E}(\chi_E, s) = Ind_{P_E}^{Spin_8^E} \chi_E \cdot \delta_{P_E}^s,$$

for $s = 4/5$ and χ_E is a finite order character associated to E. Let $E_{\chi_E}(f, s, g)$ be the Eisenstein series attached to a standard section f_s. Then the leading terms of the Laurent expansion of $E_{\chi_E}(f,s,g)$ (as f varies) at $s = 4/5$ define the embedding of Π_E into $\mathcal{A}_2(Spin_8^E)$ used in §3.

Thus, if $\pi \subset \mathcal{A}_{cusp}(G_2)$ is irreducible, it is natural to consider the following integral, first investigated by Jiang:

$$R_E(\varphi, f, \chi_E,, s) = \int_{G_2(F)\backslash G_2(\mathbb{A})} \varphi(g) \cdot E_{\chi_E}(f, \frac{s+2}{5}, g)\, dg$$

for $\varphi \in \pi$, and more generally $R_E(\varphi, f, \chi, s)$ for any character χ and $f_s \in I_{P_E}(\chi, s)$. This defines a meromorphic function of s. On unwinding, we see that $R_E(\varphi, f, \chi, s)$ is zero unless π has non-zero E-Fourier coefficient.

Here is what we would like to show:

(7.2) Conjecture: Assume that π has non-zero E-Fourier coefficient, with E Galois. Then we have the identity:

$$R_E(\varphi, f, \chi, s) = d_\chi(\varphi, f, s) \cdot L^S(s, \pi, \chi),$$

where $L^S(s, \pi, \chi)$ is the partial (degree 7) standard L-function of π twisted by χ, and d_χ is the factor corresponding to ramified and archimedean places.

This conjecture should imply a positive solution to Problem D for the cubic unipotent A-packets associated to Galois E. Indeed, we expect it to give the following stronger result:

(7.3) Conjecture: Let $\pi \subset \mathcal{A}_{cusp}(G_2)$ be irreducible. Fix a Galois cubic algebra E. Then the following statements are equivalent.

(a) $\pi \subset L^2_{\psi_E}$.

(b) π is E-distinguished.

(c) the partial L-function $L^S(\pi, s, \chi_E)$ has a pole (of order 2 if E is split, and of order 1 otherwise) at $s = 2$.

Moreover, π is CAP with respect to the Borel subgroup if and only if π satisfies the above conditions for some Galois E.

Remarks: If E is non-Galois, we still expect the equivalence of (a) and (b) above.

Acknowledgements

The first author is partially supported by NSF grant 0352682 and an AMS Centennial Fellowship. The second author is partially supported by the Israel Science Foundation.

References

[A1] J. Arthur, *On some problems suggested by the trace formula*, in Lie Groups Representations II, Lecture Notes in Math. 1041 (1983), Springer-Verlag, 1-49.

[A2] J. Arthur, *Unipotent automorphic representations: conjectures*, in Orbites Unipotentes et Representations, Asterique Vol. 171-172 (1989), Lecture Notes in Math. 1041 (1983),

[B] D. Blasius, *On multiplicities for $SL(n)$*, Israel J. Math. 88 (1994), 237-251.

[G] W. T. Gan, *Multiplicity formula for cubic unipotent Arthur packets*, to appear in Duke Math. J.

[GG] W. T. Gan and N. Gurevich, *Non-tempered A-packets of G_2: liftings from $\widetilde{SL}_2$*, to appear in American J. of Math.

[GGJ] W. T. Gan, N. Gurevich and D.-H. Jiang, *Cubic unipotent Arthur parameters and multiplicities of square-integrable automorphic forms*, Invent. Math. 149 (2002), 225-265.

[GS] W. T. Gan and G. Savin, *Endoscopic lifts from PGL_3 to G_2*, Compositio Math. 140 (2004), 793-808.

[GrS] B. H. Gross and G. Savin, *The dual pair $PGL_3 \times G_2$*, Canadian Bull. of Math. 40 (1997), No. 3, 376-384.

[HMS] J.-S. Huang, K. Magaard and G. Savin, *Unipotent representations of G_2 arising from the minimal representations of D_4^E*, J. Reine Angew Math. 500 (1998), 65-81.

[K] H. Kim, *The residual spectrum of G_2*, Canadian J. of Math. 48 (1996), No. 6, 1245-1272.

[KK] K. Konno and T. Konno, *CAP automorphic representations of $U_{E/F}(4)$ I: local A-packets*, preprint (2003).

[LS] J.-S. Li and J. Schwermer, *Construction of automorphic forms and related cohomology classes for arithmetic subgroups of G_2*, Compositio Math. 87 (1993), 45-78.

[PS] I. Piatetski-Shapiro, *On the Saito-Kurokawa lifting*, Invent. Math. 71 (1983), 309-338.

[R] J. Rogawski, *Automorphic representations of unitary groups in three variables*, Annals of Math. Studies 123 (1990), Princeton Univ. Press.

[RS] S. Rallis and G. Schiffmann, *Theta correspondence associated to* G_2, American J. of Math. 111 (1989), 801-849.

[S] G. Savin, *A class of supercuspidal representations of* $G_2(k)$, Canadian Bull. of Math. 42 (1999), No. 3, 393-400.

[So] D. Soudry, *The CAP representations of* $GSp(4, \mathbb{A})$, J. Reine Angew. Math. 383 (1988), 87-108.

[V] D. Vogan, *The unitary dual of* G_2, Invent Math. 116 (1994), 677-791.

[W1] J.-L. Waldspurger, *Correspondance de Shimura*, J. Math. Pures et Appl. 59 (1980), 1-133.

[W2] J.-L. Waldspurger, *Correspondance de Shimura et quaternions*, Forum Math. 3 (1991), 219-307.

[Z] S. Zampera, *The residual spectrum of the group of type* G_2, J. Math. Pure Appl. 76 (1997), No. 9, 805-835.

Wee Teck Gan
Department of Mathematics
University of California-San Diego
9500 Gilman Drive. La Jolla, CA 92093, U.S.A.
wgan@math.ucsd.edu

Nadya Gurevich
School of Mathematics
Ben Gurion University of the Negev
POB 653 Be'er Sheva 84105, Israel.
ngur@cs.bgu.ac.il

On the nonvanishing of the central value of the Rankin-Selberg L-functions II

David Ginzburg, Dihua Jiang, and Stephen Rallis

Abstract. We characterize the nonvanishing of the central value of the Rankin-Selberg L-functions $L(s, \pi \times \pi')$ in terms of the period of the Gross-Prasad type for the case where π is an irreducible cuspidal automorphic representation of $GL_{2r}(\mathbb{A})$ of symplectic type and π' is an irreducible cuspidal automorphic representation of $GL_{2l}(\mathbb{A})$ of orthogonal type. The other case was treated in [**GJR03b**].

1991 Mathematics Subject Classification: primary 11F67, 11F70, 22E46, 22E55.

1. Introduction

We started in [**GJR03b**] our study on the nonvanishing of the central value of the Rankin-Selberg tensor product L-functions in terms of periods of automorphic forms. In [**GJR03b**], we considered the cases where periods are defined over either a symplectic group or metaplectic double cover of a symplectic group, which was referred to as **Case (1)** in [**GJR03b**]. In this paper we will consider **Case (2)**, where the periods are defined over either an odd orthogonal group or an even orthogonal group.

Let π_1 and π_2 be irreducible unitary cuspidal automorphic representations of $GL_n(\mathbb{A})$ and $GL_m(\mathbb{A})$, respectively, where $\mathbb{A}$ is the ring of adeles attached to a number field k. In **Case (2)**, we have that (i) $n = 2r$ is even and π_1 is *symplectic* (i.e. the exterior square L-function $L(s, \pi_1, \Lambda^2)$ has a simple pole at $s = 1$), and (ii) $m = 2l$ is even and π_2 is *orthogonal* (i.e. the symmetric square L-function $L(s, \pi_2, \mathrm{Sym}^2)$ has a simple pole at $s = 1$). Our objective is to characterize the non-vanishing of the Rankin-Selberg L-function $L(s, \pi_1 \times \pi_2)$ at $s = \frac{1}{2}$ (the center of symmetry) in terms of periods attached to π_1 and π_2 over orthogonal groups. The precise definition of the period will be given in Section 2. In this case, by the Theorem of Ginzburg-Rallis-Soudry ([**GJR03b**], §1), there exist an irreducible unitary generic cuspidal automorphic representation τ of $SO_{2r+1}(\mathbb{A})$ which lifts to π_1 (since π_1 is symplectic), and an irreducible unitary generic cuspidal automorphic representation σ of $SO_{2l}(\mathbb{A})$ which lifts to π_2 (since π_2 is orthogonal).

Conjecture. *The central value of the Rankin-Selberg L-function $L(\frac{1}{2}, \pi_1 \times \pi_2)$ is non-zero if and only if there exist a relevant pair (SO'_{2r+1}, SO'_{2l}) of inner forms of (SO_{2r+1}, SO_{2l}) (see Section 2 for definition) and a pair of irreducible cuspidal automorphic representations (τ', σ') in which σ' (resp. τ') is nearly equivalent to σ (resp. τ), such that the period*

$$\mathcal{P}_{2r+1,2l}(\phi_{\tau'}, \phi_{\sigma'}) \ (r \geq l) \quad or \quad \mathcal{P}_{2l,2r+1}(\phi_{\sigma'}, \phi_{\tau'}) \ (r \leq l)$$

is non-zero. Here the period $\mathcal{P}_{2r+1,2l}(\phi_{\tau'}, \phi_{\sigma'})$ $(r \geq l)$ is defined by

$$\int_{SO'_{2l}(k)\backslash SO'_{2l}(\mathbb{A})} \phi_{\sigma'}(h) \int_{V_{r,l}(k)\backslash V_{r,l}(\mathbb{A})} \phi_{\tau'}(vh)\psi_{V_{r,l}}(v) dv dh$$

and the period $\mathcal{P}_{2l,2r+1}(\phi_{\sigma'}, \phi_{\tau'})$ $(r \leq l)$ is defined in a similar way. The details of the definition will be given in Section 2.

We sketch the relation between this conjecture and the Gross-Prasad conjecture **[GP92]**. By the Langlands functoriality from generic cuspidal automorphic representations of SO to GL (**[CKPSS03]**), one might define

$$L(s, \sigma \times \tau) := L(s, \pi_1 \times \pi_2).$$

Then the assertion on the characterization of the nonvanishing of the central value $L(\frac{1}{2}, \sigma \times \tau)$ in terms of periods attached to either (σ, τ, SO_{2l}) if $r \geq l$ or $(\sigma, \tau, SO_{2r+1})$ if $l > r$ is a conjecture of Gross and Prasad (see **[GP92]** and **[GP94]** for the global conjecture when $r = l$ or $l - 1$ and for the local conjecture in general).

When $r = 1$ and $l = 1$, the assertion was proved by Waldspurger in **[W85]**. When $l = 2$ and $r = 1$, it is a conjecture of Jacquet on the relation between the nonvanishing of the central value of the triple product L-function and the nonvanishing of the trilinear periods. For the split period case, it was proved in **[Jng98b]** and **[Jng01]**. For general period cases, it was proved completely in **[HK91]** and **[HK04]**. When $r = 2$ and $l = 2$, some special cases were studied in **[HK92]** and **[BFSP04]**. A special case of our conjecture ($r = 1$ and $l \geq 2$) was studied in **[GJR03a]**, but by a different approach.

It is important to note that for an irreducible generic cuspidal automorphic representation of SO, the image of the Langlands functorial transfer to GL is an isobaric sum. It is almost certain that the refinement of the arguments in this paper can be carried over to this generality. We will consider this in our future work. The analogue for unitary groups is a work in progress of the first and the second named authors.

In this paper, we shall use similar ideas as in **[GJR03b]** to study the above Conjecture. The following two Theorems are our main results.

Theorem A. *With the notations as in the Conjecture, if the period*

$$\mathcal{P}_{2r+1,2l}(\phi_\tau, \phi_\sigma)\ (r \geq l) \quad or \quad \mathcal{P}_{2l,2r+1}(\phi_\sigma, \phi_\tau)\ (r \leq l)$$

is non-zero, then the central value $L(\frac{1}{2}, \pi_1 \times \pi_2)$ of the Rankin-Selberg L-function $L(s, \pi_1 \times \pi_2)$ is non-zero.

The proof of Theorem A is to realize the above period as an 'inner' period of certain period of residue of Eisenstein series obtained from the pair (σ', τ'). This method has been used to study many interesting cases, about which we refer to the introduction of **[GJR03b]** for further discussions.

Theorem B. *With the notations as in the Conjecture, under certain assumptions, if the central value of the Rankin-Selberg L-function $L(\frac{1}{2}, \pi_1 \times \pi_2)$ is non-zero, then the period*

$$\mathcal{P}_{2r+1,2l}(\phi_{\tau'}, \phi_{\sigma'})\ (r \geq l) \quad or \quad \mathcal{P}_{2l,2r+1}(\phi_{\sigma'}, \phi_{\tau'})\ (r \leq l)$$

is non-zero for a pair (τ', σ').

We shall only prove Theorem B for a special case ($r \geq l = 1$) without the assumptions. For the general case, we need an assumption on the non-vanishing of certain Fourier coefficient of the residue of Eisenstein series obtained from the pair (τ', σ'), which we present the details for **Case (1)** in Section 6, **[GJR03b]**. We shall not provide the same details here for **Case (2)**. Further study of Fourier coefficients of the residual representations will be the subject matter of our on-going project.

We shall provide the general setting for **Case (2)** in Section 2, which includes the definitions of relevant pairs of orthogonal groups and the periods of automorphic forms, and other notations needed for the rest of the paper. In Section 3, we first recall basic facts of Eisenstein series and their residues, prove nonvanishing results on certain Fourier coefficients of the residues of relevant Eisenstein series, and then establish a basic identity (Theorem 3.8) which relates the 'outer' period to the 'inner' period. We remark that Theorem 3.8 deals with the residual representations with nearly generic cuspidal data. This makes the method used here potentially applicable to the investigation of residual representations without assuming the genericity of cuspidal data. This basic identity is used in Section 4 to prove Theorem A. In Section 5, we prove Theorem B for the special case $r \geq l = 1$. The general case will be considered in our future work.

During the preparation of this work, the second named author was partly supported by the NSF grant MSD-0089003 and by the Sloan Research Fellowship. The first two authors wish to express their deep admiration to Professor Stephen Rallis and thank him for his constant encouragement. Finally we would like to thank the referee for his/her comments and suggestions.

2. Periods

In this section, we give a formal definition of period integrals for automorphic forms. The remaining sections are devoted to the study of these periods for automorphic forms occurring in the discrete spectrum.

2.1. Relevant Pairs. We recall from **[GP94]** the notion of relevant pairs of orthogonal groups. Let $(V, <\cdot,\cdot>_V)$ be a quadratic space over a field k. We assume that all quadratic spaces considered in this paper are non-degenerate and of finite dimension. Let $(W, <\cdot,\cdot>_W)$ be another quadratic space, which can be embedded into V. We fix an embedding and define

$$W^{\perp} := \{v \in V \mid <v,w>_V = 0, \textit{ for all } w \in W\}.$$

Then, following Witt's Theorem, one knows that $W^{\perp}$ is a quadratic space and

$$V = W \oplus W^{\perp}. \tag{2.1}$$

We say the pair (V, W) is *relevant* if W embeds in V and $W^{\perp}$ is a split quadratic space with odd dimension over k. For more details about relevant pairs, we refer to Section 3 in **[GP94]**.

2.2. Periods of Automorphic Forms. Let k be a number field and $\mathbb{A}$ be the ring of adeles of k. Let (V, W) be a relevant pair of quadratic spaces. By definition,

$$W^{\perp} = (X + X') \oplus <a>$$

is a k-split quadratic space of dimension $2e+1$ with X and X' dual isotropic subspaces of dimension e. Let $Y = X + X'$, which is k-split and nondegenerate. We write

$$V = Y \oplus Y^{\perp}.$$

Then it is easy to see that W can be embedded into $Y^{\perp}$ with codimension one. Let $M_e = \mathrm{GL}(X) \times \mathrm{SO}(Y^{\perp})$. It is clear that M_e is a Levi subgroup of $G = \mathrm{SO}(V)$ and it is non-trivial if $e = \dim_k X > 0$. Let $P_e = M_e U_e$ be a standard parabolic subgroup of $G = \mathrm{SO}(V)$. It is proper and maximal if $e > 0$. In this paper, we include in our discussion the case that $e = \dim_k X = 0$. We have

$$0 \to \Lambda^2(X) \to U_e \to U_e^{ab} = X \otimes Y^{\perp} \to 0,$$

which is trivial if $e = 0$.

When $e = 0$, we consider the diagonal embedding of $H = \mathrm{SO}(W)$ into $G \times H$. When $e \geq 1$, we let $X_1 \subset X$ be a hyperplane and

$$\ell_1 \; : \; X \to \mathbb{G}_a$$

be a non-zero linear functional which vanishes on X_1. Let

$$\ell_W \; : \; Y^\perp \to \mathbb{G}_a$$

be a non-zero linear functional which vanishes on the hyperplane W. Consider the composite map ρ_e of the projection $U_e \to U_e^{ab} = X \otimes Y^\perp$ with

$$\ell_1 \otimes \ell_W \; : \; U_e^{ab} = X \otimes Y^\perp \to \mathbb{G}_a.$$

Since the Levi subgroup M_e normalizes U_e, the stabilizer of ρ_e in M_e is

$$\mathrm{GL}(X)_{\ell_1} \times \mathrm{SO}(W),$$

where $\mathrm{GL}(X)_{\ell_1}$ is the subgroup of $\mathrm{GL}(X)$ which fixes the linear functional ℓ_1. One can check that it is the subgroup of $\mathrm{GL}(X)$ consisting of elements of the following type

$$\begin{pmatrix} h & * \\ 0 & 1 \end{pmatrix}, \qquad h \in \mathrm{GL}_{e-1}.$$

Let $P_{1^e} = [\mathrm{GL}_1^e \times \mathrm{SO}(Y^\perp)]V_e$ be a standard parabolic subgroup of G. Then we have

$$V_e = (N_e \times I_{Y^\perp}) \ltimes U_e.$$

Let $\ell_e \; : \; N_e \to \mathbb{G}_a$ be a homomorphism which is non-trivial when restricted to each simple roots of GL_e in N_e. We define

$$\ell \; : \; V_e \to \mathbb{G}_a$$

to be a homomorphism which equals ℓ_e on N_e and equals ρ_e on U_e. It is clear that the stabilizer of ℓ in $\mathrm{SO}(Y^\perp)$ is isomorphic to $H = \mathrm{SO}(W)$. Define

$$R_e := H \ltimes V_e$$

which is an algebraic subgroup of G. We consider the natural diagonal embedding of H into $R_e \times H$.

Fix a non-trivial character ψ of $k \backslash \mathbb{A}$. We define

$$\psi_e(n) = \psi(\ell_e(n)) \tag{2.2}$$

which is a character of $N_e(\mathbb{A})$ and is trivial on the subgroup $N_e(k)$, and define

$$\psi_{\rho_e}(u) = \psi(\rho_e(u)), \tag{2.3}$$

which is a character ψ_{ρ_e} of $U_e(\mathbb{A})$ and is trivial on the subgroup $U_e(k)$. Then we define a character ψ_{V_e} of $V_e(\mathbb{A})$ by

$$\psi_{V_e}(v(n,u)) = \psi_e(n)\psi_{\rho_e}(u), \tag{2.4}$$

which is trivial on $V_e(k)$.

Now we define the period of automorphic forms. Let ϕ be an automorphic forms on $G(\mathbb{A})$ and φ be an automorphic forms on $H(\mathbb{A})$. The period of automorphic forms which we shall study is defined by the following integral

$$\mathcal{P}_{R_e}(\phi,\varphi) := \int_{H(k)\backslash H(\mathbb{A})} \varphi(h) \int_{V_e(k)\backslash V_e(\mathbb{A})} \phi(vh)\psi_{V_e}(v)dvdh. \tag{2.5}$$

It is clear that if one of ϕ and φ is cuspidal, then the period $\mathcal{P}_{R_e}(\phi,\varphi)$ converges absolutely. Otherwise, the integral needs to be defined through regularization.

From the definition of relevant pairs, the parity of $\dim_k V$ and $\dim_k W$ are different. When $\dim_k V$ is even, we denote $\dim_k V = 2l$ and $\dim_k W = 2r+1$. This implies that $r < l$. When $\dim_k V$ is odd, we denote $\dim_k V = 2r+1$ and $\dim_k W = 2l$. This implies that $r \geq l$. Hence we have $e = r - l$ if $r \geq l$; and $e = l - r - 1$ if $r < l$. We also denote the period by

$$\mathcal{P}_{R_e}(\phi,\varphi) = \begin{cases} \mathcal{P}_{2r+1,2l}(\phi,\varphi) & \textit{if } r \geq l \\ \mathcal{P}_{2l,2r+1}(\phi,\varphi) & \textit{if } r < l. \end{cases} \tag{2.6}$$

3. Periods of residues of Eisenstein series

We shall consider period $\mathcal{P}_{R_e}(\phi,\varphi)$ when both ϕ and φ are residues of Eisenstein series.

3.1. Residues of Eisenstein series. We consider orthogonal groups SO_{4r+q} with $q = 2r+1$ or $2l$. The k-split rank of SO_{4r+q} is assumed to be at least $2r$. Let $P_{2r,q}$ be the standard maximal parabolic subgroup of SO_{4r+q} with the Levi decomposition

$$P_{2r,q} = M_{2r,q} \ltimes U_{2r,q} = (\mathrm{GL}_{2r} \times \mathrm{SO}_q) \ltimes U_{2r,q}.$$

We denote by $w_{2r,q}$ the Weyl element which conjugates $P_{2r,q}$ to its opposite $P^-_{2r,q}$. The elements of $M_{2r,q}$ will be often written as

$$m = m(a,b) \in \mathrm{GL}_{2r} \times \mathrm{SO}_q. \tag{3.1}$$

Let $\pi \otimes \sigma_q$ be an irreducible generic cuspidal automorphic representation of $\mathrm{GL}_{2r} \times \mathrm{SO}_q$. We assume that the central character of π is trivial. This assumption is natural as long as we consider the existence of the residual representations that we are going to define below.

To the given generic cuspidal datum $(P_{2r,q}, \pi \otimes \sigma_q)$, one may attach an Eisenstein series $E(g, s, \phi_{\pi\otimes\sigma_q})$ on $\mathrm{SO}_{4r+q}(\mathbb{A})$. More precisely, the cuspidal automorphic representation $\pi \otimes \sigma_q$ can be realized in the space of square integrable automorphic functions $\mathrm{L}^2(Z_{M_{2r,q}}(\mathbb{A})M_{2r,q}(k)\backslash M_{2r,q}(\mathbb{A}))$, where $Z_{M_{2r,q}}$ is the center of $M_{2r,q}$. Let K be the maximal compact subgroup of SO_{4r+q} such that

$$\mathrm{SO}_{4r+q}(\mathbb{A}) = P_{2r,q}(\mathbb{A})\mathrm{K}$$

is the Iwasawa decomposition. Let $\phi_{\pi\otimes\sigma_q}$ be a $\mathrm{K}\cap M_{2r,q}(\mathbb{A})$-finite automorphic form in $\pi \otimes \sigma_q$. It can be extended as a function of $\mathrm{SO}_{4r+q}(\mathbb{A})$, so that for $g = um\mathrm{k} \in \mathrm{SO}_{4r+q}(\mathbb{A})$

$$\phi_{\pi\otimes\sigma_q}(g) = \phi_{\pi\otimes\sigma_q}(mk)$$

and for any fixed $\mathrm{k} \in \mathrm{K}$, the function

$$m \mapsto \phi_{\pi\otimes\sigma_q}(mk)$$

is a $\mathrm{K} \cap M_{2r,q}(\mathbb{A})$-finite automorphic function in $\pi \otimes \sigma_q$. We define

$$\Phi(g, s, \phi_{\pi\otimes\sigma_q}) := \phi_{\pi\otimes\sigma_q}(g) \exp < s + \rho_{P_{2r,q}}, H_{P_{2r,q}}(g) > \qquad (3.2)$$

for $g \in \mathrm{SO}_{4r+q}(\mathbb{A})$. Here the parameter s is normalized as in **[GSh88]**. In our case we have

$$\exp < s + \rho_{P_{2r,q}}, H_{P_{2r,q}}(g) >= |\det a|^{s+\frac{2r+q-1}{2}} \qquad (3.3)$$

where we write $g = um(a, b)\mathrm{k} \in \mathrm{SO}_{4r+q}(\mathbb{A})$ and $m(a, b) \in \mathrm{GL}_{2r} \times \mathrm{SO}_q$. Then the Eisenstein series is given by

$$E(g, s, \phi_{\pi\otimes\sigma_q}) = \sum_{\gamma \in P_{2r,q}(k)\backslash \mathrm{SO}_{4r+q}(k)} \Phi(\gamma g, s, \phi_{\pi\otimes\sigma_q}). \qquad (3.4)$$

The constant term of the Eisenstein series $E(g, s, \phi_{\pi\otimes\sigma_q})$ along a standard parabolic subgroup P is always zero unless $P = P_{2r,q}$. In this case, one has

$$\begin{aligned} E_{P_{2r,q}}(g, s, \phi_{\pi\otimes\sigma_q}) &= \int_{[U_{2r,q}]} E(ug, s, \phi_{\pi\otimes\sigma_q})du \qquad (3.5) \\ &= \Phi(g, s, \phi_{\pi\otimes\sigma_q}) + \mathcal{M}(w_{2r,q}, s)(\Phi(\cdot, s, \phi_{\pi\otimes\sigma_q}))(g) \end{aligned}$$

where $[U_{2r,q}] = U_{2r,q}(k)\backslash U_{2r,q}(\mathbb{A})$. We denote here by $\mathcal{M}(w_{2r,q}, s)$ the standard intertwining operator attached to the Weyl element $w_{2r,q}$, which maps the (normalized) induced representation

$$\mathrm{I}(s, \pi \otimes \sigma_q) = \mathrm{Ind}_{P_{2r,q}(\mathbb{A})}^{\mathrm{SO}_{4r+q}(\mathbb{A})}(\pi \otimes \sigma_q \otimes \exp < s, H_{P_{2r,q}}(\cdot) >) \qquad (3.6)$$

to $\mathrm{I}(-s, w_{2r,q}(\pi \otimes \sigma_q))$. It follows from the general theory of Eisenstein series of Langlands that the Eisenstein series $E(g, s, \phi_{\pi\otimes\sigma_q})$ has a pole at $s = s_0 > 0$ if and only if the term $\mathcal{M}(w_{2r,q}, s)(\Phi(\cdot, s, \phi_{\pi\otimes\sigma_q}))$ has a pole at $s = s_0$ for some holomorphic (or standard) section $\Phi(g, s, \phi_{\pi\otimes\sigma_q})$ in $\mathrm{I}(s, \pi \otimes \sigma_q)$. Since factorizable sections generate a dense subspace in $\mathrm{I}(s, \pi \otimes \sigma_q)$, it suffices to consider the factorizable sections for the existence of poles of the Eisenstein series $E(g, s, \phi_{\pi\otimes\sigma_q})$, or for the existence of poles of $\mathcal{M}(w_{2r,q}, s)(\Phi(\cdot, s, \phi_{\pi\otimes\sigma_q}))$. When the section $\Phi(\cdot, s, \phi_{\pi_2\otimes\sigma_q})$ is factorizable, i.e.

$$\Phi(\cdot, s, \phi_{\pi\otimes\sigma_q}) = \otimes_v \Phi_v(\cdot, s, \phi_{\pi_v\otimes\sigma_{q,v}}), \tag{3.7}$$

where $\Phi_v(\cdot, s, \phi_{\pi_v\otimes\sigma_{q,v}})$ is a section in $\mathrm{I}(s, \pi_v \otimes \sigma_{q,v})$ and is unramified at almost all finite local places v, the term $\mathcal{M}(w_{2r,q}, s)(\Phi(\cdot, s, \phi_{\pi\otimes\sigma_q}))$ can be expressed as an infinite product

$$\mathcal{M}(w_{2r,q}, s)(\Phi(\cdot, s, \phi_{\pi\otimes\sigma_q})) = \prod_v \mathcal{M}_v(w_{2r,q}, s)(\Phi_v(\cdot, s, \phi_{\pi_v\otimes\sigma_{q,v}})). \tag{3.8}$$

By the Langlands-Shahidi theory **[L71]** and **[Sh88]**, we have

$$\mathcal{M}(w_{2r,q}, s) = \frac{L(s, \pi \times \sigma_q) L(2s, \pi, \mathrm{r}_2)}{L(s+1, \pi \times \sigma_q) L(2s+1, \pi, \mathrm{r}_2)} \cdot \prod_v \mathcal{N}_v(w_{2r,q}, s), \tag{3.9}$$

where r_2 is the symmetric square representation $\mathrm{Sym}^2(\mathbb{C}^{2r})$ of $\mathrm{GL}_{2r}(\mathbb{C})$ if $q = 2r+1$, and the exterior square representation $\Lambda^2(\mathbb{C}^{2r})$ of $\mathrm{GL}_{2r}(\mathbb{C})$ if $q = 2l$. Here we denote by $\mathcal{N}_v(w_{2r,q}, s)$ the normalized intertwining operator

$$\mathcal{N}_v(w_{2r,q}, s) = \frac{L(s+1, \pi \times \sigma_q) L(2s+1, \pi, \mathrm{r}_2)}{L(s, \pi \times \sigma_q) L(2s, \pi, \mathrm{r}_2)} \cdot \mathcal{M}_v(w_{2r,q}, s)$$

which defines a mapping from $\mathrm{I}(s, \pi_v \otimes \sigma_{q,v})$ to $\mathrm{I}(-s, w_{2r,q}(\pi_v \otimes \sigma_{q,v}))$.

Proposition 3.1 (Theorem 4.11 [K]). The normalized local intertwining operator $\mathcal{N}_v(w_{2r,q}, s)$ is holomorphic and nonzero for the real part of s greater than or equal to $\frac{1}{2}$, i.e. for any holomorphic section $\Phi_v(\cdot, s, \phi_{\pi_v\otimes\sigma_{q,v}})$ in $\mathrm{I}(s, \pi_v \otimes \sigma_{q,v})$, as a function in s, $\mathcal{N}_v(w_{2r,q}, s)(\Phi_v(\cdot, s, \phi_{\pi_v\otimes\sigma_{q,v}}))$ is holomorphic and nonzero for the real part of s greater than or equal to $\frac{1}{2}$.

By Proposition 3.1, one deduces

Proposition 3.2. The Eisenstein series $E(g, s, \phi_{\pi\otimes\sigma_q})$ can possibly have a simple pole at $s = \frac{1}{2}$ or $s = 1$. The existence of the pole at $s = \frac{1}{2}$ or $s = 1$ of $E(g, s, \phi_{\pi\otimes\sigma_q})$ is equivalent to the existence of the pole at $s = \frac{1}{2}$ or $s = 1$ of the product of L-functions

$$L(s, \pi \times \sigma_q) L(2s, \pi, \mathrm{r}_2)$$

respectively.

Proof. By the Langlands theory of constant terms of Eisenstein series, the Eisenstein series has a pole at $s = s_0$ if and only if the constant terms of the Eisenstein series has a pole at $s = s_0$. By (3.5), this is equivalent to say that the global intertwining operator $\mathcal{M}(w_{2r,q},s)$ has a pole at $s=s_0\geq\frac{1}{2}$. From identity (3.9), if the global intertwining operator $\mathcal{M}(w_{2r,l},s)$ has a pole at $s=s_0$, then the quotient

$$\frac{L(s,\pi\times\sigma_q)L(2s,\pi,\mathrm{r}_2)}{L(s+1,\pi\times\sigma_q)L(2s+1,\pi,\mathrm{r}_2)}$$

must have a pole at $s=s_0$ since, by Proposition 3.1, the product

$$\prod_v \mathcal{N}_v(w_{2r,q},s)$$

does not vanish for $s=s_0\geq\frac{1}{2}$. Now both L-functions $L(s,\pi\times\sigma_q)$ and $L(s,\pi,\mathrm{r}_2)$ are non-zero for the real part of s greater than one. It follows that the product

$$L(s,\pi\times\sigma_q)L(2s,\pi,\mathrm{r}_2)$$

must have a pole at $s=s_0\geq\frac{1}{2}$ if $\mathcal{M}(w_{2r,l},s)$ has a pole at $s=s_0\geq\frac{1}{2}$. Conversely, if the product of L-functions

$$L(s,\pi\times\sigma_q)L(2s,\pi,\mathrm{r}_2)$$

has a pole at $s=s_0$, then the global intertwining operator $\mathcal{M}(w_{2r,q},s)$ has a pole at $s=s_0$, because in (3.9), we can always choose a particular factorizable section $\Phi(\cdot,s,\phi_{\pi\otimes\sigma_q})$ as in (3.7), so that the product

$$\prod_v \mathcal{N}_v(w_{2r,q},s)(\Phi_v(\cdot,s,\phi_{\pi_v\otimes\sigma_{q,v}}))$$

is holomorphic and non-zero at $s=s_0$. Note that in this part, we do not need Proposition 3.1.

Finally, by Theorem 7.2 in **[CKPSS03]**, the irreducible generic cuspidal automorphic representation σ_q of $\mathrm{SO}_q(\mathbb{A})$ has a Langlands functorial lift $\pi(\sigma)$, which is an irreducible unitary automorphic representation of $\mathrm{GL}_q(\mathbb{A})$ (if $q=2l$) or $\mathrm{GL}_{q-1}(\mathbb{A})$ (if $q=2r+1$), is uniquely determined by σ_q, and is of isobaric type. Hence one has

$$L(s,\pi\times\sigma_q)=L(s,\pi\times\pi(\sigma_q)).$$

It follows that the L-function $L(s,\pi\times\sigma_q)$ is holomorphic and non-vanishing at a real point $s_0>1$, and $L(s,\pi\times\sigma_q)$ has a pole at $s=1$ if and only if π is isomorphic to one of the isobaric summands of $\pi(\sigma)$. By Theorem 3.1 in **[K00]**, the complete L-function $L(s,\pi,\mathrm{r}_2)$ is holomorphic and non-vanishing at any real point $s_0>1$. By Theorem A in **[S]**, one knows that $L(s,\pi,\mathrm{r}_2)$ has a pole at $s=1$ if and only if π is the image of an irreducible unitary generic cuspidal automorphic representation τ of $\mathrm{SO}_{2r}(\mathbb{A})$ (if r_2 is the symmetric square) or $\mathrm{SO}_{2r+1}(\mathbb{A})$ (if r_2 is the exterior square), under the Langlands functorial lifting from SO_{2r} and SO_{2r+1} to GL_{2r}, respectively. Hence one knows that for the real value $s \geq \frac{1}{2}$, both L-functions $L(s, \pi \times \sigma_q)$ and $L(s, \pi, r_2)$ can have possible poles only at $s = 1$. The Proposition follows. □

In the following we are going to study some special type of Eisenstein series. Let τ be an irreducible generic unitary cuspidal automorphic representation of $SO_{2r+1}(\mathbb{A})$ which lifts to an irreducible unitary cuspidal automorphic representation $\pi(\tau)$ of $GL_{2r}(\mathbb{A})$. Let σ be an irreducible generic unitary cuspidal automorphic representation of $SO_{2l}(\mathbb{A})$ which lifts to an irreducible unitary cuspidal automorphic representation $\pi(\sigma)$ of $GL_{2l}(\mathbb{A})$. By the automorphic descent established in **[GRS01]** and **[S]**, such τ and σ exist.

Let τ' be an irreducible unitary cuspidal automorphic representation of $SO'_{2r+1}(\mathbb{A})$ (an inner form of SO_{2r+1}) nearly equivalent to τ, and σ' be an irreducible unitary cuspidal automorphic representation of $SO'_{2l}(\mathbb{A})$ (an inner form of SO_{2l}) nearly equivalent to σ. If SO'_{2l} is an inner form of SO_{2l} and SO'_{2r+1} is an inner form of SO_{2r+1}, then SO'_{6l} is the inner form of SO_{6l} such that $GL_{2l} \times SO'_{2l}$ is a Levi subgroup of SO'_{6l}, and SO'_{6r+1} is the inner form of SO_{6r+1} such that $GL_{2r} \times SO'_{2r+1}$ is a Levi subgroup of SO'_{6r+1}.

As in Proposition 3.2, we expect that both Eisenstein series $E(g, s, \phi_{\pi(\tau)\otimes\sigma'})$ and $E(g, s, \phi_{\pi(\sigma)\otimes\tau'})$ have a possible pole only at $s = \frac{1}{2}$ and the nonvanishing of the residue at $s = \frac{1}{2}$ is characterized by the nonvanishing of the value at $s = \frac{1}{2}$ of the L-function $L(s, \pi(\tau) \times \sigma')$ and $L(s, \pi(\sigma) \otimes \tau')$, respectively. We denote the residue by

$$\mathcal{E}_{\pi(\tau)\times\sigma'}(g) := \operatorname{Res}_{s=\frac{1}{2}} E(g, s, \phi_{\pi(\tau)\times\sigma'}) \tag{3.10}$$

$$\mathcal{E}_{\pi(\sigma)\otimes\tau'}(g) := \operatorname{Res}_{s=\frac{1}{2}} E(g, s, \phi_{\pi(\sigma)\otimes\tau'}). \tag{3.11}$$

One of the main results of this paper is to characterize the nonvanishing of these residues in terms of periods defined over the cuspidal support of the residues. To this end we have to introduce Eisenstein series $E(g, s, \phi_{\pi(\tau)\otimes\tau'})$ and $E(g, s, \phi_{\pi(\sigma)\otimes\sigma'})$. We show that both Eisenstein series have a non-zero residue at $s = 1$.

Lemma 3.3. With the notations as above, the following statements hold.

(1) The residue at $s = 1$ of the Eisenstein series $E(g, s, \phi_{\pi(\tau)\otimes\tau'})$ on $SO'_{6r+1}(\mathbb{A})$ is non-zero.

(2) The residue at $s = 1$ of the Eisenstein series $E(g, s, \phi_{\pi(\sigma)\otimes\sigma'})$ on $SO'_{6l}(\mathbb{A})$ is non-zero.

Proof. We shall prove Part (1), while Part(2) follows by a similar argument. By the general argument in Section 3.1, we have

$$\mathcal{M}(w_{2r,2r+1}, s) = \frac{L(s, \pi(\tau) \times \tau')L(2s, \pi(\tau), \mathrm{Sym}^2)}{L(s+1, \pi(\tau) \times \tau')L(2s+1, \pi(\tau), \mathrm{Sym}^2)} \cdot \prod_v \mathcal{N}_v(w_{2r,2r+1}, s).$$

We can choose the datum such that $\mathcal{N}_v(w_{2r,2r+1}, s)$ is non-zero at $s = 1$. Since τ' is nearly equivalent to τ and the local L-factor of $L(s, \pi(\tau) \times \tau')$ does not vanish at $s = 1$, we deduce that the L-function $L(s, \pi(\tau) \times \tau')$ has a pole at $s = 1$. Since the L-function $L(s, \pi(\tau), \mathrm{Sym}^2)$ does not vanish at $s = 2$, we know that the global intertwining operator $\mathcal{M}(w_{2r,2r+1}, s)$ has a pole at $s = 1$ with some choice of data. This implies that the Eisenstein series $E(g, s, \phi_{\pi(\tau)\otimes\tau'})$ has a pole at $s = 1$. □

We denote the residues by

$$\mathcal{E}_{\pi(\tau)\otimes\tau'}(g) = \mathrm{Res}_{s=1} E(g, s, \phi_{\pi(\tau)\otimes\tau'}) \tag{3.12}$$
$$\mathcal{E}_{\pi(\sigma)\otimes\sigma'}(g) = \mathrm{Res}_{s=1} E(g, s, \phi_{\pi(\sigma)\otimes\sigma'}). \tag{3.13}$$

3.2. Certain Fourier coefficients of the residues. We shall study the vanishing property of certain Fourier coefficients of the residue $\mathcal{E}_{\pi(\tau)\otimes\tau'}(g)$ and $\mathcal{E}_{\pi(\sigma)\otimes\sigma'}(g)$. The proof is based on the structure of the local unramified components of the residues.

Let ν be a local finite place of k and $k_\nu = F$. Let τ_ν be the local ν-component of τ and σ_ν be the local ν-component of σ. We may make assume that ν is such a local finite place such that

(1) both τ_ν and σ_ν are unramified representations of F-split SO_{2r+1} and SO_{2l}, respectively, and

(2) both τ_ν and σ_ν are generic.

We write

$$\tau_\nu = \mathrm{Ind}_{B_{\mathrm{SO}_{2r+1}}(F)}^{\mathrm{SO}_{2r+1}(F)}(\chi_\nu) \tag{3.14}$$
$$\sigma_\nu = \mathrm{Ind}_{B_{\mathrm{SO}_{2l}}(F)}^{\mathrm{SO}_{2l}(F)}(\mu_\nu) \tag{3.15}$$

where both χ_ν and μ_ν are unramified characters. Here we consider the normalized (or unitary) induced representations. Let $T_{\mathrm{SO}_{2r+1}}$ denote the maximal torus of SO_{2r+1} whose elements are written as

$$t = \mathrm{diag}(a_1, \cdots, a_r, 1, a_r^{-1}, \cdots, a_1^{-1}).$$

Then we write $\chi_\nu(t) := \prod_{i=1}^r \chi_i(a_i)$. Similarly, we write $\mu_\nu := \prod_{j=1}^l \mu_j$.

Let θ_{τ_ν} be the irreducible unramified constituent of $\mathrm{I}(1, \pi(\tau_\nu) \otimes \tau_\nu)$ and θ_{σ_ν} be the irreducible unramified constituent of $\mathrm{I}(1, \pi(\sigma_\nu) \otimes \sigma_\nu)$. Let P_{3^r} be the standard parabolic subgroup of SO_{6r+1} with Levi subgroup $(\mathrm{GL}_3)^r$ and Q_{3^l} be the standard parabolic subgroup of SO_{6l} with Levi subgroup $(\mathrm{GL}_3)^l$. We have

Lemma 3.4. Let det be the determinant of GL_3.

(1) The irreducible unramified representation θ_{τ_ν} is a constituent of the spherical unitarily induced representation
$$\mathrm{Ind}_{P_{3r}(F)}^{SO_{6r+1}(F)}(\chi_1 \circ \det \otimes \cdots \otimes \chi_r \circ \det).$$

(2) The irreducible unramified representation θ_{σ_ν} is a constituent of the spherical unitarily induced representation
$$\mathrm{Ind}_{Q_{3l}(F)}^{SO_{6l}(F)}(\mu_1 \circ \det \otimes \cdots \otimes \mu_l \circ \det).$$

Proof. We prove Part (1), while Part (2) is similar.

The representation θ_{τ_ν} is the unramified subquotient of $I(1, \pi(\tau_\nu) \otimes \tau_\nu)$. and $\tau_\nu = \mathrm{Ind}_{B_{SO_{2r+1}}(F)}^{SO_{2r+1}(F)}(\chi_\nu)$. The unramified local Langlands functorial lift of τ_ν to $GL_{2r}(F)$ is $\pi(\tau_\nu) = \mathrm{Ind}_{B_{GL_{2r}}(F)}^{GL_{2r}(F)}(\chi_\nu)$. Here χ_ν is the character of the maximal torus $T_{GL_{2r}}$ of GL_{2r} defined by
$$\chi_\nu(\mathrm{diag}(u_1, \cdots, u_r, u_{r+1}, \cdots, u_{2r})) = \prod_{i=1}^{r} \chi_i(u_i u_{2r-i+1}^{-1}).$$

Then the induced representation $I(1, \pi(\tau_\nu) \otimes \tau_\nu)$ can be written as
$$\mathrm{Ind}_{B_{SO_{6r+1}}(F)}^{SO_{6r+1}(F)}(\chi' \cdot \chi_0)$$
where
$$\begin{aligned} \chi'(t) &= \prod_{i=1}^{r} \chi_i(t_i t_{2r-i+1}^{-1}) \prod_{j=2r+1}^{3r} \chi_{j-2r}(t_j) \\ \chi_0(t) &= |t_1 \cdots t_{2r}| \end{aligned}$$
for $t = \mathrm{diag}(t_1, \cdots, t_{3r}, 1, t_{3r}^{-1}, \cdots, t_1^{-1})$ in the maximal torus $T_{SO_{6r+1}}$ of SO_{6r+1}. It is not difficult to find a Weyl element w such that
$$\mathrm{Ind}_{B_{SO_{6r+1}}(F)}^{SO_{6r+1}(F)}(\chi' \cdot \chi_0) \cong \mathrm{Ind}_{B_{SO_{6r+1}}(F)}^{SO_{6r+1}(F)}(\widetilde{\chi})$$
where the character $\widetilde{\chi}$ of $T_{SO_{6r+1}}$ is given by
$$\widetilde{\chi}(t) = \prod_{i=1}^{r} \chi_i(a_{i1} a_{i2} a_{i3}) |a_{i1} a_{i3}^{-1}|$$
for $t = \mathrm{diag}(a_{11}, a_{12}, a_{13}, \cdots, a_{r1}, a_{r2}, a_{r3}, 1, a_{r3}^{-1}, a_{r2}^{-1}, a_{r1}^{-1}, \cdots, a_{13}^{-1}, a_{12}^{-1}, a_{11}^{-1})$ in $T_{SO_{6r+1}}$. By induction in stages, Part (1) of the Lemma follows. □

Lemma 3.4 determines the structure of the local unramified component of the residues we want to study. In the following we shall show that such an unramified local component does not support certain local Fourier functional, which is related to the Fourier coefficient of the residues.

Let G be the F-split SO_{6r+1}. As in Section 2.2, we let $e \geq r+1$ and define the unipotent subgroup $V_e = V_{6r+1,e}$ of SO_{6r+1}. The elements of V_e can be written as

$$v(n,x,z) = \begin{pmatrix} n & x & z \\ & I_{6r-2e+1} & x^* \\ & & n^* \end{pmatrix} \in SO_{6r+1}, \tag{3.16}$$

where n is in N_e, the standard maximal unipotent subgroup of GL_e, $x \in Mat_{e\times(6r-2e+1)}$ and $z \in \{Z \in Mat_{e\times e} : {}^tZJ_e + J_eZ = 0\}$, where we define inductively

$$J_e = \begin{pmatrix} & & 1 \\ & J_{e-2} & \\ 1 & & \end{pmatrix}.$$

Given a vector $\zeta \in F^{6r-2e+1}$ of non-zero length, we define a character $\psi_{V_e}^{\zeta}$ of the group $V_{6r+1,e}$ by

$$\psi_{V_e}^{\zeta}(v) = \psi(n_{1,2} + \cdots + n_{e-1,e} + x_e \cdot \zeta)$$

where x_e is the last row of x.

In a similar way, for $e \geq l$, we define the unipotent subgroup $V_{6l,e}$ to be the unipotent radical of the parabolic subgroup of the split orthogonal group SO_{6l} whose Levi part is $GL_1^e \times SO_{6l-2e}$. Given a vector $\zeta \in F^{6l-2e}$ with nonzero length we define the character $\psi_{V_{6l,e}}^{\zeta}$ of $V_{6l,e}$.

Given a representation ϵ of $SO_{6r+1}(F)$ we define a functional $L_{e,\zeta}$ on the space of ϵ with quasi-invariance property that

$$L_{e,\zeta}(v\eta) = \psi_{V_{6r+1,e}}^{\zeta}(v)L_{e,\zeta}(\eta)$$

for all $v \in V_{6r+1,e}$ and all η in the space of ϵ. Similarly, we define a functional $L_{e,\zeta}$ of $V_{6l,e}$ on a representation ϵ of SO_{6l}. We now prove

Lemma 3.5. With the notations defined as above, the following statements hold.

(1) For all $e \geq r+1$, the representation θ_{τ_ν} of SO_{6r+1} has no nonzero functional $L_{e,\zeta}$ for all $\zeta \in F^{6r-2e+1}$ with nonzero length.

(2) For all $e \geq l$, the representation θ_{σ_ν} of SO_{6l} has no nonzero functional $L_{e,\zeta}$ for all $\zeta \in F^{6l-2e}$ with nonzero length.

Proof. The proof is the same in both cases and is done as in the proof of Lemma 2 in [**GRS**]. We just indicate the steps needed for the proof. From

Lemma 3.4 it is enough to show that

$$\mathrm{Ind}_{P_{3^r}(F)}^{\mathrm{SO}_{6r+1}(F)}(\chi_1 \circ \det \otimes \cdots \otimes \chi_r \circ \det)$$

has no nonzero functional $L_{e,\zeta}$. Using Mackey theory it is enough to prove that if $\gamma \in P_{3^r}\backslash \mathrm{SO}_{6r+1}/V_{6r+1,e}$ then there exists a $v \in V_{6r+1,e}$ such that $\gamma v \gamma^{-1} \in P_{3^r}$ and $\psi^{\zeta}_{V_{6r+1,e}}(v) \neq 1$. By the Bruhat theory every such γ has the form $\gamma = w u_w$ where w is a Weyl element of SO_{6r+1} and u_w is a certain upper unipotent matrix. At first we show that we can get "rid" of u_w and consider only the cases where $\gamma = w$. Then as in the proof of Lemma 2 in **[GRS]** we show that given any Weyl element w we can find a one-dimensional unipotent subgroup generated by $v \in V_{6r+1,e}$ such that $wvw^{-1} \in P_{3^r}$ and $\psi^{\zeta}_{V_{6r+1,e}}(v) \neq 1$. This finishes the proof. □

With the preparation of the local results above, we return to the global case. Let ϵ be an automorphic representation of $\mathrm{SO}'_{6r+1}(\mathbb{A})$. We define the Fourier coefficient

$$\int_{V_{6r+1,e}(k)\backslash V_{6r+1,e}(\mathbb{A})} \phi_\epsilon(vg)\psi^{\zeta}_{V_{6r+1,e}}(v)dv. \tag{3.17}$$

Similarly if ϵ is an automorphic representation on SO'_{6l} we define the Fourier coefficient

$$\int_{V_{6l,e}(k)\backslash V_{6l,e}(\mathbb{A})} \phi_\epsilon(vg)\psi^{\zeta}_{V_{6l,e}}(v)dv. \tag{3.18}$$

As a consequence of Lemma 3.5, we prove

Proposition 3.6. With the notations defined as above the following statements hold.

(1) For all $e \geq r+1$ the residual representation $\mathcal{E}_{\pi(\tau)\otimes\tau'}(g)$ of $\mathrm{SO}'_{6r+1}(\mathbb{A})$ has no nontrivial Fourier coefficient as defined by (3.17) for all $\zeta \in F^{6r-2e+1}$ with nonzero length.

(2) For all $e \geq l$ the residual representation $\mathcal{E}_{\pi(\sigma)\otimes\sigma'}(h)$ of $\mathrm{SO}'_{6l}(\mathbb{A})$ has no nontrivial Fourier coefficient as defined by (3.18) for all $\zeta \in F^{6l-2e}$ with nonzero length. Here $\mathcal{E}_{\pi(\sigma)\otimes\sigma'}(h)$ is the residue at $s=1$ of the Eisenstein series $E(h, s, \phi_{\pi(\sigma)\otimes\sigma'})$ of $\mathrm{SO}'_{6l}(\mathbb{A})$.

3.3. Regularization of periods. We study the periods of the residues of Eisenstein series defined below. If $r \geq l$, we consider

$$\begin{aligned} &\mathcal{P}_{6r+1,4r+2l}(\mathcal{E}_{\pi(\tau)\otimes\tau'}, \mathcal{E}_{\pi(\tau)\otimes\sigma'}) \\ := &\int_{[\mathrm{SO}'_{4r+2l}]} \mathcal{E}_{\pi(\tau)\otimes\sigma'}(h) \int_{[V_{6r+1,r-l}]} \mathcal{E}_{\pi(\tau)\otimes\tau'}(vh)\psi_{V_{6r+1,r-l}}(v)dvdh \end{aligned} \tag{3.19}$$

where $[H] := H(k)\backslash H(\mathbb{A})$ for any algebraic k-group H. If $r < l$, we consider

$$\begin{aligned} &\mathcal{P}_{4r+2l,6r+1}(\mathcal{E}_{\pi(\tau)\otimes\sigma'}, \mathcal{E}_{\pi(\tau)\otimes\tau'}) \\ := &\int_{[\mathrm{SO}'_{6r+1}]} \mathcal{E}_{\pi(\tau)\otimes\tau'}(g) \int_{[V_{4r+2l,l-r-1}]} \mathcal{E}_{\pi(\tau)\otimes\sigma'}(vg)\psi_{V_{4r+2l,l-r-1}}(v)dvdg. \end{aligned} \tag{3.20}$$

The convergence of these periods needs to be justified through Arthur's truncation method. We shall study the case $r \geq l$ in complete detail, while the study for case $l > r$ is similar, and the details will be omitted.

We recall that elements of $V_{6r+1,r-l}$ can be written as

$$v(n,x,z) = \begin{pmatrix} n & x & z \\ & I_{4r+2l+1} & x^* \\ & & n^* \end{pmatrix} \in \mathrm{SO}_{6r+1}. \tag{3.21}$$

Here $n \in N_{r-l}$, the standard maximal unipotent subgroup of GL_{r-l}, and x is a $(r-l) \times (4r+2l+1)$-matrix. Take $\zeta = (0_{2r+l}, a, 0_{2r+l}) \in k^{4r+2l+1}$ with $a \neq 0$. The character $\psi^{\zeta}_{V_{6r+1,r-l}}$ can be expressed as

$$\psi^{\zeta}_{V_{6r+1,r-l}}(v(n,x,z)) = \psi(n_{1,2} + \cdots + n_{r-l-1,r-l} + ax_{r-l,2r+l+1}). \tag{3.22}$$

The period $\mathcal{P}_{6r+1,4r+2l}(\mathcal{E}_{\pi(\tau)\otimes\tau'}, \mathcal{E}_{\pi(\tau)\otimes\sigma'})$ in (3.19) is

$$\int_{[\mathrm{SO}'_{4r+2l}]} \mathcal{E}_{\pi(\tau)\otimes\sigma'}(h) \int_{[V_{6r+1,r-l}]} \mathcal{E}_{\pi(\tau)\otimes\tau'}(vh)\psi^{\zeta}_{V_{6r+1,r-l}}(v)dvdh. \tag{3.23}$$

As in Section 4.1 of **[GJR03b]**, we shall regularize the above period by truncating the residue $\mathcal{E}_{\pi(\tau)\otimes\sigma'}(h)$.

In the following we denote by $P = MU = P_{2r,2l} = M_{2r,2l}U_{2r,2l}$, the maximal parabolic subgroup of $G := \mathrm{SO}'_{4r+2l}$. We identify $\mathfrak{a}_P$ with $\mathbb{R}$. Then a regular $T \in \mathfrak{a}_P$ will correspond to a real number $c \in \mathbb{R}_{>1}$, where we denote by $\mathbb{R}_{>c}$ the set of all real numbers greater than c. We denote

$$H(g) := \exp < 1, H_P(g) >= |\det m(g)| \tag{3.24}$$

for $g = um(g)\mathrm{k} \in \mathrm{SO}_{4r+2l}(\mathbb{A})$ (Iwasawa decomposition given in §3.1). Let τ_c ($c \in \mathbb{R}_{>1}$) be the characteristic function of the subset $\mathbb{R}_{>c}$.

Following **[Arthur78]** and **[Arthur80]**, the truncation of the Eisenstein series $E(g,s,\phi)$ (where $\phi = \phi_{\pi(\tau)\otimes\sigma'}$) is defined as follows:

$$\Lambda_c E(g,s,\phi) = E(g,s,\phi) - \sum_{\gamma\in P(k)\backslash G(k)} E_P(\gamma g,s,\phi)\tau_c(H(\gamma g)). \tag{3.25}$$

The constant term $E_P(g,s,\phi)$ of the Eisenstein series $E(g,s,\phi)$ along P can be expressed as (see (3.5) for definition)

$$E_P(g,s,\phi) = \Phi(g,s,\phi) + \mathcal{M}(w,s)(\Phi(\cdot,s,\phi))(g)$$

where $w = w_{2r,2l}$ and $\mathcal{M}(w,s)$ is the intertwining operator as described in §3.1. We remark that the summation in (3.25) has only finitely many terms and converges absolutely (Corollary 5.2, **[Arthur78]**). The truncated Eisenstein series can then be rewritten as follows

$$\begin{aligned}\Lambda_c E(g,s,\phi) &= \sum_{\gamma\in P(k)\backslash G(k)} \Phi(\gamma g,s,\phi)(1-\tau_c(H(\gamma g)))\\ &\quad - \sum_{\gamma\in P(k)\backslash G(k)} \mathcal{M}(w,s)(\Phi(\cdot,s,\phi))(\gamma g)\tau_c(H(\gamma g))\\ &=: \mathcal{E}_1(g)-\mathcal{E}_2(g). \end{aligned} \tag{3.26}$$

Let s_0 be a positive real number. Assume that the Eisenstein series $E(g,s,\phi)$ has a simple pole at $s = s_0$. We denote by $E_{s_0}(g,\phi)$ the non-zero residue of $E(g,s,\phi)$. Then we have

$$\begin{aligned}\Lambda_c E_{s_0}(g,\phi) &= E_{s_0}(g,\phi) - \sum_{\gamma\in P(k)\backslash G(k)} \mathcal{M}(w,s)(\Phi(\cdot,s,\phi))_{s_0}(\gamma g)\tau_c(H(\gamma g))\\ &=: E_{s_0}(g,\phi)-\mathcal{E}_3(g). \end{aligned} \tag{3.27}$$

Following the argument in **[GJR03b]**, §4.1, we obtain from (3.26) and (3.27) the formula for the period $\mathcal{P}_{6r+1,4r+2l}(\mathcal{E}_{\pi(\tau)\otimes\tau'}, E_{s_0}(g,\phi))$:

$$\begin{aligned}&\mathcal{P}_{6r+1,4r+2l}(\mathcal{E}_{\pi(\tau)\otimes\tau'}, E_{s_0}(\cdot,\phi))\\ =\ &\mathcal{P}_{6r+1,4r+2l}(\mathcal{E}_{\pi(\tau)\otimes\tau'}, \mathcal{E}_3) + \mathcal{P}_{6r+1,4r+2l}(\mathcal{E}_{\pi(\tau)\otimes\tau'}, \Lambda_c E_{s_0}(\cdot,\phi))\\ =\ &\mathcal{P}_{6r+1,4r+2l}(\mathcal{E}_{\pi(\tau)\otimes\tau'}, \mathcal{E}_3)\\ &+[\mathcal{P}_{6r+1,4r+2l}(\mathcal{E}_{\pi(\tau)\otimes\tau'}, \mathcal{E}_1) - \mathcal{P}_{6r+1,4r+2l}(\mathcal{E}_{\pi(\tau)\otimes\tau'}, \mathcal{E}_2)]_{s_0}.\end{aligned}$$

The following Proposition will be proved in the next subsection.

Proposition 3.7. For $i = 1, 2$, the following periods

$$\mathcal{P}_{6r+1,4r+2l}(\mathcal{E}_{\pi(\tau)\otimes\tau'}, \mathcal{E}_i)$$

converge absolutely for $Re(s)$ large and have meromorphic continuation to the whole complex plane.

Since both $\Lambda_c E(g,s,\phi)$ and $\Lambda_c E_{s_0}(g,\phi)$ rapidly decay in the usual sense, the periods

$$\mathcal{P}_{6r+1,4r+2l}(\mathcal{E}_{\pi(\tau)\otimes\tau'}, \Lambda_c E(\cdot,s,\phi))$$

and

$$\mathcal{P}_{6r+1,4r+2l}(\mathcal{E}_{\pi(\tau)\otimes\tau'}, \Lambda_c E_{s_0}(\cdot,\phi))$$

converge absolutely. By Proposition 3.7, both periods $\mathcal{P}_{6r+1,4r+2l}(\mathcal{E}_{\pi(\tau)\otimes\tau'}, \mathcal{E}_1)$ and $\mathcal{P}_{6r+1,4r+2l}(\mathcal{E}_{\pi(\tau)\otimes\tau'}, \mathcal{E}_2)$ converge absolutely for $Re(s)$ large and have meromorphic continuation to the complex plane. Hence we have

$$\mathcal{P}_{6r+1,4r+2l}(\mathcal{E}_{\pi(\tau)\otimes\tau'},\mathcal{E}_2)_{s_0}=\mathcal{P}_{6r+1,4r+2l}(\mathcal{E}_{\pi(\tau)\otimes\tau'},\mathcal{E}_3).$$

It follows that

$$\mathcal{P}_{6r+1,4r+2l}(\mathcal{E}_{\pi(\tau)\otimes\tau'},E_{s_0}(\cdot,\phi))=\mathcal{P}_{6r+1,4r+2l}(\mathcal{E}_{\pi(\tau)\otimes\tau'},\mathcal{E}_1)_{s_0}. \tag{3.28}$$

3.4. Period identity. We shall prove Proposition 3.7 through the following explicit calculation.

First, the convergence for the real part of s large follows as in the proof of Proposition 3.1, **[GJR01]**. The meromorphic continuation of both periods will follow from the explicit calculation.

We shall investigate the period $\mathcal{P}_{6r+1,4r+2l}(\mathcal{E}_{\pi(\tau)\otimes\tau'},\mathcal{E}_1)$ in detail, while the conclusion for the period $\mathcal{P}_{6r+1,4r+2l}(\mathcal{E}_{\pi(\tau)\otimes\tau'},\mathcal{E}_2)$ follows from the same argument. By (3.23) and (3.28), the period $\mathcal{P}_{6r+1,4r+2l}(\mathcal{E}_{\pi(\tau)\otimes\tau'},\mathcal{E}_1)$ is given by the following integral

$$\int_{[\mathrm{SO}'_{4r+2l}]}\mathcal{E}_1(g)\int_{[V_{6r+1,r-l}]}\mathcal{E}_{\pi(\tau)\otimes\tau'}(vg)\psi_{V_{6r+1,r-l}}(v)dvdg. \tag{3.29}$$

where

$$\mathcal{E}_1(g)=\sum_{\gamma\in P(k)\backslash G(k)}\Phi(\gamma g,s,\phi)(1-\tau_c(H(\gamma g))).$$

In the following we set

$$\Phi^c(g,s):=\Phi(g,s,\phi)(1-\tau_c(H(g))).$$

From the above, we know that (3.29) is absolutely convergent when the real part of s is large. Unfolding the Eisenstein series we obtain

$$\int_{P(k)\backslash\mathrm{SO}'_{4r+2l}(\mathbb{A})}\Phi^c(g)\int_{[V_{6r+1,r-l}]}\mathcal{E}_{\pi(\tau)\otimes\tau'}(vg)\psi_{V_{6r+1,r-l}}(v)dvdg. \tag{3.30}$$

Recall that U is the unipotent radical of the parabolic subgroup P. Factoring the measure $U(k)\backslash U(\mathbb{A})$, we obtain the following integral as an inner integration in (3.30)

$$\int_{[U]}\int_{[V_{6r+1,r-l}]}\mathcal{E}_{\pi(\tau)\otimes\tau'}(vug)\psi_{V_{6r+1,r-l}}(v)dvdu. \tag{3.31}$$

Define a Weyl element w of SO'_{6r+1} by

$$w = \begin{pmatrix} 0 & I_r & & & \\ I_{r-l} & 0 & & & \\ & & I_{2r+2l+1} & & \\ & & & 0 & I_{r-l} \\ & & & I_r & 0 \end{pmatrix}. \tag{3.32}$$

By the embedding of SO'_{4r+2l} into SO'_{6r+1}, we have

$$U = \{u = \begin{pmatrix} I_{r-l} & & & & \\ & I_{2r} & u_1 & u_2 & \\ & & I_{2l+1} & u_1^* & \\ & & & I_{2r} & \\ & & & & I_{r-l} \end{pmatrix} \in \mathrm{SO}'_{6r+1}\}. \tag{3.33}$$

Since U is the unipotent radical of the parabolic subgroup $P_{2r,2l}$ (of SO'_{4r+2l}), when embedded in SO'_{6r+1}, we write $u_1 \in \mathrm{Mat}_{2r\times 2l}$ as a matrix in $\mathrm{Mat}_{(2r)\times(2l+1)}$ according to the standard embedding of SO'_{2l} and SO'_{2l+1}. Here $u_2 \in \{X \in \mathrm{Mat}_{2r\times 2r} : {}^tXJ_{2r} + J_{2r}X = 0\}$. Also, we write elements in the unipotent subgroup $V_{6r+1,r-l}$ as

$$v = v(n, p, x_2, x_3, z) := \begin{pmatrix} n & p & x_2 & x_3 & z \\ & I_{2r} & 0 & 0 & x_3^* \\ & & I_{2l+1} & 0 & x_2^* \\ & & & I_{2r} & p^* \\ & & & & n^* \end{pmatrix}. \tag{3.34}$$

where $n \in N_{r-l}$, the unipotent radical consisting of all upper unipotent matrices in GL_{r-l}. Also $p, x_3 \in \mathrm{Mat}_{(r-l)\times 2r}$, $x_2 \in \mathrm{Mat}_{(r-l)\times(2l+1)}$ and $z \in \{X \in \mathrm{Mat}_{(r-l)\times(r-l)} : {}^tXJ_{r-l} + J_{r-l}X = 0\}$. We write

$$wv(n, p, x_2, x_3, z)u(u_1, u_2)w^{-1} = \ell_1(u_1, x_3, u_2)v'\hat{p} \tag{3.35}$$

where

$$\ell_1(u_1, x_3, u_2) := \begin{pmatrix} I_{2r} & 0 & u_1 & x_3 & u_2 \\ & I_{r-l} & 0 & 0 & x_3^* \\ & & I_{2l+1} & 0 & u_1^* \\ & & & I_{r-l} & 0 \\ & & & & I_{2r} \end{pmatrix}, \tag{3.36}$$

$v' \in V_{2r+1,r-l}$, the embedding of which in SO'_{6r+1} is

$$v' = \begin{pmatrix} I_{2r} & 0 & 0 & 0 & 0 \\ & n & x_2 & z & 0 \\ & & I_{2l+1} & x_2^* & 0 \\ & & & n^* & 0 \\ & & & & I_{2r} \end{pmatrix} \tag{3.37}$$

and

$$\hat{p} = \begin{pmatrix} I_{2r} & & & & \\ p & I_{r-l} & & & \\ & & I_{2l+1} & & \\ & & & I_{r-l} & \\ & & & p^* & I_{2r} \end{pmatrix}. \tag{3.38}$$

After conjugating in the above integral by w from left to right, we obtain

$$\int_{[U]} \int_{[V_{6r+1,r-l}]} \mathcal{E}_{\pi(\tau)\otimes\tau'}(\ell_1(u_1, x_3, u_2) v' \hat{p} w g) \psi^{\zeta}_{V_{6r+1,r-l}}(v) dv du. \tag{3.39}$$

Notice that from the definition of ζ, then after the conjugation by w the character $\psi^{\zeta}_{V_{6r+1,r-l}}$ agrees with $\psi^{\zeta}_{V_{2r+1,r-l}}$.

Let L_1 denote the subgroup of SO'_{6r+1} which consists of the matrices $\ell_1(u_1, x_3, u_2)$ as defined in (3.36), where, as explained above, $u_1 \in \mathrm{Mat}_{2r\times 2l}$. Let

$$L_2 = \{\ell_2(x_4, x_3, u_2) = \begin{pmatrix} I_{2r} & 0 & x_4 & x_3 & u_2 \\ & I_{r-l} & 0 & 0 & x_3^* \\ & & I_{2l+1} & 0 & x_4^* \\ & & & I_{r-l} & 0 \\ & & & & I_{2r} \end{pmatrix} \in \mathrm{SO}'_{6r+1}\}$$

where $x_4 \in \mathrm{Mat}_{2r\times(2l+1)}$ and x_3 and u_2 are as above. Then L_2/L_1 can be identified with a k-vector space of column vectors of size $2r$. In the above integral we expand along this column vector L_2/L_1 with points in $k\backslash\mathbb{A}$. We obtain

$$\sum_{\epsilon_i \in k} \int \mathcal{E}_{\pi(\tau)\otimes\tau'}(\ell_2((t, u_1), x_3, u_2) v' \hat{p} w g) \psi^{\zeta}_{V_{2r+1,r-l}}(v') \psi(\sum_{i=1}^{2r} \epsilon_i t_i) d\ell_2 dv' d\hat{p}$$

where t is a column vector whose transpose is $(t_1, \cdots, t_{2r})$ and by (t, u_1) we indicate that we added the integration of t over $(k\backslash\mathbb{A})^{2r}$ to the integration over variable u_1. The integration over all other variables are the same as before.

Let α be a row vector of size $2r$ embedded in SO'_{6r+1} as the last row of p in $\hat{p}$. Then one can find such a row $\alpha \in k^{2n}$, depending on ϵ_i and ξ such that after we conjugate it from the left to right in the above integral, change variables in v and collapse the summation over $\epsilon_i \in k$ with the adelic integration over

$(k\backslash\mathbb{A})^{2r}$ which consists of the last row of p, we obtain

$$\int_{\hat{p}_1,\hat{p}_2}\int_{[L_2]\times[V_{2r+1,r-l}]}\mathcal{E}_{\pi(\tau)\otimes\tau'}(\ell_2(x_4,x_3,u_2)v'\hat{p}_1\hat{p}_2wg)\psi^{\zeta}_{V_{2r+1,r-l}}(v')d\ell_2dv'd\hat{p}_1d\hat{p}_2, \tag{3.40}$$

where variable x_4 is in $\mathrm{Mat}_{2r\times(2l+1)}(k)\backslash\mathrm{Mat}_{2r\times(2l+1)}(\mathbb{A})$. Note that $\hat{p}$ is as defined in (3.38). In (3.40), variable p_2 is the last row of p integrated over $\mathbb{A}^{2r}$, and variable p_1 consists of all matrices in p with bottom row being zero and is integrated over points in $\mathrm{Mat}_{(r-l-1)\times(2r)}(k)\backslash\mathrm{Mat}_{(r-l-1)\times(2r)}(\mathbb{A})$. Let

$$L_3=\{\ell_3(y)=\begin{pmatrix} I_{2r} & y & & & \\ & I_{r-l} & & & \\ & & I_{2l+1} & & \\ & & & I_{r-l} & y^* \\ & & & & I_{2r}\end{pmatrix}\in \mathrm{SO}'_{6r+1}\} \tag{3.41}$$

where $y\in\mathrm{Mat}_{2r\times(r-l)}$ such that its first column is zero. Taking the Fourier expansion over $L_3(k)\backslash L_3(\mathbb{A})$ of integral (3.40) and using the integration over $\hat{p}_1$ as was done in $\hat{p}_2$, we deduce that integral (3.40) equals the following integral

$$\int_{\hat{p}}\int_{[L_3\times L_2\times V_{2r+1,r-l}]}\mathcal{E}_{\pi(\tau)\otimes\tau'}(\ell_3(y)\ell_2(x_4,x_3,u_2)v'\hat{p}wg)\psi^{\zeta}_{V_{2r+1,r-l}}(v')dyd\ell_2dv'd\hat{p}, \tag{3.42}$$

where $\hat{p}$ is integrated over $\mathrm{Mat}_{(r-l)\times 2r}(\mathbb{A})$.

Let

$$L_4=\{\ell_4(y)=\begin{pmatrix} I_{2r} & y & & & \\ & I_{r-l} & & & \\ & & I_{2l+1} & & \\ & & & I_{r-l} & y^* \\ & & & & I_{2r}\end{pmatrix} : y\in\mathrm{Mat}_{2r\times(r-l)}\}. \tag{3.43}$$

Then L_4/L_3 is a column vector which can be identified as the first column of the above matrix y. We take Fourier expansion of integral (3.42) along $(L_4/L_3)(k)\backslash(L_4/L_3)(\mathbb{A})$. Let the group GL_{2r} act on this column (L_4/L_3) via the adjoint action, which has two orbits. Here GL_{2r} is embedded in SO'_{6r+1} via the Levi subgroup $\mathrm{GL}_{2r}\times\mathrm{SO}'_{2r+1}$.

The trivial orbit will contribute

$$\int_{\hat{p}}\int_{[U_{2r,2r+1}\times V_{2r+1,r-l}]}\mathcal{E}_{\pi(\tau)\otimes\tau'}(uv'\hat{p}wg)\psi^{\zeta}_{V_{2r+1,r-l}}(v')dudv'd\hat{p}, \tag{3.44}$$

where $U_{2r,2r+1}$ is the unipotent radical of the maximal parabolic subgroup $P_{2r,2r+1}$ of SO'_{6r+1} with Levi subgroup $\mathrm{GL}_{2r}\times\mathrm{SO}'_{2r+1}$. We may write elements in $\mathrm{GL}_{2r} \times \mathrm{SO}'_{2r+1}$ as $m(a,b)$ and elements in $U_{2r,2r+1}$ as

$$u = u(y, z') = \begin{pmatrix} I_{2r} & y & z' \\ & I_{2r+1} & y^* \\ & & I_{2r} \end{pmatrix}. \tag{3.45}$$

The nontrivial orbit will contribute

$$\sum_{\gamma \in P_{2r}^0(k)\backslash \mathrm{GL}_{2r}(k)} \int_{\hat{p}} \int_{(u,v')} \mathcal{E}_{\pi(\tau)\otimes\tau'}(uv'\hat{\gamma}\hat{p}wg)\psi^{\zeta}_{V_{2r+1,r-l}}(v')\psi(y_{2r,2r+1})dudv'd\hat{p}, \tag{3.46}$$

where (u, v') is in $[U_{2r,2r+1} \times V_{2r+1,r-l}]$, and $\hat{\gamma} = m(\gamma, I_{2r+1})$ is in $\mathrm{GL}_{2r}(k) \times \mathrm{SO}'_{2r+1}(k)$. We denote by P^0_{2r} is the subgroup of GL_{2r} which fixes the character $\psi(y_{2r,2r+1})$.

From (3.37) and (3.45) an element in $U_{2r,2r+1} \times V_{2r+1,r-l}$ can be written as

$$uv' = \begin{pmatrix} I_{2r} & y_1 & y_2 & y_3 & z' \\ & n & y_4 & z & y_3^* \\ & & I_{2l+1} & y_4^* & y_2^* \\ & & & n^* & y_1^* \\ & & & & I_{2r} \end{pmatrix}. \tag{3.47}$$

The product of characters, $\psi^{\zeta}_{V_{2r+1,r-l}}(v')\psi(y_{2r,2r+1})$ equals

$$\psi((y_1)_{2r,1} + n_{1,2} + \cdots + n_{r-l-1,r-l} + a(y_4)_{r-l,l+1}), \tag{3.48}$$

according to (3.22). We are going to show that (3.46) is identically zero. To this end, we consider the Fourier expansion of (3.46) along $N_{2r}(k)\backslash N_{2r}(\mathbb{A})$, where N_{2r} is the maximal unipotent subgroup of GL_{2r} in the Levi subgroup $\mathrm{GL}_{2r} \times \mathrm{SO}'_{2r+1}$. Note that the product of N_{2r} and $U_{2r,2r+1} \times V_{2r+1,r-l}$ is equal to $V_{6r+1,3r-l}$ (see (3.16)), whose elements has form

$$v = \begin{pmatrix} n & x & z \\ & I_{2l+1} & x^* \\ & & n^* \end{pmatrix} \in \mathrm{SO}'_{6r+1}. \tag{3.49}$$

We consider the Fourier expansion, column by column, and obtain

$$\sum_{\gamma \in N_{2r}(k)\backslash \mathrm{GL}_{2r}(k)} \int_{[V_{6r+1,3r-l}]} \mathcal{E}_{\pi(\tau)\otimes\tau'}(v\hat{\gamma}\hat{p}wg)\psi^{\zeta}_{V_{6r+1,3r-l,\epsilon_1,\cdots,\epsilon_{2r-1}}}(v)dvd\hat{p}, \tag{3.50}$$

where the character $\psi^{\zeta}_{V_{6r+1,3r-l,\epsilon_1,\cdots,\epsilon_{2r-1}}}(v)$ is given by

$$\psi(\epsilon_1 n_{1,2} + \cdots + \epsilon_{2r-1}n_{2r-1,2r} + n_{2r,2r+1} + \cdots + n_{3r-l-1,3r-l} + ax_{3r-l,l+1}),$$

with $\epsilon_i \in \{0, 1\}$ and $a \in k^\times$. This is compatible with (3.22).

If there is an i such that $\epsilon_i = 0$ then the integral in (3.50) has an inner integration, which represents the constant term of $\mathcal{E}_{\pi(\tau)\otimes\tau'}$ with respect to the unipotent radical of a standard maximal parabolic subgroup not equal to $P_{2r,2r+1}$. By the cuspidal support of $\mathcal{E}_{\pi(\tau)\otimes\tau'}$, such a constant term must be zero and hence the integral in (3.50) with $\epsilon_i = 0$ must vanish. We are left with the integral in (3.50) having all $\epsilon_i = 1$. This integral represents the Fourier coefficient of $\mathcal{E}_{\pi(\tau)\otimes\tau'}$ as given in (3.17) with $e = 3r - l > r + 1$. By Part (1) of Proposition 3.6, this last integral in (3.50) must also vanish. Therefore, we prove that (3.50) is identically zero. In other words, we show that (3.31) is equal to (3.44).

In (3.44) the integration over $U_{2r,2r+1}(k)\backslash U_{2r,2r+1}(\mathbb{A})$ is the constant term of $\mathcal{E}_{\pi(\tau)\otimes\tau'}$. It follows that

$$\int_{[U_{2r,2r+1}]} \mathcal{E}_{\pi(\tau)\otimes\tau'}(ug)du = \phi_{\pi(\tau)\otimes\tau'}(g)\exp < -1 + \rho_{P_{2r,2r+1}}, H_{P_{2r,2r+1}}(g) > .$$

It is easy to check that

$$\exp < -1 + \rho_{P_{2r,2r+1}}, H_{P_{2r,2r+1}}(g) >= |\det a|^{2r-1}$$

if we write $g = um(a,b)\mathrm{k}$ using the Iwasawa decomposition

$$P_{2r,2r+1}(\mathbb{A})\mathrm{K}.$$

Hence (3.44) is equal to

$$\int_{\hat{p}}\int_{[V_{2r+1,r-l}]} |\det a(wg)|^{2r-1}\phi_{\pi(\tau)\otimes\tau'}(v'\hat{p}wg)\psi^{\varsigma}_{V_{2r+1,r-l}}(v')dv'd\hat{p}, \qquad (3.51)$$

We now return to to the explicit calculation of $\mathcal{P}_{6r+1,4r+2l}(\mathcal{E}_{\pi(\tau)\otimes\tau}, \mathcal{E}_1)$, the period defined in (3.29). In (3.30) together with (3.31), the integration over the variable g can factor through the Iwasawa decomposition, i.e.

$$\int_{M_{2r,2l}(k)U_{2r,2l}(\mathbb{A})\backslash \mathrm{SO}'_{4r+2l}(\mathbb{A})} = \int_{\mathrm{K}}\int_{[\mathrm{GL}_{2r}]\times[\mathrm{SO}'_{2r+1}]}.$$

By plugging this and (3.44) in (3.30) and changing variable in $\hat{p}$, we obtain

$$\int_{\mathrm{K}}\int_{\hat{p}}\int_{[\mathrm{GL}_{2r}\times\mathrm{SO}'_{2l}]} \Phi^c(m(a,b)\mathrm{k})|\det a|^{-(r+l)} \qquad (3.52)$$
$$\int_{[V_{2r+1,r-l}]} \phi_{\pi(\tau)\otimes\tau'}(v'm(a,b)\hat{p}w\mathrm{k})\psi^{\varsigma}_{V_{2r+1,r-l}}(v')dv'd\hat{p}dmdk.$$

Recall that

$$\begin{aligned}\Phi^c(m(a,b)\mathrm{k},s) &= \Phi(m(a,b)\mathrm{k},s,\phi)(1-\tau_c(H(a)))\\ &= \phi_{\pi(\tau)\otimes\sigma'}(m(a,b)\mathrm{k})|\det a|^{s+\frac{2r+2l-1}{2}}(1-\tau_c(H(a))).\end{aligned}$$

Then (3.52) can be re-written as

$$\int_{\mathrm{K}}\int_{\hat{p}}\int_{[\mathrm{GL}_{2r}]}|\det a|^{s-\frac{1}{2}}(1-\tau_c(H(a)))\int_{[\mathrm{SO}'_{2l}]}\phi_{\pi(\tau)\otimes\sigma'}(m(a,b)\mathrm{k}) \tag{3.53}$$
$$\int_{[V_{2r+1,r-l}]}\phi_{\pi(\tau)\otimes\tau'}(v'm(a,b)\hat{p}w\mathrm{k})\psi^{\varsigma}_{V_{2r+1,r-l}}(v')dv'dmd\hat{p}d\mathrm{k}.$$

In (3.53) we set

$$\mathcal{F}^{\psi^{\varsigma}}(\phi_{\pi(\tau)\otimes\tau'})(g):=\int_{[V_{2r+1,r-l}]}\phi_{\pi(\tau)\otimes\tau'}(v'g)\psi^{\varsigma}_{V_{2r+1,r-l}}(v')dv'. \tag{3.54}$$

We consider the Langlands decomposition for $\mathrm{GL}_{2r}(\mathbb{A})$

$$\mathrm{GL}_{2r}(\mathbb{A})=\mathrm{GL}_{2r}(\mathbb{A})^1\cdot A^+,$$

and set $\mathrm{PM}_{2r,2l}=(Z_{\mathrm{GL}_{2r}}(\mathbb{A})\mathrm{GL}_{2r}(k)\backslash\mathrm{GL}_{2r}(\mathbb{A}))\times(\mathrm{SO}'_{2l}(k)\backslash\mathrm{SO}'_{2l}(\mathbb{A}))$. Let d be the degree of the number field k over $\mathbb{Q}$. Then (3.53) equals

$$\mathrm{vol}(\mathbb{A}^1/k^\times)\int_{\mathbb{R}^+}|t|^{2rd(s-\frac{1}{2})}(1-\tau_c(t))dt^\times\int_{\mathrm{K}}\int_{\hat{p}} \tag{3.55}$$
$$\int_{\mathrm{PM}_{2r,2l}}\phi_{\pi(\tau)\otimes\sigma'}(m\mathrm{k})\mathcal{F}^{\psi^{\varsigma}}(\phi_{\pi(\tau)\otimes\tau'})(m\hat{p}w\mathrm{k})dmd\hat{p}d\mathrm{k}.$$

It is easy to see that

$$\int_{\mathbb{R}^+}|t|^{2rd(s-\frac{1}{2})}(1-\tau_c(t))dt^\times=\frac{c^{2rd(s-\frac{1}{2})}}{2rd(s-\frac{1}{2})}$$

which has a simple pole at $s=\frac{1}{2}$. In (3.54) we are left with

$$\int_{\mathrm{K}\times\mathrm{Mat}_{r-l,2r}(\mathbb{A})\times\mathrm{PM}_{2r,2l}}\phi_{\pi(\tau)\otimes\sigma'}(m\mathrm{k})\mathcal{F}^{\psi^{\varsigma}}(\phi_{\pi(\tau)\otimes\tau'})(m\hat{p}w\mathrm{k})dmd\hat{p}d\mathrm{k} \tag{3.56}$$

which is holomorphic in s. Hence we obtain from the above explicit calculation and (3.28) the main identity which relates the 'outer' period to the 'inner' period.

Theorem 3.8. When $s\neq\frac{1}{2}$, the period

$$\mathcal{P}_{6r+1,4r+2l}(\mathcal{E}_{\pi(\tau)\otimes\tau'},E_s(\cdot,\phi_{\pi(\tau)\otimes\sigma'}))$$

is identically zero. When $s=\frac{1}{2}$, the period

$$\mathcal{P}_{6r+1,4r+2l}(\mathcal{E}_{\pi(\tau)\otimes\tau'},\mathcal{E}_{\pi(\tau)\otimes\sigma'})$$

is equal to

$$\frac{\mathrm{vol}(\mathbb{A}^1/k^\times)}{2rd}\int_{\mathrm{K}\times\mathrm{Mat}_{r-l,2r}(\mathbb{A})\times\mathrm{PM}_{2r,2l}}\phi_{\pi(\tau)\otimes\sigma'}(m\mathrm{k})\mathcal{F}^{\psi^{\varsigma}}(\phi_{\pi(\tau)\otimes\tau'})(m\hat{p}w\mathrm{k})dmd\hat{p}d\mathrm{k}$$

where $\mathrm{PM}_{2r,2l} = (Z_{\mathrm{GL}_{2r}}(\mathbb{A})\mathrm{GL}_{2r}(k)\backslash\mathrm{GL}_{2r}(\mathbb{A})) \times (\mathrm{SO}'_{2l}(k)\backslash\mathrm{SO}'_{2l}(\mathbb{A}))$.

Remark 3.9. From formulas (3.55) and (3.56), the period

$$\mathcal{P}_{6r+1,4r+2l}(\mathcal{E}_{\pi(\tau)\otimes\tau'},\mathcal{E}_1)$$

has meromorphic continuation to the whole complex plane.

This completes the proof of Proposition 3.7 for this period. The statement for the period

$$\mathcal{P}_{6r+1,4r+2l}(\mathcal{E}_{\pi(\tau)\otimes\tau'},\mathcal{E}_2)$$

follows from the same argument (as in **[GJR01]**). Hence the proof for Proposition 3.7 is now completed.

Remark 3.10. We can use the same argument to prove Theorem 3.8 for the case of $l > r$. In this case, we can also use the following 'outer' period

$$\int_{[\mathrm{SO}'_{4l+2r+1}]} \mathcal{E}_{\pi(\sigma)\otimes\tau'}(g) \int_{[V_{6l,l-r-1}]} \mathcal{E}_{\pi(\sigma)\otimes\sigma'}(vg)\psi_{V_{6l,,l-r-1}}(v)dvdg$$

where $\pi(\sigma)$ is the Langlands transfer of σ to GL_{2l}. We omit the details.

4. The proof of Theorem A

In this section we prove Theorem A which gives a sufficient condition for $L(\frac{1}{2},\pi_1\times\pi_2)$ to be non-zero.

Let π_1 be an irreducible unitary cuspidal automorphic representation of $\mathrm{GL}_{2r}(\mathbb{A})$ such that the exterior square L-function $L(s,\pi_1,\Lambda^2)$ has a pole at $s=1$, and let τ be an irreducible unitary generic cuspidal automorphic representation of $\mathrm{SO}_{2r+1}(\mathbb{A})$ which has a weak lift to π_1. Let π_2 be an irreducible unitary cuspidal automorphic representation of $\mathrm{GL}_{2l}(\mathbb{A})$ with the properties that the symmetric square L-function $L(s,\pi_2,\mathrm{Sym}^2)$ has a pole at $s=1$ and let σ be an irreducible unitary generic cuspidal automorphic representation of $\mathrm{SO}_{2l}(\mathbb{A})$ which has a weak lift to π_2. Note that the existence of τ and σ is established by the automorphic descent method ([**GRS99c**], [**GRS01**], and [**S**]). By the work of [**CKPSS03**], [**JS03**] and [**JS04**], such weak lifts are Langlands functorial.

We first discuss the cases more general than what Theorem A asserts. Let τ' be an irreducible unitary generic cuspidal automorphic representation of $\mathrm{SO}'_{2r+1}(\mathbb{A})$ (an inner form of SO_{2r+1}), which is nearly equivalent to τ, and σ' be an irreducible unitary generic cuspidal automorphic representation of $\mathrm{SO}'_{2l}(\mathbb{A})$ (an inner form of SO_{2l}) which is nearly equivalent to σ.

Proposition 4.1. If there exists a pair (τ, σ') as above such that the period

$$\mathcal{P}_{2r+1,2l}(\phi_\tau, \phi_{\sigma'}) \ (r \geq l) \quad or \quad \mathcal{P}_{2l,2r+1}(\phi_{\sigma'}, \phi_\tau) \ (r < l)$$

does not vanish for some given $\phi_{\sigma'} \in \sigma'$ and $\phi_\tau \in \tau$, then the residue $\mathcal{E}_{\pi(\tau)\otimes\sigma'}$ at $s = \frac{1}{2}$ of the Eisenstein series $E(g, s, \phi_{\pi(\tau)\otimes\sigma'})$ does not vanish.

We shall prove this Theorem by using Theorem 3.8 for the case $r \geq l$. The case $r < l$ follows by a similar argument. Recall from (2.5) that the period $\mathcal{P}_{2r+1,2l}(\phi_\tau, \phi_{\sigma'})$ is defined by

$$\mathcal{P}_{2r+1,2l}(\phi_\tau, \phi_{\sigma'}) = \int_{[\mathrm{SO}'_{2l}]} \phi_{\sigma'}(g) \int_{[V_{2r+1,r-l}]} \phi_\tau(vg)\psi^\varsigma_{V_{2r+1,r-l}}(v)dvdg. \quad (4.1)$$

We consider the cuspidal data $(P_{2r,2l}, \pi(\tau) \otimes \sigma')$ and $(P_{2r,2r+1}, \pi(\tau) \otimes \tau)$ of $\mathrm{SO}_{4r+2l}(\mathbb{A})$ and $\mathrm{SO}_{6r+1}(\mathbb{A})$, respectively. Note that σ' may not be generic. We may assume that

$$\phi_{\pi(\tau)\otimes\sigma'} = \phi_{\pi(\tau)} \otimes \phi_{\sigma'} \quad (4.2)$$

$$\phi_{\pi(\tau)\otimes\tau} = \overline{\phi}_{\pi(\tau)} \otimes \phi_\tau \quad (4.3)$$

where $\overline{\phi}_{\pi(\tau)}$ is the complex conjugate of $\phi_{\pi(\tau)}$. Recall from (3.2) that

$$\Phi(g, s, \phi_{\pi(\tau)\otimes\sigma'}) = \phi_{\pi(\tau)\otimes\sigma'}(g) \exp < s + \rho_{P_{2r,2l}}, H_{P_{2r,2l}}(g) >$$

$$\Phi(g, s, \phi_{\pi(\tau)\otimes\tau}) = \phi_{\pi(\tau)\otimes\tau}(g) \exp < s + \rho_{P_{2r,2r+1}}, H_{P_{2r,2r+1}}(g) > .$$

We need the following Proposition to finish the proof of Proposition 4.1.

Proposition 4.2. If the period

$$\mathcal{P}_{2r+1,2l}(\phi_\tau, \phi_{\sigma'})$$

does not vanish for some given $\phi_{\sigma'} \in \sigma'$ and $\phi_\tau \in \tau$, then integral (3.56)

$$\int_{K\times \mathrm{Mat}_{r-l,2r}(\mathbb{A})\times \mathrm{PM}_{2r,2l}} \phi_{\pi(\tau)\otimes\sigma'}(mk)\mathcal{F}^{\psi^\varsigma}(\phi_{\pi(\tau)\otimes\tau'})(m\hat{p}nk)dmd\hat{p}dk$$

does not vanish for the corresponding data defined in (4.2) and (4.3), and for $\phi_{\pi(\tau)} \in \pi(\tau)$, where $\mathcal{F}^{\psi^\varsigma}(\phi_{\pi(\tau)\otimes\tau'})$ is defined as in (3.54).

Proof. The proof is the same as that of Proposition 5.3 in **[GJR03b]**. We omit the details here. □

By Proposition 4.2 and Theorem 3.8, if the period

$$\mathcal{P}_{2r+1,2l}(\phi_\tau, \phi_{\sigma'})$$

does not vanish for some given $\phi_{\sigma'} \in \sigma'$, $\phi_\tau \in \tau$, and $\phi_{\pi(\tau)} \in \pi(\tau)$, then the period

$$\mathcal{P}_{6r+1,4r+2l}(\mathcal{E}_{\pi(\tau)\otimes\tau}, \mathcal{E}_{\pi(\tau)\otimes\sigma'})$$

does not vanish for the corresponding data. In particular, this implies that the residue

$$\mathcal{E}_{\pi(\tau)\otimes\sigma'}$$

does not vanish for the given data. This proves Proposition 4.1.

It is worthwhile noting that Proposition 4.1 provides a characterization of the existence of residue in terms of a period of the cuspidal datum without assuming the genericity of the cuspidal support of the residue.

To link the existence of the residue $\mathcal{E}_{\pi(\tau)\otimes\sigma'}$ with the nonvanishing of the central value $L(\frac{1}{2}, \pi(\tau) \times \sigma')$, we have to use Formula (3.9). The key point here is Proposition 3.2, which has been proved only when σ' is generic.

In the following we assume that $\sigma' = \sigma$ is generic. Then from Proposition 3.2 the product of L-functions

$$L(s, \pi(\tau) \times \sigma)L(2s, \pi(\tau), \Lambda^2)$$

has a simple pole at $s = \frac{1}{2}$. Since the exterior square L-function

$$L(s, \pi(\tau), \Lambda^2)$$

has a simple pole at $s = 1$, the product L-function

$$L(s, \pi(\tau) \times \sigma)$$

does not vanish at $s = \frac{1}{2}$. Since $\pi_2 = \pi(\sigma)$ and $\pi_1 = \pi(\tau)$, we obtain that the tensor product L-function

$$L(s, \pi_1 \times \pi_2)$$

does not vanish at $s = \frac{1}{2}$.

Since in Theorem A, both τ and σ are generic, the above argument gives a complete proof of Theorem A.

5. The proof of Theorem B

In this section we prove Theorem B for the case $r \geq l = 1$. Let π_1 be an irreducible unitary cuspidal automorphic representation of $\mathrm{GL}_{2r}(\mathbb{A})$ such that the exterior square L-function $L(s, \pi_1, \Lambda^2)$ has a pole at $s = 1$ and let π_2 be an irreducible unitary cuspidal automorphic representation of $\mathrm{GL}_2(\mathbb{A})$ such that the symmetric square L-function $L(s, \pi_2, \mathrm{Sym}^2)$ has a pole at $s = 1$. We assume that $\pi_1 = \pi(\tau)$ is a lift from an irreducible unitary generic cuspidal automorphic representation of $\mathrm{SO}_{2r+1}(\mathbb{A})$ and $\pi_2 = \pi(\sigma)$ is a lift from an irreducible unitary automorphic representation of $\mathrm{SO}_2(\mathbb{A})$ (here σ is a character $\mathrm{SO}_2(k)\backslash\mathrm{SO}_2(\mathbb{A})$). Let SO'_{2r+1} be an inner form of SO_{2r+1} and SO'_2 be an inner form of SO_2. Assume that $(\mathrm{SO}'_{2r+1}, \mathrm{SO}'_2)$ is a relevant pair.

Theorem 5.1 ($r \geq l = 1$). If the central value of the Rankin-Selberg L-function $L(\frac{1}{2}, \pi_1 \times \pi_2)$ is non-zero, then there is a pair (τ', σ') of $(\mathrm{SO}'_{2r+1}, \mathrm{SO}'_2)$ which is nearly equivalent to (τ, σ) such that the period $\mathcal{P}_{2r+1,2}(\phi_{\sigma'}, \phi_{\tau'})$ is not identically zero.

To prove this Theorem, we have to consider two cases. First is the case when SO_2 is the split orthogonal group. In this case $\mathrm{SO}_2 \cong \mathrm{GL}_1$ and the only relevant pair is $(\mathrm{SO}_{2r+1}, \mathrm{GL}_1)$ where SO_{2r+1} is the split orthogonal group. In this case the period $\mathcal{P}_{2r+1,2}(\phi_{\sigma'}, \phi_{\tau'})$ is just the Norodvorsky integral [N75] evaluated at $s = 1/2$. Since the Norodvorsky integral represents the L-function $L(s, \pi_1 \times \pi_2)$. The Theorem follows in this case.

The second case is when SO_2 is anisotropic. For this case we consider the Eisenstein series $E(g, s, \phi_{\pi(\tau)\otimes\sigma})$ on SO'_{4r+2} as studied in Section 3.1. Note that the quasisplit orthogonal group SO'_{4r+2} has a Levi subgroup isomorphic to $\mathrm{GL}_{2r} \times \mathrm{SO}_2$. By Proposition 3.2, we know that the residue at $s = \frac{1}{2}$ of $E(g, s, \phi_{\pi(\tau)\otimes\sigma})$ does not vanish if the L-function

$$L(s, \pi_1 \times \pi_2) = L(s, \pi(\tau) \times \sigma)$$

is non-zero at $s = \frac{1}{2}$. As before, we denote the residue by $\mathcal{E}_{\pi(\tau)\otimes\sigma}$. The proof of the Theorem in this case is based on the study of certain Bessel model of the residue $\mathcal{E}_{\pi(\tau)\otimes\sigma}$. We need the information on the structure of the local unramified component of the residue $\mathcal{E}_{\pi(\tau)\otimes\sigma}$ before we investigate the global Bessel coefficients of $\mathcal{E}_{\pi(\tau)\otimes\sigma}$.

Fix a local finite place ν of k and set $k_\nu = F$. Assume that the local ν-components of τ and σ are unramified, the relevant pair $(\mathrm{SO}_{2r+1}, \mathrm{SO}_2)$ is split over F. Since τ_ν is generic, we may write

$$\tau_\nu = \mathrm{Ind}_{B_{\mathrm{SO}_{2r+1}}(F)}^{\mathrm{SO}_{2r+1}(F)}(\chi)$$

and then the lift to $\mathrm{GL}_{2r}(F)$ of τ_ν is

$$\pi(\tau_\nu) = \mathrm{Ind}_{B_{\mathrm{GL}_{2r}}(F)}^{\mathrm{GL}_{2r}(F)}(\chi)$$

where the character χ is defined as in the proof of Lemma 3.4. Let μ denote an unramified character of $\mathrm{SO}_2(F) = \mathrm{GL}_1(F)$. We denote by $P_{2^r,1}$ the parabolic subgroup of SO_{4r+2} whose Levi part is $\mathrm{GL}_2^r \times \mathrm{GL}_1$. Define a character $\mu \otimes \chi$ of $\mathrm{GL}_2^r(F) \times \mathrm{GL}_1(F)$ to be

$$\mu \otimes \chi(h_1, \cdots, h_r, a) = \prod_{i=1}^{r} \chi_i(\det h_i)\mu(a),$$

where $h_i \in \mathrm{GL}_2(F)$ and $a \in \mathrm{GL}_1(F)$. We have

Lemma 5.2. The local unramified component of the residue $\mathcal{E}_{\pi(\tau)\otimes\sigma}$ is the unique unramified quotient of $\mathrm{Ind}_{P_{2^r,1}(F)}^{\mathrm{SO}_{4r+2}(F)}(\mu \otimes \chi)$.

Proof. This is proved exactly as in Proposition 1, **[GRS99b]**. □

We now return to the global situation and consider the Bessel coefficients of the residue $\mathcal{E}_{\pi(\tau)\otimes\sigma}$. Let U denote the upper unipotent subgroup of SO'_{4r+2}, the elements of which can be written as

$$u = \begin{pmatrix} n & x & z \\ & I_2 & x^* \\ & & n^* \end{pmatrix} \in \mathrm{SO}'_{4r+2}$$

were $n \in N_{2r}$, the maximal upper unipotent subgroup of GL_{2r}, and $z \in \{X \in \mathrm{Mat}_{2r\times 2r} : {}^tXJ_r + J_rX = 0\}$. Assume that SO'_{4r+2} is defined in terms of the symmetric matrix

$$\begin{pmatrix} & & J_{2r} \\ & \begin{pmatrix} \alpha & \\ & \beta \end{pmatrix} & \\ J_{2r} & & \end{pmatrix}$$

where we define inductively $J_{2r} = \begin{pmatrix} & & 1 \\ & J_{2r-2} & \\ 1 & & \end{pmatrix}$, and $\begin{pmatrix} \alpha & \\ & \beta \end{pmatrix}$ defines SO_2. Then if the two columns of x are $\alpha\ell_1$ and $\beta\ell_2$, then the rows of x^* are $-(J_r\ell_1)^t$ and $-(J_r\ell_2)^t$.

For $1 \le e \le 2r$ let V_e be the subgroup of U which consists of all matrices of the form

$$v(v_1, v_2, v_3) = \begin{pmatrix} v_1 & v_2 & v_3 \\ & I_{4r-2e+2} & v_2^* \\ & & v_1^* \end{pmatrix} \in U,$$

where $v_1, v_3 \in \mathrm{Mat}_{e\times e}, v_2 \in \mathrm{Mat}_{e\times(4r-2e+2)}$. Notice that $U = V_{2r}$. Let ψ be an additive character of $k\backslash\mathbb{A}$. Let $\xi \in k^{4r-2e+2}$ be a vector with nonzero length. We define a character ψ_e^ξ of V_e by

$$\psi_e^\xi(v(v_1, v_2, v_3)) = \psi((v_1)_{1,2} + \cdots + (v_1)_{e-1,e} + (v_2)_e \cdot \xi)$$

where $(v_1)_{i,j}$ is the (i,j)- entry of v_1, $(v_2)_e$ is the e-th row of v_2 and $(v_2)_e \cdot \xi$ is the inner product of $(v_2)_e$ with ξ.

Let ϵ be an automorphic representation on $\mathrm{SO}'_{4r+2}(\mathbb{A})$. For all $1 \le e \le 2r$ we define

$$\mathcal{B}_e(\epsilon, \xi)(g) = \int_{V_e(F)\backslash V_e(\mathbb{A})} \phi_\epsilon(vg)\psi_e^\xi(v)dv$$

for $\phi_\epsilon \in \epsilon$, and refer to it as the e-th Bessel coefficient of ϕ_ϵ. We have

Proposition 5.3. Suppose that $r < e \leq 2r$. Then the the e-th Bessel coefficient of the residue $\mathcal{E}_{\pi(\tau)\otimes\sigma}$, i.e. the integral $\mathcal{B}_e(\mathcal{E}_{\pi(\tau)\otimes\sigma}, \xi)$ vanishes for all choice of data and all $\xi \in k^{4r-2e+2}$ which have nonzero length.

Proof. The proof is based on the structure of the local unramified component of the residue $\mathcal{E}_{\pi(\tau)\otimes\sigma}$. In other words, it is enough to show that at one local place the local component of the residue $\mathcal{E}_{\pi(\tau)\otimes\sigma}$ has no such functionals. From Lemma 5.2 we can find a local unramified place ν such that $(\mathcal{E}_{\pi(\tau)\otimes\sigma})_\nu$ is the unramified quotient of $\mathrm{Ind}_{P_{2^r,1}(F)}^{\mathrm{SO}_{4r+2}(F)}(\mu \otimes \chi)$ $(F = k_\nu)$. It suffices to show that $\mathrm{Ind}_{P_{2^r,1}(F)}^{\mathrm{SO}_{4r+2}(F)}(\mu \otimes \chi)$ has no such functionals. This is done in the same way as in Lemma 3.5 and the references cited there. Namely, we consider the double cosets $P_{2^r,1}\backslash \mathrm{SO}_{4r+2}/V_e$ and show that if γ is any representative of a double coset, then there exists $v \in V_e$ such that $\gamma v \gamma^{-1} \in P_{2^r,1}$ and $\psi_e^\xi(v) \neq 1$. This proves the proposition. □

Next we consider the r-th Bessel coefficient of the residue $\mathcal{E}_{\pi(\tau)\otimes\sigma}$, i.e. $\mathcal{B}_r(\mathcal{E}_{\pi(\tau)\otimes\sigma}, \xi)(g)$. Depending on ξ this coefficient is an automorphic function on $\mathrm{SO}(r+1, r)(\mathbb{A})$ or $\mathrm{SO}(r+2, r-1)(\mathbb{A})$. We have

Proposition 5.4. As an automorphic function in g, the r-th Bessel coefficient $\mathcal{B}_r(\mathcal{E}_{\pi(\tau)\otimes\sigma}, \xi)(g)$ is cuspidal.

Proof. This is done in a similar way as in Theorem 8 pp. 839–844, [**GRS99c**]. Indeed, when one computes the constant terms of $\mathcal{B}_r(\mathcal{E}_{\pi(\tau)\otimes\sigma}, \xi)(g)$ along various maximal unipotent radicals of SO_{2r+1} one obtains integrations which either contain constant terms of the residue $\mathcal{E}_{\pi(\tau)\otimes\sigma}$ or contain as inner integration the e-th Bessel coefficients of the residue $\mathcal{E}_{\pi(\tau)\otimes\sigma}$ for $r < e \leq 2r$. By Proposition 5.3 these e-th Bessel coefficients are zero. As for the constant terms obtained, one can check that none of them are along the unipotent radical of $P_{2r,2}$. Hence, by cuspidality of $\pi(\tau)$, they are all zero. □

The following nonvanishing property is the key to prove Theorem 5.1.

Proposition 5.5. There exists a $\xi_0 \in k^{2r+2}$ with nonzero length such that the r-th Bessel coefficient $\mathcal{B}_r(\mathcal{E}_{\pi(\tau)\otimes\sigma}, \xi_0)(g)$ of the residue $\mathcal{E}_{\pi(\tau)\otimes\sigma}$ is nonzero for some choice of data.

Proof. Let $Q_{2^r,2}$ be the standard parabolic subgroup of SO'_{4r+2} whose Levi part is $\mathrm{GL}_2^r \times \mathrm{SO}_2$. Let $U_{2^r,2}$ denote its unipotent radical, the elements of which can be expressed as

$$u = \begin{pmatrix} I_2 & x_1 & & * & & & & \\ & I_2 & \ddots & & & * & & \\ & & \ddots & x_{r-1} & & & & \\ & & & I_2 & y & z & & \\ & & & & I_2 & y^* & & \\ & & & & & I_2 & & * \\ & & & & & & \ddots & \\ & & & & & & & I_2 \end{pmatrix} \in \mathrm{SO}'_{4r+2}, \tag{5.1}$$

where $x_i, y \in \mathrm{Mat}_{2\times 2}$ and $z = \begin{pmatrix} t & 0 \\ 0 & -t \end{pmatrix}$. Define below a character $\psi_{U_{2^r,2}}$ of $U_{2^r,2}(\mathbb{A})$, which is trivial on $U_{2^r,2}(k)$

$$\psi_{U_{2^r,2}}(u) = \psi(\mathrm{tr}(x_1) + \cdots + \mathrm{tr}(x_{n-1})).$$

The subgroup of $U_{2^r,2}$ which consists of the (y,z) variables is a Heisenberg group H_5 with five variables. It is also clear that there exists a canonical group homomorphism from $U_{2^r,2}$ to H_5, which we denote by ℓ.

Let $g \in \mathrm{SL}_2$ be embedded in GL_2^r diagonally. Then it is easy to see that SL_2 preserves the character $\psi_{U_{2^r,2}}$ and hence we can define

$$f(g) := \int_{U_{2^r,2}(k)\backslash U_{2^r,2}(\mathbb{A})} \mathcal{E}_{\pi(\tau)\otimes\sigma}(ug)\widetilde{\theta}_\varphi^\psi(\ell(u)g)\psi_{U_{2^r,2}}(u)du$$

where $\widetilde{\theta}_\varphi^\psi$ is the theta function on $\widetilde{\mathrm{Sp}}_4(\mathbb{A})$ and $g \in \mathrm{SL}_2(\mathbb{A})$. Since the copy of SL_2 is part of the dual pair $(\mathrm{SL}_2, \mathrm{SO}_2)$ in Sp_4 it splits under the cover.

Clearly $f(w_0 g) = f(g)$ where w_0 is the Weyl element of $\mathrm{SL}_2(k)$. To prove the Theorem we will prove that under the assumption that $\mathcal{B}_r(\mathcal{E}_{\pi(\tau)\otimes\sigma}, \xi)$ is zero for all choice of data, the function $f(g)$ enjoys the following two properties:

(1) $f(g)$ is not identically zero.

(2) $f(\begin{pmatrix} b & \\ & b^{-1} \end{pmatrix} g) = |b|^2 f(g)$ for all $b \in \mathbb{A}^\times$.

This is clearly a contradiction. In fact, by (2) and the automorphicity of $f(g)$, we have

$$|b|^2 f(g) = f(\begin{pmatrix} b & \\ & b^{-1} \end{pmatrix} g) = f(w_0 \begin{pmatrix} b & \\ & b^{-1} \end{pmatrix} g) = f(\begin{pmatrix} b^{-1} & \\ & b \end{pmatrix} w_0 g) = |b|^{-2} f(g).$$

If $f(g)$ is not identically zero, we derive a contradiction. This will prove the proposition.

In order to show that the function $f(g)$ has properties (1) and (2), we unfold the theta function in the above integral and we obtain

$$\int_{\mathbb{A}^2}\int_{U^0_{2r,2}(k)\backslash U^0_{2r,2}(\mathbb{A})} \mathcal{E}_{\pi(\tau)\otimes\sigma}(u(x_1,x_2)^\wedge g)(\omega_\psi(g)\phi)(x_1,x_2)\psi_{U^0_{2r,2}}(u)dudx_i$$

where $U^0_{2r,2}$ is the subgroup of $U_{2r,2}$ where the bottom row of y, as denoted in (5.1), is zero. By $(x_1,x_2)^\wedge$ we denote the matrix u as defined in (5.1) where (x_1,x_2) is the bottom row of y, and all other variables in u are zero. We integrate x_i over $\mathbb{A}^2$. Finally

$$\psi_{U^0_{2r,2}}(u)=\psi_{U_{2r,2}}(u)\psi(t)$$

where $z=\begin{pmatrix} t & 0\\ 0 & -t\end{pmatrix}$ as denoted in (5.1).

We now concentrate on the integral

$$\int_{U^0_{2r,2}(k)\backslash U^0_{2r,2}(\mathbb{A})}\mathcal{E}_{\pi(\tau)\otimes\sigma}(um)\psi_{U^0_{2r,2}}(u)du.$$

Let w be the Weyl element of SO'_{4r+2} defined by $w[i,2i-1]=w[r+i,2r+2i-1]=w[2r+1,2r+1]=w[2r+2,2r+2]=1$. This determines w uniquely. Conjugating the above integral by w from left to right we obtain

$$\int \mathcal{E}_{\pi(\tau)\otimes\sigma}(\begin{pmatrix} v_1 & v_2 & v_3\\ & I_2 & v_2^*\\ & & v_1^*\end{pmatrix}\begin{pmatrix} I_{2n} & & \\ p & I_2 & \\ q & p^* & I_{2n}\end{pmatrix} wm)\psi_r(v_1)dv_idpdq$$

where v_1 is integrated over $N_{2r}(k)\backslash N_{2r}(\mathbb{A})$ (N_{2r} is the maximal upper unipotent subgroup of GL_{2r}) and

$$\psi_r(v_1)=\psi((v_1)_{1,2}+\cdots+(v_1)_{2r-1,2r}).$$

The variable v_2 is integrated over $\mathrm{Mat}_{2r\times 2}(k)\backslash\mathrm{Mat}_{2r\times 2}(\mathbb{A})$ such that the last r rows are zero. The variables v_3 and q are integrated over all matrices in

$$\{X\in \mathrm{Mat}_{2r\times 2r}\ :\ {}^tXJ_{2r}+J_{2r}X=0 \textit{ and } X_{i,j}=0 \textit{ for all } i\geq j\}.$$

Finally, p is integrated over all matrices in $\mathrm{Mat}_{2\times 2r}$ such that $p_{i,j}=0$ for all $j\leq r+1$. These variables are integrated over the $\mathbb{A}$-rational points modulo the k-rational points of the corresponding varieties. Using the assumption that $\mathcal{B}_r(\mathcal{E}_{\pi(\tau)\otimes\sigma},\xi)$ is zero for all nonzero length vectors $\xi\in k^{2r+2}$ and apply similar Fourier expansions as in the proof of Proposition 7.1 (**[GJR03b]**) starting at integral (7.6), we deduce that the above integral equals

$$\int \mathcal{E}^{U_{2r,2}}_{\pi(\tau)\otimes\sigma}(\begin{pmatrix} v_1 & & \\ & I_2 & \\ & & v_1^*\end{pmatrix}\begin{pmatrix} I_{2r} & & \\ p & I_2 & \\ q & p^* & I_{2r}\end{pmatrix} wm)\psi_r(v_1)dv_1dpdq.$$

Here $\mathcal{E}^{U_{2r,2}}_{\pi(\tau)\otimes\sigma}$ is the constant term along $U_{2r,2}$ which is the unipotent radical of $Q_{2r,2}$. The matrices v_1, p and q are defined as before but p and q are integrated over the adelic points. Combining this with the Weil representation we thus obtain

$$f(g) = \int \mathcal{E}^{U_{2r,2}}_{\pi(\tau)\otimes\sigma}\left(\begin{pmatrix} v_1 & & \\ & I_2 & \\ & & v_1^* \end{pmatrix}\begin{pmatrix} I_{2r} & & \\ p & I_2 & \\ q & p^* & I_{2r} \end{pmatrix} w(x_1,x_2)^\wedge g\right)$$
$$(\omega_\psi(g)\phi)(x_1,x_2)\psi_r(v_1)d(\ldots).$$

The torus of SL_2 is embedded in SO'_{4r+2} as

$$\mathrm{diag}(b, b^{-1}, \cdots, b, b^{-1}, 1, 1, b, b^{-1}, \cdots, b, b^{-1}).$$

When conjugating by w we obtain $\mathrm{diag}(bI_{2r}, I_2, b^{-1}I_{2r})$ which is the center of the Levi part of $Q_{2r,2}$. Hence $f(g)$ has a left transformation property under this torus. When we conjugate across the $(x_1,x_2)^\wedge$, p and q integration we obtain from a change of variables, a factor of $|b|^{-2r^2}$. From the Weil representation we get a factor of $|b|^2$ and from $\mathcal{E}^{U_{2r,2}}_{\pi(\tau)\otimes\sigma}$ we get $|b|^{2r^2}$. Overall we are left with $|b|^2$ and hence (2) holds for $f(g)$. As for the nonvanishing of $f(g)$, we argue as in Lemmas 1 and 2 on p.895–896, **[GRS99a]** and deduce that $f(g)$ is not zero for some choice of data. □

We now finish the proof of Theorem 5.1. From Propositions 5.4 and 5.5 we deduce that there is a $\xi_0 \in k^{2r+2}$ with nonzero length such the r-th Bessel coefficient $\mathcal{B}_r(\mathcal{E}_{\pi(\tau)\otimes\sigma}, \xi_0)(g)$ is a nonzero cuspidal automorphic function on $\mathrm{SO}'_{2r+1}(\mathbb{A})$. Let τ' be any irreducible summand of the cuspidal automorphic representation of $\mathrm{SO}'_{2r+1}(\mathbb{A})$ generated by all r-th Bessel coefficients $\mathcal{B}_r(\mathcal{E}_{\pi(\tau)\otimes\sigma}, \xi_0)(g)$. It then follows that the integral

$$\int_{\mathrm{SO}'_{2r+1}(k)\backslash \mathrm{SO}'_{2r+1}(\mathbb{A})} \int_{V_r(k)\backslash V_r(\mathbb{A})} \phi_{\tau'}(g)\mathcal{E}_{\pi(\tau)\otimes\sigma}(vg)\psi_r^{\xi_0}(v)dvdg$$

is nonzero for some choice of data. Replacing the residue by the Eisenstein series itself we deduce that the integral

$$\int_{\mathrm{SO}'_{2r+1}(k)\backslash \mathrm{SO}'_{2r+1}(\mathbb{A})} \int_{V_r(k)\backslash V_r(\mathbb{A})} \phi_{\tau'}(g)\mathcal{E}_{\pi(\tau)\otimes\sigma}(vg, s)\psi_r^{\xi_0}(v)dvdg$$

is not zero for the real part of s large.

Unfolding this integral as in **[GPSR97]** we deduce that $\mathcal{P}_{2r+1,2l}(\phi_\sigma, \phi_{\tau'})$ is not zero for some choice of data.

Finally, arguing as in Proposition 5 **[GRS99a]** we deduce that τ' is nearly equivalent to τ. This finishes the proof of Theorem 5.1 and hence the proof of Theorem B.

References

[Arthur78] Arthur, J. *A trace formula for reductive groups. I. Terms associated to classes in $G(Q)$*. Duke Math. J. 45 (1978), no. 4, 911–952.

[Arthur80] Arthur, J. *A trace formula for reductive groups. II. Applications of a truncation operator*. Compositio Math. 40 (1980), no. 1, 87–121.

[BFSP04] Boecherer, S.; Furusawa, M.; Schulze-Pillot, R. *On the global Gross-Prasad conjecture for Yoshida liftings*. Contributions to automorphic forms, geometry, and number theory, 105–130, Johns Hopkins Univ. Press, Baltimore, MD, 2004.

[CKPSS01] Cogdell, J.; Kim, H.; Piatetski-Shapiro, I.; Shahidi, F. *On lifting from classical groups to* GL_N. Inst. Hautes Études Sci. Publ. Math. No. 93 (2001), 5–30.

[CKPSS03] Cogdell, J.; Kim, H.; Piatetski-Shapiro, I.; Shahidi, F. *Functoriality for the classical groups*. Publ. Math.IHES. 99(2004), 163–233.

[GSh88] Gelbart, S.; Shahidi, F. *Analytic properties of automorphic L-functions*. Perspectives in Mathematics, 6. Academic Press, Inc., Boston, MA, 1988.

[GJR01] Ginzburg, D.; Jiang, D.; Rallis, S. *Nonvanishing of the central critical value of the third symmetric power L-functions*. Forum Math. 13 (2001), no. 1, 109–132.

[GJR03a] Ginzburg, D.; Jiang, D. and Rallis, S. *Periods of residue representations of* SO_{2l}. Manuscripta Math. 113(2004), 319–358.

[GJR03b] Ginzburg, D.; Jiang, D.; Rallis, S. *On the nonvanishing of the central value of the Rankin-Selberg L-function*. Journal of AMS. 17 (2004), 679–722.

[GPSR97] Ginzburg, D.; Piatetski-Shapiro, I.; Rallis, S. *L Functions for the Orthogonal group*. Memoirs of AMS 611 (1997).

[GRS99a] Ginzburg, D.; Rallis, S; Soudry, D. *On the Correspondence between cuspidal representations of* GL_{2n} *and* $\widetilde{\mathrm{Sp}}_{2n}$. J. of AMS, Vol. 12 (1999), 849-907.

[GRS99b] Ginzburg, D.; Rallis, S; Soudry, D. *Lifting Cusp Forms on* GL_{2n} *to* $\widetilde{\mathrm{Sp}}_{2n}$*: The Unramified Correspondence*. Duke Math. J. Vol. 100, No. 2 (1999), 243–266.

[GRS99c] Ginzburg, D.; Rallis, S; Soudry, D. *On explicit lifts of cusp forms from* GL_m *to classical groups*. Ann. of Math. 150 (1999), 807–866.

[GRS01] Ginzburg, D.; Rallis, S; Soudry, D. *Generic automorphic forms on* $\mathrm{SO}(2n+1)$*: functorial lift to* $\mathrm{GL}(2n)$*, endoscopy, and base change*. Internat. Math. Res. Notices 2001, no. 14, 729–764.

[GRS] Ginzburg, D.; Rallis, S; Soudry, D. *Constructions of CAP representations for symplectic groups using the descent method*. This proceedings.

[GP92] Gross, B.; Prasad, D. *On the decomposition of a representation of* SO_n *when restricted to* SO_{n-1}. Canad. J. Math. 44 (1992), no. 5, 974–1002.

[GP94] Gross, B.; Prasad, D. *On irreducible representations of* $SO_{2n+1} \times SO_{2m}$. Canad. J. Math. 46 (1994), no. 5, 930–950.

[HK91] Harris, M.; Kudla, S. *The central critical value of a triple product L-function.* Ann. of Math. (2) 133 (1991), no. 3, 605–672.

[HK92] Harris, M.; Kudla, S. *Arithmetic automorphic forms for the nonholomorphic discrete series of* GSp(2). Duke Math. J. 66 (1992), no. 1, 59–12.

[HK04] Harris, M.; Kudla, S. *On a conjecture of Jacquet.* Contributions to automorphic forms, geometry, and number theory, 355–371, Johns Hopkins Univ. Press, Baltimore, MD, 2004.

[Jng98a] Jiang, D. *G_2-periods and residual representations.* J. Reine Angew. Math. 497 (1998), 17–46.

[Jng98b] Jiang, D. *Nonvanishing of the central critical value of the triple product L-functions.* Internat. Math. Res. Notices 1998, no. 2, 73–84.

[Jng01] Jiang, D. *On Jacquet's conjecture: the split period case.* Internat. Math. Res. Notices 2001, no. 3, 145–163.

[JS03] Jiang, D.; Soudry, D., *The local converse theorem for* $SO(2n+1)$ *and applications.* Ann. of Math. 157(2003), 743–806.

[JS04] Jiang, D.; Soudry, D., *Generic Representations and the Local Langlands Reciprocity Law for p-adic* $SO(2n+1)$. Contributions to automorphic forms, geometry, and number theory, 457–519, Johns Hopkins Univ. Press, Baltimore, MD, 2004.

[K00] Kim, H. *Langlands-Shahidi methods and poles of automorphic L-functions II.* Israel J. of Math. 117(2000), 261–284.

[K] Kim, H. *On local L-functions and normalized intertwining operators.* To appear in Canad. J. Math.

[L71] Langlands, R. *Euler products.* A James K. Whittemore Lecture in Mathematics given at Yale University, 1967. Yale Mathematical Monographs, 1. Yale University Press, New Haven, Conn.-London, 1971.

[N75] Norodvorsky, M. *Fonction J pour des groupes orthogonaux.* (French) C. R. Acad. Sci. Paris Sér. A-B 280 (1975), Ai, A1421–A1422.

[Sh88] Shahidi, F. *On the Ramanujan conjecture and the finiteness of poles of certain L-functions.* Ann. of Math. (2)(1988)(127), 547-584.

[S] Soudry, D. *Langlands functoriality from classical groups to* GL_n. To appear in Asterisque.

[W85] Waldspurger, J.-L. *Sur les valeurs de certaines fonctions L automorphes en leur centre de symétrie.* (French) Compositio Math. 54 (1985), no. 2, 173–242.

David Ginzburg
School of Mathematical Sciences
Sackler Faculty of Exact Sciences
Tel Aviv University
Ramat-Aviv, 69978 Israel
ginzburg@post.tau.ac.il

Dihua Jiang
School of Mathematics
University of Minnesota
206 Church Street. S.E.
Minneapolis, MN 55455, USA
dhjiang@math.umn.edu

Stephen Rallis
Department of Mathematics
The Ohio State University
Columbus, OH 43210. USA
haar@math.ohio-state.edu

Construction of CAP representations for symplectic groups using the descent method

David Ginzburg, Stephen Rallis, and David Soudry

Abstract. In this paper we use a descent method to construct examples of CAP representations for the symplectic group. The main step involves a Fourier–Jacobi coefficient of a residue of an Eisenstein series. We give conditions when the lift is cuspidal and show that it is always nonzero.

1. Introduction

It is important in the theory of automorphic representations to understand the structure of cuspidal representations of a given reductive group G. For the group GL_n it is well known that every cuspidal representation is generic. However, for other groups this is not the case. The notion of a CAP (Cuspidal Associated with Parabolic) representation was first introduced in **[PS]**. Basically, these cuspidal representations are nearly equivalent to induced representations. It follows from **[J-S]** that there are no such representations for the group GL_n.

To understand the general predictions, given an induced representation induced from cuspidal data, one can associate with it an Eisenstein series. The general conjecture is that a CAP representation will exist precisely at the points where the corresponding Eisenstein series can have a pole. So far all examples constructed were of this nature.

The first construction of CAP representations was done in **[PS]** for the group of type C_2. Since then many other examples were found in various groups, cf. **[G-R]**, **[I]**, **[R-S]**, **[S]**. The main theme of these constructions was the use of a certain kernel function which was constructed from some small representation, usually a residue of an Eisenstein series.

In this paper we will use the so called descent method, which was introduced in **[G-R-S1]**, in order to construct examples of CAP representations on the symplectic group and on its double cover. Our method is inductive. We start with a generic cuspidal representation σ on the double cover $\widetilde{Sp}_{2n}(\mathbb{A})$. We make two assumptions on σ. First we assume that its lift to GL_{2n} produces a cuspidal representation denoted by τ. This assumption can be weakened. In fact we need only assume that τ is generic. The second assumption is that σ has a missing $\psi_{U_{n,2n}}$ Fourier coefficient (see Assumption 1 in Section 4). This assumption is crucial; without it we can not guarantee that our construction will produce a cuspidal representation. Using σ we construct a CAP representation π_2 on the group $Sp_{4n}(\mathbb{A})$. This representation will be a CAP representation with respect to $Ind_{P(\mathbb{A})}^{Sp_{4n}(\mathbb{A})}\tau\delta_P^s$, where P is the Siegel parabolic subgroup of Sp_{4n} and s is a certain fixed complex point. The construction of π_2 uses a residue of an Eisenstein series on the group $\widetilde{Sp}_{6n}(\mathbb{A})$. The process can be realized in the following diagram

$$\begin{array}{ccccccccc} & & \widetilde{Sp}_{6n} & & Sp_{8n} & & \cdots \\ & \nearrow & \downarrow & \nearrow & \downarrow & \nearrow & \\ \widetilde{Sp}_{2n} & & Sp_{4n} & & \widetilde{Sp}_{6n} & & \end{array} \tag{1.1}$$

We give some details. Let σ be as above. Given an additive character ψ of $F\backslash\mathbb{A}$, we denote by $\tau_\psi(\sigma)$ its lift to $\mathrm{GL}_{2n}(\mathbb{A})$. This lift is given by first using the theta correspondence to pass to the odd orthogonal group and then using the results of **[C-K-PS-S]**. We shall assume that $\tau_\psi(\sigma)$ is cuspidal. With this data we construct an Eisenstein series $\widetilde{E}_{\tau_\psi(\sigma),\sigma}(g,s)$ defined on the group $\widetilde{Sp}_{6n}(\mathbb{A})$. This Eisenstein series has a residue and in sections 2 and 3 we study the properties of this residue. (In these sections we study these properties for a certain more general family of Eisenstein series). Our main result is Proposition 3.6, which describes the Fourier coefficients of this residue. This proposition relies heavily on Key Lemma 3.3, where we study the unramified constituent of the residue representation. Having done all this, we use the residue representation as a kernel to define our lifting. We form the Fourier-Jacobi coefficient of the residue and descend to the group $Sp_{4n}(\mathbb{A})$. Under a certain assumption on the cuspidal representation σ we prove that the $Sp_{4n}(\mathbb{A})$ representation we obtain is indeed a non-zero cuspidal representation. As mentioned above this is the starting point for our induction. In section 4 we introduce the general setup for our construction, and state our main results, Theorems A and B. We prove them in section 5.

This first step is the only case where our construction produces a CAP with respect to a maximal parabolic subgroup. The other examples will all be CAP's with respect to non-maximal parabolics. Next we proceed by induction and we construct examples of CAP representations on the group Sp_{4nk} with respect to the parabolic subgroup P whose Levi part is isomorphic to $\mathrm{GL}_{2n}\times\cdots\times\mathrm{GL}_{2n}$. As before, in order to guarantee that our lift is cuspidal we need to make a general assumption on a vanishing of a certain Fourier coefficient (see Assumptions 2a and 2b in Section 4). Unfortunately, we don't know how to prove that this assumption is indeed satisfied. All assumptions and the general inductive process are explained in Section 4.

Recently, a different method was used in [I] in order to construct examples of CAP representations for the group Sp_{8k} with respect to the parabolic subgroup P whose Levi part is isomorphic to $\mathrm{GL}_2 \times \cdots \times \mathrm{GL}_2$. Our construction generalizes these results (using a different method) but, as explained above, it is somewhat weaker since we don't know how to prove the above assumptions.

Finally, in the last section, we study the converse problem. We start with a CAP representation on the group Sp_{4n} with respect to the Siegel parabolic subgroup and, under a certain assumption on its Fourier coefficients, we associate with it a generic cuspidal representation on $\widetilde{Sp}_{2n}(\mathbb{A})$. We then prove that this representation has a missing $\psi_{U_{n,2n}}$ Fourier coefficient.

The first and third authors would like to express their deep admiration and gratitude to Steve Rallis, for his generosity in sharing his beautiful ideas and for his friendship.

Acknowledgement. The research of the first and third named authors was supported by the Israel Science Foundation, founded by the Israel Academy of Sciences and Humanities.

2. Notation and definitions

We start by introducing the notion of CAP representations which we shall construct. Let τ be a cuspidal representation on $\mathrm{GL}_{2n}(\mathbb{A})$. For any $k \geq 1$ let $Q_k = (\mathrm{GL}_{2n} \times \ldots \times \mathrm{GL}_{2n})U(Q_k)$ denote the standard parabolic subgroup of Sp_{4nk} whose Levi part is a direct product of k copies of GL_{2n}. In general, if P is any parabolic subgroup, we shall denote by $U(P)$ its unipotent radical. Let P_r, for $1 \leq r \leq k$ denote the standard maximal parabolic subgroup of Sp_{4nk} defined by $P_r = (\mathrm{GL}_{2nr} \times \mathrm{Sp}_{4n(k-r)})U(P_r)$. For the induced representation $I(\tau, \bar{s}) = \mathrm{Ind}_{Q_k(\mathbb{A})}^{\mathrm{Sp}_{4nk}(\mathbb{A})}(\tau \otimes \cdots \otimes \tau)\delta_{Q_k}^{\bar{s}}$, (we always use non-normalized induction) we denote by $E_\tau(g, \bar{s})$ a corresponding Eisenstein series. Here $\delta_{Q_k}^{\bar{s}} = \prod_{r=1}^{k} \delta_{P_r}^{s_r}$ where $s_r \in \mathbb{C}$. Conjecturally, the poles of $E_\tau(g, \bar{s})$ are determined by

$$\prod_{r=1}^{k} L^S(\tau, \rho_r, \zeta_r) L_\tau^S(\bar{s}) \ .$$

Here, if $1 \leq r \leq k-1$,

$$L^S(\tau, \rho_r, \zeta_r) = L^S(\tau \otimes \tau, (4nk - 2rn + 1)s_r - 2n),$$

and, for $r = k$,

$$L^S(\tau, \rho_k, \zeta_k) = L^S(\tau, (2nk+1)s_k - (n+1/2)) \cdot L^S(\tau, \Lambda^2, 2(2nk+1)s_k - (2n+1)) \ .$$

Also, $L_\tau^S(\overline{s})$ is a product of L-functions of the same type as above (i.e. tensor product, standard and exterior square) evaluated at some linear combinations of ζ_r with positive coefficients. Let $\overline{s}_1 = (s_{1,1}, \ldots, s_{k,1})$ where $s_{r,1} = (2n+1)/(4nk-2rn+1)$ if $1 \le r \le k-1$ and $s_{k,1} = (n+1)/(2nk+1)$. At this point the tensor product L-function has a pole provided τ is selfdual. If in addition $L^S(\tau, \Lambda^2, s)$ has a pole at $s=1$ and $L^S(\tau, 1/2) \neq 0$ then $E_\tau(g, \overline{s})$ will have a multi-residue at $\overline{s} = \overline{s}_1$. But we shall not assume that. We have

Definition 2.1. Let $\epsilon = \otimes_\nu \epsilon_\nu$ be an irreducible automorphic representation of $\mathrm{Sp}_{4nk}(\mathbb{A})$. We say that ϵ is associated with $(Q_k, \tau, \overline{s}_1)$ if for almost all ν the representation ϵ_ν is isomorphic to the unramified constituent of $I(\tau_\nu, \overline{s}_1) = \mathrm{Ind}_{Q_k(F_\nu)}^{\mathrm{Sp}_{4nk}(F_\nu)}(\tau_\nu \otimes \cdots \otimes \tau_\nu)\delta_{Q_k}^{\overline{s}_1}$. Here τ_ν is the local component of τ at ν. If, in addition, ϵ is a cusp form we call it a CAP representation with respect to $(Q_k, \tau, \overline{s}_1)$. Note that

$$I(\tau, \overline{s}_1) = Ind_{Q_k(\mathbb{A})}^{Sp_{4nk}(\mathbb{A})}(\tau|\det|^{k-1/2} \otimes \tau|\det|^{k-3/2} \otimes \cdots \otimes \tau|\det|^{1/2})\,\delta_{Q_k}^{1/2}.$$

Remark 2.2. If the residue of $E_\tau(g, \overline{s})$ exists then ϵ is a nearly equivalent representation to that residue.

We need a similar definition for the double cover of the symplectic group. Let σ be a cusp form on $\widetilde{\mathrm{Sp}}_{2n}(\mathbb{A})$. For any $k \ge 1$ let $L_k = (\mathrm{GL}_{2n} \times \cdots \times \mathrm{GL}_{2n} \times \mathrm{Sp}_{2n})U(L_k)$ denote the standard parabolic subgroup of $\mathrm{Sp}_{2n(2k+1)}$ whose Levi part is a direct product of k copies of GL_{2n}. Let R_j, for $1 \le j \le k$, denote the standard maximal parabolic subgroup of $\mathrm{Sp}_{2n(2k+1)}$ defined by $R_j = (\mathrm{GL}_{2n_j} \times \mathrm{Sp}_{2n(2k-2j+1)})U(R_j)$. Let $\tilde{L}_k(\mathbb{A})$ denote the inverse image of $L_k(\mathbb{A})$ in $\widetilde{\mathrm{Sp}}_{2n(2k+1)}(\mathbb{A})$. To the induced representation $I(\tau, \sigma, \overline{s}) = \mathrm{Ind}_{\tilde{L}_k(\mathbb{A})}^{\widetilde{\mathrm{Sp}}_{2n(2k+1)}(\mathbb{A})}(\tau \otimes \cdots \otimes \tau \otimes \sigma)\,\delta_{\tilde{L}_k}^{\overline{s}}$ we denote by $\tilde{E}_{\tau,\sigma}(g, \overline{s})$ a corresponding Eisenstein series. Here we understand $\tau \otimes \cdots \otimes \tau \otimes \sigma$ as a representation of $\tilde{L}_k(\mathbb{A})$ as follows. First, it is trivial on $U(L_k)(\mathbb{A})$. Given $(a_1, \ldots, a_k, h) \in \mathrm{GL}_{2n}(\mathbb{A}) \times \cdots \times \mathrm{GL}_{2n}(\mathbb{A}) \times \mathrm{Sp}_{2n}(\mathbb{A})$ the action of $\tau \otimes \cdots \otimes \tau \otimes \sigma$ is given by

$$((a_1, \ldots, a_k, h), \zeta) \mapsto \zeta \prod_{1 \le i < j \le k} (\det a_i, \det a_j) \prod_{i=1}^{k} (\det a_i, x(h))\, \gamma_\psi(\det a_i)$$
$$\times \otimes_{i=1}^{k} \tau(a_i) \otimes \sigma(h),$$

where $x(h)$ is as defined in [R]. Here $(,)$ is the usual Hilbert symbol and $\gamma_\psi(\ )$ is the Weil factor associated with a fixed additive non-trivial character ψ of $F \backslash \mathbb{A}$. Also, we have $\delta_{\tilde{L}_k}^{\overline{s}} = \prod_{j=1}^{k} \delta_{R_j}^{s_j}$ where $s_j \in \mathbb{C}$. Conjecturally, the poles of

$\widetilde{E}_{\tau,\sigma}(g,\bar{s})$ are determined by

$$\prod_{r=1}^{k-1} L^S(\tau\otimes\tau,\zeta_r)L^S_\psi(\tau\otimes\sigma,\zeta_k)L^S_{\tau,\sigma,\psi}(\bar{s})$$

where $\zeta_r = (4nk-2n(r-1)+1)s_r - 2n$ if $1\le r\le k-1$ and $\zeta_k = (2nk+2n+1)s_k - (2n+1/2)$. Also $L^S_{\tau,\sigma,\psi}(\bar{s})$ is a product of L-functions of the same type (i.e. tensor products) and also the symmetric square of τ evaluated at some linear combinations of $\bar{s}$ with positive coefficients. Let $\bar{s}_2 = (s_{2,1},\ldots,s_{2,k})$ where $s_{2,r} = (2n+1)/(4nk-2nr+2n+1)$ if $1\le r\le k-1$ and $s_{2,k} = (4n+3)/2(2nk+2n+1)$. As in the linear case, we make the following definition.

Definition 2.3. Let $\epsilon = \otimes_\nu\epsilon_\nu$ be an irreducible automorphic representation of $\widetilde{\mathrm{Sp}}_{2n(2k+1)}(\mathbb{A})$. We say that ϵ is associated with $(L_k,\tau,\sigma,\bar{s}_2)$ if for almost all ν the representation ϵ_ν is isomorphic to the unramified constituent of $I(\tau_\nu,\sigma_\nu,\bar{s}_2) = \mathrm{Ind}_{\widetilde{L}_k(F_\nu)}^{\widetilde{\mathrm{Sp}}_{2n(2k+1)}(F_\nu)}(\tau_\nu\otimes\cdots\otimes\tau_\nu\otimes\sigma_\nu)\delta_{\widetilde{L}_k}^{\bar{s}_2}$. Here τ_ν is as in definition 2.1 and σ_ν is the local component of σ at ν. If, in addition, ϵ is a cusp form we call it a CAP representation with respect to $(L_k,\tau,\sigma,\bar{s}_2)$. Note that

$$I(\tau,\sigma,\bar{s}_2) = \mathrm{Ind}_{\widetilde{L}_k(\mathbb{A})}^{\mathrm{Sp}_{2n(2k+1)}(\mathbb{A})}(\tau|\det|^k\otimes\tau|\det|^{k-1}\otimes\cdots\otimes\tau|\det|\otimes\sigma)\delta_{L_k}^{1/2}.$$

3. On certain Fourier coefficients

In this section we shall study certain Fourier coefficients of certain representations which we will use later on.

We first start with some local results. Let F be a local non archimedean field. Let σ be an irreducible unramified representation of $\widetilde{\mathrm{Sp}}_{2n}(F)$. Assume that σ is a constituent of $\mathrm{Ind}_{\widetilde{B}(\mathrm{Sp}_{2n})}^{\widetilde{\mathrm{Sp}}_{2n}(F)}(\chi\circ\gamma_\psi)\delta_{B(\mathrm{Sp}_{2n})}^{1/2}$ where $(\chi\circ\gamma_\psi)(t,\zeta) = \zeta\prod_{i=1}^{n}\chi_i(t_i)\gamma_\psi(t_1\cdots t_n)$. Here χ_i are unramified characters of F^*, γ_ψ is the Weil factor and $\zeta\in\{\pm1\}$. Also $t = \mathrm{diag}(t_1,\ldots,t_n,t_n^{-1},\ldots,t_1^{-1})$. Let τ be an unramified representation of $\mathrm{GL}_{2n}(F)$. We shall assume that τ is the local ψ lift of σ. By that we mean that τ is the unramified constituent of $\mathrm{Ind}_{B(\mathrm{GL}_{2n})}^{\mathrm{GL}_{2n}(F)}\chi\delta_{B(\mathrm{GL}_{2n})}^{1/2}$. Let θ_τ denote the unramified constituent of $I(\tau,\bar{s}_1)$ (This induced representation was defined in Section 2). Similarly, let $\widetilde{\theta}_{\tau,\sigma}$ denote the unramified constituent of $I(\tau,\sigma,\bar{s}_2)$. Let $\widehat{Q}_k$ denote the standard parabolic subgroup of Sp_{4nk} defined by $\widehat{Q}_k = (\mathrm{GL}_{2k}\times\cdots\times\mathrm{GL}_{2k})U(\widehat{Q}_k)$ where GL_{2k} occurs n times. Let $\widehat{L}_k$ denote the standard parabolic subgroup of $\mathrm{Sp}_{2n(2k+1)}$ defined by $\widehat{L}_k = (\mathrm{GL}_{2k+1} \times \cdots \times \mathrm{GL}_{2k+1})\, U(\widehat{L}_k)$ where GL_{2k+1} occurs n times.

We have the following:

Lemma 3.1. *1) Assume that τ above is a selfdual (unramified) representation of $\mathrm{GL}_{2n}(F)$. Then the representation θ_τ is the unramified constituent of the representation*

$$\mathrm{Ind}_{\widehat{Q}_k(F)}^{\mathrm{Sp}_{4nk}(F)} (\chi_1(\det) \otimes \cdots \otimes \chi_n(\det))\, \delta_{\widehat{Q}_k}^{1/2}.$$

Here χ_i are the unramified characters associated with χ.

2) Assume that τ is the local ψ-lift of σ. Then the representation $\widetilde{\theta}_{\tau,\sigma}$ is the unramified constituent of the representation

$$\mathrm{Ind}_{\widetilde{\widehat{L}}_k(F)}^{\widetilde{\mathrm{Sp}}_{2n(2k+1)}(F)} (\chi_1 \circ \gamma_\psi(\det) \times \cdots \times \chi_n \circ \gamma_\psi(\det)\mu)\, \delta_{\widehat{L}_k}^{1/2}.$$

Here μ is a product of Hilbert symbols which depend only on the determinant.

Proof. We shall prove 2). The proof of 1) is similar. Recall that

$$I(\tau,\sigma,\bar{s}_2) = \mathrm{Ind}_{\widetilde{L}_k(F)}^{\widetilde{\mathrm{Sp}}_{2n(2k+1)}(F)} (\tau \otimes \cdots \otimes \tau \otimes \sigma)\, \delta_{\widetilde{L}_k}^{\bar{s}_2}.$$

Let B be the Borel subgroup of $\mathrm{Sp}_{2n(2k+1)}$ and $\widetilde{B}$ its full inverse image in $\widetilde{\mathrm{Sp}}_{2n(2k+1)}(F)$. We parameterize the maximal torus T of $\mathrm{Sp}_{2n(2k+1)}$ as

$$t = \mathrm{diag}(t_1, \ldots, t_k, t_{k+1}, t_k^*, \ldots, t_1^*)$$

where $t_i = \mathrm{diag}(t_{i,1}, \ldots, t_{i,2n})$ for $1 \le i \le k$ and

$$t_{k+1} = \mathrm{diag}(t_{k+1,1}, \ldots, t_{k+1,n}, t_{k+1,n}^{-1}, \ldots, t_{k+1,1}^{-1}).$$

Then we have

$$I(\tau,\sigma,\bar{s}_2) = \mathrm{Ind}_{\widetilde{B}}^{\widetilde{\mathrm{Sp}}_{2n(2k+1)}(F)} (\mu_\chi \delta_{L_k}^{-1/2} \delta_{\widetilde{L}_k}^{\bar{s}_2} \widetilde{\mu})\, \delta_B^{1/2}$$

where $\mu_\chi \delta_{L_k}^{-1/2} \delta_{\widetilde{L}_k}^{\bar{s}_2} \widetilde{\mu}$ is a genuine character of $\widetilde{T}$ and on the T component it is defined as follows. First

$$\mu_\chi(t) = \prod_{i=1}^{k+1} \chi(t_i)$$

where

$$\chi(t_i) = \prod_{j=1}^{n} \chi_j(t_{i,j} t_{i,2n-j+1}^{-1}),$$

and

$$\chi(t_{k+1}) = \prod_{j=1}^{n} \chi_j(t_{k+1,j}).$$

This follows from the assumption that τ is a lift of σ and hence τ is selfdual. Next, we have

$$\delta_{L_k}^{1/2}(t) = \prod_{i=1}^{k} |\det t_i|^{-2nk+2n(i-1)-1/2}$$

and

$$\delta_{\overline{L}_k}^{\overline{s}_2}(t) = \prod_{i=1}^{k} |\det t_i|^{(2n+1)(k-i)+(4n+3)/2}.$$

Finally we have

$$\widetilde{\mu}(t) = \prod_{i<j}(\det t_i, \det t_j) \prod_{i=1}^{k}(\det t_i, x(t_{k+1}))\gamma_\psi(\det t_i)\gamma_\psi(t_{k+1})$$

where $x(t_{k+1}) = t_{k+1,1} \cdots t_{k+1,n}$ and $\gamma_\psi(t_{k+1}) = \gamma_\psi(t_{k+1,1} \cdots t_{k+1,n})$. Using the identity $\gamma_\psi(a)\gamma_\psi(b) = (a,b)\gamma_\psi(ab)$ implies that

$$\widetilde{\mu}(t) = \gamma_\psi(x(t_{k+1}) \prod_{i=1}^{k} \det t_i).$$

Hence

$$I(\tau, \sigma, \overline{s}_2) = \mathrm{Ind}_{\widetilde{B}}^{\widetilde{\mathrm{Sp}}_{2n(2k+1)}(F)} (\mu_\chi \mu_1 \widetilde{\mu}) \delta_B^{1/2}$$

where $\mu_1(t) = \prod_{i=1}^{k} |\det t_i|^{k+1-i}$. Let w denote the following Weyl element of $\mathrm{Sp}_{2n(2k+1)}$. For all $1 \le i \le 2k+1$ and $0 \le j \le n-1$ set $w[(2k+1)j+i, 2n(i-1)+j+1] = 1$. This specifies w in the first $n(2k+1)$ rows and hence determines w uniquely. Thus $I(\tau, \sigma, \overline{s}_2)$ and $\mathrm{Ind}_{\widetilde{B}}^{\widetilde{\mathrm{Sp}}_{2n(2k+1)}(F)} {}^{w}(\mu_\chi \mu_1 \widetilde{\mu})\, \delta_B^{1/2}$ have the same semisimplification (and in particular share the same unramified constituent) where for a given character ν of T we denote ${}^{w}\nu(t) = \nu(w^{-1}tw)$. This twist does not contribute any cocycle to the computations. Parameterize T as

$$t = \mathrm{diag}(a_1, a_2, \ldots, a_n, a_n^*, \ldots, a_2^*, a_1^*)$$

where now a_i is a diagonal matrix of order $2k+1$. Then we have

$${}^{w}(\mu_\chi \mu_1 \mu)(t) = \gamma_\psi(\prod_{i=1}^{n} \det a_i) \prod_{i=1}^{n} \chi_i(\det a_i)\, \delta_{B_k}^{1/2}(a_i)$$

where B_k is the Borel subgroup of GL_{2k+1}. Thus $I(\tau, \sigma, \bar{s}_2)$ and

$$\mathrm{Ind}_{\widetilde{B}}^{\widetilde{\mathrm{Sp}}_{2n(2k+1)}(F)}(\chi_1 \circ \gamma_\psi \delta_{B_k} \times \cdots \times \chi_n \circ \gamma_\psi \delta_{B_k} \mu)\, \delta_{\widehat{L}_k}^{1/2}$$

have the same semisimplification where $\mu(t) = \prod_{i<j} (\det a_i, \det a_j)$. Since the identity representation is a quotient of $\mathrm{Ind}_{B_k(F)}^{\mathrm{GL}_{2k+1}(F)} \delta_B$ and since $\chi_i \circ \gamma_\psi$ depends only on the determinant it follows that the representation

$$\mathrm{Ind}_{\widetilde{\widehat{L}}_k(F)}^{\widetilde{\mathrm{Sp}}_{2n(2k+1)}(F)}(\chi_1 \circ \gamma_\psi(\det) \times \cdots \times \chi_n \circ \gamma_\psi(\det)\mu)\, \delta_{\widehat{L}_k}^{1/2} \tag{3.1}$$

is a subquotient of $I(\tau, \sigma, \bar{s}_2)$. Thus $\widetilde{\theta}_{\tau,\sigma}$, the unramified constituent in $I(\tau, \sigma, \bar{s}_2)$, is a quotient of (3.1) and from this our result follows. □

Next, we wish to show that θ_τ (resp. $\widetilde{\theta}_{\tau,\sigma}$) does not support a certain functional. Let U^p denote the unipotent radical of the standard parabolic subgroup of Sp_{2r} whose Levi part is $\mathrm{GL}_1^p \times \mathrm{Sp}_{2(r-p)}$. Define

$$U_p = \{u = (u_{ij}) \in U^p : u_{p,j} = 0; p+1 \le j \le 2r-p\}\ .$$

For a given $a \in F^*$ we define a character $\psi_{p,a}$ of U_p by

$$\psi_{p,a}(u) = \psi(u_{1,2} + \cdots + u_{p-1,p} + a u_{p,2r-p+1})\ .$$

Given a representation π of $\mathrm{Sp}_{2r}(F)$ (resp. $\widetilde{\mathrm{Sp}}_{2r}(F)$) we consider functionals $\ell_{p,a}$ on the space of π satisfying $\ell_{p,a}(\pi(u)v) = \psi_{p,a}(u)\ell_{p,a}(v)$ for all $u \in U_p$ and $v \in V(\pi)$.

Remark 3.2. As in [M-W] we can associate to any unipotent class $\mathcal{O}$ of Sp_{2r} a functional on a given representation. The functional $\ell_{p,a}$ corresponds to the unipotent class $((2p)1^{2(r-p)})$. We will say more about this later.

We now prove

Key Lemma 3.3. *Let τ and σ be as in Lemma 3.1. The representation θ_τ (resp. $\widetilde{\theta}_{\tau,\sigma}$) of the group $\mathrm{Sp}_{4nk}(F)$ (resp. $\widetilde{\mathrm{Sp}}_{2n(2k+1)}(F)$) has no nonzero functionals $\ell_{n+1,a}$ for all $a \in F^*$.*

Proof. The proof is similar for both cases. We shall consider the Sp_{4nk} case. Due to Lemma 3.1 it is enough to show that $\mathrm{Ind}_{\widehat{Q}_k(F)}^{\mathrm{Sp}_{4nk}(F)}(\chi_1(\det) \times \cdots \times \chi_n(\det))\delta_{\widehat{Q}_k}^{1/2}$ does not have a nonzero $\ell_{n+1,a}$ functional. Indeed, showing this means that the Jacquet module of the full induced representation, with respect to U_{n+1} and the character $\psi_{n+1,a}$, is zero, and due to exactness of Jacquet functors, this will imply that our Jacquet module is trivial on all constituents of the full induced representation, and in particular, on the spherical one. Using

Mackey theory one needs to prove the following. Let $\gamma \in \widehat{Q}_k\backslash \mathrm{Sp}_{4nk}/U_{n+1}$. Then there exists $u \in U_{n+1}$ such that $\gamma u\gamma^{-1} \in \widehat{Q}_k$ and $\psi_{n+1,a}(u) \neq 1$. This follows from the fact that the character $\chi_1(\det) \times \cdots \times \chi_n(\det)$ is trivial on any unipotent element in $\widehat{Q}_k$. Let $\widehat{u}_{i,i+1}(r) = I_{4nk} + re'_{i,i+1}$ if $1 \le i \le n$ and $\widehat{u}_{n+1,4nk-n}(r) = I_{4nk} + re_{n+1,4nk-n}$. Here $e_{i,j}$ is the $4nk \times 4nk$ matrix with one at the i,j entry and zero elsewhere and $e'_{i,j} = e_{ij} - e_{4nk-i,4nk-j+1}$. Then from the definition of $\psi_{n+1,a}$ we deduce that it is nontrivial on each of the matrices $\widehat{u}_{i,i+1}(r)$ and $\widehat{u}_{n+1,4nk-n}(r)$ where $1 \le i \le n$ and on any other one-dimensional unipotent subgroup which corresponds to a positive root in Sp_{4nk}, the character $\psi_{n+1,a}$ is trivial. Let $\gamma \in \widehat{Q}_k\backslash \mathrm{Sp}_{4nk}/U_{n+1}$. Then by the Bruhat decomposition we can write $\gamma = wu_w$ where w is a Weyl element of Sp_{4nk} and u_w is an upper unipotent matrix in $\mathrm{Sp}_{4nk-2n}(F)$ which is embedded in $\mathrm{Sp}_{4nk}(F)$ in the obvious way. Thus

$$u_w = \begin{pmatrix} I_n & & \\ & \overline{u}_w & \\ & & I_n \end{pmatrix}, \qquad \overline{u}_w \in \mathrm{Sp}_{4nk-2n}(F) .$$

From the above discussion we deduce that it is enough to show that either $\gamma\widehat{u}_{i,i+1}(r)\gamma^{-1} \in \widehat{Q}_k(F)$ for some $1 \le i \le n$ or $\gamma\widehat{u}_{n+1,4nk-n}(r)\gamma^{-1} \in \widehat{Q}_k(F)$. First we want to get rid of u_w. To do so we claim that given u_w and $\widehat{u}_{i,i+1}(r)$ or $\widehat{u}_{n+1,4nk-n}(r)$ there exists $u' \in U_{n+1}$ such that $u_w u' u_w^{-1} = \widehat{u}_{i,i+1}(r)$ or $u_w u' u_w^{-1} = \widehat{u}_{n+1,4nk-n}(r)$ and such that $\psi_{n+1,a}(u') = \psi_{n+1,a}(\widehat{u}_{i,i+1}(r))$ or $\psi_{n+1,a}(u') = \psi_{n+1,a}(\widehat{u}_{n+1,4nk-n}(r))$. Indeed all $\widehat{u}_{i,i+1}(r)$ and $\widehat{u}_{n+1,4nk-n}(r)$ commute with u_w unless $i = n$. Thus we can choose them to be u'. For $\widehat{u}_{n,n+1}(r)$ one can find a matrix

$$u' = \begin{pmatrix} I_{n-1} & 0 & 0 & 0 & 0 \\ & 1 & t & * & 0 \\ & & I & t^* & 0 \\ & & & 1 & 0 \\ & & & & I_{n-1} \end{pmatrix}$$

where $t = (r, t_2, \ldots, t_{4nk-2n}) \in \mathrm{Mat}_{1\times(4nk-2n)}(F)$ such that $u_w u' u_w^{-1} = \widehat{u}_{n,n+1}(r)$. Clearly then $\psi_{n+1,a}(u') = \psi_{n+1,a}(\widehat{u}_{n,n+1}(r))$. All this means that it is enough to prove that given a Weyl element w of Sp_{4nk} then either $w\widehat{u}_{i,i+1}w^{-1} \in \widehat{Q}_k(F)$ for some $1 \le i \le n$ or that $w\widehat{u}_{n+1,4nk-n}w^{-1} \in \widehat{Q}_k(F)$.

Let $\overline{\widehat{Q}}_k = (\mathrm{GL}_{2k} \times \cdots \times \mathrm{GL}_{2k})Z$ denote the opposite parabolic subgroup to $\widehat{Q}_k$ in Sp_{4nk}. Thus $\overline{\widehat{Q}}_k$ is the transpose of $\widehat{Q}_k$ and Z is a lower unipotent subgroup. Thus in matrices we have

$$Z = \left\{ \begin{pmatrix} I & & & & \\ * & I & & & \\ * & * & I & & \\ \vdots & \vdots & & \ddots & \\ * & * & * & \cdots & I \end{pmatrix} \in \mathrm{Sp}_{4nk} \right\}$$

We think of Z as block matrices of size $2k \times 2k$. Let $z_{ij} \in \mathrm{Mat}_{2k\times 2k}$. For all $2 \le i \le n$ and $1 \le j \le i-1$ we define the element $\widehat{z}_{ij}$ of Z by

$$\widehat{z}_{ij} = \begin{pmatrix} I & & & & & & & & & \\ & \ddots & & & & & & & & \\ & & I & & & & & & & \\ & & \vdots & \ddots & & & & & & \\ & & z_{ij} & \cdots & I & & & & & \\ & & & & & \ddots & & & & \\ & & & & & & I & & & \\ & & & & & & \vdots & \ddots & & \\ & & & & & & z_{ij}^* & \cdots & I & \\ & & & & & & & & & \ddots & \\ & & & & & & & & & & I \end{pmatrix} \in \mathrm{Sp}_{4nk}.$$

Denote by $\widehat{Z}_{ij}$ the subgroup of Z which consists of these elements $\widehat{z}_{ij}$. Similarly we define $\widehat{z}_{ij}$ and $\widehat{Z}_{ij}$ for $n+1 \le i \le 2n$ and $1 \le j \le 2n-i+1$. However for $\widehat{z}_{i,2n-i+1}$, when $n+1 \le i \le 2n$, to be in Sp_{4nk} we need to have that $z_{i,2n-i+1} \in \mathrm{Mat}^0_{2k\times 2k} = \{A \in \mathrm{Mat}_{2k\times 2k} : A^t J_{2k} = J_{2k} A\}$ (recall that J_m is the $m \times m$ matrix which has ones on the other diagonal and zeros elsewhere). In this way Z is generated by all $\widehat{Z}_{ij}$ where i and j are specified above. Notice that in the terms of this block notation Z is an n step nilpotent group.

Assume that w is a Weyl element of Sp_{4nk} such that $w\widehat{u}_{i,i+1}w^{-1} \in Z$ and $w\widehat{u}_{n+1,4nk-n}w^{-1} \in Z$ for all $1 \le 1 \le n$. We shall derive a contradiction. Since $\widehat{u}_{n+1,4nk-n}$ corresponds to a short root it follows that $w\widehat{u}_{n+1,4nk-n}w^{-1} = \widehat{z}_{i_1,2n-i_1+1} \in \widehat{Z}_{i_1,2n-i_1+1}$ for some $n+1 \le i_1 \le 2n$. Let $z_{i_1,2n-i_1+1}(\ell_1,\ell_2)$ denote the (ℓ_1,ℓ_2) entry of $z_{i_1,2n-i_1+1}$ and let $\widehat{z}_{i_1,2n-i_1+1}(\ell_1,\ell_2)$ denote the embedding of this entry in Sp_{4nk}. By multiplying w from the left with a suitable Weyl element of $\widehat{Q}_k$ we may assume that $w\widehat{u}_{n+1,4nk-n}w^{-1} = \widehat{z}_{i_1,2n-i_1+1}(1,2k)$. Next consider $w\widehat{u}_{n,n+1}w^{-1}$. Notice that $\widehat{u}_{n,n+1}$ and $\widehat{u}_{n+1,4nk-n}$ do not commute. Since the only matrices in Z which do not commute with $\widehat{z}_{i_1,2n-i_1+1}$ are $\widehat{z}_{2n-i_1+1,j}$ for $1 \le j \le 2n-i_1$, it follows that $w\widehat{u}_{n,n+1}w^{-1} = \widehat{z}_{2n-i_1+1,i_2} \in \widehat{Z}_{2n-i_1+1,i_2}$ for some $1 \le i_2 \le 2n-i_1$. Adjusting w if needed we may assume that $w\widehat{u}_{n,n+1}w^{-1} = \widehat{z}_{2n-i_1+1,i_2}(2k,2k)$. Next consider $\widehat{u}_{n-1,n}$. It commutes with $\widehat{u}_{n+1,4nk-n}$ but not with $\widehat{u}_{n,n+1}$. The only matrices in Z which do not commute with $\widehat{z}_{2n-i_1+1,i_2}$ are either $\widehat{z}_{i_1,2n-i_1+1}$ or $\widehat{z}_{i_2,j}$ for some $1 \le j \le i_2-1$. Hence $w\widehat{u}_{n-1,n}w^{-1} \in \widehat{Z}_{i_1,2n-i_1+1}$ or $w\widehat{u}_{n-1,n}w^{-1} \in \widehat{Z}_{i_2,j}$ for some $1 \le j \le i_2 - 1$. Notice that as matrices in Sp_{4nk} the variables (in F) of $\widehat{u}_{n-1,n}, \widehat{u}_{n,n+1}$ and $\widehat{u}_{n+1,2nk-n}$ lie in distinct rows and columns. This means that the same holds after the conjugation by w. Hence, since $w\widehat{u}_{n+1,4nk-n}w^{-1} = \widehat{z}_{i_1,2n-i_1+1}(1,2k)$, we get that if $w\widehat{u}_{n-1,n}w^{-1} = \widehat{z}_{i_1,2n-i_1+1}(\ell_1,\ell_2)$ then $\ell_1 \neq 1$ and $\ell_2 \neq 2k$. However, simple matrix multiplication implies that such a matrix $\widehat{z}_{i_1,2n-i_1+1}(\ell_1,\ell_2)$ commutes with $\widehat{z}_{2n-i_1+1,i_2}(2k,2k)$ which is impossible since $\widehat{u}_{n-1,n}$ and $\widehat{u}_{n,n+1}$ do not commute. Thus $w\widehat{u}_{n-1,n}w^{-1} \in \widehat{z}_{i_2,j}$ for some $1 \le j \le i_2-1$. Continuing this process we deduce that the matrices $w\widehat{u}_{i,i+1}w^{-1}$ and $w\widehat{u}_{n+1,4nk-n}w^{-1}$ for $1 \le i \le n$ all lie in distinct blocks $\widehat{z}_{\ell_1,\ell_2}$. Since the roots $\widehat{u}_{i,i+1}, \widehat{u}_{n+1,4nk-n}$ for $1 \le i \le n$ form a set of simple roots of $\mathrm{Sp}_{2(n+1)}$, then the unipotent subgroup generated by these roots is an $n+1$ step nilpotent group. As mentioned above, in terms of $2k \times 2k$ blocks, Z is an n step nilpotent group. Since we showed that w conjugates the above $n+1$ roots to distinct blocks this is impossible and hence we derive a contradiction. This completes the proof of Lemma 3.3. □

For the rest of this section let F be a global field. The groups U_p of Sp_{2r} and the character $\psi_{p,a}$ of U_p are now considered over the adeles of F. Here $a \in F^*$. Let π be an automorphic representation of $\mathrm{Sp}_{2r}(\mathbb{A})$. We start with

Lemma 3.4. *Suppose that the integral*

$$I_{p,a}(g) = \int\limits_{U_p(F)\backslash U_p(\mathbb{A})} \varphi_\pi(ug)\,\psi_{p,a}(u)\,du$$

is zero for all choice of data (φ_π is in the space of π). Then the integral

$$I_{p+1,a}(g) = \int\limits_{U_{p+1}(F)\backslash U_{p+1}(\mathbb{A})} \varphi_\pi(ug)\,\psi_{p+1,a}(u)\,du$$

is zero for all choice of data.

Proof. The integral $I_{p,a}(g)$ defines an automorphic function on $\mathrm{Sp}_{2(r-p)}(\mathbb{A})$. Let V be the unipotent radical of the standard maximal parabolic subgroup of $\mathrm{Sp}_{2(r-p)}$ whose Levi part is $\mathrm{GL}_1 \times \mathrm{Sp}_{2(r-p-1)}$. The integral

$$\int\limits_{V(F)\backslash V(\mathbb{A})} I_{p,a}(vg)\,dv \tag{3.2}$$

is a special case of similar integrals which were computed in **[G-R-S1]** Theorem 8 p. 838. It is shown there that $I_{p+1,a}(g)$ is a Fourier coefficient of integral

(3.2). By our assumption (3.2) is zero for all choice of data. Thus the result follows. □

To study certain Fourier coefficients we make the following definitions. Let H be a reductive group and π and automorphic representation of $H(\mathbb{A})$. As in **[G-R-S5]** and also in **[M-W]** one can associate with any unipotent class of H a set of Fourier coefficients. We have

Definition 3.5. We shall denote by $\mathcal{O}_H(\pi)$ the set of all unipotent classes of H with the following property. A unipotent class $\mathcal{O} \in \mathcal{O}_H(\pi)$ if for all unipotent classes $\widetilde{\mathcal{O}} > \mathcal{O}$ the representation π does not have a nonzero Fourier coefficient corresponding to $\widetilde{\mathcal{O}}$.

Let ϵ be an irreducible automorphic representation of $\mathrm{Sp}_{4nk}(\mathbb{A})$ (resp. $\widetilde{\mathrm{Sp}}_{4nk+2n}(\mathbb{A})$) associated with $(Q_k, \tau, \bar{s}_1)$ (resp. $(L_k, \tau, \sigma, \bar{s}_2)$). (See Section 2 for definitions.) We further make the assumption that for almost all ν, $\epsilon_\nu \simeq \theta_\tau$ (resp. $\epsilon_\nu \simeq \widetilde{\theta}_{\tau,\sigma}$).

Let U_r denote the unipotent radical subgroup of the parabolic subgroup of $\mathrm{Sp}_{4nk}(\mathbb{A})$ (resp. $\widetilde{\mathrm{Sp}}_{4nk+2n}(\mathbb{A})$) whose Levi part is $\mathrm{GL}_1^r \times Sp_{2l}$ where l equals $4nk - 2r$ or $4nk + 2n - 2r$. For $u \in U_r$ we define the additive character $\psi_{U_r}(u) = \psi(u_{1,2} + \cdots + u_{r-1,r})$.

Applying this we will now prove

Proposition 3.6. *Let $\mathcal{O}$ be a unipotent class of Sp_{4nk} (resp. Sp_{4nk+2n}) which is greater or not related to $((2n)^{2k})$ (resp. $((2n)^{2k+1})$). Assume that for all $r > n$ the integrals*

$$\int_{U_r(F)\backslash U_r(\mathbb{A})} \phi_\epsilon(ug)\psi_{U_r}(u)\,du$$

are zero for all choice of data. Then ϵ has no nonzero Fourier coefficient corresponding to $\mathcal{O}$.

Remark 3.7. Later we will show that for a certain representation ϵ on $\mathrm{Sp}_{4nk}(\mathbb{A})$ we have $\mathcal{O}_{\mathrm{Sp}_{4nk}}(\epsilon) = ((2n)^{2k})$ and for a certain representation ϵ on $\widetilde{\mathrm{Sp}}_{4nk+2n}$ we have $\mathcal{O}_{\mathrm{Sp}_{4nk+2n}}(\epsilon) = ((2n)^{2k+1})$.

Proof. We prove the Sp_{4nk} case. The other one is similar. Let $\mathcal{O} = (d_1 d_2 \cdots d_\ell)$ be a unipotent class of Sp_{4nk}, where we assume that $d_i \geq d_{i+1}$, such that $\mathcal{O}$ is greater or not related to $((2n)^{2k})$. Assume ϵ has a nonzero Fourier coefficient corresponding to $\mathcal{O}$. Clearly $d_1 > 2n$. If d_1 is odd then by Lemma 2.4 in **[G-R-S5]** there is a $2p > d_1 > 2n$ such that ϵ has a nonzero Fourier coefficient with respect to $((2p)1^{4nk-2p})$. Indeed, the proof of Lemma 2.4 in **[G-R-S5]** does not require that ϵ be a cuspidal representation. All that is needed is that the above integrals will be identically zero. Similarly if d_1 is even then by Lemma 2.6 in [G-R-S5] such a p exists also. One can check that the Fourier coefficient corresponding to the unipotent class $((2p)1^{4nk-2p})$ is given by the integral $I_{p,a}(g)$ where $a \in F^*$. It follows from Lemma 3.3 that ϵ has no nonzero Fourier coefficient corresponding to $((2n+2)1^{4nk-2n-2})$. Thus from Lemma 3.4 it follows that ϵ has no nonzero Fourier coefficient corresponding to $((2p)1^{4nk-2p})$ for all $p \geq n+1$. This is a contradiction. Thus the proposition follows. □

4. The setup for the construction

In this section we will explain the basic notation and setup for the construction of the CAP representations. Here F is a number field.

We recall some notation and definitions used in Section 3. Let $V_{p,2r}$ denote the unipotent radical of the standard parabolic subgroup of Sp_{2r} whose Levi part is $\mathrm{GL}_1^p \times \mathrm{Sp}_{2(r-p)}$. Define

$$U_{p,2r} = \{v = (v_{ij}) \in V_{p,2r} : v_{p,j} = 0; p+1 \leq j \leq 2r-p\} .$$

These two groups, $V_{p,2r}$ and $U_{p,2r}$, were denoted in Section 3 by U^p and U_p. We define characters $\psi_{V_{p,2r}}$ and $\psi^a_{U_{p,2r}}$ of $V_{p,2r}(\mathbb{A})$ and $U_{p,2r}(\mathbb{A})$ respectively as follows. Fix a character ψ of $F\backslash\mathbb{A}$. First, for $v = (v_{ij}) \in V_{p,2r}(\mathbb{A})$ we define

$$\psi_{V_{p,2r}}(v) = \psi(v_{1,1} + \cdots + v_{p-1,p}) .$$

Next, for $u = (u_{ij}) \in U_{p,2r}(\mathbb{A})$ and $a \in F^*$ let

$$\psi^a_{U_{p,2r}}(u) = \psi(u_{1,2} + \cdots + u_{p-1,p} + au_{p,2r-p+1}) .$$

Given an automorphic representation π on $\mathrm{Sp}_{2r}(\mathbb{A})$ or on $\widetilde{\mathrm{Sp}}_{2r}(\mathbb{A})$ we can associate with $\psi^a_{U_{p,2r}}$ the Fourier coefficient (φ_π is in the space of π)

$$\int\limits_{U_{p,2r}(F)\backslash U_{p,2r}(\mathbb{A})} \varphi_\pi(ug)\psi^a_{U_{p,2r}}(u)\, du .$$

We will say that π has a missing $\psi^a_{U_{p,2r}}$ Fourier coefficient if the above integral is zero for all choice of data. It can happen that π has a missing $\psi^a_{U_{p,2r}}$ Fourier coefficient but a nonzero Fourier coefficient with respect to $\psi^b_{U_{p,2r}}$ for some $b \in F^*$. Clearly, if this is the case, then a and b are in different square classes of F^*.

Let us remark that the above Fourier coefficient corresponds to the unipotent class $((2p)1^{2(r-p)})$.

The construction we will define shortly is inductive. To start it let σ be an irreducible cuspidal generic representation of $\widetilde{\mathrm{Sp}}_{2n}(\mathbb{A})$. We shall write $\psi_{U_{p,2r}}$ for $\psi^a_{U_{p,2r}}$ when $a = 1$. We make the following assumption on σ.

Assumption 1: *The representation σ has a missing $\psi_{U_{n,2n}}$ Fourier coefficient. In other words σ has a missing Whittaker Fourier coefficient with respect to $\psi_{U_{n,2n}}$.* □

Write $\sigma = \otimes_\nu \sigma_\nu$ and $\psi = \otimes \psi_\nu$. Assume that when σ_ν is unramified then σ_ν occurs in $\mathrm{Ind}_{B(F_\nu)}^{\mathrm{Sp}_{2n}(F_\nu)}(\chi_\nu \circ \gamma_{\psi_\nu})\delta_B^{1/2}$ where B is the Borel subgroup of Sp_{2n}, χ_ν is an unramified character of $B(F_\nu)$ and γ_{ψ_ν} is the Weil factor.

Definition 4.1. Let $\tau = \otimes_\nu \tau_\nu$ be a cuspidal representation of $\mathrm{GL}_{2n}(\mathbb{A})$. We say that τ is a ψ weak lift of σ if for almost all ν we have

$$\tau_\nu = \mathrm{Ind}_{B(\mathrm{GL}_{2n})}^{\mathrm{GL}_{2n}(F_\nu)} \chi_\nu \delta_{B(\mathrm{GL}_{2n})}^{1/2}.$$

In this case we write $\tau_\psi(\sigma)$ for τ.

Since σ is generic we can lift it to a generic cuspidal representation π of $\mathrm{SO}_{2n+1}(\mathbb{A})$ using the theta correspondence with respect to some character $\psi' \neq \psi$. By **[C-K-PS-S]**and **[G-R-S4]** π has a weak lift to an irreducible representation τ' of $\mathrm{GL}_{2n}(\mathbb{A})$, which is well described. Let $\psi' = \psi^a$, $a \in F^*$, and let χ_a be the quadratic character to a. Then it is easy to see that $\tau_\psi(\sigma) = \tau' \otimes \chi_a$ is a ψ-weak lift of σ to $\mathrm{GL}_{2n}(\mathbb{A})$,

Remark 4.2. 1) The compatibility between $\psi_{U_{n,2n}}$ and the fact that we take the ψ lift is important for our construction.

2) The assumption that $\tau_\psi(\sigma)$ is cuspidal is not essential. We can also handle the case when $\tau_\psi(\sigma)$ is generic but not cuspidal. However, in this case the cuspidality assumption as stated in Assumptions 2 below need to be modified.

We will construct a sequence of representations $\{\pi_m(\sigma)\}$ where $\pi_1(\sigma) = \sigma$. If $m = 2k-1$ then $\pi_{2k-1}(\sigma)$ will be a representation on $\widetilde{\mathrm{Sp}}_{2n(2k-1)}(\mathbb{A})$ and if $m = 2k$ then $\pi_{2k}(\sigma)$ will be a representation on $\mathrm{Sp}_{4nk}(\mathbb{A})$. To construct $\pi_m(\sigma)$ we will use $\pi_{m-1}(\sigma)$. For the construction of $\pi_m(\sigma)$ and in order to prove that this is cuspidal we will have to make an assumption on $\pi_{m-1}(\sigma)$ which we shall soon explain.

At this point we wish to explain the inductive process. Assume we have constructed a representation $\pi'_{2k-1}(\sigma)$ on $\widetilde{\mathrm{Sp}}_{2n(2k-1)}(\mathbb{A})$ with the property that $\pi'_{2k-1}(\sigma)$ is a CAP representation with respect to $(L_{k-1}, \tau_\psi(\sigma), \sigma, \bar{s}_2)$. (See Definition 2.3 in Section 2).

We make the following

Assumption 2a: *There is a nearly equivalent irreducible CAP representation to $\pi'_{2k-1}(\sigma)$ which has a missing $\psi_{U_{n,2n(2k-1)}}$ Fourier coefficient. We denote this representation by $\pi_{2k-1}(\sigma)$.* □

Remark 4.3. In the above notation, Assumption 1 is just Assumption 2a with $k = 1$. This is the base of our induction. Thus we assume such cusp forms do exist.

With these assumptions we will now explain how to construct the representation $\pi_{2k}(\sigma)$ on the group $Sp_{4nk}(\mathbb{A})$. Let $P_{2n(2k+1)}$ denote the standard parabolic subgroup of $Sp_{2n(2k+1)}$ whose Levi part is $GL_{2n} \times Sp_{2n(2k-1)}$. To the induced representation $\mathrm{Ind}_{\widetilde{P}_{2n(2k+1)}(\mathbb{A})}^{\widetilde{Sp}_{2n(2k+1)}(\mathbb{A})}(\tau_\psi(\sigma) \otimes \pi_{2k-1}(\sigma))\delta^s_{P_{2n(2k+1)}}$ we associate an Eisenstein series $\widetilde{E}_{\tau_\psi(\sigma),\pi_{2k-1}(\sigma)}(g, s)$. We will prove

Lemma 4.4. *The Eisenstein series $\widetilde{E}_{\tau_\psi(\sigma),\pi_{2k-1}(\sigma)}(g, s)$ has a simple pole at $s = \zeta_1$ where $\zeta_1 = (4nk + 2k + 1)/2(4nk + 1)$. (At this point the residue representation is a quotient of* $\mathrm{Ind}_{\widetilde{P}_{2n(2k+1)}(\mathbb{A})}^{\widetilde{Sp}_{2n(2k+1)}(\mathbb{A})}(\tau_\psi(\sigma)|\det|^k \otimes \pi_{2k-1}(\sigma)) \cdot \delta^{1/2}_{P_{2n(2k+1)}})$.

Let $\widetilde{E}_{\tau_\psi(\sigma),\pi_{2k-1}(\sigma)}(g) = \mathrm{Res}_{s=\zeta_1} \widetilde{E}_{\tau_\psi(\sigma),\pi_{2k-1}(\sigma)}(g, s)$. We construct the following function

(4.1)
$$f(h) = \int_{V_{n,2n(2k+1)}(F)\backslash V_{n,2n(2k+1)}(\mathbb{A})} \widetilde{\theta}^\psi_\phi(\ell(v)h)\widetilde{E}_{\tau_\psi(\sigma),\pi_{2k-1}(\sigma)}(vh)\psi_{V_{n,2n(2k+1)}}(v)\, dv$$

Here $\widetilde{\theta}^\psi_\phi$ is the theta function on $\widetilde{Sp}_{4nk}(\mathbb{A})$ corresponding to the additive character ψ and ϕ is a Schwartz function. Also ℓ denotes the homomorphism from $V_{n,2n(2k+1)}$ onto the Heisenberg group H_{4nk+1} in $4nk + 1$ variables. Clearly $f(h)$ is an automorphic function on $Sp_{4nk}(\mathbb{A})$. Let $\pi'_{2k}(\sigma)$ denote the representation of $Sp_{4nk}(\mathbb{A})$ generated by the functions $f(h)$ defined in (4.1). Our main result is the following:

Theorem A. *Assume that $\pi_{2k-1}(\sigma)$ has missing $\psi_{U_{n,2n(2k-1)}}$ Fourier coefficient. (See Assumption 2a). Then the representation $\pi'_{2k}(\sigma)$ is a nonzero CAP representation with respect to $(Q_k, \tau_\psi(\sigma), \overline{s}_1)$.*

This assumptions of missing Fourier coefficient is natural in these types of constructions. Even though it is not clear to us how to construct in general examples of cuspidal representations which have certain missing Fourier coefficients, such examples do exist at least for low rank groups. See **[PS]** for examples in $\widetilde{SL}_2(\mathbf{A})$.

To construct $\pi_{2k+1}(\sigma)$ we apply a similar construction. We make the following

Assumption 2b: *There is a nearly equivalent irreducible (CAP) representation to $\pi'_{2k}(\sigma)$ which has a missing $\psi_{U_{n,4nk}}$ Fourier coefficient. We denote this representation by $\pi_{2k}(\sigma)$.* □

Let $P_{4n(k+1)}$ denote the standard maximal parabolic subgroup of $\mathrm{Sp}_{4n(k+1)}$ whose Levi part is $\mathrm{GL}_{2n} \times \mathrm{Sp}_{4nk}$. To the induced representation

$$\mathrm{Ind}_{P_{4n(k+1)}(\mathbb{A})}^{\mathrm{Sp}_{4n(k+1)}(\mathbb{A})}(\tau_\psi(\sigma) \otimes \pi_{2k}(\sigma))\delta^s_{P_{4n(k+1)}}$$

we associate an Eisenstein series $E_{\tau_\psi(\sigma),\pi_{2k}(\sigma)}(g,s)$. As in Lemma 4.4 we will prove the following:

Lemma 4.5. *The Eisenstein series $E_{\tau_\psi(\sigma),\pi_{2k}(\sigma)}(g,s)$ has a simple pole at $s = \zeta_2$ where $\zeta_2 = (2kn - n + k)/(4nk - 2n + 1)$.*

Denote $E_{\tau_\psi(\sigma),\pi_{2k}(\sigma)}(g) = \mathrm{Res}_{s=\zeta_2} E_{\tau_\psi(\sigma),\pi_{2k}(\sigma)}(g,s)$. This time we construct functions on $\widetilde{\mathrm{Sp}}_{2n(2k+1)}(\mathbb{A})$ given by
(4.2)

$$\tilde{f}(h) = \int\limits_{V_{n,4n(k+1)}(F)\backslash V_{n,4n(k+1)}(\mathbb{A})} \tilde{\theta}^\psi_\phi(\ell(v)h) E_{\tau_\psi(\sigma),\pi_{2k}(\sigma)}(vh)\psi_{V_{n,4n(k+1)}}(v)\, dv.$$

The notation is the same as in (4.1). Here $\tilde{\theta}^\psi_\phi$ is the theta function on $\widetilde{\mathrm{Sp}}_{2n(2k+1)}(\mathbb{A})$ corresponding to the character ψ. We define $\pi'_{2k+1}(\sigma)$ to be the representation of $\widetilde{\mathrm{Sp}}_{2n(2k+1)}(\mathbb{A})$ generated by the functions $\tilde{f}(h)$ defined in (4.2). We will prove the following:

Theorem B. *Assume that $\pi_{2k}(\sigma)$ has a missing $\psi_{U_{n,4nk}}$ Fourier coefficient (see Assumption 2b) then $\pi_{2k+1}(\sigma)$ is a nonzero CAP representation with respect to $(L_k, \tau_\psi(\sigma), \sigma, \bar{s}_2)$. (see Definition 2.3 in Section 2).*

5. The proofs of Theorems A and B

The proofs of both Theorems are similar. We shall provide the details concerning the construction of $\pi'_{2k}(\sigma)$ using $\pi_{2k-1}(\sigma)$.

Proof of Lemma 4.4: This is quite standard. Let $U(P_{2n(2k+1)})$ be the unipotent radical subgroup of $P_{2n(2k+1)}$. Computing the constant term of

$$\tilde{E}_{\tau_\psi(\sigma),\pi_{2k-1}(\sigma)}(g,s)$$

we obtain a sum of two terms. First we obtain the section which is holomorphic for all s and then we also obtain the intertwining operator corresponding to the long Weyl element modulo $P_{2n(2k+1)}$. The poles of the intertwining operator, and hence the poles of the Eisenstein series, are determined, (up to bad primes) after cancellation by

$$\frac{L^S(\tau_\psi(\sigma)\times\tau_\psi(\sigma),(4nk+1)(s-1/2)-k+1)}{L^S(\tau_\psi(\sigma)\times\tau_\psi(\sigma),(4nk+1)(s-1/2)+k)} \times \frac{L^S(\tau,sym^2,(4nk+1)(2s-1))}{L^S(\tau,sym^2,(4nk+1)(2s-1)+1)}.$$

Here we used the fact that $\pi_{2k-1}(\sigma)$ is a CAP with respect to $(L_{k-1},\tau_\psi(\sigma),\sigma,\overline{s}_2)$. Clearly this L-function has a simple pole at $\zeta_1=(4nk+2k+1)/2(4nk+1)$. Since data can be chosen at bad primes to be nonvanishing at s_1, we get that ζ_1 is a pole of the Eisenstein series. □

We denote

$$\widetilde{E}_{\tau_\psi(\sigma),\pi_{2k-1}(\sigma)}(g)=\mathrm{Res}_{s=\zeta_1}\widetilde{E}_{\tau_\psi(\sigma),\pi_{2k-1}(\sigma)}(g,s)$$

and write $\widetilde{E}_{\tau_\psi(\sigma),\pi_{2k-1}(\sigma)}$ for the representation itself. We shall now prove

Proposition 5.1. *Assume that* $\mathcal{O}_{\mathrm{Sp}_{2n(2k-1)}}(\pi_{2k-1}(\sigma))=((2n)^{2k-1})$. *Then*

$$\mathcal{O}_{\mathrm{Sp}_{2n(2k+1)}}(\widetilde{E}_{\tau_\psi(\sigma),\pi_{2k-1}(\sigma)})=((2n)^{2k+1}).$$

Proof. Before proving the proposition, let us mention that since σ is generic, then the assumption stated in the proposition is true for $k=1$. Arguing as in Lemma 3.1 part 2) we deduce that if $(\widetilde{E}_{\tau_\psi(\sigma),\pi_{2k-1}(\sigma)})_\nu$ is unramified at the place ν then this local representation is the unramified quotient of

$$\mathrm{Ind}_{\widetilde{L}_k(F_\nu)}^{\widetilde{\mathrm{Sp}}_{2n(2k+1)}(F\nu)}(\chi_1\circ\gamma_{\psi_\nu}(\det)\times\cdots\times\chi_n\circ\gamma_{\psi_\nu}(\det)\mu)\,\delta_{\widetilde{L}_k}^{1/2}.$$

In particular this implies that $\widetilde{E}_{\tau_\psi(\sigma),\pi_{2k-1}(\sigma)}$ is associated with $(L_k,\tau_\psi(\sigma),\sigma,\overline{s}_2)$. To apply Proposition 3.6 we need to verify that for all $r>n$ the integrals

$$\int\limits_{U_r(F)\backslash U_r(\mathbb{A})}\phi_\epsilon(ug)\psi_{U_r}(u)\,du$$

are zero for all choice of data. Here $\epsilon=\widetilde{E}_{\tau_\psi(\sigma),\pi_{2k-1}(\sigma)}$. To do that we follow the same Fourier expansion as in the proof of Lemma 2.2 in **[G-R-S5]**. This will show that the above integral can be expressed as sum of integrals which are of two types. First, we will obtain Fourier expansions corresponding to unipotent classes $\mathcal{O}=((2p)1^{4nk+2n-2p})$ where $p>n$. Using Lemmas 3.3 and 3.4 we deduce that these terms are zero. Second, we obtain integrals of the form

$$\int_{U_p(F)\backslash U_p(\mathbb{A})} \phi_\epsilon(ug)\psi_{U_p}(u)\,du$$

where $p > n$. These integrals will be zero unless $p = mn$ for some m. Indeed, this follows from the induction data from which the residue $\widetilde{E}_{\tau_\psi(\sigma),\pi_{2k-1}(\sigma)}$ is constructed. If $p = mn$ then the assumption that $r > n$ implies that $m > 1$. In this case, from the definition of the induction data, we may realize the integral over $U_p(F)\backslash U_p(\mathbb{A})$ as the Whittaker coefficient of a certain residue of an Eisenstein series on GL_{mn}. However, since $m > 1$ the residue of this Eisenstein series cannot have a Whittaker model and hence it is zero.

From all this we deduce that $\widetilde{E}_{\tau_\psi(\sigma),\pi_{2k-1}(\sigma)}$ does not have a nonzero Fourier coefficient with respect to any unipotent class which is greater or not related to $((2n)^{2k+1})$. To complete the proof we need to show that $\widetilde{E}_{\tau_\psi(\sigma),\pi_{2k-1}(\sigma)}$ has a nonzero Fourier coefficient corresponding to the unipotent class $((2n)^{2k+1})$.

We shall now describe in matrices the relevant Fourier coefficients. For $r = 2k-1$ or $2k+1$, let Z_{2nr} denote the unipotent radical of the standard parabolic subgroup of Sp_{2nr} whose Levi part is

$$\underbrace{\mathrm{GL}_r \times \cdots \times \mathrm{GL}_r}_{n}.$$

We can identify the commutator quotient $Z_{2nr}/[Z_{2nr}, Z_{2nr}]$ with the group of matrices

$$M_r = \Big(\bigoplus_{i=1}^{n-1} \mathrm{Mat}_{r\times r}\Big) \oplus \mathrm{Mat}^0_{r\times r}$$

where

$$\mathrm{Mat}^0_{m\times m} = \{A \in \mathrm{Mat}_{m\times m} : {}^tAJ_m = J_mA\}.$$

Given $b_i \in F^*$ where $1 \le i \le r$, we set $b = (b_1, \ldots, b_r)$. To define an additive character on $Z_{2nr}(\mathbb{A})$, we just need to specify it on $M_r(\mathbb{A})$. Given $z \in Z_{2nr}$ we let $(z(1), \ldots, z(n-1), y)$ be the corresponding element in the quotient M_r. We define a character $\psi_{Z_{2nr},b}$ of $Z_{2nr}(F)\backslash Z_{2nr}(\mathbb{A})$ by

$$\psi_{Z_{2nr},b}(z) = \psi(trz(1) + \cdots + trz(n-1) + b_1y_{1,r} + \cdots + b_ry_{r,1})\,.$$

For given $b \in (F^*)^{2k-1}$ and $c \in (F^*)^{2k+1}$, the unipotent subgroups $Z_{2n(2k-1)}$ and $Z_{2n(2k+1)}$ with characters $\psi_{Z_{2n(2k-1)},b}$ and $\psi_{Z_{2n(2k+1)},c}$ define Fourier coefficients corresponding to $((2n)^{2k-1})$ and $((2n)^{2k+1}$, respectively. Thus, we need to prove that if the integral

$$\int_{Z_{2n(2k-1)}(F)\backslash Z_{2n(2k-1)}(\mathbb{A})} \varphi_{\pi_{2k-1}(\sigma)}(zg)\psi_{Z_{2n(2k-1)},b}(z)\,dz \tag{5.1}$$

is nonzero for some choice of data then the integral

$$(5.2)\qquad \int\limits_{Z_{2n(2k+1)}(F)\backslash Z_{2n(2k+1)}(\mathbb{A})} \widetilde{E}_{\tau_\psi(\sigma),\pi_{2k-1}(\sigma)}(zg)\psi_{Z_{2n(2k+1)},c}(z)\,dz$$

is nonzero for some choice of data. Here $\varphi_{\pi_{2k-1}(\sigma)}(g)$ is a function in the space of $\pi_{2k-1}(\sigma)$. As might be expected the vector $c \in (F^*)^{2k+1}$ will depend on $b \in (F^*)^{2k-1}$. Indeed, let $c = (1, -1, b_1, \ldots, b_{2k-1})$. We shall assume that integral (5.2) is zero for all choice of data (for the c we fixed above) and derive a contradiction. Let m denote the matrix $\begin{pmatrix} 1 & -1 \\ -1/2 & -1/2 \end{pmatrix}$. Denote

$$t = \operatorname{diag}(m, I_{2k-1}, m, I_{2k-1}, \ldots, m, I_{2k-1}, I_{2k-1}, m^*, \ldots)$$

where m occurs $2n$ times. Let

$$w = \operatorname{diag}(\nu, \ldots, \nu, \nu^*, \ldots, \nu^*) \quad \nu = \begin{pmatrix} 1 & 0 & 0 \\ 0 & 0 & I_{2k-1} \\ 0 & 1 & 0 \end{pmatrix}.$$

Conjugating (5.2) by the matrix t and then by w we obtain that the integral

$$(5.3)\qquad \int \widetilde{E}_{\tau_\psi(\sigma),\pi_{2k-1}(\sigma)}(zg)\psi_c(z)\,dz$$

is zero for all choice of data. Here z is integrated as in (5.2) and

$$\psi_c(z) = \psi(trz(1) + \cdots + trz(n-1) + y_{1,1} + b_1 y_{2,2k} + \cdots + b_{2k-1} y_{2k,2}) .$$

Next we define a Weyl element w of $\mathrm{Sp}_{2n(2k+1)}$ as follows. For $1 \le i \le 2n$, let $w[i, (2k+1)i - 2k] = 1$, and for all $1 \le i \le 2k-1$ and $0 \le j \le n-1$ let $w[2n + (2k-1)j + i, (2k+1)j + i + 1) = 1$. This defines w in the first $n(2k+1)$ rows which determines w completely as a matrix in $\mathrm{Sp}_{2n(2k+1)}$. Conjugating (5.3) by w we obtain the integral

(5.4)

$$\int \widetilde{E}_{\tau_\psi(\sigma),\pi_{2k-1}(\sigma)} \left[\begin{pmatrix} v_1 & x_1 & x_2 \\ 0 & v_2 & x_1^* \\ 0 & 0 & v_1^* \end{pmatrix} \begin{pmatrix} I & & \\ r_1 & I & \\ r_2 & r_1^* & I \end{pmatrix} g \right] \psi_c((v_1, v_2))\,dv_i\,dx_i\,dr_i$$

is zero for all choices of data. Here all variables are integrated over the groups we shall define next, with points in $F\backslash\mathbb{A}$. First v_1 is integrated over all upper unipotent matrices in GL_{2n}. Next $v_2 \in Z_{2n(2k-1)}$ and

$$\psi_c((v_1, v_2)) = \psi_{Z_{2n(2k-1)},b}(v_2) \cdot \psi(v_1(1,2) + \cdots + v_1(2n-1, 2n)).$$

The variable $x_1 \in \mathrm{Mat}'_{2n\times 2n(2k-1)} = \{X \in \mathrm{Mat}_{2n\times 2n(2k-1)} : X(j,i) = 0$ for all $1 \le i \le j(2k-1);\ 1 \le j \le 2n\}$. The variable r_1 is integrated over all matrices of the form $J_{2n} x J_{2n(2k-1)}$ where $x \in \mathrm{Mat}'_{2n\times 2n(2k-1)}$. Here J_ℓ is the

$\ell \times \ell$ matrix defined by

$$J_\ell = \begin{pmatrix} & & 1 \\ & \cdot^{\cdot^{\cdot}} & \\ 1 & & \end{pmatrix}.$$

Finally

$$x_2, r_2 \in \mathrm{Mat}''_{2n\times 2n} = \{x \in \mathrm{Mat}^0_{2n\times 2n} : x(i,j) = 0 \text{ for all } i \geq j\}.$$

Next we proceed as in **[G-R-S2]** in the proof of Theorem 1 p. 889–894. In other words, we use a series of Fourier expansion and the assumption of the vanishing of (5.4) for all choice of data to deduce that the integral

$$(5.5) \qquad \int \widetilde{E}_{\tau_\psi(\sigma),\pi_{2k-1}(\sigma)}\left[\begin{pmatrix} v_1 & t_1 & t_2 \\ & v_2 & t_1^* \\ & & v_1^* \end{pmatrix} g\right] \psi_c((v_1, v_2))\, dv_i\, dt_i$$

is zero for all choice of data. Here v_1 and v_2 are integrated as before, t_1 is integrated over all $\mathrm{Mat}_{2n\times 2n(2k-1)}$ and $t_2 \in \mathrm{Mat}^0_{2n\times 2n}$ (points in $F\backslash\mathbb{A}$). To derive that we also used an analogue of **[G-R-S2]** Lemma 2 p. 896. The integration over t_1 and t_2 yields the constant term of $\widetilde{E}_{\tau_\psi(\sigma),\pi_{2k-1}(\sigma)}$ along $U(P_{2n(2k+1)})$. As in **[G-R-S2]** Lemma 1 we deduce that the vanishing of (5.5) for all choice of data implies either the vanishing of the Whittaker coefficient of $\tau_\psi(\sigma)$ or the vanishing of (5.1) for all choice of data. By our assumptions both possibilities are false and hence we derived a contradiction. Thus, integral (5.2) is nonzero for some choice of data and the proposition is proved. □

At this point we can prove Theorem A. Recall that $\pi'_{2k}(\sigma)$ is the automorphic representation of $\mathrm{Sp}_{4nk}(\mathbb{A})$ which is generated by all functions $f(h)$ as defined in Section 4 formula (4.1). That is, by all functions of the form

$$f(h) = \int\limits_{V_{n,2n(2k+1)}(F)\backslash V_{n,2n(2k+1)}(\mathbb{A})} \widetilde{\theta}^{\psi}_{\phi}(\ell(v)h)\widetilde{E}_{\tau_\psi(\sigma),\pi_{2k-1}(\sigma)}(vh)\psi_{V_{n,2n(2k+1)}}(v)\, dv\,.$$

We shall prove Theorem A in three steps. First we will study the cuspidality of $\pi'_{2k}(\sigma)$. Then its nonvanishing and finally we will consider the unramified correspondence. We start with

Proposition 5.2. *Assume that $\pi_{2k-1}(\sigma)$ has a missing $\psi_{U_{n,2n(2k-1)}}$ Fourier coefficient and assume that $\mathcal{O}_{\mathrm{Sp}_{2n(2k-1)}}(\pi_{2k-1}(\sigma))$ is not zero. Then $\pi'_{2k}(\sigma)$ is a cusp form of* $\mathrm{Sp}_{4nk}(\mathbb{A})$.

Proof. For any $0 \leq m \leq 2nk-1$ let L_m denote the unipotent radical of the standard maximal parabolic subgroup of Sp_{4nk} whose Levi part is $\mathrm{GL}_{2nk-m} \times$

Sp_{2m}. In matrices we have

$$L_m = \left\{ \begin{pmatrix} I_{2nk-m} & x & y \\ & I_{2m} & x^* \\ & & I_{2nk-m} \end{pmatrix} \in Sp_{4nk} \right\}.$$

Thus we need to prove that

$$\int\limits_{L_m(F)\backslash L_m(\mathbb{A})} f(\ell h)\, d\ell \tag{5.6}$$

is zero for all choice of data and all $0 \le m \le 2nk - 1$. Let $V^-_{n,2n(2k+1)}$ be the subgroup of $V_{n,2n(2k+1)}$ defined as

$$V^-_{n,2n(2k+1)} = \left\{ v = (v_{ij}) \in V_{n,2n(2k+1)} : v_{n,j} = 0 \text{ for all } n+1 \le j \le 2nk+n \right\}.$$

Unfolding the theta function as in **[G-R-S1]** p. 844–847 we see that in order to prove the vanishing of (5.6) we need to show that

$$\int \widetilde{E}_{\tau_\psi(\sigma),\pi_{2k-1}(\sigma)}(v\ell h)\psi_{U_{n,2n(2k+1)}}(v)\, dv\, d\ell \tag{5.7}$$

is zero for all choice of data. Here ℓ is integrated over $L_m(F)\backslash L_m(\mathbb{A})$ and

$$v \in V^-_{n,2n(2k+1)}(F)\backslash V^-_{n,2n(2k+1)}(\mathbb{A}) .$$

Also we view $\psi_{U_{n,2n(2k+1)}}$ as a character of $V^-_{n,2n(2k+1)}(\mathbb{A})$ by extending it trivially on

$$V^-_{n,2n(2k+1)}(\mathbb{A})\backslash U_{n,2n(2k+1)}(\mathbb{A}).$$

By definition, $U_{n,2n(2k+1)}$ is a subgroup of $V^-_{n,2n(2k+1)}$. Let

$$w_m = \begin{pmatrix} & I_r & & & \\ I_n & & & & \\ & & I_{2m} & & \\ & & & & I_n \\ & & & I_r & \end{pmatrix}, \qquad r = 2nk - m.$$

Conjugating inside (5.7) by w_m we need to consider the integral

(5.8)

$$\int \widetilde{E}_{\tau_\psi(\sigma),\pi_{2k-1}(\sigma)} \left[\begin{pmatrix} I_r & x_1 & x_2 \\ 0 & v_2 & x_1^* \\ 0 & 0 & I_r \end{pmatrix} \begin{pmatrix} I_r & & \\ z & I & \\ 0 & z^* & I_r \end{pmatrix} \right] \psi_{U_{n,2(n+m)}}(v_2)\, dx_i\, dv_2\, dz .$$

Here $x_2 \in Mat^0_{r\times r}$ where $r = 2nk - m$. x_1 is integrated over all $r \times 2(n+m)$ matrices whose n first columns are zero. z is integrated over all $2(n+m) \times r$ matrices whose last $n + 2m + 1$ rows are zero. The matrix v_2 is integrated over $V^-_{n,2(n+m)}$. All integrations are over the corresponding groups with points in

$F\backslash\mathbb{A}$. Next we use Fourier expansions in (5.8) along the $n-1$ columns in x_1 from the second to the n-th. Using the z integration we collapse summation and integration and deduce that (5.8) equals

(5.9)

$$\int \widetilde{E}_{\tau_\psi(\sigma),\pi_{2k-1}(\sigma)}\left[\begin{pmatrix} I_r & x_1 & x_2 \\ 0 & v_2 & x_1^* \\ 0 & 0 & I_r \end{pmatrix}\begin{pmatrix} I_r & & \\ z & I & \\ 0 & z^* & I_r \end{pmatrix}\right]\psi_{U_{n,2(n+m)}}(v_2)\,dx_i\,dv_2\,dz$$

Here x_1 is integrated over all $r \times 2(n+m)$ whose first column is zero. z is integrated over the same group as in (5.8) but with adelic points. All other variables are integrated as in (5.8). Clearly we can ignore the z integration. In (5.9) we can now use the usual expansion column by column starting from the $r+1$-th column and going down. Since GL_p acts on a column of size p with two orbits we deduce that (5.9) will be zero for all choice of data if and only if each of the following integrals

$$(5.10) \qquad \int \widetilde{E}_{\tau_\psi(\sigma),\pi_{2k-1}(\sigma)}\left[\begin{pmatrix} v_1 & x_1 & x_2 \\ 0 & v_2 & x_1^* \\ 0 & 0 & v_1^* \end{pmatrix}\right]\psi_{U_{n,2(n+m)}}(v_2)\psi_\epsilon(v_1,x_1)\,dv_i\,dx_i$$

is zero for all choice of data. Here x_1 is integrated over all $r \times 2(n+m)$ matrices and v_1 is integrated over all upper unipotent matrices of size $r \times r$. All integrations are with respect to points in $F\backslash\mathbb{A}$. Also

$$\psi_\epsilon((v_1,x_1)) = \psi(\epsilon_1 v_1(1,2) + \cdots + \epsilon_{r-1}v_1(r-1,r) + \epsilon_r x_1(r,1))$$

where $\epsilon_i \in \{0,1\}$ and $v_1(i,j)$ is the (i,j)-th entry of v_1. Since $v_2 \in V^-_{n,2(n+m)}$, then combining all variables, we obtain the integration over the group $V^-_{2nk+n-m,2n(2k+1)}$ with the above specified character. Recall that the residue corresponds to induction from cuspidal data of the maximal parabolic whose Levi part is $\mathrm{GL}_{2n} \times \mathrm{Sp}_{2n(2k-1)}$. This means that integration over any other unipotent radical subgroup with a trivial character on it, will give zero. Thus we need to consider two cases. First let $\epsilon_i = 1$ for all $1 \le i \le r$. In this case the contribution of (5.10) will be

$$\int \widetilde{E}_{\tau_\psi(\sigma),\pi_{2k-1}(\sigma)}(v)\psi_{U_{2nk+n-m,2n(2k+1)}}(v)\,dv$$

where v is integrated over $V^-_{2nk+n-m,2n(2k+1)}(F)\backslash V^-_{2nk+n-m,2n(2k+1)}(\mathbb{A})$. This integration corresponds to the unipotent class $(2(2nk+n-m)1^{2m})$. But this unipotent class is either not related or greater than $((2n)^{2k+1})$. By our assumption $\mathcal{O}_{Sp_{2n(2k-1)}}(\pi_{2k-1}(\sigma))$ is not zero. Hence, from Proposition 5.1 it follows that $(\widetilde{E}_{\tau_\psi(\sigma),\pi_{2k-1}(\sigma)}) = ((2n)^{2k+1})$ and we get zero.

The other possible contribution to (5.10) can come from the integral

$$\int \widetilde{E}_{\tau_\psi(\sigma),\pi_{2k-1}(\sigma)}\left[\begin{pmatrix} v_1 & x_1 & x_2 \\ 0 & v_2 & x_1^* \\ 0 & 0 & v_1^* \end{pmatrix}\right] \psi_{U_{n,2n(2k-1)}}(v_2)\widetilde{\psi}(v_1)\, dv_i\, dx_i$$

where v_1 is integrated over all $2n\times 2n$ upper unipotent matrices, v_2 is integrated over $V^-_{n,2n(2k-1)}$ and the x_1 and x_2 integration combine to the unipotent radical subgroup of the parabolic $P_{2n(2k+1)} = (\mathrm{GL}_{2n} \times \mathrm{Sp}_{2n(2k-1)})\, U(P_{2n(2k+1)})$. All integrations are over the corresponding groups with points in $F\backslash\mathbb{A}$. Finally

$$\widetilde{\psi}(v_1) = \psi(v_1(1,2) + \cdots + v_1(2n-1,2n))\ .$$

This integral is zero since we obtain

$$\int_{U_{n,2n(2k-1)}(F)\backslash U_{n,2n(2k-1)}(\mathbb{A})} \varphi_{\pi_{2k-1}(\sigma)}(u)\, \psi_{U_{n,2n(2k-1)}}(u)\, du$$

as inner integration. But this integral is zero since we assumed that $\pi_{2k-1}(\sigma)$ has a missing $\psi_{U_{n,2n(2k-1)}}$–Fourier coefficient. This completes the proof of the cuspidality. □

Next we study the nonvanishing of the lifting.

Proposition 5.3. $\mathcal{O}_{\mathrm{Sp}_{4nk}}(\pi'_{2k}(\sigma)) = ((2n)^{2k})$.

Proof. One can easily check that the integration defining the function $f(h)$ (and hence the representation $\pi'_{2k}(\sigma)$) is a unipotent integration corresponding to the unipotent class $((2n)1^{4nk})$. Suppose that $\pi'_{2k}(\sigma)$ has a nonzero Fourier coefficient corresponding to $\mathcal{O}$ where $\mathcal{O}$ is a unipotent class of Sp_{4nk} which is either greater or not related to $((2n)^{2k})$. Arguing as in the proof of Proposition 3.6, it follows that there is $p > n$ such that $\pi'_{2k}(\sigma)$ has a nonzero Fourier coefficient corresponding to the unipotent class $((2p)1^{4nk-2p})$. In other words we assume the integral

$$\int_{U_{p,4nk-2p}(F)\backslash U_{p,4nk-2p}(\mathbb{A})} f(uh)\, \psi^a_{U_{p,4nk-2p}}(u)\, du$$

is nonzero for some choice of data. (In **[G-R-S5]** Lemma 2.5 this type of integral was denoted by $((2n)1^{4nk}) \circ ((2p)1^{4nk-2p})$). It follows from **[G-R-S5]** Lemma 2.5 that if $p > n$ then there exists a $d_1 > 2n$ such that if $\mathcal{O} = (d_1 \cdots d_s)$ then $\widetilde{E}_{\tau_\psi(\sigma),\pi_{2k-1}(\sigma)}$ has a nonzero Fourier coefficient corresponding to the unipotent class $\mathcal{O}$. Here, as in the proof of Proposition 5.1 one has to use the fact that the integrals

$$\int_{U_r(F)\backslash U_r(\mathbb{A})} \phi_\epsilon(ug)\psi_{U_r}(u)\, du$$

are all zero if $r > n$ where $\epsilon = \widetilde{E}_{\tau_\psi(\sigma),\pi_{2k-1}(\sigma)}$. This contradicts Proposition 5.1. To complete the proof all we have to show is that $\pi'_{2k}(\sigma)$ has a nonzero Fourier coefficient corresponding to the unipotent class $((2n)^{2k})$.

To do that we define the unipotent subgroup Z_{4nk} as in the proof of Proposition 5.1. In other words Z_{4nk} is the unipotent radical of the standard parabolic subgroup of Sp_{4nk} whose Levi part is $\mathrm{GL}_{2k} \times \cdots \times \mathrm{GL}_{2k}$. We identify $Z_{4nk}/[Z_{4nk}, Z_{4nk}]$ with $\left(\bigoplus_{i=1}^{n-1} \mathrm{Mat}_{2k\times 2k}\right) \oplus \mathrm{Mat}^0_{2k\times 2k}$ and, as in Proposition 5.1, we define a character $\psi_{Z_{4nk},d}$ of Z_{4nk} as

$$\psi_{Z_{4nk},d}(z) = \psi(trz(1) + \cdots + trz(n-1) + d_1 z_{1,2k} + \cdots + d_{2k} z_{2k,1})$$

where $d = (d_1, \ldots, d_{2k}) \in (F^*)^{2k}$. To complete the proof we need to show that for some d the integral

$$\int\limits_{Z_{4nk}(F^*)\backslash Z_{4nk}(\mathbb{A})} f(zh)\psi_{Z_{4nk},d}(z)\, dz \tag{5.11}$$

is nonzero for some choice of data. In Proposition 5.1 we proved that integral (5.2) is nonzero for some choice of data with $c = (1, -1, b_1, \ldots, b_{2k-1})$. We start with that integral. Let β denote the following Weyl element of $\mathrm{GL}_{n(2k+1)}$. The ones in the first n rows of β are at entries $(i, (2k+1)i - 2k)$ for $1 \le i \le n$. Then we have $\beta[n + 2kj + i, 2kj + i + j + 1] = 1$ for all $1 \le i \le 2k$ and $0 \le j \le n-1$. All other entries of β are zero. We conjugate inside (5.2) by the Weyl element $w = \begin{pmatrix} \beta & 0 \\ 0 & \beta^* \end{pmatrix}$ of Sp_{4nk}. Hence we conclude that the integral

(5.12)

$$\int \widetilde{E}_{\tau_\psi(\sigma),\pi_{2k-1}(\sigma)} \left[\begin{pmatrix} v & x & y \\ 0 & z & x^* \\ 0 & 0 & v^* \end{pmatrix} \begin{pmatrix} I_n & & \\ r & I & \\ 0 & r^* & I \end{pmatrix} \right] \psi_c((v,z,y))\, dv\, dz\, dx\, dy\, dr$$

is nonzero for some choice of data. Here v is integrated over all upper unipotent matrices of size $n \times n$, $y \in \mathrm{Mat}^0_{n\times n}$ and $z \in Z_{4nk}$. The matrix x is integrated over all $n \times 4nk$ matrices described as follows. Write $x = (x_1, x_2)$ where $x_i \in \mathrm{Mat}_{n\times 2nk}$. Then x_2 is integrated over all $\mathrm{Mat}_{n\times 2nk}$ and x_1 is integrated over

$$X_1 = \{x_1 \in \mathrm{Mat}_{n\times 2nk} : x_1(i,j) = 0 \text{ for all } 1 \le i \le n; 1 \le j \le 2ki\}\,.$$

Finally the matrix r is integrated over

$$\{r \in \mathrm{Mat}_{4nk\times n} : r = (J_1 x_1 J_{2nk})^t \text{ where } x_1 \in X_1\}.$$

All variables are integrated in their groups with points in $F\backslash\mathbb{A}$. We also have

$$\psi_c((v,z,y)) = \psi_{Z_{4nk},d}(z)\psi(v_{1,2} + \cdots + v_{n-1,n} + y_{n,1}),$$

where $d = (-1, b_1, \ldots, b_{2k-1})$.

Expanding (5.12) along the "missing" roots of x_1 and arguing in a similar way as in **[G-R-S1]** Theorem 8 p. 844–847 we obtain that the integral (5.13)

$$\int \widetilde{E}_{\tau_\psi(\sigma),\pi_{2k-1}(\sigma)}\left[\begin{pmatrix} v & x & y \\ 0 & z & x^* \\ 0 & 0 & v^* \end{pmatrix}\begin{pmatrix} I_n & & \\ r & I & \\ 0 & r^* & I \end{pmatrix}\right]\psi_c((v,z,y))\,dv\,dz\,dx\,dy\,dr$$

is nonzero for some choice of data. Here $x=(x_1,x_2)$ where now x_1 is integrated over all $n\times 2nk$ matrices with zero bottom row and r is integrated as above but with adelic points. All other integrations are as in (5.12). Notice that the v,x,y integration combines to an integration over $V^-_{n,2n(2k+1)}(F)\backslash V^-_{n,2n(2k+1)}(\mathbb{A})$ and ψ_c restricted to (v,y) produces the character $\psi_{U_{n,2n(2k+1)}}$ on $V^-_{n,2n(2k+1)}$ where we extend it trivially from $U_{n,2n(2k+1)}$ to $V^-_{n,2n(2k+1)}$. As in **[G-R-S2]** Lemma 1 p. 895 we may ignore the r integration and deduce that (5.13) is nonzero if and only if the integral

$$(5.14)\qquad \int \widetilde{E}_{\tau_\psi(\sigma),\pi_{2k-1}(\sigma)}\left(u\begin{pmatrix} I_n & & \\ & z & \\ & & I_n \end{pmatrix}\right)\psi_{U_{n,2n(2k+1)}}(u)\,\psi_{Z_{4nk},d}(z)\,du\,dz$$

is nonzero for some choice of data. Here u is integrated over

$$V^-_{n,2n(2k+1)}(F)\backslash V^-_{n,2n(2k+1)}(\mathbb{A})$$

and z is as before. Using the argument in **[G-R-S5]** which is given right at the end of Section 1 we deduce that (5.14) is nonzero if and only if (5.11) is nonzero. Thus the proposition follows. □

To complete the proof of Theorem A we need to establish the local unramified correspondence.

In the next proposition if $\sigma=\otimes\sigma_\nu$ and σ_ν is unramified then assume that σ_ν is the unramified constituent of $\mathrm{Ind}_{B(F_\nu)}^{\widetilde{\mathrm{Sp}}_{2n(2k-1)}(F_\nu)}(\chi\circ\gamma_{\psi_\nu})\delta_B^{1/2}$. For $t\in T$, the maximal torus of B, we write $\chi(t)=\prod_{i=1}^{n}\chi_i(t_i)$ where χ_i are unramified. Assume that $\pi_{2k}(\sigma)=\otimes(\pi_{2k}(\sigma))_\nu$. We prove

Proposition 5.4. *At a place ν as above, $(\pi'_{2k}(\sigma))_\nu$ is the unramified constituent of*

$$\mathrm{Ind}_{\widehat{Q}_k}^{\mathrm{Sp}_{4nk}(F_\nu)}(\chi_1(\det)\times\cdots\times\chi_n(\det))\delta_{\widehat{Q}_k}^{1/2}.$$

(See Lemma 3.1 for the notation.)

Proof. The proof is similar to the proof of Lemma 3.3. Denote the representation (3.1) by $\rho^{(n,k)}_{\chi,\psi}$. Denote, for short, $F=F_\nu$, $U_n=U_{n,2n(2k+1)}(F)$,

$H = H_{4nk+1}(F)$. Denote by C the center of H. (We embed H in the unipotent radical $V_{n,2n(2k+1)}(F)$.) Let $\omega_{\psi^{-1}}$ denote the Weil representation (corresponding to ψ^{-1}) of $\widetilde{\mathrm{Sp}}_{4nk}(F) \ltimes H$ (embedded in the inverse image of the Levi part of the standard parabolic subgroup Q_n of $\mathrm{Sp}_{2n(2k+1)}(F)$, whose unipotent radical is $V_{n,2n(2k+1)}(F)$). The main point of the proof is to consider the Jacquet module $J_{C\backslash H}(J_{U_n,\psi_{U_n}}(\rho_{\chi,\psi}^{(n,k)} \otimes \omega_{\psi^{-1}})$ as a module over $\mathrm{Sp}_{4nk}(F)$, and show that it has a unique unramified constituent, which is the unramified constituent of $\mathrm{Ind}_{\widehat{Q}_k}^{\mathrm{Sp}_{4nk}(F)}(\chi_1(\det) \otimes \cdots \otimes \chi_n(\det))\delta_{\widehat{Q}_k}^{1/2}$. We first consider $J_{U_n,\psi_{U_n}}(\rho_{\chi,\psi}^{(n,k)}) = J_{U_n,\psi_{U_n}}(\mathrm{Res}_{\widetilde{Q}_n} \rho_{\chi,\psi}^{(n,k)})$. As in Lemma 3.3, $\mathrm{Res}_{\widetilde{Q}_n} \rho_{\chi,\psi}^{(n,k)}$ has a normal series, whose subquotients are parameterized by $\widehat{L}_k \backslash \mathrm{Sp}_{2n(2k+1)}(F)/Q_n$. We may take representatives w which are Weyl elements. Such an element w is not relevant (in the sense that it has a zero contribution to the Jacquet module, with respect (U_n, ψ_{U_n})) if there is $u \in U_n$, such that $\psi_{U_n}(u) \neq 1$, and $wuw^{-1} \in \widehat{L}_k$. Note that for v_0 in the unipotent radical of Q_n, and

$$\widehat{g}_0 = \begin{pmatrix} I_n & & \\ & g & \\ & & I_n \end{pmatrix} \in Q_n,$$

we have $\widehat{g}_0 v_0 U_n v_0^{-1} \widehat{g}_0^{-1} = U_n$, and hence, if there is $u \in U_n$, such that $wuw^{-1} \in \widehat{L}_k$, then

$$w\widehat{g}_0 v_0 (v_0^{-1} \widehat{g}_0^{-1} u \widehat{g}_0 v_0) v_0^{-1} \widehat{g}_0^{-1} w^{-1} = wuw^{-1} \in \widehat{L}_k$$

and clearly,

$$\psi_{U_n}(v_0^{-1} \widehat{g}_0^{-1} u \widehat{g}_0 v_0) = \psi_{U_n}(u) \ ,$$

so that $u' = v_0^{-1} \widehat{g}_0^{-1} u \widehat{g}_0 v_0 \in U_n$ satisfies

$$(w\widehat{g}_0 v_0) u' (w\widehat{g}_0 v_0)^{-1} \in \widehat{L}_k \text{ and } \psi_{U_n}(u') \neq 1 \ .$$

A relevant Weyl element w must satisfy

$$wu(t)w^{-1} \in Z$$

for each one parameter subgroup $u(t)$ in U_n, such that $\psi_{U_n}(u(t)) \not\equiv 1$. Here Z is the unipotent radical of the parabolic subgroup of $\mathrm{Sp}_{2n(2k+1)}(F)$, opposite to $\widehat{L}_k$. Using notation similar to that of Lemma 3.3, since $\widehat{u}_{n,4nk+n+1}$ corresponds to a short root, $w\widehat{u}_{n,4nk+n+1}w^{-1} \in \widehat{Z}_{i_1,2n-i_1+1}$, for some $n+1 \leq i_1 \leq 2n$. We may assume, as in Proposition 5.1, that $w\widehat{u}_{n,unk+1}w^{-1} = \widehat{z}_{i_1,2n-i_1+1}(1, 2k+1)$. Now consider $\widehat{u}_{n-1,n}$ (which does not commute with $\widehat{u}_{n,4nk+n+1}$). The only "blocks" of Z which don't commute with $\widehat{z}_{i_1,2n-i_1+1}$ are $\widehat{z}_{2n-i_1+1,j}$, $1 \leq j \leq 2n - i_1$. Hence $w\widehat{u}_{n-1,n}w^{-1} = \widehat{z}_{2n-i_1+1,i_2}$, for some $1 \leq i_2 \leq 2n - i_1$. We may assume, as in Lemma 3.3, that $w\widehat{u}_{n-1,n}w^{-1} = \widehat{z}_{2n-i_1+1,i_2}(2k+1, 2k+1)$. Next, as in Lemma 3.3, we get that $w\widehat{u}_{n-2,n-1}w^{-1} \in \widehat{Z}_{i_2,i_3}$ for some $1 \leq i_3 \leq i_2 - 1$, and we may assume that $w\widehat{u}_{n-2,n-1}w^{-1} = \widehat{z}_{i_2,i_3}(2k+1,2k+1)$. Continuing in this way, we get that $w\widehat{u}_{\ell,\ell+1}w^{-1}$, for $\ell = n-1, n-2, \ldots, 2, 1$ is respectively $\widehat{z}_{2n-i_1+1,i_2}(2k+1,2k+1), \widehat{z}_{i_2,i3}(2k+1,2k+1), \ldots, \widehat{z}_{i_{n-1},i_{n-2}}(2k+1,2k+1)$ where

$$n \geq 2n - i_1 + 1 > i_2 > i_3 > \cdots > i_{n-1} \geq 2 \, .$$

This forces $i_{n-1} = 2, i_{n-2} = 3, \ldots, i_2 = n-1, i_1 = n+1$. It is now easy to see that modulo $\widehat{L}_k$ from the left and Q_n from the right there is a unique such Weyl element w. Then one can check that $w^{-1}\widehat{L}_k w \cap Q_n = L_{n,k}$ consists of the elements

$$(5.15) \qquad \begin{pmatrix} t & 0 & u & 0 \\ & g & x & u' \\ & 0 & g^* & 0 \\ & & & t^* \end{pmatrix}$$

in $\mathrm{Sp}_{2n(2k+1)}(F)$, where $t = \mathrm{diag}(a_1, \ldots, a_n)$, $\begin{pmatrix} g & x \\ 0 & g^* \end{pmatrix} \in \widehat{Q}_k \subset \mathrm{Sp}_{4nk}(F)$ and if we write u as a matrix (u_{ij}), $1 \leq i,j \leq n$, where u_{ij} are rows of $2k$ coordinates, then $u_{ij} = 0$ for $1 \leq j < n+1-i$. Thus

$$J_{U_n,\psi_{U_n}}(\rho^{(n,k)}_{\chi,\psi}) \cong J_{U_n,\psi_{U_n}}(\mathrm{Ind}^{\tilde{Q}_n}_{\tilde{L}_{n,k}} \alpha_{\chi,\psi})$$

where $\alpha_{\chi,\psi}$ is the character which takes $(h,1)$, for h of the form (5.15) to

$$\prod_{i=1}^{n} (|a_i^{-1}| \, |\det g_i|)^{(2m-2i+1)(k+\frac{1}{2})+1/2} \cdot \gamma_\psi(a_i^{-1}\det g_i)\chi_i(a_i^{-1}\det g_i)) \\ \times \prod_{1\leq i<j\leq n} (a_i^{-1}\det g_i, a_j^{-1}\det g_j),$$

where

$$g = \begin{pmatrix} g_1 & & * \\ & \ddots & \\ & & g_n \end{pmatrix},$$

with $g_i \in \mathrm{GL}_{2k}(F)$. We have an isomorphism of $\tilde{Q}' = \widetilde{\mathrm{Sp}}_{4nk}(F) \ltimes H$ modules

$$J_{U_n,\psi_{U_n}}(\mathrm{Ind}^{\tilde{Q}_n}_{\tilde{L}_{n,k}} \alpha_{\chi,\psi}) \cong J_{C,\psi}(\mathrm{Ind}^{\tilde{Q}'}_{\tilde{L}'} \beta_{\chi,\psi}),$$

where $(g;(x,t)) \in \mathrm{Sp}_{4nk}(F)\cdot H$ is embedded in Q_n as

$$\begin{pmatrix} I_{n-1} &&&&& \\ & 1 & x & t && \\ && g & x' && \\ &&& 1 && \\ &&&& I_{n-1} \end{pmatrix},$$

L' is the subgroup of $\mathrm{Sp}_{4nk}(F)\cdot H$ which consists of elements of the form

$$\begin{pmatrix} 1 & 0 & \cdots & 0 & * & \cdots & * & 0 \\ & g_1 & & * & & & * & * \\ && \ddots &&&&& \vdots \\ &&& g_n &&&& * \\ &&&& g_n^* && * & 0 \\ &&&&& \ddots && \vdots \\ &&&&&& g_1^* & 0 \\ &&&&&&& 1 \end{pmatrix}, \quad g_i \in \mathrm{GL}_{2k}(F), \tag{5.16}$$

and $\beta_{\chi,\psi}$ takes $(h,1)$, for h of the form (5.16), to

$$\prod_{i=1}^{n-1} |\det g_i|^{i-n} \prod_{i=1}^{n}(|\det g_i|^{(2n-2i+1)(k+\frac{1}{2})+1/2}\gamma_\psi(\det g_i)\chi_i(\det g_i)) \\ \times \prod_{1\le i<j\le n} (\det g_i, \det g_j)\ .$$

The isomorphism above is induced by the map

$$T(f)(h_1') = \int\limits_{U_n\cap L_{n,k}\backslash U_n} \psi_{U_n}^{-1}(u) f\left(u \begin{pmatrix} I_{n-1} && \\ & h' & \\ && I_{n-1} \end{pmatrix}, 1\right) du$$

for $h' \in Q' = \mathrm{Sp}_{4nk}(F)\cdot H$, and f in the space of $\mathrm{Ind}_{\tilde{L}_{n,k}}^{\tilde{Q}_n} \alpha_{\chi,\psi}$. Thus

$$J_{C\backslash H}(J_{U_n,\psi_{U_n}}(\rho_{\chi,\psi}^{(n,k)}) \otimes \omega_{\psi^{-1}}) \cong J_{C\backslash H}(J_{C,\psi}(\mathrm{Ind}_{\tilde{L}'}^{\tilde{Q}'} \beta_{\chi,\psi}) \otimes \omega_{\psi^{-1}}) \tag{5.17}$$

as $\mathrm{Sp}_{4nk}(F)$-modules. It is easy to see that the right hand side of (5.17) has a finite composition series with a unique unramified constituent. Consider the functional

$$R : \mathrm{Ind}_{\tilde{L}'}^{\tilde{Q}'} \beta_{\chi,\psi} \otimes \omega_{\psi^{-1}} \longrightarrow \mathbb{C}$$

defined by

$$R(f\otimes\phi)=\int\limits_{F^{2nk+1}} f\left(\begin{pmatrix}1 & x & 0 & z\\ & I_{2nk} & 0 & 0\\ & & I_{2nk} & x'\\ & & & 1\end{pmatrix},1\right)\omega_{\psi^{-1}}(x,0;z)\phi(0)\,dx\,dz$$

$$=\int\limits_{F^{2nk+1}} f\left(\begin{pmatrix}1 & x & 0 & z\\ & I_{2nk} & 0 & 0\\ & & I_{2nk} & x'\\ & & & 1\end{pmatrix},1\right)\psi^{-1}(z)\phi(x)\,dx\,dz\ .$$

Here $\phi\in S(F^{2nk})$. R clearly factors through $J_{C,\psi}(\mathrm{Ind}_{\tilde{L}'}^{c\tilde{Q}'}\beta_{\chi,\psi})\otimes\omega_{\psi^{-1}}$. For

$$g=\begin{pmatrix}g_1 & & *\\ & \ddots & \\ 0 & & g_n\end{pmatrix},\ g_i\in \mathrm{GL}_{2k}(F),$$

R satisfies,

$$R(\rho(\begin{pmatrix}1 & 0 & * & 0\\ & g & * & *\\ & & g^* & 0\\ & & & 1\end{pmatrix},1)f\otimes\omega_{\psi^{-1}}((\begin{pmatrix}g & *\\ 0 & g^*\end{pmatrix},1)\phi))$$

$$=\delta_{\widehat{Q}_k}^{1/2}\begin{pmatrix}g & *\\ 0 & g^*\end{pmatrix}\prod_{i=1}^{n}\chi_i(\det g_i)R(f,\phi)\ .$$

Here ρ denotes the right translation. Thus R defines an $\mathrm{Sp}_{4nk}(F)$-map

$$J_{C\backslash H}(J_{C,\psi}(\mathrm{Ind}_{\tilde{L}'}^{\tilde{Q}'}\beta_{\chi,\psi})\otimes\omega_{\psi^{-1}})\longrightarrow \mathrm{Ind}_{\widehat{Q}_k}^{\mathrm{Sp}_{4nk}(F)}(\chi_1(\det)\otimes\cdots\otimes\chi_n(\det))\cdot\delta_{\widehat{Q}_k}^{1/2}$$

It is easy to see that this map is nontrivial on the unramified constituent, and hence the unramified constituent of the $\mathrm{Sp}_{4nk}(F)$ module

$$J_{C\backslash H}(J_{U_n,\psi_{U_n}}(\rho_{\chi,\psi}^{(n,k)})\otimes\omega_{\psi^{-1}})$$

is the unramified constituent of $\mathrm{Ind}_{\widehat{Q}_k}^{\mathrm{Sp}_{4nk}(F)}(\chi_1(\det)\otimes\cdots\otimes\chi_n(\det))\delta_{\widehat{Q}_k}^{1/2}$. □

6. On CAP representations with respect to the Siegel parabolic subgroup

In Section 5, we showed how to construct CAP representations on symplectic groups. To do that we started with an irreducible generic cuspidal representation σ of $\widetilde{Sp}_{2n}(\mathbb{A})$ and we made the assumption that σ has a missing $\psi_{U_{n,2n}}$

Fourier coefficient (see Assumption 1). It is natural to ask if the converse statement is true.

In this section we want to study this question. We shall only consider CAP representations with respect to the Siegel parabolic subgroup of $\mathrm{Sp}_{4n}(\mathbb{A})$.

To start, let τ be a selfdual cuspidal representation of $\mathrm{GL}_{2n}(\mathbb{A})$ such that $L^S(\pi, \wedge^2, s)$ has a simple pole at $s = 1$. Let π be a cuspidal representation of $\widetilde{\mathrm{Sp}}_{4n}(\mathbb{A})$. We will say that π has a Fourier coefficient with respect to the unipotent class $((2n)^2)$ if the integral as defined in (5.11) with $k = 1$ is not zero for some choice of data.

We prove:

Theorem C. *Let π be a CAP representation of $G = Sp_{4n}(\mathbb{A})$ with respect to $(Q_1, \tau, \bar{s}_1)$ (See Definition 2.1). Suppose that $\mathcal{O}_G(\pi) = ((2n)^2)$ (see Definition 3.5). Then there exists a generic cuspidal representation σ of $\widetilde{Sp}_{2n}(\mathbb{A})$ such that σ has a missing $\psi_{U_{n,2n}}$ Fourier coefficient and such that σ has a ψ weak lift to τ.*

Proof. From the assumptions it follows that integral (5.11) is nonzero for some choice of data where $d = (d_1, d_2)$. Conjugating, if needed, by a diagonal element in GSp_{4n} we obtain a cusp form which is nearly equivalent to π such that $d_1 = -1$. Thus we can assume that π has a nonzero Fourier coefficient as defined in (5.11) with $d_1 = -1$.

At this point we use the descent method in a similar way as in **[G-R-S1]**. To be more specific let U denote the unipotent radical of the parabolic subgroup of Sp_{4n} whose Levi part is $\mathrm{GL}_1^n \times Sp_{2n}$. We define a character ψ_U of U as follows. If $u = (u_{i,j})$ then $\psi_U(u) = \psi(u_{1,2} + \cdots + u_{n-1,n})$.

We define

$$\tilde{f}(g) = \int\limits_{U(F)\backslash U(\mathbb{A})} \theta_\phi^{\psi^{-1}}(l(u)g)\varphi_\pi(ug)\psi_U(u)\,du.$$

Here l defines the projection from U to the Heisenberg group with $2n + 1$ variables. (It was denoted by j_k in **[G-R-S1]** (2.13) page 836.) Also θ denotes the usual theta function on the double cover of Sp_{2n}. Thus as we vary the data we get a representation σ on $\widetilde{Sp}_{2n}(\mathbb{A})$ by taking all right translations of the functions $\tilde{f}$. The fact that σ is not zero follows from our assumption that π has a nonzero Fourier coefficient with respect to the unipotent class $((2n)^2)$. Indeed, as in Lemma 2.6 in **[G-R-S5]** one can easily deduce that π has a nonzero Fourier coefficient with respect to the unipotent class $((2n)1^{2n})$ for a suitable character. This last unipotent class corresponds to the above integral.

Also, from Lemma 2.6 in **[G-R-S5]**, it follows that σ is generic. Indeed, in the notation of that lemma we can deduce that π has a nonzero Fourier coefficient with respect to $((2n)1^{2n}) \circ (2n)$. But this just means that π is generic.

Next we show that σ is cuspidal. This follows as in **[G-R-S1]**. As is shown there, when we compute each constant term of $\tilde{f}$ with respect to any maximal unipotent subgroup of $\widetilde{Sp}_{2n}$ we obtain Fourier coefficients which corresponds to unipotent class which are bigger then $((2n)^2)$. By our assumption they are all zero. Hence σ is cuspidal.

To show that σ has a missing $\psi_{U_{n,2n}}$ we proceed as in **[G-R-S2]** Section 5. Indeed, one can check that the proof of Theorem 2 in **[G-R-S2]** Section 5 is applicable in our case as well. Thus we can derive the formula (5.26) in **[G-R-S2]** in our case. However, since π is a cuspidal representation then the right hand side of (5.26) is zero in our case. This means that σ has a missing $\psi_{U_{n,2n}}$ Fourier coefficient.

Finally, to show that σ has a ψ weak lift to τ we just need to apply the unramified computations as described in **[G-R-S3]**. This completes the proof of the theorem. □

References

[C-K-PS-S] J. Cogdell, H. Kim, I. Piatetski-Shapiro, F. Shahidi, *On Lifting from Classical Groups to* GL_N, IHES Pub. Math. **93** (2001), 5–30.

[G-R] S. Gelbart and J. Rogawski, *L-functions and Fourier-Jacobi coefficients for the unitary group U(3)*, Invent. Math., **105** (1991), 445–472.

[G-R-S1] D. Ginzburg, S. Rallis, D. Soudry, *On explicit lifts of cusp forms from* GL_m *to classical groups*, Annals of Math., **150** (1995), 807–866.

[G-R-S2] D. Ginzburg, S. Rallis, D. Soudry, *On the Correspondence between cuspidal representations of* GL_{2n} *and* $\widetilde{Sp}_{2n}$, J. of AMS, **12** (1999), 849–907.

[G-R-S3] D. Ginzburg, S. Rallis, D. Soudry, *Lifting Cusp forms on* GL_{2n} *to* $\widetilde{Sp}_{2n}$ *: The Unramified Correspondence*, Duke Math. J., **100** (1999) p. 243–266.

[G-R-S4] D. Ginzburg, S. Rallis, D. Soudry, *Generic Automorphic Forms on* $SO(2n+1)$*: Functorial Lift to* GL(2n)*, Endoscopy and Base Change*, Inter. Math. Res. Notices, **14** (2001), 749–764.

[G-R-S5] D. Ginzburg, S. Rallis, D. Soudry, *On Fourier Coefficients of Automorphic Forms of Symplectic Groups*, manuscripta math. 111, (2003), 1-16.

[I] T. Ikeda, *On the lifting of Elliptic cusp forms to Siegel cusp forms of degree* 2n, Annals of Math., **154** (2001), 641–681.

[J-S] H. Jacquet, J. Shalika, *On Euler products and the classification of automorphic representations I, II*, Amer. J. Math., **103** (1981) 499–558, 777–815.

[M-W] C. Moeglin, J.L. Waldspurger, *Modéles de Whittaker dégénérés pour des groupes p-adiques*, Math. Zeit., **196** (1987), 47–452.

[PS] I. Piatetski-Shapiro, *On the Saito–Kurokawa Lifting*, Invent. Math., **71** (1983), 309–338.

[R-S] S. Rallis and G. Schiffmann, *θ–correspondences associated to G_2*, Amer. J. Math., **111** (1988), 801–849.

[R] R. Ranga Rao, *On some explicit formulas in the theory of Weil representation*, Pacific J. Math., **157** (1993), 335–371.

[S] D. Soudry, *The CAP representations of* GSp(4, $\mathbb{A}$), J. reine angew. Math., **383** (1988), 87–108.

David Ginzburg
School of Mathematical Sciences
Sackler Faculty of Exact Sciences
Tel Aviv University
Ramat-Aviv, 69978 Israel

Steve Rallis
Department of Mathematics
The Ohio State University
Columbus, OH 43210, USA

David Soudry
School of Mathematical Sciences
Sackler Faculty of Exact Sciences
Tel Aviv University
Ramat-Aviv, 69978 Israel

The Rallis inner product formula and p-adic L-functions

Michael Harris[*], *Jian-Shu Li*[†] and *Christopher M. Skinner*[‡]

To Steve Rallis, the oracle

1. Introduction

This article is a report on work in progress, whose goal is to show in certain cases that congruences satisfied by normalized special values of L-functions of geometric automorphic forms on unitary groups give rise to non-trivial elements of certain Galois cohomology groups. More precisely, letting p be an odd prime number, the Galois cohomology groups, or generalized Selmer groups, can be viewed as modules over appropriate Iwasawa algebras, and, possibly up to p-power torsion, the associated p-adic L-functions divide the characteristic power series, in the sense of Iwasawa theory, of the corresponding Selmer groups. Divisibility in both directions is a version of the main conjecture of Iwasawa theory.

Divisibility results of this kind were first obtained by Ribet for class groups of cyclotomic fields. Ribet observed that divisibility of normalized values of Dirichlet L-functions by p gave rise to congruences modulo p between holomorphic Eisenstein series and cusp forms on $GL(2)$. Here two holomorphic modular forms f and g are considered congruent if their Fourier expansions at the cusp at infinity have algebraic integer coefficients $a_n(f) \equiv a_n(g) \pmod{p}$ for all $n \geq 0$. Ribet's technique was vastly generalized by Mazur and Wiles, who obtained the first proof of Iwasawa's main conjecture for cyclotomic fields, and Wiles extended these arguments to abelian extensions of totally real fields.

[*]Institut de Mathématiques de Jussieu, U.M.R. 7586 du CNRS. Membre, Institut Universitaire de France

[†]Supported in part by RGC-CERG grants HKUST6126/00P, HKUST6115/02P, HKUST601805, and grant from the Chinese Academy of Science

[‡]Partially supported by a grant from the National Science Foundation and fellowships from the Sloane Foundation and the David and Lucile Packard Foundation

More recently, this technique has been applied successfully by Urban to congruences of Siegel modular forms, obtaining the first proof of an Iwasawa-type main conjecture for p-adic L-functions of degree greater than 1 – specifically, the symmetric square p-adic L-functions for elliptic modular forms. Admitting some expected consequences of the stable trace formula for certain unitary groups, Skinner and Urban have used similar techniques to prove the main conjecture for the standard p-adic L-function associated to an elliptic modular form for a large class of such forms. The principle underlying these constructions is the following: the normalized value of the L-function in question appears as the constant term of a certain holomorphic Eisenstein series. In very favorable cases, vanishing of the constant term modulo p^r implies congruence of the Eisenstein series with a cusp form modulo p^r. This in turn implies isomorphism of the semisimplified reductions modulo p^r of the Galois representations associated to the cusp form and the Eisenstein series, and following Ribet this isomorphism gives rise to the non-trivial Galois cohomology class, as will be explained in §8. Very favorable cases are unfortunately very rare, however, and the greatest challenge in this approach, as well as in the one described below, is to make the argument work in relatively unfavorable cases.

Our approach is based on a different principle. We work with the standard L-function $L(s,\pi)$ of an automorphic representation π of the unitary group of a hermitian vector space V' of dimension n over a CM field $\mathcal{K}$. These L-functions can and will be viewed as the L-functions of the corresponding automorphic representation $BC(\pi)$ of $GL(n,\mathcal{K})$ obtained by stable base change. Suppose there is an $n+1$-dimensional positive-definite hermitian vector space V over $\mathcal{K}$ such that the theta lift $\Theta(\pi)$ of π to $U(V)$ is non-trivial. Then the *Rallis inner product formula* identifies the value at $s=1$ of $L(s,\pi)$, or more precisely of the twist $L(s,\pi,\chi)$ where χ belongs to a certain family of Hecke characters of $\mathcal{K}$, as the ratio of the square norm of the theta lift $<\theta(f),\theta(f)>$ of a vector $f\in\pi$ by the square norm $<f,f>$ of f, multiplied by a finite product of bad local zeta integrals and explicit correction factors. By an *endoscopic automorphic representation* of $U(V)$ (more precisely, an endoscopic representation of type $(n,1)$), we will understand an automorphic representation of the form $\Theta(\pi)$; *stable automorphic representations* of $U(V)$ will be automorphic representations not of this form. A principle originating in work of Hida affirms that p-divisibility of $<f,f>$ (resp. $<\theta(f),\theta(f)>$) corresponds to congruences modulo p between forms in π (resp. $\Theta(\pi)$) and forms in other representations π' of $U(V')$ (resp. $U(V)$). Thus p-divisibility of the ratio $\frac{<\theta(f),\theta(f)>}{<f,f>}$ corresponds, in very favorable cases, to congruences modulo p between forms in $\Theta(\pi)$ and forms in *stable* automorphic representations of $U(V)$. Applying Ribet's construction to these congruences one then obtains Galois cohomology classes as before.

Much work is required in order to sort out what we mean by a very favorable case, not to mention making the technique work more generally. We actually need to study congruences in the Iwasawa algebra, which can be viewed as a power series ring over a p-adic integer ring in up to n-variables. This means we need to construct the p-adic L-functions for continuous families – Hida families – of p-adic automorphic forms on $U(V')$ and $U(V)$. We also need to obtain a version of the Rallis inner product formula adapted to these Hida families. These results are nearly complete, at least for the one- and two-dimensional families that suffice for our purposes, though some questions remain concerning the behavior of bad local zeta integrals. The most serious problem has to do with the p-adic properties of the theta lift. Roughly speaking, p-adic divisibility of the inner product $< \theta(f), \theta(f) >$ is only useful if we can guarantee that $\theta(f)$ is p-adically integral but not itself divisible by p. This problem can be resolved when $n = 2$. It is likely that we can obtain conditional results for general n, but this is not yet completely clear.

Finally, the normalization of the p-adic L-function depends on the choice of a good integral structure on the cohomology of the Shimura variety attached to the unitary group $U(V')$. This is well understood when V' is positive-definite, or when $n = 2$, but otherwise is completely mysterious. We will restrict our attention to the case of positive-definite V', with some remarks about the case $n = 2$. When $\mathcal{K}$ is an imaginary quadratic field, which we assume henceforward, the case $n = 2$ includes the case of p-adic L-functions of elliptic modular forms over $\mathbb{Q}$, base changed to $\mathcal{K}$.

The idea of studying the congruences between stable and endoscopic forms on unitary groups was first proposed by Richard Taylor in a message to one of the authors in 1993. The connection to the theta correspondence, and specifically to the Rallis inner product formula, was first observed in 2000 or 2001. Throughout the project we have benefited from the advice and suggestions of many mathematicians, whom we will acknowledge in due course, as our articles are completed. However, we must immediately express our gratitude to Haruzo Hida, without whose advice and encouragement this work would have been simply impossible.

It gives us great pleasure to dedicate this article to Steve Rallis. Steve's generosity is exceptional and his insight has been a constant source of inspiration. The "Rallis inner product formula" of the title originated in Steve's work on the Howe duality correspondence and incorporates his results obtained with Piatetski-Shapiro on the standard L-functions of classical groups, on the one hand, and the results of his collaboration with Kudla on the extended Siegel-Weil formula, on the other hand, in its version for unitary groups due to Ichino. The applications of these results to arithmetic are just beginning to be explored. Yet all this work is at least ten years old. Since then Steve has, if anything, been even busier. The possibility is real that arithmetic will never catch up.

2. L-functions of unitary groups

For simplicity, we will work with unitary groups with base field $\mathbb{Q}$, though everything goes over with little change to general totally real fields. Thus $\mathcal{K}$ is an imaginary quadratic field, given with a fixed embedding as a subfield of $\mathbb{C}$, V' and V are hermitian spaces of dimension n and $n+1$, respectively, over $\mathcal{K}$. We take V to be positive-definite. In the applications V' will usually be positive-definite as well, but in the present report we assume the signature of the hermitian form on V' to be (a,b), with $a+b=n$. We let $H = U(V)$, $G = U(V')$, viewed as algebraic groups over $\mathbb{Q}$, and let $K_\infty \simeq U(a) \times U(b)$ be a maximal compact subgroup of $G(\mathbb{R})$. Let $\varepsilon_{\mathcal{K}}$ be the quadratic character attached to the extension $\mathcal{K}/\mathbb{Q}$ and c the non-trivial automorphism of $\mathcal{K}$ (complex conjugation). We fix a prime p, assumed to split in $\mathcal{K}$.

For the moment we keep V and H in reserve and restrict our attention to automorphic representations π of G unramified at p. The representation π_∞ will be assumed to belong to the discrete series, and to have cohomology with values in a one-dimensional representation (τ, W) of $G(\mathbb{R}) = U(a,b)$:

$$H^{ab}(Lie(G)(\mathbb{R}), K_\infty; \pi_\infty \otimes W) \neq 0 \qquad (2.1)$$

(since π_∞ is in the discrete series, all the cohomology is concentrated in the middle dimension ab). There are more general coefficient systems but we will ignore these.

Via twisted (unitary) stable base change, π can be identified – unconditionally for $n \leq 3$ [Rog], conditionally [1] for general n [HL] – with an automorphic representation $BC(\pi) = \pi_{\mathcal{K}}$ of $GL(n,\mathcal{K})$, unramified at p, which is *cohomological* at the unique archimedean place and which satisfies

$$BC(\pi)^\vee \xrightarrow{\sim} BC(\pi)^c. \qquad \text{(polarization)}$$

Arithmetically, this corresponds – again, under some conditions for $n > 3$ – to a continuous representation $\sigma : Gal(\overline{\mathbb{Q}}/\mathcal{K}) \to GL(A)$ on an n-dimensional $\overline{\mathbb{Q}}_p$-vector space A which is unramified outside a finite set and is *crystalline* at p. This can be viewed either as the Galois representation attached to $BC(\pi)$ or to π; there is no difference. But the former has an easier interpretation: if $BC(\pi) = \otimes_v BC(\pi)_v$, where v runs over places of $\mathcal{K}$, prime to p, then it follows from [HT] that $\sigma = \sigma(BC(\pi))$ is characterized by

$$\sigma\,|_{\mathfrak{G}_v} \xrightarrow{\sim} \mathcal{L}(BC(\pi)_v) \qquad (2.2)$$

up to semisimplification, where v is any finite prime of $\mathcal{K}$ not dividing p, $\mathfrak{G}_v$ is a decomposition group at v and $\mathcal{L}$ is the local Langlands correspondence.

[1] One has to assume π is supercuspidal or Steinberg at two places of $\mathbb{Q}$ that split in $\mathcal{K}$.

appropriately normalized, taking irreducible representations of $GL(n,\mathcal{K}_v)$ to n-dimensional representations of $\mathfrak{G}_v$. [2]

The Galois representation σ is realized, up to an explicit abelian twist, on the cohomology of some Shimura variety, not necessarily attached to the group G. [3] Suppose for the moment that the coefficient representation τ is trivial. Then the weight of the p-adic representation σ, determined by Frobenius eigenvalues at unramified primes, is $n-1$, the Hodge-Tate weights are $\{0,1,\ldots,n-1\}$, and the representation is *regular* in the following sense: each Hodge-Tate weight has multiplicity one (over the coefficient field). This regularity hypothesis, which remains valid for general τ (the list of Hodge-Tate weights depends on τ) is the principal restriction imposed by the use of Shimura varieties.

We continue to assume τ trivial. The polarization condition has one degree of freedom, namely twisting by a Hecke character χ of $\mathcal{K}$ which is *anticyclotomic* in the sense that χ factors through the map $a \mapsto a/a^c$. To χ we can associate a weight k (basically $\chi(z) = (z/z^c)^{-k}$) and a p-adic incarnation $\sigma_p(\chi)$, so that the p-adic representation $\sigma \otimes \sigma_p(\chi)$ is pure of weight $w = n-1+k$. The L-function with arithmetic normalization:

$$
\begin{aligned}
& L(s - \tfrac{n-1+k}{2}, BC(\pi), \chi) && (GL(n) \text{ version}) \\
(2.3) \qquad & = L(s - \tfrac{n-1+k}{2}, \pi, \chi) && (U(n) \text{ version}) \\
& = L(s, \sigma \otimes \sigma_p(\chi)) && (\text{Galois version })
\end{aligned}
$$

then has critical special values, in the sense of Deligne, for most k, the nature of the periods depending on k. In the first line of (2.3) we are working with the principal L-function of $GL(n)$, whereas in the second line we are working with the Langlands L-function attached to π and the standard representation of the L-group of $U(n)$. Suppose $n=2$, for instance, in which case $BC(\pi)$ is the base change to $\mathcal{K}$ of the representation generated by an elliptic modular newform f of weight 2. Then if $k=0$ (so $w=1$) $L(s,\sigma\otimes\sigma_p(\chi))$ has a critical value at the center $s=1$ of Birch-Swinnerton-Dyer type whose associated Deligne period is roughly the Petersson norm $< f, f >$. If $k=3$, so $w=4$, $L(s,\sigma\otimes\sigma_p(\chi))$ has a critical value at the near-central point $s=3$ whose Deligne period depends only on χ – i.e., on CM periods – up to algebraic factors. For general n, if V' is supposed positive-definite, but k is fixed and τ is trivial, we only obtain the

[2]We also need information at primes dividing p, specifically that the p-adic absolute values of the normalized Satake parameters correspond to the Newton polygon of the Frobenius operator on crystalline cohomology. At least in the situations considered in [HL] the Katz-Messing theorem applies to provide what we need. For intersection cohomology of general Shimura varieties this should follow from work in progress of A. Nair [N].

[3]In general one may only obtain the restrictions of σ to the Galois groups of a sufficiently general family of quadratic extensions of $\mathcal{K}$ directly on the cohomology of Shimura varieties; cf. [HT], Theorem VII.1.9.

analogue of the latter case with $k = s = n+1$; in particular, the critical value $L(n+1, \sigma \otimes \sigma_p(\chi))$ is an algebraic multiple of a fixed CM period independent of π.

Fixing k eliminates the archimedean degree of freedom but χ can still vary in a one-dimensional p-adic (anticyclotomic) family. Henceforward we incorporate k into the representation τ – i.e., we drop the assumption that τ be trivial. We take L-functions in the unitary normalization, so that $Re(s) = \frac{1}{2}$ is the axis of symmetry, and $s = 1$ is the possible pole for L-functions of tempered automorphic representations of $GL(n)$. We write $L(s, \pi, \chi) = L(s, BC(\pi), \chi)$ as above. In what follows, we let $\chi = \otimes_v \chi_v$ vary among Hecke characters satisfying:

$$\chi\,|_{\mathbb{A}_{\mathbb{Q}}^{\times}} = \varepsilon_{\mathcal{K}}^{n+1}; \quad |\chi_\infty| = 1. \tag{2.4}$$

If χ_0 and χ are two characters satisfying (2.4), then $\chi \cdot \chi_0^{-1}$ is *anticyclotomic* in the sense defined above, so that a family of χ satisfying (2.4) can be viewed as the product of a fixed base point χ_0 by the family of all anticyclotomic Hecke characters. We write $\chi_f^p = \otimes_{v \neq p, \infty} \chi_v$, and view χ as a family of characters of finite order with values in $\bar{\mathbb{Q}}^\times \subset \mathbb{C}_p^\times$, where $\mathbb{C}_p$ is the p-adic completion of $\bar{\mathbb{Q}}_p$. Consider the family $X_{\mathcal{K}}$ of *p-adically continuous* characters satisfying (2.4):

$$\chi : \mathbb{A}_{\mathcal{K}}^\times \to \mathbb{C}_p^\times$$

such that the restriction of χ_f^p to the product of the local units away from p is fixed (i.e., χ_f^p is fixed up to inertial equivalence). This can be viewed as the set of $\mathbb{C}_p$-valued points of a rigid-analytic space over $\mathbb{Q}_p$. Our results concern the p-adic L-function – which we need to construct for $n > 2$ – corresponding to the values $L(1, \pi, \chi)$ as a function of χ. We can also let π vary in an appropriate Hida family. [4]

The p-adic L-function, like the standard L-function on which it is based, is defined by means of the doubling method of Piatetski-Shapiro and Rallis, studied in some detail, though insufficient for our purposes, by Shimura in two recent books ([GPR], [Sh1], [Sh2]; see also [Gar]). Let $-V'$ be the space V' with its hermitian form multiplied by -1, let $2V' = V' \oplus (-V')$, and define two totally isotropic subspaces of $2V'$ by $V'^{,d} = \{(v, v) \mid v \in V'\}$, $V'_d = \{(v, -v) \mid v \in V'\}$. Let $P \subset U(2V')$ be the stablizer of $V'^{,d}$; then the stabilizer $M \subset P$ of V'_d is a Levi component, canonically isomorphic to $GL(V'^{,d}) \simeq GL(V')$. Let $m : GL(V') \to M$ denote the canonical isomorphism. We write $P = M \cdot U_P$; the exponential map identifies U_P with the group Her_n of $n \times n$-hermitian matrices over $\mathcal{K}$, viewed as an algebraic vector group over $\mathbb{Q}$.

[4] Especially in §3 and §4, we will make use, without explanation, of notions from Hida's theory of (nearly) ordinary families of modular forms. Most of the results we use are contained in [Hi 02]. There is no simple introduction to Hida theory in higher dimensions, but the reader can refer to [Hi IHP] for background.

The inclusion $V' \oplus (-V') \subset 2V'$ defines a natural embedding $G \times G \hookrightarrow U(2V')$. The group $U(2V')$ is quasi-split, and $U(2V')(\mathbb{R})$ can be identified with $U(n,n)$. Choose a maximal compact subgroup $K_{\infty,2} \subset U(2V')(\mathbb{R})$ so that $K_{\infty,2} \cap (G \times G)(\mathbb{R}) = K_{\infty,G} \times K_{\infty,G}$.

Let χ be any unitary Hecke character of $\mathcal{K}$, viewed as a character of $M(\mathbb{A}) \cong GL(V')(\mathbb{A})$ via composition with $\det(\bullet)$. Consider the induced representation of $U(2V')(\mathbb{A})$:

$$I(\chi, s) = \operatorname{Ind}(\chi|\cdot|_{\mathcal{K}}^{s}) \xrightarrow{\sim} \otimes_v I_v(\chi_v|\cdot|_v^s),$$

the induction being normalized; the local factors I_v, as v runs over places of $\mathbb{Q}$, are likewise defined by normalized induction. At archimedean places we assume our sections to be $K_{\infty,2}$-finite. For a section $f(h;\chi,s) \in I(\chi,s)$ we form the Eisenstein series

$$E_f(h;\chi,s) = \sum_{\gamma \in P(\mathbb{Q})\backslash U(2V')(\mathbb{Q})} f(\gamma h;\chi,s).$$

This series is absolutely convergent for $\operatorname{Re}(s) > \frac{n-1}{2}$, and it can be continued to a meromorphic function on the entire plane.

Let (π, H_π) be a cuspidal automorphic representation of G, $(\pi^\vee, H_{\pi^\vee})$ its contragredient, which we assume given with compatible isomorphisms of $G(\mathbb{A})$-modules

$$\pi \xrightarrow{\sim} \otimes_v \pi_v, \quad \pi^\vee \xrightarrow{\sim} \otimes_v \pi_v^\vee,$$

the tensor products taken over places v of $\mathbb{Q}$, with π_∞ an admissible $(\mathfrak{g}_\infty, K_{\infty,G})$-module of cohomological type, with lowest K_∞-type τ_∞. For each v we let $(\bullet,\bullet)_{\pi_v}$ denote the canonical bilinear pairing $\pi_v \otimes \pi_v^\vee \to \mathbb{C}$.

Let $f(h;\chi,s)$ be a section, as above, $\varphi \in H_\pi$, $\varphi' \in H_{\bar\pi}$, and let $\varphi'_\chi(g) = \varphi'(g)\chi^{-1}(\det g')$. We define the zeta integral:

$$(2.5)\quad Z(s,\varphi,\varphi',f,\chi) = \int_{(G\times G)(\mathbb{Q})\backslash (G\times G)(\mathbb{A})} E_f((g,g');\chi,s)\varphi(g)\varphi'_\chi(g')\,dg\,dg',$$

with $dg = dg'$ the Tamagawa measure on $G(\mathbb{A})$.

The theory of this function was worked out (for trivial χ) by Li [L92] and more generally in [HKS,§6]. We make the following hypotheses:

Hypotheses 2.6.

(a) There is a finite set of finite places S_f of $\mathbb{Q}$ such that, for any non-archimedean $v \notin S_f$, the representations π_v, the characters χ_v, and the fields $\mathcal{K}_w$, for w dividing v, are all unramified.

(b) The section f admits a factorization $f = \otimes_v f_v$, and f_∞ is an eigenvector for a one-dimensional representation of $K_{\infty,2}$.

(c) The functions φ, φ' admit factorizations $\varphi = \varphi_{S_f} \otimes_{v \notin S_f} \varphi_v$, $\varphi' = \varphi_{S_f} \otimes_{v \notin S_f} \varphi'_v$.

(d) For $v \notin S_f$ non-archimedean, the local vectors f_v, φ_v, and φ'_v, are the normalized spherical vectors in their respective representations, with $(\varphi_v, \varphi'_v)_{\pi_v} = 1$.

(e) The vector φ_∞ (resp. φ'_∞) is a non-zero highest (resp. lowest) weight vector in τ_∞ (resp. in $\tau_\infty^\vee$), such that $(\varphi_\infty, \varphi'_\infty)_{\pi_\infty} = 1$.

We let $S = \{\infty\} \cup S_f$. Define

$$d_n^S(s,\chi) = \prod_{r=0}^{n-1} L^S(2s+n-r, \varepsilon_K^{n-1+r}) = \prod_{v \notin S} d_{n,v}(s,\chi),$$

$$Q_{V'}^0(\varphi \otimes \varphi') = \int_{G(\mathbb{Q})\backslash G(\mathbb{A})} \varphi(g)\varphi'(g)dg;$$

(2.7) $$Z_S(s,\varphi,\varphi',f,\chi) = \prod_{v \in S} \int_{G(\mathbb{Q}_v)} f_v((g_v,1);\chi,s)(\pi_v(g_v)\varphi,\varphi')dg_v.$$

The integrals (2.5, 2.7) converge absolutely in a right halfplane and admit meromorphic continuations to all s, and we have the following identity of meromorphic functions on $\mathbb{C}$:

Basic Identity of Piatetski-Shapiro and Rallis

(2.8) $$d_n^S(s,\chi)Z(s,\varphi,\varphi',f,\chi) = Q_{V'}^0(\varphi \otimes \varphi')Z_S(s,\varphi,\varphi',f,\chi)L^S(s+\frac{1}{2},\pi,\chi).$$

Here and throughout the superscript S means that we omit the Euler factors at the places in S.

Let $X(n,n)$ be the tube domain associated to the unitary group $U(n,n)$; there is a homogenous isomorphism

$$U(2V')/K_{\infty,2} = U(n,n)/U(n) \times U(n) \xrightarrow{\sim} X(n,n).$$

A well-known procedure associates to the Eisenstein series E_f a function $\mathbf{E}_f(z,u_f,s)$, meromorphic in s, on $X(n,n) \times U(2V')(\mathbb{A}_f) \times \mathbb{C}$, with values in $\mathbb{C}$ (because f_∞ belongs to a one-dimensional $K_{\infty,2}$-type), which satisfies a familiar automorphy relation with respect to the left action of $U(2V')(\mathbb{Q})$. In what follows we call $\mathbf{E}_f$ the *classical Eisenstein series* associated to E_f.

When V' is positive definite the zeta integral (2.5) is just a finite sum of values of the classical Eisenstein series $\mathbf{E}_f$. When $n = 1$, the basic identity reduces to the expression of the Hecke L-function on the right hand side as a finite sum of special values of the normalized elliptic modular Eisenstein series on the left. This expression is familiar to number theorists for its role in the proof of Damerell's theorem and its generalizations by Shimura, the starting point for Katz' construction of p-adic L-functions of Hecke characters of CM fields. The generalization of Katz' construction is described in the following section.

In general, with the appropriate normalization and choices of data f_v, φ_v, and φ'_v at primes $v \in S_f$, the basic identity yields the following result:

Theorem 2.9. *The expression*

$$\pi^{-\frac{n(n+3)}{2}} Q^0_{V'}(\varphi \otimes \varphi') Z_S(\frac{1}{2}, \varphi, \varphi', f, \chi) L^S(1, \pi, \chi)$$

is an algebraic number.

In order for this to hold, the zeta integral has to be identified with a natural cup product pairing in coherent cohomology, the Eisenstein series has to be $\overline{\mathbb{Q}}$-rational in the sense of having algebraic Fourier coefficients, and the functions φ, φ' have to be identified with $\overline{\mathbb{Q}}$-rational antiholomorphic modular forms. These conditions are discussed in sections 3 and 4 below (cf. condition (3.4)(ii), for example). The power of π in the denominator is an algebraic multiple of the factor $d_n^S(\frac{1}{2}, \chi)$. We prefer not to be precise about choices of measures so the power of π is somewhat spurious. Later we will implicitly assume that measures have been chosen so that the archimedean factor $Z_\infty(\frac{1}{2}, \varphi, \varphi', f, \chi)$ is an algebraic number (even a p-unit); any discrepancy is of course incorporated into the definition of $Q^0_{V'}$. The non-archimedean bad local factors in $Z_S(\frac{1}{2}, \varphi, \varphi', f, \chi)$ can easily be arranged to be algebraic.

Remark. When φ and φ' are antiholomorphic and $\overline{\mathbb{Q}}$-rational in the sense of coherent cohomology, the period $Q^0_{V'}(\varphi \otimes \underline{\varphi'})$ can be rewritten as the reciprocal of the Petersson inner product of two $\overline{\mathbb{Q}}$-rational *holomorphic* modular forms. Theorem 2.9 is thus an expression of a critical value of the motivic L-function $L(1, \pi, \chi)$ in terms of periods of arithmetic modular forms, as in work of Shimura (see [Sh1,Sh2] for results on classical groups) and [H97], which adopts the present point of view.

3. p-adic Eisenstein measures

Henceforward we assume π_p is unramified. We also assume $S_f = S_{\mathcal{K}} \coprod S_{\pi,\chi} \cup \{p\}$, where $S_{\mathcal{K}}$ is the set of primes ramified in $\mathcal{K}$ and $S_{\pi,\chi}$ is the set of primes at which either π or χ is ramified. We make the hypothesis

Hypothesis 3.1. *Every prime in $S_{\pi,\chi}$ splits in $\mathcal{K}$.*

In the cases we consider the classical Eisenstein series $\mathbf{E}_f(z, u_f, s)$ is a holomorphic modular form at the point $s = \frac{1}{2}$, corresponding to the value $L(1, \pi, \chi)$ in the basic identity. When φ and φ' correspond to (anti)-holomorphic modular forms on the Shimura variety associated to G, the basic identity, combined with rationality properties of the Fourier coefficients of holomorphic Eisenstein series, thus provides an expression of the special values $L(1, \pi, \chi)$, as χ varies with χ_∞ fixed, as algebraic multiples of a fixed period factor:

$$L(1, \pi, \chi) = L^{alg}(1, \pi, \chi) \cdot p(\pi, \chi_\infty), \tag{3.2}$$

where $p(\pi, \chi_\infty)$ is essentially the Petersson norm $< \varphi, \varphi' >$, up to elementary factors [Sh1,H97].

The philosophy of p-adic L-functions is summarized by the following two sentences, the first of which is highly misleading:

(i) As χ varies over a p-adic family X, for instance the $X_{\mathcal{K}}$ defined above, one expects the values $L^{alg}(1, \pi, \chi)$ to vary p-adically analytically, and thus to extend to an analytic function on X with values in $\mathbb{C}_p$.

(ii) The properties of this analytic function depend strongly on the p-adic absolute values of the Satake parameters of the unramified representation π_p.

When one actually tries to construct p-adic L-functions, one immediately realizes that the values $L^{alg}(1, \pi, \chi)$ need to be modified in order for something like (i) to be true. The modification involves multiplication by p-adic and archimedean correction factors, whose conjectural form is proposed by Coates in [Co]. We have chosen to keep the real components of π and χ fixed, so the archimedean correction factors can be ignored. In the ultimate theory π_∞ and χ_∞ will vary and the archimedean correction terms – mainly products of factorials – will have to be addressed.

On the other hand, we allow χ_p to be arbitrarily ramified, so p-adic correction factors are essential. This amounts to including the prime p in the set $S_{\pi,\chi}$ of bad primes, and making specific choices of local data that allow explicit calculation of the zeta integral

$$Z_p(s, \varphi_p, \varphi'_p, f_p, \chi_p) = \int_{G(\mathbb{Q}_p)} f_p((g_p, 1); \chi, s)(\pi_p(g_p)\varphi_p, \varphi'_p)dg_p \tag{3.3}$$

at the point $s = 1$. The choices of vectors φ_p, φ'_p are more or less imposed by Hida theory. Let $P \subset GL(n)$ be a standard parabolic subgroup, of parabolic rank ℓ, say. In the most general setting, what Hida theory provides is a p-adic manifold X_P of dimension ℓ and a countable Zariski dense subset $X_P^{alg} \subset X_P(\mathbb{C}_p)$ of points corresponding to triples $(\pi_x, \varphi_x, \chi_x)$, with π_x an automorphic representation of G of fixed infinity type and fixed central character, $\varphi_x \in \pi_x$ a factorizable vector, and χ_x a Hecke character of $\mathcal{K}$ satisfying (2.4). The pairs

(φ_x, χ_x) extend to an analytic family (φ, χ) of some sort of p-adic modular forms on $G(\mathbb{A})$. The problem is to find $f_p(x) = f_p(\varphi_x, \chi_x)$ such that the corresponding Eisenstein series also extend to a p-adic analytic map from X_P to p-adic modular forms on $U(2V')$.

One of the rules of the game is that the f_v, for $v \neq p$, are constant as functions on X_P. Specifically:

(3.4)

(i) When $v \notin S$ hypothesis (2.6)(d) remains true.
(ii) When $v \in S_{\pi,\chi}$ is prime to p, f_v is chosen to make all zeta integrals constant and algebraic.
(iii) The section f_∞ is chosen to make the classical Eisenstein series $\mathbf{E}_f(z, u_f, s)$ holomorphic.[5]
(iv) When $v \in S_K$, we fix f_v.
(v) Finally, we will assume below that $f = \otimes f_v$ is a Siegel-Weil section for some $(n+1)$ – dimensional hermitian space V.

Only f_p needs to vary. Condition (v) is required for application of the Rallis inner product formula and will be explained in §6; it is also used to obtain the effective formulas for Fourier coefficients of Proposition 3.8.

At this point we need to provide at least a rough definition of a p-adic modular form and of a p-adic family of modular forms, on $U(2V')$ as well as on G. For $U(2V')$ the definition is relatively simple. We restrict our attention to Eisenstein series. The classical Eisenstein series $\mathbf{E}_f(z, u_f) = \mathbf{E}_f(z, u_f, \frac{1}{2})$ – we only work with Eisenstein series holomorphic at $s = \frac{1}{2}$, corresponding to the value of the L-function at $s = 1$ – is completely determined by its Fourier expansion. Let $\mathbf{E}_f^* = \pi^{-\frac{n(n+3)}{2}} d_n^S(\frac{1}{2}, \chi)\mathbf{E}_f$. Then the Fourier expansion of $\mathbf{E}_f^*$ is

(3.5)
$$\mathbf{E}_f^*(z, u_f) = \sum_{\beta \in Her_n(\mathbb{Q})} c_{f,\beta}(u_f) q^\beta(z), z \in X(n,n) \quad , u_f \in M(\mathbb{A}^S) \times \prod_{v \in S_f} K_v.$$

Here $K_v \in GL(V_v')$ is a special maximal compact subgroup (by strong approximation one does not need all values of the argument $u_f \in U(2V')(\mathbb{A}_f)$ to define $\mathbf{E}_f^*$, the values in $M(\mathbb{A}^S) \times \prod_{v \in S_f} K_v$ suffice). The symbol $q^\beta = q^\beta(z)$ denotes the function $z \mapsto e^{2\pi i \alpha Tr(\beta \cdot z)}$, where $\alpha \in \mathbb{Q}^\times$, assumed to be a p-unit, is a free parameter corresponding to the choice of an additive character. Holomorphy and the Koecher principle imply that $\mathbf{E}_{f,\beta}^* = 0$ unless β is positive semidefinite. Our choices of f_p provide a stronger vanishing condition:

(3.6)
$$c_{f,\beta} \neq 0 \Rightarrow \det \beta \neq 0;$$

[5]More generally, when π_∞ is allowed to vary, one will want $\mathbf{E}_f(z, u_f, s)$ to have holomorphic restriction to the Shimura variety associated to $G \times G$. We actually work with unitary similitude groups in order to define Shimura varieties, but in fact the theory of the standard L-functions only depends on the unitary groups.

as well as a factorization condition [6]

$$c_{f,\beta}(u_f) = c_{\infty,\beta} \cdot \prod_{v \neq \infty} c_{f_v,\beta,v}(u_v). \tag{3.7}$$

Here u_v is the v-component of u_f and the term $c_{\infty,\beta}$ is a function of β alone.

Since we are working with $\mathbf{E}^*$ rather than $\mathbf{E}$, the normalized coefficient in (3.7) is a finite Euler product. Let $\Lambda \subset H_n(\mathbb{Q})$ denote the lattice of hermitian matrices with entries in $\mathcal{O}_K$, Λ^* the dual lattice

$$\Lambda^* = \{B \in Her_n(\mathbb{Q}) \mid tr(B\Lambda) \subset \mathcal{O}_K\}.$$

For any v let T_v denote the characteristic function of Λ_v^*.

Proposition 3.8. *(a) For $A \in GL(V_v')$, $v \notin S$,*

$$c_{f_v,\beta,v}(m(A)) = T_v(\alpha^t \bar{A}\beta A), v \notin S.$$

(b) Let $c^0_{f,\beta} = c_\infty(\beta) \cdot \prod_{v \in S_{\pi,\chi}, v \neq p} c_{f_v,\beta,v}$. Under hypotheses (3.4) one can find lattices $L_v \in Her_n(\mathbb{Q}_v)$ and a constant $\gamma_0 \in \overline{\mathbb{Q}}$ such that, for all $m(A) \in \prod_{v \in S_{\pi,\chi} \cup \infty, v \neq p} K_v$, $c^0_{f,\beta}(m(A)) = \gamma_0 \cdot \det(\beta) \cdot \chi \circ \det(A)$ for $\beta \in \prod_{v \in S_{\pi,\chi}} L_v^*$ and equals zero otherwise.

Here L_v^* is the dual lattice.

The formulas in Proposition 3.8 can easily be reinterpreted in terms of classical holomorphic modular forms on the tube domain attached to $U(n,n)$. Together with the conditions on the p-adic local factors to be discussed below, they imply that the Fourier coefficients vanish for β outside some lattice, and that the dependence on β can be controlled. Similar formulas were obtained years ago by Shimura; see [Sh1, Sh2] for his more recent results. Shimura's hypotheses are somewhat different from ours and he does not restrict his attention to Siegel-Weil Eisenstein series.

We only really care about the p-adic valuation of $c_{f,\beta}(u_f)$, and Proposition 3.8 tells us that the contributions of primes outside $\{p\} \cup S_K$ to this valuation are *bounded.* Moreover, the constant γ_0 is a product of explicit local constants. The archimedean contribution is a p-unit if $p > n+1$ and if $\det \beta$ is a p-unit, which will be true for our choice of f_p. The other local contributions are indices of open compact subgroups in $G(\mathbb{Q}_v)$, whose p-adic valuations can be determined easily. At the time of writing we do not have good choices of f_v for $v \in S_K$. However, we can again guarantee at least that the corresponding factor $c^K_{f,\beta}(u_f) = \prod_{v \in S_K} c_{f_v,\beta,v}(u_f)$ has p-adic absolute value bounded above and below, and we will ignore these factors for the time being.

It remains to consider the factor $c_{f_p,\beta,p}$, which depends on explicit choices of local factors f_p. Consider a partition $n = n_1 + \cdots + n_\ell$, corresponding to

[6] Compare [Sh1, Proposition 19.2]. Shimura's hypothesis $\mathfrak{c} \neq \mathfrak{g}$, which has to be read together with the definition (18.4.8, 18.4.9) and various other definitions in [Sh1], guarantees the factorization condition for the Eisenstein series he considers.

a standard parabolic subgroup $P \subset GL(n)$. Let $M_n(\mathbb{Q}_p)$ denote the space of $n \times n$-matrices with values in $\mathbb{Q}_p$, and let $K_P \subset GL(n, \mathbb{Z}_p) \subset M_n(\mathbb{Q}_p)$ be the standard parahoric subgroup corresponding to the parabolic P. Let $\nu = (\nu_1, \ldots, \nu_\ell)$ be an ℓ-tuple of characters of $\mathbb{Z}_p^\times$ of finite order, with values in $\overline{\mathbb{Q}}^\times \subset \mathbb{C}^\times$. The group K_P consists of matrices $Z = (Z_{ij})_{1 \leq i,j \leq \ell}$, written according to the chosen partition, with $Z_{ii} \in GL(n_i, \mathbb{Z}_p)$ for all i, with above-diagonal entries p-integral, and below-diagonal entries divisible by p. We define a function $\phi_\nu : K_P \longrightarrow \overline{\mathbb{Q}}^\times$ by $\phi_\nu(Z) = \prod_{1 \leq i \leq \ell} \nu_i(\det Z_{ii})$. We extend ϕ_ν by zero to a function, still denoted ϕ_ν, on $M_n(\mathbb{Q}_p)$.

Let w_1, w_2 denote the two primes of $\mathcal{K}$ dividing p. The local component χ_p of the Hecke character maps $\mathcal{K}_p^\times = \mathcal{K}_{w_1}^\times \times \mathcal{K}_{w_2}^\times \xrightarrow{\sim} \mathbb{Q}_p^\times \times \mathbb{Q}_p^\times$ to $\overline{\mathbb{Q}}^\times$. By restriction to the first factor (more precisely, by fixing w_1 rather than w_2) χ_p becomes a character of $\mathbb{Z}_p^\times$ of finite order. The group

$$M(\mathbb{Q}_p) = GL(V_p') \xrightarrow{\sim} GL(V_{w_1}') \times GL(V_{w_2}') \xrightarrow{\sim} GL(n, \mathbb{Q}_p) \times GL(n, \mathbb{Q}_p),$$

and elements of $M(\mathbb{Z}_p)$ will be denoted $m(A, B)$ for $A, B \in GL(n, \mathbb{Z}_p)$. There is a choice of basis implicit in the constructions below.

Proposition 3.9. *For every $\ell + 1$-tuple $(\nu_1, \ldots, \nu_\ell, \chi_p)$ of characters of $\mathbb{Z}_p^\times$ of finite order, there exists a section $f_p(\nu, \chi_p; s) \in I_p(\chi_p, s)$ such that, setting $s = \frac{1}{2}$,*

$$c_{f_p(\nu,\chi_p),\beta,p}(m(A, B)) = \chi_p(\det(BA^{-1}))\phi_\nu({}^tA\beta B^{-1})$$

for any $A, B \in GL(n, \mathbb{Z}_p)$, provided β is p-integral with determinant a p-unit, and vanishes otherwise.

The construction of f_p was inspired by Katz' construction [Katz] when $n = 1$. It involves tensoring ϕ_ν with its Fourier transform and viewing the result as a Schwartz-Bruhat function Φ_ν for $M(n, 2(n+1))$, hence a datum for the local Siegel-Weil formula for the theta lift from $U(n+1)(\mathbb{Q}_p) = GL(n+1, \mathbb{Q}_p)$ to $U(2V')(\mathbb{Q}_p) = GL(2n, \mathbb{Q}_p)$. The details are quite complicated, but we would like to point out that, although the partial Fourier transform does not preserve p-integrality, the resulting local coefficient in (3.9) is p-integral. We write $f(\nu, \chi) = \otimes_{v \neq p} f_v \otimes f_p(\nu, \chi_p)$, $\mathbf{E}^*_{f^p,\chi,\nu} = \mathbf{E}^*_{f(\nu,\chi)}$.

One can define a p-adic modular form on $U(2V')$ to be a formal expression of the form (3.5), where now the q^β are symbols satisfying the obvious relations $q^\beta q^{\beta'} = q^{\beta+\beta'}$, and the coefficients $c_{f,\beta}(u_f)$ are $\mathbb{C}_p$-valued functions on $M(\mathbb{A}_f) \cap (M(\mathbb{A}^S) \times M(\mathbb{Z}_p) \times \prod_{v \in S_f, v \neq p} K_v)$ with p-adic absolute values uniformly bounded in β as well as u_f. In order to qualify as a p-adic modular form, the formal series has to arise as the p-adic limit of Fourier expansions of genuine holomorphic modular forms on $U(2V')$, whose weights and levels are allowed to vary. This condition, first introduced by Serre, was given a rigid-analytic interpretation by Katz, and it is this interpretation that allows us to restrict p-adic modular forms to the Shimura varieties attached to $G \times G$, and hence to give a p-adic interpretation of the doubling method.[7]

We stress that, in the setting of Shimura varieties of higher dimension, the foundations of the theory of p-adic modular forms are entirely due to Hida (especially [Hi00, Hi02, Hi IHP]). Moreover, the possibility of applying these results to generalizing Katz' construction of p-adic L-functions is explicitly stated in Hida's articles, and – at least when V' is positive-definite – the construction of these p-adic L-functions will appear in a forthcoming article of Hida.

Let $\mathcal{M}(2V')$ denote the $\mathbb{C}_p$-Banach space of p-adic modular forms on $U(2V')$. Let $Z^\ell = (\mathbb{Z}_p^\times)^{\ell+1}$ and let $\mathbf{X}_P$ denote the set of continuous $\mathbb{C}_p^\times$-valued characters on Z^ℓ. The set $\mathbf{X}_P$ has the structure of the set of $\mathbb{C}_p$-valued points of a rigid-analytic variety over $\mathbb{Q}_p$ whose subset $\mathbf{X}_P^{alg}$ of characters of finite order is Zariski-dense. A *measure* $d\mu$ on Z^ℓ with values in $\mathcal{M}(2V')$ is a finitely additive continuous map from the Banach space of continuous $\mathbb{C}_p$-valued functions F on Z^ℓ to $\mathcal{M}(2V')$:

$$C(Z^\ell, \mathbb{C}_p) \ni \phi \mapsto \int_{Z^\ell} \phi d\mu \in \mathcal{M}(2V').$$

Such a measure is determined uniquely by its restriction to $\mathbf{X}_P \subset C(Z^\ell, \mathbb{C}_p)$, and even by its restriction to the Zariski-dense subset $\mathbf{X}_P^{alg}$.

Theorem 3.10. *Fix $f^p = \otimes_{v \neq p} f_v$ as above. There is a (unique) p-adic measure $d\mu_{Eis}(f^p)$ on Z^ℓ with values in $\mathcal{M}(2V')$ such that, for any $(\nu, \chi) \in \mathbf{X}_P^{alg}$,*

$$\int_{Z^\ell} (\nu, \chi) d\mu_{Eis}(f^p) = \mathbf{E}^*_{f^p, \chi, \nu}.$$

This is an immediate consequence of the above results on Fourier coefficients. Indeed, (3.5) and (3.6) show that the Fourier coefficients factorize as a product of local coefficients; Proposition 3.8 shows that for fixed f^p, the product of local coefficients prime to p is bounded uniformly in β and u_f; and Proposition 3.9 asserts that the local coefficient at p varies p-adically continuously on $\mathbf{X}_P^{alg}$. Let $F = (\phi, \xi) : Z^{\ell+1} \to \mathbb{C}_p$ be a continuous function where ϕ depends on the first ℓ variables and ξ on the last (so $F(z_1, \ldots, z_\ell, x) = \phi(z_1, \ldots, z_\ell)\xi(x)$). Define a function $\phi_F : I_P \to \mathbb{C}_p$ by composing ϕ with the natural map

$$I_P \to \mathbb{Z}_p^\ell; \quad Z \mapsto (\det Z_{11}, \ldots, \det Z_{\ell,\ell}).$$

[7] Note that p-adic integrality of the Fourier coefficients is not in general preserved by the action of the full group $U(2V')(\mathbb{Q}_p)$, but only by the subgroup that stabilizes the "cusp at infinity." In the theories of Katz and Hida this is interpreted in terms of the geometry of the Igusa varieties.

Let $T_{(L_v)}(\beta, A) : Her_n(\mathbb{Q}) \times GL(V'_{\mathbb{A}_f^p}) \rightarrow \mathbb{C}_p$ be the function that takes the value 1 if $\alpha^t \bar{A}_v \beta A_v \in L_v^*$ for all $v \notin S$, $\beta \in L_v^*$ for all $p \neq v \in S_{\pi,\chi}$, and β is integral at p and $\det(\beta) \in \mathbb{Z}_p^\times$, and otherwise takes the value 0. Then

$$\int_{Z^\ell} F \ \ d\mu_{Eis}(f^p) = \sum_\beta c_\beta(u) q^\beta$$

where $c_\beta(u)$ is the function of $u \in GL(V'_{\mathbb{A}_f}) = M(\mathbb{A}_f)$, $u_p = (A_p, B_p) \in M(\mathbb{Z}_p)$ and $u_v \in K_v$ for all $p \neq v \in S_f$, given by

(3.11)

$$c_\beta(u) = \gamma_0 \det(\beta) \xi(\det u^p_{S_{\pi,\chi}}) \chi_p(\det(B_p A_p^{-1})) \phi_F({}^t A_p \beta B_p^{-1}) \times T_{(L_v)}(\beta, u) c_{\beta,\mathcal{K}}(u),$$

where $u^p_{S_{\pi,\chi}} = \prod_{p \neq v \in S_{\pi,\chi}} u_v$ and $c_{\beta,\mathcal{K}}(u) = \prod_{v \in S_\mathcal{K}} c_{\beta,v}(u_v)$.

4. p-adic L-functions for unitary groups

The rigid-analytic variety $\mathbb{X}_P$ introduced above breaks up as the disjoint union of $(p-1)^{\ell+1}$ connected components, indexed by the distinct characters of the torsion subgroup $((\mathbb{Z}/p\mathbb{Z})^\times)^{\ell+1} \subset Z^\ell$. We fix such a character $\nu^0 = (\nu_1^0, \ldots, \nu_\ell^0, \chi^0)$ and the associated connected component $\mathbb{X}_P^0 \subset \mathbb{X}_P$. For our purposes, a Hida family of type P on $U(V')$ will correspond to

(1) A rigid-analytic variety $\mathbb{Y}$ and a finite flat morphism $wt : \mathbb{Y} \rightarrow \mathbb{X}_P^0$;

(2) An analytic family (φ_y, χ_y) over $\mathbb{Y}$ of pairs where φ_y is a p-adic modular form on $U(V')$ and $\chi_y : \mathbb{Z}_p^\times \rightarrow \mathbb{C}_p^\times$ is a continuous character whose restriction to the torsion subgroup $(\mathbb{Z}/p\mathbb{Z})^\times$ is χ^0;

(3) When $wt(y) = (\nu, \chi) \in \mathbb{X}_P^{alg} \cap \mathbb{X}_P^0$, $\chi_y = \chi$ and φ_y is a holomorphic automorphic form on $U(V')$, belonging to an irreducible automorphic representation π_y, of archimedean type π_∞ (as in §2) and of type ν at p (see below);

(4) The family φ_y is P-nearly ordinary in Hida's sense (see below).

Of coures, if (φ_y, χ_y) is any Hida family then one obtains another by multiplying by any holomorphic rigid-analytic function f on $\mathbb{Y}$: $(f(y)\varphi_y, \chi_y)$. We call a Hida family *primitive* if it can not be obtained from another by multiplication by a non-constant f.

We will not define p-adic modular forms on a unitary group of general signature except to say that (i) they are uniform limits of classical holomorphic modular forms of variable level and/or weight which are $\overline{\mathbb{Q}}$-*rational* in a sense compatible with Shimura's theory of canonical models, and (ii) they can be defined without recourse to q-expansions. We refer to the points y such that $wt(y) \in \mathbb{X}_P^{alg}$ as *classical points* and the associated φ_y as *classical modular forms*. The character ν in (3) is assumed to restrict to $\nu^0 = (\nu_1^0, \ldots, \nu_\ell^0)$ on the torsion subgroup. A form is of *type* ν *at* p if it belongs to an irreducible automorphic representation π of $U(V')(\mathbb{A})$ with π_p in the principal series, and if it is an eigenvector for $P(\mathbb{Z}_p) \subset I_P \subset U(V')(\mathbb{Q}_p) = GL(n, \mathbb{Q}_p)$ with eigenvalue $Z \mapsto \nu(Z) = \nu_1(\det(Z_{11})) \cdots \nu_\ell(\det(Z_{\ell,\ell}))$. It is also invariant under a certain open compact subgroup $K_\nu \subset I_P$ containing the kernel of ν in $P(\mathbb{Z}_p)$ - in particular, K_ν contains the $\mathbb{Z}_p$-points $R_u P(\mathbb{Z}_p)$ of the unipotent radical of P. In many cases ν extends to a character of K_ν and (K_ν, ν) is a *semisimple type* in the sense of Bushnell-Kutzko, but this is (unfortunately!) only the case when the conductors of the ν_i are monotonic (increasing or decreasing; so $\ell \leq 2$ is OK).

Finally, the P-ordinarity condition is a condition on the inducing character of the principal series π_p. Suppose $\alpha = \alpha_1 \otimes \cdots \otimes \alpha_n : B \rightarrow \overline{\mathbb{Q}}^\times$ is a character of the standard Borel subgroup of $GL(n, \mathbb{Q}_p)$, each α_i a character of $\mathbb{Q}_p^\times$, and π_p is the principal series representation unitarily induced from α. In practice π_p will be essentially tempered, so the principal series is necessarily irreducible. Condition (3) implies that the α_i can be ordered in such a way that their restrictions $\tilde{\alpha}_i$ to $\mathbb{Z}_p^\times$ satisfy

$$\tilde{\alpha}_j = \nu_j, \quad n_1 + \cdots + n_{i-1} < j \leq n_1 + \cdots + n_i.$$

Fix an embedding of $\overline{\mathbb{Q}}$ in $\mathbb{C}_p$, i.e. a p-adic valuation $|\cdot|_p$ on $\overline{\mathbb{Q}}$, and let $\mu_j = |\alpha_j(p)|_p$. Condition (4) above is that

$$\sum_{j=1}^{n_1+n_2+\cdots+n_i} \mu_j = \sum_{j=1}^{n_1+n_2+\cdots+n_i} j - 1 + k \quad, i = 1, \ldots, \ell \tag{4.1}$$

where k is a fixed constant (this is the same k we fixed to be $n+1$ in §2). A condition of this sort is the starting point of Hida's deformation theory and corresponds to a condition on the p-adic Galois representation associated to π, but we will have no use for its specific form here. We note that (4.1) together with condition (3) imposes a partial order on the characters α_j: one can permute the α_j's within the n_i-th block $(n_1 + \cdots + n_{i-1} < j \leq n_1 + \cdots + n_i)$ for any i, but no permutations are allowed between blocks.

We henceforward assume that the partition $n = n_1 + \cdots + n_\ell$ associated to P is a refinement of the partition $n = a + b$ associated to the signature of V'; i.e. that $a = n_1 + \cdots + n_m$, $b = n_{m+1} + \cdots + n_\ell$ for some index m. Let $\mathbb{Y}^{alg} = (wt)^{-1}(\mathbb{X}^{0,alg})$, so that every $y \in \mathbb{Y}^{alg}$ defines an irreducible automorphic representation π_y of infinity type π_∞. One wants to guarantee that the map from $\mathbb{Y}^{alg}$ to automorphic representations is injective, and in particular we want a way to single out vectors $\varphi_y \in \pi_y$, up to scalar multiples. At places outside S we assume φ_y is an unramified vector. At the infinite place

φ_y is determined by holomorphicity. Every $v \in S(\pi, \chi)$ splits in $\mathcal{K}$, hence $\pi_{y,v}$ is a representation of $GL(n, \mathbb{Q}_v)$, and we assume the condition of being a semi-simple type vector determines the v-component of φ_y uniquely. This is possible as long as π_v is fully induced from a supercuspidal representation of some Levi component, provided we remember to interpret φ_y as a vector valued modular form with values in the space dual to the (tensor product over $v \in S(\pi, \chi)$ of) the type spaces. We have no good choice for $v \in S_{\mathcal{K}}$ unless π_v is spherical for some special maximal compact subgroup – and even then there are two conjugacy classes of special maximal compacts if n is odd, so we have to choose one – so in this exposition we simply ignore this issue and pretend φ_y is fixed up to scalar multiples.

Finally, and most importantly, Hida has shown that conditions (3) and (4) determine the p-component of φ_y up to scalars, and we can identify it explicitly as a (K_ν, ν)-type when the latter is a type. The conditions guarantee that φ_y is a factorizable vector for any $y \in \mathbb{Y}^{alg}$. We determine vectors $\varphi'_y \in \pi_y^\vee$ by the dual conditions and the requirement that the Petersson norms be the same. We would like to apply the basic identity (2.8). However, since we have defined Hida families so that φ_y, $y \in \mathbb{Y}^{alg}$, is holomorphic (at least when $ab \neq 0$), we apply the basic identity not to this pair but to the corresponding pair $\bar{\varphi}_y, \bar{\varphi}_{y'}$, where the former is *antiholomorphic* and the latter is defined again by the dual conditions and both are normalized so that $Q_V^0(\bar{\varphi}_y \otimes \bar{\varphi}'_y) = Q_V^0(\varphi_y \otimes \varphi'_y)$.[8] The condition that $\bar{\varphi}_y$ be antiholomorphic implies, under our conventions (which we have not specified here), that $\bar{\varphi}_{y'}$ is holomorphic on the Shimura variety naturally attached to the hermitian space V' and therefore is again antiholomorphic on the Shimura variety attached to the hermitian space $-V'$. The zeta integral (2.5), applied to the Eisenstein series constructed in §3, is then naturally an integral of the restriction to $G \times G$ of a holomorphic Eisenstein series against an anti-holomorphic automorphic form on $G \times G$. This integral can be interpreted as a cup-product in coherent cohomology, and this is how one expresses special values of the L-function as algebraic multiples of certain periods (cf. [H97]).

When $ab = 0$ holomorphy (resp. antiholomorphy) is replaced by the condition of taking values in a certain finite-dimensional representation of $U(n)$ (resp. the dual representation).

Corresponding to Proposition 3.8, we have that, with an appropriate choice of measures,

Proposition 4.2. *The factor $Z_S^p(1, \bar{\varphi}_y, \bar{\varphi}'_y, f, \chi)$ is an algebraic number whose p-adic absolute value is uniformly bounded above and below as a function of y and χ.*

[8] This has the effect of replacing $L(s, \pi, \chi)$ by $L(s, \bar{\pi}, \chi)$ in the Basic Identity. To avoid changing all the formulas we make the simplifying hypothesis that π_f is isomorphic to its complex conjugate.

(The superscript $'p'$ denotes the ommision of the relevant factors at p.) One actually knows more. If we avoid small primes the archimedean contribution is a p-adic unit, and for the choices made (without explanation) in (3.4)(ii), the factors corresponding to primes in $S(\pi,\chi)$ are just group indices again. Any variation is again due to $S_{\mathcal{K}}$.

It remains to determine the local zeta integral at p. We let $P_{a,b}$ be the standard parabolic subgroup of $GL(n)$ corresponding to the partition $n = a+b$. We write $\pi_p = Ind_{P_{a,b}(\mathbb{Q}_p)}^{GL(n,\mathbb{Q}_p)} \pi_a \otimes \pi_b$ (normalized induction) where π_a is a principal series representation of $GL(a,\mathbb{Q}_p)$ and π_b a principal series representation of $GL(b,\mathbb{Q}_p)$. In what follows, $\ell = 1$ is only allowed if $U(V')(\mathbb{R})$ is definite; i.e. if $P_{a,b} = GL(n)$.

Proposition 4.3. *Write*

$$Z_a(1,\pi) = \varepsilon_p(0,\pi_a\otimes\chi,\psi_p^{-1})\frac{L_p(1,\pi_a^\vee\otimes\chi^{-1})}{L_p(0,\pi_a\otimes\chi)}$$

$$Z_b(1,\pi) = \varepsilon_p(0,\pi_b^\vee\otimes\chi^{-1},\psi_p^{-1})\frac{L_p(1,\pi_b\otimes\chi)}{L_p(0,\pi_b^\vee\otimes\chi^{-1})}.$$

Suppose $\ell = 2$. Then for any $y \in \mathbb{Y}^{alg}$

$$Z_p(\frac{1}{2},\bar{\varphi}_y,\bar{\varphi}'_y,f_p,\chi_p) = A(a,b)p^{N(\pi_p)}\cdot Z_a(1,\pi)\cdot Z_b(1,\pi), \tag{4.3.1}$$

where $A(a,b) \in \mathbb{Q}$ is an explicit constant depending only on a and b and $N(\pi_p) \in \mathbb{Z}$ depends only on the conductors of the characters ν_i and $\nu_i\nu_j^{-1}$. The formula (4.3.1) remains true for any ℓ, for a Zariski dense set of points $y \in \mathbb{Y}^{alg}$.

The shape of the local factor (4.3.1) is roughly as predicted by Coates [Co], although we have not checked that the constants $A(a,b)$ and (especially) the power $N(\pi_p)$ of p corresponds exactly to his normalization. For purposes of comparison, the standard L-factor at p is just

$$L_p^{U(V')}(s,\pi,\chi) = L_p^{GL(n)}(s,\pi,\chi)L_p^{GL(n)}(s,\pi^\vee,\chi^{-1}).$$

The p-adic normalization process, sometimes called *p-stabilization*, selects half the Euler factors, places the others (with s replaced by $1-s$) in the denominator, and for good measure multiplies the whole by a product of ε factors (products of Gauss sums). Here the presence of the denominator corresponds precisely to the partial Fourier transform used to define the section f_p, and appears in the course of applying the Godement-Jacquet functional equation for the standard L-function of $GL(n)$.

Finally, Hida theory provides a way to pair the Eisenstein measure of the preceding with the Hida family φ_y (or rather $\varphi_y\otimes\varphi'_y$) in such a way that at points in $\mathbb{Y}^{alg}$ the result is the global zeta integral involving $\bar{\varphi}_y$ and $\bar{\varphi}'_y$. Let

$$L^p(1,\pi_y,\chi_y) = L^S(1,\pi,\chi)Z_S^p(1/2,\bar{\varphi}_y,\bar{\varphi}'_y,f,\chi_y).$$

The final result (see the remarks following Theorem 2.9) is:

Theorem 4.4. *There is an analytic function (generalized measure)*

$$L_p(1,\bullet) = \varphi * d\mu_{Eis} : \mathbb{Y} \to \mathbb{C}_p$$

such that

$$L_p(1,y) = A(a,b)q^{N(\pi_{y,p})} \cdot Z_a(1,\pi_y) \cdot Z_b(1,\pi_y)L^p(1,\pi_y,\chi_y)/\Omega_y$$

for all $y \in \mathbb{Y}^{alg}$.

Here Ω_y is a period essentially equal to the quotient of the Petersson norm of φ_y by a *congruence number* measuring congruences between φ_y and other forms on G.

The notation $L_p(1,y)$ indicates the expectation that the p-adic L-function extends to an analytic function in the first argument ($s=1$) but we have not considered values other than $s=1$. We also write $L_p(1,\pi,\chi) = L_p(1,y)$ if $\pi = \pi_y$, $\chi = \chi_y$.

(4.5) *Hecke algebras in Hida's theory*

Rigid analytic geometry is not really necessary for Hida's theory but it provides a useful geometric picture of p-adic variation. Let Λ_P and $H_{\mathbb{Y}}$ be the rings of functions on $\mathbb{X}^0$ and $\mathbb{Y}$, respectively. Hida actually works with integral versions of these rings, in which case the former is an Iwasawa algebra in rank(P) variables and the latter, finite and torsion-free over the former, is a p-adic arithmetic Hecke algebra. In our rigid-analytic setting, the rings Λ_P and $H_{\mathbb{Y}}$ are obtained from Hida's rings upon tensoring with $\mathbb{Q}_p$. We have constructed $L_p(1,y)$ to be an element in $H_{\mathbb{Y}} \otimes \mathbb{C}_p$ (and even in $H_{\mathbb{Y}}$).

The irreducible components of $\mathbb{Y}$ correspond to Hida families, whereas the connected components of $\mathbb{Y}$ correspond to collections of Hida families admitting congruences modulo maximal ideals in Λ_P. We will make more precise the notion of such congruences when we return to these considerations in §7.

5. Endoscopic forms and the Rallis inner product formula

Formulas are different depending on the parity of n, and to shorten the exposition, we will henceforth assume n to be *even*. Let χ be an (adelic) Hecke character of $\mathcal{K}$ satisfying (2.4), so

$$\chi \mid_{\mathbb{A}_{\mathbb{Q}}^\times} = \varepsilon_{\mathcal{K}}.$$

Recall that we have fixed a definite hermitian space V at the beginning of the text, with unitary group H. Let $\pi \times \xi$ be an automorphic representation of $G \times U(1)$. The Arthur conjectures predict the existence of an automorphic representation $\Pi_\chi(\pi \times \xi)$ of H, with the property that $BC(\Pi_\chi(\pi \times \xi))$ is induced from the representation $(BC(\pi) \otimes \chi \circ \det) \times BC(\xi)$ of $GL(n) \times GL(1)$:

$$L(s, BC(\Pi_\chi(\pi \times \xi))) = L(s, BC(\pi) \otimes \chi \circ \det) L(s, BC(\xi)). \tag{5.1}$$

Here recall that $BC(\xi)$ is the character $z \mapsto \xi(z/z^c)$, where c denotes complex conjugation. The $\Pi_\chi(\pi \otimes \xi)$ are the endoscopic representations of the title (more properly attached to the elliptic endoscopic group $U(n) \times U(1)$). Note that $\Pi_\chi(\pi \times \xi)$ comes along with a conjectural multiplicity (proved for $n = 2$) which may be zero, but in general is not. Also $\Pi_\chi(\pi \times \xi)$ is actually a packet, but in the examples we consider there will be only one representation. Indeed, $H(\mathbb{R})$ is compact, so L-packets at ∞ are singletons. [9]

The theta correspondence constructs these representations "by hand". The tensor product $V' \otimes V$ has a non-degenerate hermitian form $<,>_{V'\otimes V}$, and the choice of a non-zero element $\daleth \in \mathcal{K}$ with $tr_{\mathcal{K}/\mathbb{Q}}(\daleth) = 0$ transforms $<,>_{V'\otimes V}$ into a skew-hermitian form $\daleth^{-1} <,>_{V'\otimes V}$. Thus $\mathcal{V} = R_{\mathcal{K}/\mathbb{Q}}(V' \otimes V)$ can be viewed as a symplectic vector space over $\mathbb{Q}$ of dimension $4n(n+1)$, and there is a natural map $G \times H = U(V') \times U(V) \rightarrow Sp(\mathcal{V})$. For our purposes, the metaplectic cover $Mp(\mathcal{V})(\mathbb{A})$ is the extension of $Sp(\mathcal{V})(\mathbb{A})$ by the circle. The character χ introduced above can be used to define a splitting $G(\mathbb{A}) \times H(\mathbb{A}) \rightarrow Mp(\mathcal{V})(\mathbb{A})$ (cf. [Kudla]). Given χ as above and an additive adelic character ψ, which has to have the right sign at ∞, one can then define the metaplectic representation $\omega_{\psi,\chi}$ of $G(\mathbb{A}) \times H(\mathbb{A})$ and, for any automorphic representation π of G, the theta lift $\Theta_{\psi,\chi}(\pi)$ as an automorphic representation of H.

Assume π is in the domain of the theta correspondence; i.e., that $\Theta_{\psi,\chi}(\pi)$ is not identically zero. As explained below, this can be guaranteed by a global condition (which will be automatically satisfied) and by local conditions (essentially at primes in $S_{\mathcal{K}}$). The local components of $\Theta_{\psi,\chi}(\pi)$ at places outside S can be determined in terms of those of π and χ, by a procedure introduced by Rallis in [Rallis HDC]. We have

$$\Theta^S_{\psi,\chi}(\pi) = \Pi^S_{\chi^{-1}}(\check{\pi} \times 1) \tag{5.2}$$

in the sense that the left-hand side of (5.2) satisfies (5.1) at unramified places. That is,

$$L^S(s, \Theta_{\psi,\chi}(\pi)) = L^S(s, \check{\pi} \otimes \chi^{-1}) \zeta^S_{\mathcal{K}}(s) = L^S(s, \pi \otimes \chi) \zeta^S_{\mathcal{K}}(s). \tag{5.3}$$

[9] Local L-packets in general arise at places that do not split in $\mathcal{K}$. We retain the hypothesis 3.1 that if v is an unramified inert place then π_v is spherical and χ_v is unramified; therefore packets are singletons at such places as well. Here as previously, we don't really know what to say about places v ramified in $\mathcal{K}$, but the explicit construction of endoscopic representations described in this section yields well-defined representations of $H(\mathbb{Q}_v)$.

We may thus *define* $\Pi_{\chi^{-1}}(\check{\pi} \times 1)$ to be $\Theta_{\psi,\chi}(\pi)$.

Not every $\Pi_{\chi^{-1}}(\pi \times \xi)$ occurs as such a theta-lift; $\Theta_{\psi,\chi}(\pi) = 0$ unless π_∞ belongs to the domain of the local theta correspondence for $U(a,b)\times U(n+1,0)$. For example, suppose $n = 2$ and V' is positive-definite, so $a = 2$, $b = 0$, and π is of the form $\pi^0 \otimes \beta$ where π^0_∞ is the trivial representation and β is a character of $U(1)$. Then there exists a classical holomorphic new form f of weight 2 such that

$$L(s,\pi^0) = L(s,f)L(s,f,\varepsilon_{\mathcal{K}})$$

(i.e., $\pi^0 = \pi(f)_{\mathcal{K}}$ where $\pi(f)$ is the automorphic representation attached to f). More generally, if $\theta(?)$ is the holomorphic modular form on $GL(2,\mathbb{Q})$ attached to the Hecke character ? (a binary theta function) then

$$(L(s,\Theta_{\psi,\chi}(\pi)) =)L(s,\pi\otimes\chi)\zeta_{\mathcal{K}}(s) = L(s,f,\theta(BC(\beta)\cdot\chi))\zeta_{\mathcal{K}}(s)$$

where the first term on the right-hand side is the Rankin-Selberg product. Then $\Theta_{\psi,\chi}(\pi) = 0$ unless $\theta(BC(\beta)\cdot\chi)$ is a holomorphic modular form of weight at least 4. In that case, $s = 1$ is a critical value of the Rankin-Selberg product and Shimura's theorem implies that

$$L(1,f,\theta(BC(\beta)\cdot\chi))/(2\pi i)^* < \theta(BC(\beta)\cdot\chi), \theta(BC(\beta)\cdot\chi) > \quad \in \ \overline{\mathbb{Q}}.$$

Moreover, the Petersson norm in the denominator can be expressed in terms of periods of elliptic curves with CM by $\mathcal{K}$.

6. The Rallis inner product formula

Let $\varphi \in \pi$ be as in §4 and let χ be a splitting character as above. We use the notation $\theta_{\psi,\chi,\Phi_0}(\varphi)$ to designate the theta lift of φ to an element of $\Theta_{\psi,\chi}(\pi)$, an automorphic form on the definite unitary group $U(V)$. As indicated, the notation depends on a Schwartz-Bruhat function Φ_0, which in turn depends on a choice of polarization of the symplectic vector space $\mathcal{V} = R_{\mathcal{K}/\mathbb{Q}}(V'\otimes V)$. The construction becomes more canonical when we consider a pair of forms $\varphi \in \pi, \varphi' \in \pi^\vee$; their theta lift

$$\theta_{\psi,\chi,\Phi}(\varphi\otimes\varphi') \in \Theta_{\psi,\chi}(\pi)\otimes\Theta_{\psi,\chi}(\pi)^\vee$$

is defined in terms of a choice of maximal isotropic subspace X of the skew-hermitian space $2(V'\otimes V)$, and for X one can choose the diagonal subspace $(V'\otimes V)^d$, which we identify with $V'\otimes V$.

This same polarization defines a theta lift from functions on $U(V)$ to functions on $U(2V')$; i.e., there is a seesaw diagram

$$\begin{array}{ccc} & U(2V') & U(V)\times U(V) \\ (6.1) & \uparrow & \uparrow \\ & U(V')\times U(V') & U(V) \end{array}$$

Seesaw duality yields the following expression for the Petersson inner product of theta lifts to $U(V)$:

$$(6.2) \qquad Q_V(\theta_{\psi,\chi,\Phi}(\varphi\otimes\varphi')) \stackrel{def}{=} \int_{U(V)(\mathbb{Q})\backslash U(V)(\mathbb{A})} \theta_{\psi,\chi,\Phi}(\varphi\otimes\varphi')(h)dh$$

$$= \int_{U(V')\times U(V')(\mathbb{Q})\backslash U(V')\times U(V')(\mathbb{A})} \theta_{\psi,\chi,\Phi}(1)(g_1,g_2)\cdot\bar{\varphi}(g_1)\otimes\bar{\varphi}'(g_2)dg_1dg_2.$$

Here $\bar{\varphi}$ and $\bar{\varphi}'$ are as in the discussion preceding Proposition 4.2. For appropriate choices of a Schwartz-Bruhat function Φ, the theta lift $\theta_{\psi,\chi,\Phi}(\varphi\otimes\varphi')$ can be factored as a product of forms on the two copies of $U(V')$, and the first integral can be written as a genuine inner product. The $\theta_{\psi,\chi,\Phi}(1)$ appearing in the second integral denotes the theta lift of the constant function 1 to $U(2V')$.

At this stage it is more natural to work with a fixed choice of Φ. Indeed, expression (6.2) becomes the *Rallis inner product formula* when the function $\theta_{\psi,\chi,\Phi}(1)$ can be identified with a Siegel Eisenstein series on $U(2V')$. This identification is provided by the extended Siegel-Weil formula, proved in the present case by Ichino [Ich], following the methods introduced by Kudla and Rallis in their work on Eisenstein series on symplectic groups. For appropriate choices of Φ we obtain precisely [10] the value at $s=\frac{1}{2}$ of the Eisenstein series attached to the section $f=\otimes_v f_v$ defined in §3; this is the meaning of condition (3.4)(v). We now apply formula (6.2) to a Hida family depending on the parameter y, as in §4. The precise formula is then

$$(6.3) \qquad \begin{aligned} &d_n^S(\tfrac{1}{2},\chi_y)Q_V(\theta_{\psi,\chi_y,\Phi}(\varphi_y\otimes\varphi_y')) \\ &\quad = d_n^S(\tfrac{1}{2},\chi_y)\int E_f((g_1,g_2);\chi,\tfrac{1}{2})\bar{\varphi}_y(g_1)\otimes\bar{\varphi}_y'(g_2)dg_1dg_2 \\ &\quad = Q_{V'}^0(\varphi_y\otimes\varphi_y')L_p(1,y)\Omega_y, \end{aligned}$$

where the integral is over $U(V')\times U(V')(\mathbb{Q})\backslash U(V')\times U(V')(\mathbb{A})$ and Ω_y is the period appearing in Theorem 4.4.. The appearance of $L_p(1,y)$ suggests that the Rallis inner product formula has a variant in terms of p-adic L-functions. This is indeed the case, but the precise formulation requires some foundational work.

[10] Up to multiplication by $\frac{1}{2}$, which we incorporate into the constants.

p-adic families of endoscopic forms.

On $G \times G$ we work with holomorphic modular forms, which, as we have already intimated, have a natural $\overline{\mathbb{Q}}$-rational structure compatible with the theory of canonical models. We can do the same on $H \times H$, but since $H(\mathbb{R})$ is compact, the locally symmetric variety attached to H and an open compact subgroup $K \subset H(\mathbb{A}_f)$ is the finite set ${}_KSh(H) = H(\mathbb{Q})\backslash H(\mathbb{A})/H(\mathbb{R}) \times K$. The theta lift $\Theta_{\psi,\chi}(\pi)$ to H of an automorphic representation π of G of our fixed infinity type π_∞ is an automorphic representation Π with Π_∞ fixed and, with our choice of τ, necessarily of dimension 1. We may thus view forms in Π as sections of a local system $\mathcal{L}$ on the finite set ${}_KSh(H)$, where $\mathcal{L}$ depends only on τ and has a natural $\overline{\mathbb{Q}}$-rational structure, and even an integral structure over the ring $\bar{\mathbb{Z}}$ of algebraic integers in $\overline{\mathbb{Q}}$. (Of course $\mathcal{L}$ is constant on connected components of ${}_KSh(H)$, which are just points!) We let $Sh(H)$ be the profinite set $\varprojlim_K {}_KSh(H)$. [11]

It is an important fact that the theta lift $\theta_{\psi,\chi,\Phi}(\varphi \otimes \varphi')$ almost never preserves arithmeticity. If $\varphi \otimes \varphi'$ is a holomorphic modular form on $G \times G$ which is $\overline{\mathbb{Q}}$-rational and belongs to an irreducible automorphic representation $\pi \otimes \pi^\vee$, then one can define without much difficulty an almost certainly transcendental factor $q_\pi \in \mathbb{C}^\times$, well-defined up to $\overline{\mathbb{Q}}^\times$, such that

$$q_\pi^{-1}\theta_{\psi,\chi,\Phi}(\varphi \otimes \varphi') \in H^0(Sh(H), \mathcal{L})(\overline{\mathbb{Q}}). \tag{6.4}$$

Unfortunately, unless either (i) $n = 2$ or (ii) $ab = 0$ we know no sensible way to define a variant of q_π that is well-defined up to p-integers. More precisely, as long as π is unramified at p (which will not generally be the case!) there are natural choices of q_π so that $q_\pi^{-1}\theta_{\psi,\chi,\Phi}(\varphi \otimes \varphi') \in H^0(Sh(H), \mathcal{L})(\bar{\mathbb{Z}}_{(p)})$, but no choice is known that does not compromise valuable information about congruences between holomorphic forms on $G \times G$, except in cases (i) or (ii). We will say no more about this issue, which for practical purposes amounts to restricting our attention to cases (i) and (ii), even though much of what we say below is true more generally.

The Eisenstein measure $d\mu_{Eis}$ introduced in Theorem 3.10, depends on the p-adic family of sections $f_p = f_p(\nu, \chi)$ described after Proposition 3.9, which in turn depends on a certain collection of Schwartz-Bruhat functions Φ_ν on $\mathcal{V}(\mathbb{Q}_p) \simeq M(n, 2(n+1))(\mathbb{Q}_p)$ depending on characters ν of finite order. We have not checked enough details to remove the question mark in the following Proposition 6.5.(?), but it is undoubtedly true. Note that the period Ω_y appears squared in the denominator.

[11] This is one place where we need to take care to work with similitude groups rather than unitary groups, but there is no real difficulty and we ignore the issue in this exposition.

Proposition 6.5. (?) *Let (φ_y, χ_y) be a Hida family over $\mathbb{Y}$, as in §4. Let $\Phi_\nu \in S(M(n, 2(n+1))(\mathbb{Q}_p))$ and $\Phi^p \in S(\mathcal{V}(\mathbb{A}_f^p))$ be the Schwartz-Bruhat functions used to define the Eisenstein measure $d\mu_{Eis}$. Then there is a Hida family $F^\theta(y) = F^\theta_{\varphi_y \otimes \varphi'_y, \chi_y}$ of p-adic forms in $H^0(Sh(H) \times Sh(H), \mathcal{L} \otimes \mathcal{L}^\vee)(\mathbb{C}_p)$, parametrized by $\mathbb{Y}$, such that*

$$F^\theta(y) = (d_n^S(1/2, \chi_y)/\Omega_y^2)\theta_{\psi, \chi_y, \Phi_{wt(y)}}(\varphi_y \otimes \varphi'_y)$$

for all $y \in \mathbb{Y}^{alg}$.

Once one admits the above Proposition 6.5.(?) the following assertion is not difficult:

Proposition 6.6. *There is a rigid meromorphic function g_θ on $\mathbb{Y}$ such that $F(y) = g_\theta(y) F^\theta(y)$ is a primitive Hida family.*

If (φ_y, χ_y) is a Hida family relative to a standard parabolic associated to a partition $n = n_1 + \cdots n_\ell$, then $F(y)$ is a Hida family relative to the standard parabolic P associated to the partition $n + 1 = n_1 + \cdots + n_\ell + 1$.

Much of our work at present is concerned with finding conditions under which g_θ can be taken to equal 1. The reason for this is explained in the following section. Meanwhile, we can now state a p-adic version of the Rallis inner product formula. Define

$$Q^{alg}_{V'}(\varphi_y \otimes \varphi'_y) = \Omega_y^{-1} Q^0_{V'}(\varphi_y \otimes \varphi'_y)$$

whenever $y \in \mathbb{Y}^{alg}$.

Proposition 6.7. *The pairings Q_V and $Q^{alg}_{V'}$ extend to natural pairings on p-adic modular forms on $U(V) \times U(V)$ and $U(V') \times U(V')$, respectively, such that*

$$\begin{aligned} Q_V(F(y)) &= g_\theta(y) Q_V(F^\theta(y)) \\ &= g_\theta(y) Q^{alg}_{V'}(\varphi_y \otimes \varphi'_y) L_p(1, y). \end{aligned}$$

The pairing $Q^{alg}_{V'}$ is essentially the usual Petersson pairing, but normalized by a period so as to be algebraic. For example, when $n = 2$ and forms on $U(V')$ can be identified with classical modular forms, $Q^{alg}_{V'}$ is just the pairing constructed (either implicitly or explicitly) and used by Hida in his papers on p-adic congruence modules and p-adic Rankin products (cf. [Hi90]). A similar remark applies to Q_V (since $U(V)$ is definite, the pairing is already algebraic).

Remark 6.8. We note in passing that there is no natural way to separate φ_y from φ'_y in the above discussion, mainly because the doubling method cannot be applied to pairs $\varphi_1 \otimes \varphi_2$, with $\varphi_1 \in \pi_1$, $\varphi_2 \in \pi_2$, where φ_2 is congruent to a form in $\pi_1^\vee$ but is not actually in $\pi_1^\vee$. This is the source of serious problems whose resolution brings into play the deep structure theory of arithmetic Hecke algebras.

7. Inner products and congruences

To construct congruences, we use a variant of an idea introduced by Hida. Fix a level subgroup $K = \prod_q K_q$ with K_p determined by the level of the character χ, K_q hyperspecial maximal at all inert primes, and some other local conditions if necessary. The space $M_K(\mathcal{L}, \mathbb{C}) = H^0({}_KSh(H), \mathcal{L})$ of K-fixed vectors in the space of all automorphic forms on H of type Π_∞ contains a natural $\mathcal{O} = \mathcal{O}_{\mathcal{K},p}$-lattice $L(\mathcal{L})$ (if Π_∞ were trivial it would have a $\mathbb{Z}$-lattice, namely the $\mathbb{Z}$-valued functions). The Petersson inner product on this lattice, with respect to the measure on the finite set ${}_KSh(H)$ giving each point mass 1, is a unimodular hermitian form (more precisely, we work with the perfect pairing between $L(\mathcal{L})$ and $L(\mathcal{L}^{-1})$).

Now we write

$$(7.1) \qquad H^0({}_KSh(H), \mathcal{L}) = H^0({}_KSh(H), \mathcal{L})_{end} \oplus H^0({}_KSh(H), \mathcal{L})_{st}$$

ignoring endoscopic forms of other types (this can be arranged by adding local conditions). There is correspondingly an inclusion

$$L(\mathcal{L}) \supset L(\mathcal{L})_{end} \oplus L(\mathcal{L})_{st}$$

where $L(\mathcal{L})_? = H^0({}_KSh(G), \mathcal{L})_? \cap L(\mathcal{L})$. This inclusion is an equality if and only if there are no pairs $f_{end} \in L(\mathcal{L})_{end}, f_{st} \in L(\mathcal{L})_{st}$ of p-primitive forms such that $p^{-1}[f_{end} - f_{st}] \in L(\mathcal{L})$, i.e. if and only if there are no congruences between endoscopic and stable forms of the form

$$f_{end} \equiv f_{st} \pmod p.$$

Hida's observation, made in the setting of classical modular forms, is that this is true if and only if the restriction to $L(\mathcal{L})_{end}$ of the hermitian pairing on $L(\mathcal{L})$ is again unimodular [Hi81].

We now allow the level subgroup K_p to vary, always assuming $K_p \supset R_uP(\mathbb{Z}_p)$. Applying Hida's constructions, we obtain modules of P-nearly ordinary forms $L_{P-ord}(\mathcal{L}) \supset L_{P-ord}(\mathcal{L})_{end} \oplus L_{P-ord}(\mathcal{L})_{st}$ of finite type over Λ_P, the ring of functions on $\mathbb{X}_0$. The torsion Λ_P-module

$$\mathcal{C} = L_{P-ord}(\mathcal{L})/(L_{P-ord}(\mathcal{L})_{st} \oplus L_{P-ord}(\mathcal{L})_{end})$$

measures congruences between stable and endoscopic forms.

Hida's constructions also give a Hida family of modular forms on H over some $\mathbb{Y}$ such that $H_{\mathbb{Y}}$ (a Hecke algebra; see (4.5)) acts faithfully on $L_{P-ord}(\mathcal{L})$

and hence acts on $\mathcal{C}$. (To make all this true we have to consider Hida families on H consisting simply of p-adic forms, not pairs of p-adic forms and characters.) Let

$$I = Ann_{H_{\mathbb{Y}}}(\mathcal{C}).$$

Let (φ_y, χ_y) over $\mathbb{Y}_0$ be an irreducible Hida family of modular forms on G. Let $R = H_{\mathbb{Y}_0}$. In general this is finite and integral over Λ_P, but for the sake of simplicity we will assume that $R = \Lambda_P$. Recall that we can view $L_p(1, \bullet)$ as an element of R (see (4.5)). For the appropriate choice of the prime-to-p levels K_q, if we admit Proposition 6.5 then the endoscopic Hida family F of Proposition 6.6 is associated to an irreducible component of $\mathbb{Y}$, which identifies R as a Λ_P-quotient of $H_{\mathbb{Y}}$. In fact, this identifies $\mathbb{Y}_0$ as a component of $\mathbb{Y}$. The considerations introduced in the first paragraph of this section apply to the unimodular pairing

$$L_{P-ord}(\mathcal{L})_{\mathfrak{P}} \otimes L_{P-ord}(\mathcal{L})_{\mathfrak{P}} \rightarrow \Lambda_{P,\mathfrak{P}}, \tag{7.2}$$

where $\mathfrak{P} \subset \Lambda_P$ is any height one prime. In particular, the discriminant of the restriction of the pairing (7.2) to $L_{P-ord}(\mathcal{L})_{end,\mathfrak{P}} \otimes L_{P-ord}(\mathcal{L})_{end,\mathfrak{P}}$ divides $IR_{\mathfrak{P}}$. Concretely, we obtain the following theorem (admitting Proposition 6.5).

Theorem 7.3. $IR \subseteq (g_\theta(\bullet)L_p(1,\bullet))$.

Assume we can eliminate the function g_θ from the above statement. Then the theorem asserts that the p-adic L-function divides the ideal measuring congruences between endoscopic and stable forms (really Hida families). In the next section we will briefly explain how such congruences give rise to non-trivial classes in Galois cohomology, as predicted by the appropriate generalization of Iwasawa's main conjecture.

In order to eliminate g_θ it suffices to show that the normalized theta lifts $F^\theta_{\varphi_y\otimes\varphi'_y,\chi_y}$ are already p-primitive (integral, but not divisible by any fractional power of p). The methods available for proving the arithmeticity of $F^\theta_{\varphi_y\otimes\varphi'_y,\chi_y}$ – i.e., for determining the transcendental factors q_{π_y} introduced in (6.4) – can be applied to prove the desired primitivity, provided one knows that sufficiently many special values of auxiliary L-functions are not divisible by p. It appears that one can derive what one needs from results of Vatsal in the case $n = 2$.

(7.4) *Hecke algebras on H and congruences*

Let $\mathfrak{P} \subset R = \Lambda_P$ be any prime ideal containing IR. It follows that localization at $\mathfrak{P}$ of the quotient of $H_{\mathbb{Y}}$ corresponding to all the families of stable forms (i.e., the quotient acting faithfully on $L_{P-ord}(\mathcal{L})_{st}$) is non-trivial. This can be interpreted as meaning that there is a component of $\mathbb{Y}$ corresponding to a family of stable forms that intersects the endoscopic component $\mathbb{Y}_0$ "at the prime $\mathfrak{P}$."

(7.5) *Theorem* (7.3) *as a statement about forms on G*

As written, Theorem 7.3 asserts that the p-adic L-function of a Hida family on $G = U(V')$ controls congruences between endoscopic and stable forms on $H = U(V)$. The applications in the next section concern the Galois representations attached to the Hida family on G. Thus neither the input (the Hida family on G) nor the output (the construction of Galois cohomology classes) makes mention of the group H, and one would like to be able to eliminate H and the hermitian space V from the statement. In other words:

Question 7.5.1. *Does the typical classical point $y = (\pi_y, \chi_y)$ in the Hida family admit a non-zero lifting to $H = U(V)$, for some positive-definite hermitian space V of dimension $n+1$?*

As Waldspurger was the first to observe, and as Rallis explained in great generality, there are both global and local obstructions. The Rallis inner product formula accounts for both obstructions: if $Q_V(\theta_{\psi,\chi,\Phi}(\varphi\otimes\varphi')) \neq 0$ for some choice of data then the theta lift is non-trivial. It suffices to restrict our attention to Hida families such that π_y is *tempered* for all (classical) y, and for which no π_y is itself a theta lift from some $U(W)$ with $\dim W < n$. Under our running assumption that π_y admits a base change to $GL(n)_{\mathcal{K}}$, the L-function $L(s,\pi,\chi)$ does not vanish at $s=1$, hence there is no global obstruction. The local obstructions are given by the product of the local terms in the Rallis inner product formula. There is a genuine local obstruction at the archimedean prime, since we have to take V positive-definite, but the local component $\pi_{y,\infty}$ has been chosen to eliminate this obstruction. It is shown in [HKS] that there is a non-archimedean local obstruction at a prime p if and only p is inert or ramified in $\mathcal{K}$ and the local theta lift of π_p (determined by χ_p, as well as the additive character ψ_p and $\beth$) to $U(V)(\mathbb{Q}_p)$ is trivial. It is also shown in [loc. cit.] that for at least one choice of local space V_p of dimension $n+1$ up to isometry – there are two isometry classes – the theta lift of π_p to V_p is non-trivial. Thus Question (7.5.1) comes down to whether the set of local V_p's to which π_p lifts can be realized as the local components of a given global positive-definite hermitian space over $\mathcal{K}$. In principle there is a parity obstruction; however, at least when n is even, "most" π_p lift to either isometry class, which means that "most" π (with the given π_∞) lift to some global $U(V)$. Finally, it is not difficult to see that, by varying the choice of χ (away from p) one can always arrange that the theta lift to a given global positive-definite V of π is non-trivial.

All this is of little consolation, however, if one's interests are served by a specific pair (π,χ). At present we do not know the answer to Question (7.5.1), which has something to do with the nature of the values of "incoherent" Eisenstein series at the point $s = \frac{1}{2}$, which (in principle) are not explained by the Siegel-Weil formula.

8. Applications to Selmer groups

We keep to the notation of section 7. We assume that associated to the Hida family (φ_y, χ_y) over $\mathbb{Y}_0$ is an analytic family of n-dimensional Galois representations $\sigma : Gal(\bar{\mathcal{K}}/\mathcal{K}) \to GL_n(R)$ such that $\sigma(y) = \sigma(\pi_y \otimes \chi_y)$, the latter being the n-dimensional representation corresponding to $BC(\pi) \otimes \chi$ as in section 2. At this point, we make a number of assumptions on p, S, and σ, the most important of which is that the mod p reduction of some (and therefore all) $\sigma(\pi_y \otimes \chi_y)$ in our family is *absolutely irreducible*.

Let $\mathfrak{P} \subset R = \Lambda_P$ be any height one prime containing I. Let $\sigma_{\mathfrak{P}}$ be σ viewed as a representation into $GL_n(R_{\mathfrak{P}})$ and let $\bar{\sigma}_{\mathfrak{P}}$ be the reduction of $\sigma_{\mathfrak{P}}$ modulo $\mathfrak{P}$. This is a representation into $GL_n(k(\mathfrak{P}))$, where $k(\mathfrak{P}) = R_{\mathfrak{P}}/\mathfrak{P}$ is the residue field. Our assumptions imply that $\bar{\sigma}_{\mathfrak{P}}$ is irreducible.

From Theorem (7.3) it follows that there is some irreducible component $\mathbb{Y}'$ corresponding to a Hida family of stable forms that intersects the endoscopic family $\mathbb{Y}_0$ at $\mathfrak{P}$ (see (7.4)) (recall that for these families we have dropped the extra character). Again, for simplicity we shall assume that $R' = H_{\mathbb{Y}'}$ (a quotient of $H_{\mathbb{Y}}$) is also isomorphic to Λ_P. Then $\mathfrak{P}$ determines a prime of R' and we assume that there is an $n+1$-dimensional representation $\sigma' : Gal(\bar{\mathcal{K}}/\mathcal{K}) \to GL_{n+1}(R')$ such that its specializations at points in $\mathbb{Y}'^{alg}$ are the $n+1$-dimensional representations (conjecturally) associated to the representations $BC(\pi')$ as in section 2. The existence of σ' is already more difficult to arrange than that of σ, since the methods of [HL] only apply to representations that are locally supercuspidal or Steinberg at some finite place split in $\mathcal{K}$, and this will in general not be the case for the points in $\mathbb{Y}'^{alg}$, precisely because they are congruent to endoscopic representations. However, when $n+1 \leq 3$ the existence of the representations $\sigma'(y)$ has been shown by Blasius and Rogawski [BR], and they are expected to exist in general.

Define $\sigma'_{\mathfrak{P}}$ and $\bar{\sigma}'_{\mathfrak{P}}$ as we did $\sigma_{\mathfrak{P}}$ and $\bar{\sigma}_{\mathfrak{P}}$. The properties of these representations connecting Hecke operators and eigenvalues of Frobenius elements then implies that

$$(\bar{\sigma}'_{\mathfrak{P}})^{ss} = \bar{\sigma}_{\mathfrak{P}} \oplus \mathbf{1},$$

where the "ss" denotes semi-simplification. But we actually get more. The representations associated to stable forms are expected to have "large" image. When this is so, after possibly replacing $\sigma'_{\mathfrak{P}}$ with a conjugate over the field of fractions of $R_{\mathfrak{P}}$, we have that

$$\bar{\sigma}'_{\mathfrak{P}} = \begin{pmatrix} \bar{\sigma}_{\mathfrak{P}} & * \\ & \mathbf{1} \end{pmatrix},$$

with the "$*$" in the upper-right corner non-trivial. This "$*$" determines a non-trivial class in $H^1(\mathcal{K}, \bar{\sigma}_{\mathfrak{P}}^{-1})$. With more work this class can be shown to have nice ramification properties at the primes above p, and its existence can be reinterpreted as saying that localization at $\mathfrak{P}$ of the Bloch-Kato/Greenberg Selmer group associated to an integral version of σ has length ≥ 1 as a $R_{\mathfrak{P}}$-module.

An argument originally due to Wiles, as generalized by Urban, shows that by working with all stable components congruent to the endoscopic component $\mathbb{Y}_0$ modulo $\mathfrak{P}$ it is possible to prove that the length of the Selmer group localized at $\mathfrak{P}$ is at least $ord_{\mathfrak{P}}(IR_{\mathfrak{P}})$.

If we assume the function g_θ of Proposition 6.6 to be a scalar, then the congruence ideal I is divisible by the p-adic L-function $L_p(1, \bullet)$, and therefore the length of the Selmer group localized at any height one prime $\mathfrak{P}$ of Λ_P is at least $\mathrm{ord}_{\mathfrak{P}}(L_p(1, \bullet))$. Provided we can justify all the assumptions we have made up to this point !

References

[BR] D. Blasius and J. Rogawski, Tate classes and arithmetic quotients of the two-ball, in R. P. Langlands, Jr., and D. Ramakrishnan, eds., *The zeta functions of Picard modular surfaces* Montréal: Univ. Montréal, (1992) 421–444.

[Co] J. Coates, Motivic p-adic L-functions, in J. Coates and M. J. Taylor, eds., *L-functions and Arithmetic* London Mathematical Society Lecture Note Series, **153** Cambridge: Cambridge University Press (1991), 141-172.

[Gar] P. B. Garrett, Integral representations of Eisenstein series and L-functions, in *Number theory, trace formulas and discrete groups (Oslo, 1987)*, Boston: Academic Press (1989), 241-264.

[GPR] S. Gelbart, I. Piatetski-Shapiro, and S. Rallis, Explicit constructions of automorphic L-functions, *Lecture Notes in Math.*, **1254** (1987).

[H97] M. Harris, L-functions and periods of polarized regular motives, *J. reine angew. Math.*, **483** (1997), 75-161.

[HKS] M. Harris, S. Kudla, and W. J. Sweet, Theta dichotomy for unitary groups, *J. Am. Math. Soc.*, **9** (1996) 941-1004.

[HL] M. Harris and J.-P. Labesse, Conditional base change for unitary groups, manuscript (2003).

[HT] M. Harris and R. Taylor, The Geometry and Cohomology of Some Simple Shimura Varieties, *Annals of Math. Studies*, **151**, Princeton: Princeton University Press (2001).

[Hi81] H. Hida, Congruence of cusp forms and special values of their zeta functions, *Invent. Math.*, **63** (1981) 225-261.

[Hi90] H. Hida, Le produit de Petersson et de Rankin p-adique. *Séminaire de Thorie des Nombres, Paris 1988-1989*, 87–102, Progr. Math., 91, Birkhuser Boston, Boston, MA (1990).

[Hi00] H. Hida, Irreducibility of generalized Igusa towers, manuscript (2000); available at http://www.math.ucla.edu/~hida.

[Hi02] H. Hida, Control theorems of coherent sheaves on Shimura varieties of PEL type, *J. Inst. Math. Jussieu*, **1** (2002) 1-76.

[Hi IHP] H. Hida, p-adic automorphic forms on reductive groups, Notes of a course at the IHP in 2000, *Astérisque* (to appear); available at http://www.math.ucla.edu/~hida.

[Ich] A. Ichino, A regularized Siegel-Weil formula for unitary groups (manuscript, 2001-2003).

[Katz] N. Katz, p-adic L-functions for CM fields, *Invent. Math.*, **49** (1978) 199-297.

[Kudla] S. Kudla, Splitting metaplectic covers of dual reductive pairs, *Israel J. Math.*, **87** (1994) 361-401.

[L92] J.-S. Li, Non-vanishing theorems for the cohomology of certain arithmetic quotients, *J. reine angew. Math.*, **428** (1992), 177-217.

[N] A. Nair, Intersection cohomology, Shimura varieties, and motives, Manuscript (2003).

[Rallis HDC] S. Rallis, On the Howe duality conjecture, *Compositio Math.*, **51** (1984) 333-399.

[Rog] J. Rogawski, Automorphic Representations of Unitary Groups in Three Variables, *Annals of Math. Studies*, **123**, Princeton: Princeton University Press (1990).

[Sh1] G. Shimura, Euler products and Eisenstein series, *CBMS Regional Conference Series in Mathematics*, **93**, Providence, R.I.: American Mathematical Society (1997).

[Sh2] G. Shimura, Arithmeticity in the theory of automorphic forms, *Mathematical Series and Monographs*, **82**, Providence, R.I.: American Mathematical Society (2000).

Michael Harris
UFR de Mathématique
Université Paris 7
2 Pl. Jussieu
75251 Paris cedex 05. FRANCE
harris@math.jussieu.fr

Jian-Shu Li
Department of Mathematics
HKUST and Zhejiang University
Clear Water Bay. Hong Kong
matom@ust.hk

Christopher M. Skinner
Department of Mathematics
University of Michigan
Ann Arbor, MI 48109, USA
cskinner@math.lsa.umich.edu

A guide to the relative trace formula

Hervé Jacquet

Abstract. We discuss several examples of the relative trace formula.

1991 Mathematics Subject Classification: primary 11F55; secondary 11F67, 11F72, 11S37.

1. Introduction: relative Kusnietsov formulas

Recently there have been many examples of **relative trace formulas**. We will review some of the examples, hoping to provide a guide through the burgeoning literature.

Rallis and his collaborators are responsible for a substantial part of this literature. This lecture is dedicated to him. In my view, Rallis is to the field of automorphic forms what Donizetti[1] is to operatic music.

We start with what one could call the **relative Kusnietsov trace formula**.

Let G and G' be two algebraic reductive groups defined over a number field F and quasi-split. Let N be a maximal unipotent subgroup of G, θ a character of $N(F_{\mathbb{A}})$ trivial on $N(F)$. Let N', θ' be similarly defined for G'. We assume that θ is in general position. In addition, we let H be a subgroup of G' defined over F and χ a character of $H(F_{\mathbb{A}})$ trivial on $H(F)$. We consider the usual kernel functions. We let f and f' be smooth functions of compact support on $G(F_{\mathbb{A}})$ and $G'(F_{\mathbb{A}})$ respectively. We set:

$$K_f(x,y) = \sum_{\gamma \in G(F)} f(x^{-1}\gamma y)\,,$$

$$K_{f'}(x,y) = \sum_{\gamma' \in G'(F)} f'(x^{-1}\gamma' y)\,.$$

Then the geometric side of the relative trace formula reads:

$$\int\int K_f(u_1,u_2)\overline{\theta}(u_1)\theta(u_2)du_1du_2 = \int\int K_{f'}(h,u')\overline{\chi}(h)\theta'(u')du'\,; \qquad (1)$$

[1]Gaetano Donizetti composed more than 60 operas.

this identity is satisfied when f and f' are related by a certain linear correspondence. The simplest case is the case where the correspondence is **geometric**. This means the following. The left hand side of (1) can be computed as a sum of orbital integrals:

$$\sum_{\xi} \int \int f(u_1 \xi u_2) \theta(u_1 u_2) du_1 du_2 .$$

The sum is over all orbits (or double cosets) $N(F)\xi N(F)$, the integral being over the quotient of the product $N(F_{\mathbb{A}}) \times N(F_{\mathbb{A}})$ by the stabilizer of ξ; in order for the orbital integral to make sense, we have to assume that the character $\theta(u_1 u_2)$ is trivial on this stabilizer; equivalently, we have to assume that $\theta(u_1 u_2)$ depends only on the product $u_1 \xi u_2$. We shall say that ξ (or its double coset) is **relevant** if this condition is satisfied. Thus the sum is over the set of all relevant double cosets $N(F)\xi N(F)$.

Similarly, the right hand side of (1) can be computed as a sum of orbital integrals of the form

$$\int \int f'(h\xi' u') \chi(h) dh \theta'(u') du' ,$$

the sum being over all relevant double cosets $H(F)\xi' N'(F)$. One expects to have a bijection $N(F)\xi N(F) \leftrightarrow H(F)\xi' N'(F)$ between the set of relevant double cosets. The geometric linear relation is that

$$\int \int f(u_1 \xi u_2) \theta(u_1 u_2) du_1 du_2 = \int \int f'(h\xi' u') \overline{\chi}(h) dh \theta'(u') du'$$

each time $N(F)\xi N(F) \leftrightarrow H(F)\xi' N'(F)$.

More precisely, the functions f and f' are product of local functions f_v and f'_v and the above conditions is more correctly formulated as a set of equalities, one for each place v,

$$\int \int f_v(u_1 \xi u_2) \theta_v(u_1 u_2) du_1 du_2 =$$

$$\gamma_v(\xi_v, \xi'_v) \int \int f'_v(h\xi' u') \chi_v(h) dh \theta'_v(u') du' .$$

The factor $\gamma_v(\xi_v, \xi'_v)$ is a **transfer** factor, to be defined appropriately. We shall say that the functions f_v and f'_v **match** if this relation is satisfied each time $N(F_v)\xi_v N(F_v) \leftrightarrow H(F_v)\xi'_v N'(F_v)$. The following assertions have to be proved. Every f_v matches a f'_v and conversely (**smooth matching**). At almost all finite places v, the characteristic function of the (suitable) maximal compact subgroups K_v and K'_v match (**fundamental lemma**). The product of the local transfer factors $\prod_v \gamma_v(\xi, \xi')$ is one when ξ and ξ' are rational. One expects to have at almost all places an homomorphism of the Hecke algebras:

either

$$b : \mathcal{H}(G_v, K_v) \to \mathcal{H}(G'_v, K'_v)$$

or

$$b' : \mathcal{H}(G'_v, K'_v) \to \mathcal{H}(G_v, K_v)\,;$$

Hecke functions which correspond to one another by this homomorphism match.

This setup is quite general. For instance, as will be apparent from the examples discussed below, one expects such a trace formula identity (with a geometric correspondence) to hold true **each time** H is the group of fixed points of an automorphism of order 2 of a group G', the group G being appropriately defined.

Now let ϕ be an automorphic form on G and ϕ' be an automorphic form on G'. We will set:

$$\mathcal{W}(\phi) = \int_{N(F)\backslash N(F_{\mathbb{A}})} \phi(u)\overline{\theta(u)}du\,, \tag{2}$$

$$\mathcal{W}'(\phi') = \int_{N(F)'\backslash N(F_{\mathbb{A}})'} \phi'(u)\overline{\theta'}(u)du\,, \tag{3}$$

$$\mathcal{P}(\phi') = \int_{H(F)\backslash H(F_{\mathbb{A}})} \phi'(h)\overline{\chi}(h)dh\,. \tag{4}$$

We will say that an automorphic representation π' of G' is **distinguished** by (H, χ) (or $\mathcal{P}$) if $\mathcal{P}$ is not identically 0 on the space of π'. Similarly we will say that π' is **distinguished** by θ' if the restriction of $\mathcal{W}'$ to the space of π' is non-zero. Similarly we shall say that an automorphic representation π of G is **distinguished** by θ if the restriction of $\mathcal{W}$ to the space of π is non-zero. In the examples investigated so far, the character θ is in general position but the character θ' needs not be.

One expects to be able to associate to every automorphic (cuspidal) representation π' of G' distinguished by (H, χ) and θ' an automorphic representation π of G distinguished by θ and conversely such that:

$$\sum_i \mathcal{W}(\pi(f)\phi_i)\overline{\mathcal{W}(\phi_i)} = \sum_j \mathcal{P}(\pi'(f')\phi'_j(h))\overline{\mathcal{W}'(\phi'_j)}\,. \tag{5}$$

Here, in the left hand side, ϕ_i is an orthonormal basis of the space of π, f is a smooth function of compact support on $G(F_{\mathbb{A}})$. Likewise for the right hand side. The functions f and f' are related by the linear correspondence and the representations π and π' are related by the principle of **functoriality**. In fact, the group G may be a two fold cover of an algebraic group, so the principle of functoriality has to be extended to that case.

Formally, the left hand side of (5) computes a trace. Indeed, let A be the operator such that

$$\langle A\phi_1, \phi_2\rangle = \mathcal{W}(\phi_1)\overline{\mathcal{W}}(\phi_2).$$

Then the left hand side is the trace of the composition operator $A\pi(f)$. In particular, the left hand side does not depend on the choice of the basis. Likewise for the right hand side. Thus the related identities (1) and (5) appear as an extension of the duality between characters and conjugacy classes which is at the heart of the standard trace formula.

However, in the recent work of Rallis and his collaborators, another kind of relative trace formula appears where the functions f and f' are related by a linear correspondence of a more general type. The correspondence is expected to have several formal properties. In particular, it should be local, that is, writing the functions f and f' as product, for every place v, the functions (f_v, f'_v) should be related by a linear correspondence.

Note also that the double coset $G^{\Delta}\backslash(G \times G)/(N \times N)$ is the same as $N\backslash G/N$. This suggests that more generally, one should consider more relative trace formulas comparing $H_1\backslash G_1/N_1$ and $H_2\backslash G_2/N_2$ where H_1 and H_2 are the groups of fixed points of involutions in G_1 and G_2 respectively.

2. Relation with the Weil representation

The ongoing work of Rallis and Mao suggests that the machinery of the Weil representation will lead to many examples of relative trace formulas of the type (5) (or suitable generalizations). As a matter of fact, many early examples of the relative trace formula were suggested but the work of Waldspurger and others, which, in turn, was based on the Weil representation.

For clarity, I now restrict the discussion to one case studied in detail by Furusawa ([**8**]). I am very grateful to Erez Lapid for making his note on the subject available to me ([**29**] (1)). Lapid gives the following global interpretation of the work of Rallis and Mao. Here G is the two fold cover of the symplectic group in $2n$ variables $Sp(n)$ and G' is the split orthogonal group in $2n+1$ variables. Thus $Sp(n)$ operates on a symplectic vector space Y and G' on a quadratic vector space X. The Weil representation $\omega(g', g)$ is a representation of the product $G'(F_{\mathbb{A}}) \times G(F_{\mathbb{A}})$ on the space of Schwartz functions on $X \otimes Y_0(F_{\mathbb{A}})$, where Y_0 is a maximal isotropic subspace of Y. The Θ kernel is defined by

$$\Theta_{\Phi}(g', g) = \sum_{\xi \in X \otimes Y_0(F)} \omega(g', g)\Phi(\xi).$$

In fact Furusawa considers two models of the Weil representation ω^i, $i = 1, 2$. In the first model the orthogonal group G' operates linearly and in the second model the symplectic group G operates linearly. Thus we have in fact two equivalent representations ω^i, $i = 1, 2$.

Furusawa considers a subgroup H of G' (**Bessel subgroup**) with a character χ with the following property. Let ϕ' be an automorphic form on G'. Let Φ be a Schwartz function and set

$$\kappa_\Phi^1(\phi')(g) = \int \Theta_\Phi^1(g', g)\phi'(g')dg'.$$

Then

$$\mathcal{W}(\kappa_\Phi^1(\phi')) = \int_{R_\epsilon(F_\mathbb{A})\backslash G'(F_\mathbb{A})} \omega^1(g', 1)\Phi(\epsilon)\mathcal{P}(\rho(g)\phi')dg'. \tag{6}$$

Here ϵ is a suitable vector, R_ϵ is the stabilizer of ϵ in G' and $\mathcal{P}$ the period integral associated with (H, χ). Now choose a function f' on G' such that

$$\int_{R_\epsilon(F_\mathbb{A})} f'(rg')dr = \omega^1(g', 1)\Phi(\epsilon). \tag{7}$$

We ignore the fact that f' may not be a function of compact support. We also take for granted that any function may be obtained in this way. This should follow from the work of Rallis and Mao. Then

$$\mathcal{W}(\kappa_\Phi^1(\phi')) = \mathcal{P}(\rho(f')\phi'). \tag{8}$$

On the other hand let ϕ be a (genuine) automorphic form on G. Let Ψ be a Schwartz function and set

$$\kappa_\Psi^2(\phi)(g') = \int \Theta_\Psi^2(g', g)\phi(g)dg.$$

Then

$$\overline{\mathcal{W}'}(\kappa_\Psi^2(\phi)) = \int_{R_\eta(F_\mathbb{A})\backslash G(F_\mathbb{A})} \omega(1, g)\Psi(\eta)\overline{\mathcal{W}}(\rho(g)\phi)dg.$$

Here η is a suitable vector and R_η is its stabilizer in G. Let f be a function such that

$$\int_{R_\eta(F_\mathbb{A})} f(rg)dr = \omega^2(1, g)\Psi(\eta). \tag{9}$$

Then

$$\overline{\mathcal{W}'}(\kappa_\Psi^2(\phi)) = \overline{\mathcal{W}}(\rho(f)\phi). \tag{10}$$

Now let Φ and Ψ be Schwartz functions which correspond to one another in the isomorphism between ω^1 and ω^2. Let f' and f be functions satisfying (7) and (9) respectively. The corresponding kernels are equal: $\Theta_\Phi^1 = \Theta_\Psi^2$. It follows that

$$\int\int \Theta^1_\Phi(u',u)\theta'(u')du'\overline{\theta}(u)du = \int\int \Theta^2_\Psi(u',u)\theta'(u')du'\overline{\theta}(u)du\,. \qquad (11)$$

To continue, let us argue formally. Thus consider an orthonormal basis (ϕ') of the space of all automorphic forms on G' and consider the sum

$$\sum_{\phi'} \mathcal{W}(\kappa^1_\Phi(\phi'))\overline{\mathcal{W}'(\phi')}\,.$$

This is equal to

$$\sum_{\phi'} \int \Theta^1_\Phi(g',u)\phi'(g')dg'\overline{\theta}(u)du\overline{\int \phi'(u')\overline{\theta'}(u')du'}\,.$$

Since

$$\sum_{\phi'} \int \Theta^1_\Phi(g',u)\phi'(g')dg'\overline{\phi'(u')} = \Theta^1_\Phi(u',u)$$

this is actually equal to

$$\int\int \Theta^1_\Phi(u',u)\theta'(u')du'\overline{\theta}(u)du\,.$$

Similarly let us consider an orthonormal basis (ϕ) of the space of all genuine forms on G. Then

$$\sum_{\phi} \overline{\mathcal{W}'(\kappa^2_\Psi(\phi))}\overline{\overline{\mathcal{W}}(\phi)} =$$

$$\int\int \Theta^2_\Psi(u',u)\theta'(u')du'\overline{\theta}(u)du\,.$$

It follows from (11) that

$$\sum_{\phi'} \mathcal{W}(\kappa^1_\Phi(\phi'))\overline{\mathcal{W}'(\phi')} = \sum_{\phi} \overline{\mathcal{W}'(\kappa^2_\Psi(\phi))}\overline{\overline{\mathcal{W}}(\phi)}$$

or

$$\sum_{\phi'} \mathcal{P}(\rho(f')\phi')\overline{\mathcal{W}'(\phi')} = \sum_{\phi} \mathcal{W}(\rho(f)\phi)\overline{\mathcal{W}(\phi)}\,. \qquad (12)$$

At this point, the identity is only heuristic. However, one should be able to prove rigourously the following. Let π' an irreducible automorphic representation of G' and let π be its image under the θ correspondence. Then

$$\sum_{\phi'} \mathcal{P}(\pi'(f')\phi')\overline{\mathcal{W}'(\phi')} = \sum_{\phi} \mathcal{W}(\pi(f)\phi)\overline{\mathcal{W}(\phi)} \qquad (13)$$

where the sum is now over an orthonormal basis of π and π' respectively. We will discuss the possible use of this formula in a different context in the next section.

The identity (12) suggests that the identity (1) is satisfied. Indeed Mao and Rallis prove rigourously that it is so. Furthermore, they show that the correspondence between f and f' is geometric.

Further work of Mao and Rallis suggests these considerations apply to generalized dual pairs.

3. Relation with the descent

There are several examples where G' is a linear group and (H, χ) detects the existence of a pole for a Langlands $L-$ function. That is, a cuspidal representation π' of G' is distinguished by (H, χ) if and only if the (partial) $L-$function attached to π' has a pole at $s = 1$. Then π' should be the functorial image of a packet of representations of a classical group G, or more simply, the functorial image of a generic representation π of G. Conversely, every π should give rise to a π'. Ginzberg-Rallis-Soudry's technique of **descent** reconstructs π in terms of the residue of an appropriate Eisenstein series built out of π'.

It is natural to try to prove these results in terms of a relative trace formula (5). In the three examples below preliminary work had indicated the existence of such a formula with a geometric correspondence. Recently Mao and Rallis have improved on the backward method to establish the existence of a relative trace formula based on a correspondence which is not (at least not obviously) geometric .

Let G be as before the two fold cover of the symplectic group in $2n$ variables and let G' be the group $GL(2n)$. Let H be the group $GL(n) \times GL(n)$ viewed as a subgroup of G'. The character χ is trivial. The case $n = 1$ is a special case of the discussion in the previous section. In his thesis ([**32**] (2)) Mao checked an identity for the case $n = 2$ which would be the fundamental lemma of the trace formula in question. The $L-$ function at hand is the product $L(\frac{s}{2}, \pi')L(s, \bigwedge^2 \pi')$. At any rate, one possible application is as follows. Call $B(f)$ the distribution on the left hand side and $B'(f')$ the distribution on the right hand side of (5). The theory of representations of $L-$ functions gives a way to decompose $B'(f')$ into a product of local distributions $B'_v(f'_v)$: the local distribution is defined by

$$B'_v(f'_v) = \sum \mathcal{P}_v(\pi'_v(f'_v)\phi'_i)\mathcal{W}'_v(\phi'_i)$$

where $\mathcal{P}_v$ is a linear form invariant under H_v and $\mathcal{W}'_v$ a linear form transforming under the character θ'_v. Roughly speaking, the decomposition takes the form:

$$B'(f') = c \prod_v B'_v(f'_v)$$

where c is an explicit constant containing $L(1/2, \pi')$ and other factors. From the comparison, one will get a similar decomposition for $B(f)$. This will be a generalization of the fact that the square of the Fourier coefficients of a modular form of half integral weight is related in a simple way to $L(1/2, \pi)$. See **[2]** for a formulation of the case $n = 1$ along those lines. A similar discussion applies to the case considered in the previous section (**[29]** (1)).

Another example is the case where G is the quasi-split unitary group in $n-$variables relative to a quadratic extension E/F, $G' = GL(n, E)$ and $H = GL(n, F)$. The relevant $L-$function is the appropriate Asai $L-$function (**[4]** (1)). The case $n = 3$ was treated in **[6]** and **[10]**. In positive characteristic the techniques of Ngo (**[34]**) can be used to prove the fundamental lemma (Ngo, private communication).

Still another example is the case where $G = SO(2n+1)$ (or a closely related group) and $G' = GL(2n)$, the group H is the Shalika subgroup of matrices of the form

$$h = \begin{pmatrix} g & 0 \\ 0 & g \end{pmatrix} \begin{pmatrix} 1_n & X \\ 0 & 1_n \end{pmatrix}$$

and

$$\chi(h) = \psi(\mathrm{tr}(X)) .$$

A representation π' of $GL(2n)$ is distinguished by (H, χ) if and only if its exterior square $L-$function has a pole at $s = 1$. Conjecturally such π' is the functorial image of a packet of representations of G (or simply a generic representation π). The purpose of **[7]** (1) was to obtain evidence that this could be proven in terms of a relative trace formula. In fact the case $n = 2$ can be treated directly in terms of the Weil representations (without a relative trace formula)(**[19]**).

4. Factorization of periods

In all the previous examples the period $\mathcal{P}$ had a precise interpretation. In other words, the properties of the right hand side of (5) were well understood. The goal was to use (5) to investigate the properties of the left hand side. In the next example, whose study was initiated by Ye Yangbo in his thesis (**[36]**, (1)), the reverse is true. Namely let E/F be a quadratic extension and let $G' = GL(n, E)$, $G = GL(n, F)$. We denote by $z \mapsto \overline{z}$ the Galois conjugation. The quadratic base change associates to every cuspidal automorphic representation π of G an automorphic representation π' of $GL(n, E)$ which is in general cuspidal. It is conjectured that an automorphic cuspidal representation π' of G' is a base change if and only if it is distinguished by some form of the unitary group ([**25**]). Thus to study this case, in the right hand side of (5), one must consider a sum over all unitary groups.

It is then convenient to adopt a slightly different point of view. Let $S(n\times n)$ be the space of invertible $n \times n$ Hermitian matrices. The group $G(E) = GL(n, E)$ operates on $S(n \times n)$ by $s \mapsto {}^t\overline{g}sg$. If Ψ is in $\mathcal{C}^\infty(S(n \times n, F_{\mathbb{A}}))$ we construct a function $\Theta_\Psi(g)$ on $G(E)\backslash G(E_{\mathbb{A}})$ by

$$\Theta_\Psi(g) = \sum_{\xi\in S(n\times n,F)} \Psi({}^t\overline{g}\xi g)\,.$$

The invariant space spanned by the functions Θ_Ψ is the **automorphism spectrum** of the space of Hermitian matrices. The cuspidal automorphism representations which appear in the spectrum are exactly the cuspidal representations π which are **distinguished** by the stabilizer H of some point in $S(n \times n, F)$; thus H is a unitary group; as before, the distinction means that there is ϕ in the space of π such that

$$\int_{H(F)\backslash H(F_{\mathbb{A}})} \phi(h)dh \neq 0\,.$$

To prove the conjecture it is natural to compare with another similarly defined space of functions on $(G(F)\backslash G(F_{\mathbb{A}}))^2$. The group $GL(n, F) \times GL(n, F)$ operates on $GL(n, F)$ by $s \mapsto {}^tg_1sg_2$. To every function Φ in $\mathcal{C}^\infty(GL(n, F_{\mathbb{A}}))$ we associate a function $\Theta_\Phi(g_1, g_2)$ defined by

$$\Theta_\Phi(g_1, g_2) = \sum_{\xi\in GL(n,F)} \Phi({}^tg_1\xi g_2)\,.$$

We consider the invariant space spanned by these functions. The automorphism cuspidal representations $\pi = \pi_1 \otimes \pi_2$ which appear in the space are exactly those distinguished by the twisted diagonal subgroup $\{({}^tg^{-1}, g)\}$, that is, those π for which π_1 is contravariant to π_2.

As above, we let N be the space of triangular matrices with unit diagonal and θ a generic character of $N(F_{\mathbb{A}})$ trivial on $N(F)$. Then (1) is replaced by the identity

$$\int_{N(E)\backslash N(E_{\mathbb{A}})} \Theta_\Psi(n)\theta(n\overline{n})dn = \int_{N(F)\backslash N(F_{\mathbb{A}})^2} \Theta_\Phi(n_1, n_2)\theta(n_1n_2)dn_1dn_2\,.$$

There is a simple bijection of relevant orbits. In particular, the diagonal matrices with entries in F form a common system of representatives for the (relevant) orbits of maximum dimension.

We now describe the notion of **local matching** in the context of a quadratic unramified extension of local non Archimedean fields. Thus we let E/F be such an extension. We let η be the corresponding (unramified) quadratic character of $F^\times$. We assume the residual characteristic is not 2. We denote by $v(\bullet)$ the valuation of F. We let q be the cardinality of the residual field of F and set $|\, x\, |_F = q^{-v(x)}$. We let ψ be an additive character of F whose conductor is the ring of integers $\mathcal{O}_F$ of F. We let $\mathcal{P}_F$ be the maximal ideal in $\mathcal{O}_F$ and ϖ a generator of $\mathcal{P}_F$. We denote by dx the self dual Haar measure on F. We let N_n be the group of upper triangular matrices with unit diagonal in $GL(n)$. We define a character θ_n or simply $\theta : N_n(F) \to \mathbb{C}^\times$ by

$$\theta(u) = \psi\left(\sum_i u_{i,i+1}\right) .$$

Locally, it is best to consider orbital integrals for smooth functions of compact support on $M(n \times n, F)$. Let Φ be a such a function. We define the **Kloosterman integral**

$$\Omega(\Phi, \psi : a) = \int_{N_n(F)\times N_n(F)} \Phi({}^t u_1 a u_2)\theta(u_1 u_2) du_1 du_2$$

where

$$a = \operatorname{diag}(a_1, a_2, a_3, \ldots, a_n)$$

is a diagonal matrix with

$$a_i \in F^\times \,,\, 1 \leq i \leq n-1 \,,\; a_n \in F \,,$$

and du is the Haar measure such that by

$$\int_{N_n(\mathcal{O}_F)} du = 1 \,.$$

We often write $\Omega(\Phi, \psi : a)$ as a function of n variables

$$\Omega(\Phi, \psi : a_1, a_2, a_3, \ldots, a_n) \,.$$

Likewise, we define a character $u \mapsto \theta(u\overline{u})$ of $N_n(E)$ by

$$\theta(u\overline{u}) = \psi\left(\sum_i (u_{i,i+1} + \overline{u}_{i,i+1})\right) .$$

We let $H(n \times n, E/F)$ be the space of Hermitian matrices. Let Ψ be a smooth function of compact support on $H(n \times n, E/F)$. We define the **relative Kloosterman integral**

$$\Omega(\Psi, E/F, \psi : a) := \int_{N_n(E)} \Psi({}^t\overline{u} a u)\theta(u\overline{u}) du$$

where a is as above and du is the Haar measure such that

$$\int_{N_n(\mathcal{O}_E)} du = 1\,.$$

We say that Φ **matches** Ψ for ψ (see [**16**](7), (8))) and we write $\Phi \overset{\psi}{\leftrightarrow} \Psi$ if

$$\Omega(\Phi, \psi : a) = \gamma(a)\Omega(\Psi, E/F, \psi : a)\,,$$

where

$$\gamma(a) = \eta(a_1)\eta(a_1 a_2)\cdots\eta(a_1 a_2 \cdots a_{n-1})\,.$$

This identity implies similar identities for the other orbital integrals. The fundamental lemma takes then the following form ([**34**] for positive characteristic, [**16**] (10) for the general case). Let Φ_n be the characteristic function of $M(n \times n, \mathcal{O}_F)$ and Ψ_n be the characteristic function of $M(n\times n, \mathcal{O}_E) \cap H(n \times n, E/F)$. Then $\Phi_n \overset{\psi}{\leftrightarrow} \Psi_n$, that is,

$$\Omega(\Phi_n, \psi : a) = \gamma(a)\Omega(\Psi_n, E/F, \psi : a)\,.$$

The spectral expansion of the relative trace formula is given in [**29**],(2). In addition to a proof of the conjectural characterization of the base change, the relative trace formula at hand can be used to prove that the period integral over H (with H possibly replaced by the similitude group) factorizes into a product of local integrals.

5. Square of period integrals

In many cases it is conjectured that the square of a period integral $\mathcal{P}$ is, roughly speaking, the value of an L–function at a special point (e.g. the center of symmetry). Perhaps this conjecture can be formulated precisely and established in terms of a relative trace formula of the form:

$$\sum_i \mathcal{P}(\pi(f)\phi_i)\overline{\mathcal{P}(\phi_i)} = \sum_j \mathcal{P}_1(\pi'(f')\phi'_j)\overline{\mathcal{P}_2(\phi'_j)}\,. \tag{14}$$

For a simple case see [**18**]. This should apply to the case of a period $\mathcal{W}$ (See previously quoted references [**2**] and [**29**] (10).). Again in this setup one would know how to factorize the distribution on the right as a product of local distributions times the special value of an L–function and the goal would be to obtain a similar factorization of the left-hand side.

Still another variant should hold when H is the group of fixed points of an automorphism of order 2 and χ is trivial. The putative relative trace formula takes the form

$$\sum_i \mathcal{P}(\pi'(f)\phi'_i)\overline{\mathcal{P}(\phi'_i)} = \sum_j \int \pi(f)\phi_j(x)\overline{\phi_j(x)}E(x)dx\,. \tag{15}$$

Here E is a suitable Eisenstein series or a closely related function. Roughly speaking the group G is related to G' by the condition that there be a bijection between the double cosets of H in G' and the conjugacy classes of G. See **[21]**, **[22]**, **[30]**. To make the statement more precise, let us consider again the situation discussed in the previous section. Thus π' is an automorphic cuspidal representation of $GL(n,E)$ which is the base change of an automorphic representation π of $GL(n,F)$. The period $\mathcal{P}$ is the one defined by integration over a unitary group (or possibly a similitude group). The Eisenstein series has the following form. Set

$$E_\Phi(x,s) = \int_{F_{\mathbb{A}}^\times/F^\times} \sum_{\xi\in F^n-0} \Phi(\xi t g) \mid t \mid^{ns} d^\times t \mid \det g \mid^s \,.$$

Let η be the quadratic character of F attached to E. Then an integral of the form:

$$\int \pi(f)\phi_j(x)\overline{\phi_j(x)}\eta(\det x)E_\Phi(x,s)dx$$

is an holomorphic multiple of $L(s,\pi\otimes\eta,\tilde{\pi})$. Since π is not equivalent to $\eta\otimes\pi$, this integral is holomorphic even at $s=1$. Thus the right hand side of (15) is to be interpreted as being the value at $s=1$ of

$$\sum_j \int \pi(f)\phi_j(x)\overline{\phi_j(x)}\eta(x)E_\Phi(x,s)dx\,.$$

The ultimate consequence of such a formula should be that the left hand side should be expressible as the product of local distributions times $L(1,\pi\otimes\eta,\tilde{\pi})$ (and other explicit factors).

A simpler variation on the same idea is the case of two pairs (G,H) and (G',H') where H,H' are the groups of fixed points of involutive automorphisms of G and G' and, in an appropriate sense, (G,H) is an inner form of (G',H'). The trace formula in question takes the form

$$\sum_i \mathcal{P}(\pi(f)\phi_i)\overline{\mathcal{P}(\phi_i)} = \sum_j \mathcal{P}'(\pi'(f')\phi'_j)\overline{\mathcal{P}'(\phi'_j)}\,. \tag{16}$$

It can be used to compare distinction for representations π and π' corresponding to one another. See **[17]**, **[28]**, **[12]**, **[5]**.

For the local analogue of the condition of distinction see **[12]**, **[13]**, **[14]**.

References

[1] J. ARTHUR AND L. CLOZEL, *Simple algebras, base change, and the advanced theory of the trace formula*, Ann. Math. Studies, N0 **120**, Princeton Univ. Press 1989.

[2] E.M. BARUCH, Z. MAO, (1) *Bessel identities in the Waldspurger correspondence over a p-adic field.*, Amer. J. Math. 125 (2003), no. 2, 225–288. (2) *Bessel identities in the Waldspurger correspondence, The Archimedean Theory* preprint. (3) *Central values of Automorphic L-functions*, preprint.

[3] D. BUMP, S. FRIEDBERG, *The exterior square automorphic L-functions on* GL(n), Festschrift in honor of I. I. Piatetski-Shapiro on the occasion of his sixtieth birthday, Part II (Ramat Aviv, 1989), 47–65, Israel Math. Conf. Proc., 3, Weizmann, Jerusalem, 1990.

[4] Y. FLICKER (1) *Automorphic forms with anisotropic periods on a unitary group*, J. Algebra 220 (1999), no. 2, 636–663. (2) *On distinguished representations*, J. Reine Angew. Math. 418 (1991), 139–172. (3) *Cyclic automorphic forms on a unitary group*, J. Math. Kyoto Univ. 37 (1997), no. 3, 367–439.

[5] Y. FLICKER, J. HAKIM, *Quaternionc distinguished representations*, Americ. J. Math. 116 (1994), 683-736.

[6] S. FRIEDBERG, SOLOMON, S. GELBART, H. JACQUET, J. ROGAWSKI *Représentations génériques du groupe unitaire à trois variables* (French) [Generic representations for the unitary group in three variables], C. R. Acad. Sci. Paris Sr. I Math. 329 (1999), no. 4, 255–260.

[7] S. FRIEDBERG, H. JACQUET, (1) *The fundamental Lemma for the Shalika subgroup of $GL(4)$*, Memoirs AMS, Vol **124** no 594, 1996. (2) *Linear periods*, J. Reine Angew. Math. 443 (1993), 91–139.

[8] M. FURUSAWA, *On the theta lift from SO_{2n+1} to $\tilde{Sp}_n$*. J. Reine Angew. Math. 466 (1995), 87–110.

[9] M. FURUSAWA, J. SHALIKA (1) *On inversion of the Bessel and Gelfand transforms*, Trans. Amer. Math. Soc. 354 (2002), no. 2, 837–852 (electronic). (2) *The fundamental lemma for the Bessel and Novodvorsky subgroups of* GSp(4). II. C. R. Acad. Sci. Paris Sr. I Math. 331 (2000), no. 8, 593–598. (3) *The fundamental lemma for the Bessel and Novodvorsky subgroups of* GSp(4), C. R. Acad. Sci. Paris Sr. I Math. 328 (1999), no. 2, 105–110.

[10] S. GELBART, H. JACQUET, J. ROGAWSKI, *Generic representations for the unitary group in three variables*, Israel J. Math. 126 (2001), 173–237.

[11] J. GUO (1) *Uniqueness of generalized Waldspurger model for* GL($2n$), Pacific J. Math. 180 (1997), no. 2, 273–289. (2) *On a generalization of a result of Waldspurger*, Canad. J. Math. 48 (1996), no. 1, 105–142. (3) *On the positivity of the central critical values of automorphic L-functions for* GL(2), Duke Math. J. 83 (1996), no. 1, 157–190.

[12] J. HAKIM (1) *Admissible distributions on p-adic symmetric spaces*, J. Reine Angew. Math. 455 (1994), 1–19. (2) *Character relations for distinguished rep-

resentations, Amer. J. Math. 116 (1994), no. 5, 1153–1202. (3) *Howe/Kirillov theory for p-adic symmetric spaces*, Proc. Amer. Math. Soc. 121 (1994), no. 4, 1299–1305. (4)*Distinguished p-adic representations*, Duke Math. J. 62 (1991), no. 1, 1–22. (5) *Sur le caractre relatif*, (French) [On the relative character] C. R. Acad. Sci. Paris Sr. I Math. 310 (1990), no. 5, 243–244.

[13] J. HAKIM, Z. MAO. (1) *Cuspidal representations associated to* (GL(n), O(n)) *over finite fields and p-adic fields*, J. Algebra 213 (1999), no. 1, 129–143. (2) *Supercuspidal representations of* GL(n) *distinguished by a unitary subgroup*, Pacific J. Math. 185 (1998), no. 1, 149– 162.

[14] J. HAKIM, F. MURNAGHAN (1) *Supercuspidal representations of* GL(n) *distinguished by a unitary group*, Compositio Math. 133 (2002), no. 2, 199–244. (2) *Two types of distinguished supercuspidal representations*, Int. Math. Res. Not. 2002, no. 35, 1857–1889. (3) *Globalization of distinguished supercuspidal representations of* GL(n), Canad. Math. Bull. 45 (2002), no. 2, 220–230.

[15] G. HARDER, R.P. LANGLANDS, M. RAPOPORT *Algebraische Zyklen auf Hilbert-Blumenthal-Flchen*,(German) [Algebraic cycles on Hilbert-Blumenthal surfaces] J. Reine Angew. Math. 366 (1986), 53–120.

[16] H. JACQUET (1) *Sur un résultat de Waldspurger*, (French) [On a result of Waldspurger] Ann. Sci. cole Norm. Sup. (4) 19 (1986), no. 2, 185–229. (2) *Sur un résultat de Waldspurger. II*, (French) [On a result of Waldspurger. II] Compositio Math. 63 (1987), no. 3, 315–389. (3) *On the nonvanishing of some L-functions*, Proc. Indian Acad. Sci. Math. Sci. 97 (1987), no. 1-3, 117–155 (1988). (4) *The continuous spectrum of the relative trace formula for $GL(3)$ over a quadratic extension*, Israel J. Math. **89** (1995), 1-59. (5) *A theorem of density for Kloosterman integrals*, Asian J. Math. **2** (1998), 759-778. (6) *Transfert lisse d'intégrales de Kloosterman*, C.R. Acad. Sci. Paris. **335** (2002), p. 229-302. (7) *Smooth transfer of Kloosterman integrals*, Duke Math. J., 120 (2003), 121-152. (8) *Facteurs de Transfert pour les intégrales de Kloosterman*, [Transfer factors for Kloosterman integrals] C. R. Math. Acad. Sci. Paris 336 (2003), no. 2, 121–124. (9) *Factorization of period integrals*, J. Number Theory 87 (2001), no. 1, 109–143. (10) *Kloosterman identities over a quadratic extension*, to appear in Annals Math.

[17] H. JACQUET, K. LAI, *A relative trace formula*, Compositio Math. 54 (1985), 243–310.

[18] H. JACQUET, NAN CHEN *Positivity of quadratic base change L-functions*, Bull. Soc. Math. France 129 (2001), no. 1, 33–90.

[19] H. JACQUET, I. PITATESKI-SHAPIRO, J. SHALIKA *Unwritten manuscript.*

[20] H. JACQUET AND S. RALLIS (1) *Kloosterman integrals for skew symmetric matrices*, Pacific J. Math. 154 (1992), no. 2, 265–283. (2) *Uniqueness of linear periods*, Compositio Math. 102 (1996), no. 1, 65–123. (3) *Symplectic periods*, J. Reine Angew. Math. 423 (1992), 175–197.

[21] H. JACQUET, K. LAI, S. RALLIS, *A trace formula for symmetric spaces*, Duke Math. J. 70 (1993), 305–372.

[22] H. JACQUET, E. LAPID, S. RALLIS, *A spectral identity for skew symmetric matrices*, in A supplemental volume to the Americ. Math. J.. Contributions to Automorphic Forms, Geometry and Number Theory. To appear.

[23] H. JACQUET, E. LAPID, J. ROGAWSKI *Periods of automorphic forms*, J. Amer. Math. Soc. 12 (1999), no. 1, 173–240.

[24] H. JACQUET, J. SHALIKA *Exterior square L-functions*, in Automorphic forms, Shimura varieties, and L-functions, Vol. II (Ann Arbor, MI, 1988), 143–226, Perspect. Math., 11, Academic Press, Boston, MA, 1990.

[25] H. JACQUET, Y. YE (1) *Relative Kloosterman integrals for $GL(3)$*, Bull. Soc. Math. France **120** (1992), 449–558. (2) *Distinguished Representations and quadratic base change for $GL(3)$*, Trans. Amer. Math. Soc. **348** (1996), 913-939. (3) H. JACQUET AND Y. YE, *Germs of Kloosterman integrals for $GL(3)$*, Trans. Amer. Math. Soc. **351** (1999), 1227-1255.

[26] D. JIANG (1) *Nonvanishing of the central critical value of the triple product L-functions*, Internat. Math. Res. Notices 1998, no. 2, 73–84. (2) *On Jacquet's conjecture: the split period case*, Internat. Math. Res. Notices 2001, no. 3, 145–163.

[27] D. JIANG, Z. MAO, S. RALLIS *A relative Kuznietsov trace formula on G_2*, Manuscripta Math. 99 (1999), no. 3, 411–423.

[28] K.F. LAI *Algebraic cycles on compact Shimura surface*, Math. Z. 189 (1985), no. 4, 593–602.

[29] E. LAPID (1) *Untitled preprint.* (2) *On the fine expension of Jacquet trace formula*, preprint.

[30] E. LAPID, S. RALLIS, (1) *Positivity of $L(\frac{1}{2}, \pi)$ for symplectic representations*, C. R. Math. Acad. Sci. Paris 334 (2002), no. 2, 101–104. (2) *On the non-negativity of $L(1/2, \pi)$ for $SO(2n+1)$*, to appear in Annals Math. (3) *A spectral identity between symmetric spaces*, preprint.

[31] E. LAPID AND J. ROGAWSKI, (1) *Periods of Eisenstein series: the Galois case*, to appear in the Duke Math. J. (2) *Periods of Eisenstein series*, C. R. Acad. Sci. Paris Sr. I Math. 333 (2001), no. 6, 513–516. (3) *Stabilization of periods of Eisenstein series and Bessel distributions on* GL(3) *relative to* U(3), Doc. Math. 5 (2000), 317–350 (electronic).

[32] Z. MAO, (1) *A fundamental lemma for metaplectic correspondence*, J. Reine Angew. Math. 496 (1998), 107–129. (2) *Kloosterman sums, and Sali sums*. J. Number Theory 65 (1997), no. 2, 316–320. (3) *Relative Kloosterman integrals for* GL(3), III. Canad. J. Math. 45 (1993), no. 6, 1211–1230. (4) *Sur les sommes de Sali relatives.* [Relative Salie sums] C. R. Acad. Sci. Paris Sr. I Math. 316 (1993), no. 12, 1257–1262. (5) *Relative Kloosterman integrals for the unitary group.* C. R. Acad. Sci. Paris Sr. I Math. 315 (1992), no. 4, 381–386.

[33] Z. MAO AND S. RALLIS, (1) *Howe duality and the trace formula*, Pacific J. Math. 191 (1999), no. 2, 329–346. (2) *A trace formula for dual pairs*, Duke Math. J. 87 (1997), no. 2, 321–341. (3) *The induced Weil representation and the Shalika period*, Pacific J. Math. 191 (1999), no. 2, 313–327. (4) *Jacquet modules*

of the oscillator representation and families of relative trace identities, preprint. (5) *Lifting from $\tilde{Sp}_n$ to $GL(2n)$*, Preprint. (6) *Global method in relative trace identities* in A supplemental volume to the Americ. Math. J.. Contributions to Automorphic Forms, Geometry and Number Theory. To appear.

[34] B.C. NGO, (1) *Le lemme fondamental de Jacquet et Ye en caractéristique positive*, Duke Math. J. **96** (1999), 473-520. (2) *Faisceaux pervers, homomorphisme de changement de base et lemme fondamental de Jacquet et Ye*, Ann. Sci. Ecole Norm. Sup. **32** (1999), 619-679.

[35] J,-L., WALDSPURGER (1) *Sur les valeurs de certaines fonctions L automorphes en leur centre de symétrie*, (French) [Values of certain automorphic *L*-functions at their center of symmetry] Compositio Math. 54 (1985), no. 2, 173–242. (2) *Quelques propriétés arithmétiques de certaines formes automorphes sur* GL(2), (French) [Some arithmetical properties of certain automorphic forms on GL(2)] Compositio Math. 54 (1985), no. 2, 121–171. (3) *Correspondances de Shimura*, (French) [Shimura correspondences] Proceedings of the International Congress of Mathematicians, Vol. 1, 2 (Warsaw, 1983), 525–531, PWN, Warsaw, 1984. (4) *Sur les coefficients de Fourier des formes modulaires de poids demi-entier*, (French) [On the Fourier coefficients of modular forms of half-integral weight] J. Math. Pures Appl. (9) 60 (1981), no. 4, 375–484. (4) *Correspondance de Shimura*, (French) [The Shimura correspondence] J. Math. Pures Appl. (9) 59 (1980), no. 1, 1–132. (5) *Correspondances de Shimura et quaternions*, (French) [Shimura correspondences and quaternions] Forum Math. 3 (1991), no. 3, 219–307.

[36] Y. YE, (1) Ye *Kloosterman integrals and base change for* GL(2), J. Reine Angew. Math. 400 (1989), 57–121. (2) *The fundamental lemma of a relative trace formula for $GL(3)$*, Compositio Math. **89** (1993), 121-162. (3) *An integral transform and its applications*, Math. Ann. **30** (1994), 405-417.

Hervé Jacquet
Department of Mathematics
Columbia University
2990 Broadway, 509 Mathematics Building
Mail Code: 4406, New York, NY 10027, USA
hj@math.columbia.edu

On first occurrence in the local theta correspondence

Stephen S. Kudla and Stephen Rallis†*

Abstract. This paper discusses a conservation conjecture for the first occurrence indices. Such indices record the first occurrence of an irreducible admissible representation π of a fixed group G in the local theta correspondence as the second member of a dual pair (G, G') runs over the isometry groups in a Witt tower. For example, for a fixed nonarchimedean local field F and for a fixed quadratic character χ of $F^\times$, there are two Witt towers $\{V_m^+\}$ and $\{V_m^-\}$, where the quadratic space $V_m^\pm$ has dimension m, character χ and Hasse invariant ± 1. If $m_\chi^\pm(\pi)$ is the dimension of the space $V_m^\pm$ for which the irreducible admissible representation π of $\mathrm{Sp}_n(F)$ first occurs in the theta correspondence for the dual pair $(Sp_n(F), O(V_m^\pm))$, then we conjecture that

$$m_\chi^+(\pi) + m_\chi^-(\pi) = 4n + 4.$$

We prove this conjecture in many cases, in particular, when π is a supercuspidal representation.

1991 Mathematics Subject Classification: primary 11F27; secondary 22E50.

0. Introduction

A basic step in understanding the local theta correspondence is the description of the behavior of the correspondence as one group in the reductive dual pair varies in a Witt tower, **[18]**. The combination of this Witt tower technique, with the method of 'doubling', **[4]**, **[18]**, and the local 'doubling' zeta integrals, **[16]**, **[17]**, lies behind much of our qualitative understanding of the correspondence. In this paper[1], in the case of a nonarchimedean local field, we conjecture a 'conservation relation' between the first occurrence indices in a pair of related Witt towers and prove it in many cases.

*Partially supported by NSF Grant DMS-0200292 and by a Max-Planck Research Prize from the Max-Planck Society and Alexander von Humboldt Stiftung.

†Partially supported by NSF Grant DMS 0140422.

[1] which has languished in a drawer for nearly 10 years, due to the idolence of the first author

Suppose that F is a nonarchimedean local field of characteristic zero. Two quadratic spaces V and V' over F are said to lie in the same Witt tower if they are isomorphic after the addition of a suitable number of hyperbolic planes. The Witt tower of V is then determined by the discriminant character $\chi = \chi_V$ and the Hasse invariant $\epsilon = \epsilon(V) = \pm 1$, and, of course, the parity of the dimension of V. If χ, a quadratic character of $F^\times$, and a parity have been fixed, there are two associated Witt towers $\{V_m^+\}$ and $\{V_m^-\}$, where $\chi_{V_m^\pm} = \chi$, $\epsilon(V_m^\pm) = \pm 1$, and $\dim_F(V_m^\pm) = m$. Let $G'_n = \mathrm{Sp}(W_n)$ be the symplectic group of rank n over F, in the case of even parity, or $\mathrm{Mp}(W_n)$, its metaplectic extension, in the case of odd parity. Let $\mathrm{Irr}(G'_n)$ be the set of isomorphism classes of irreducible admissible (genuine) representations of G'_n. For $\pi \in \mathrm{Irr}(G'_n)$, the first occurrence index $m_\chi^\pm(\pi)$ is the smallest dimension m for which π occurs in the local theta correspondence for the dual pair $(G'_n, O(V_m^\pm))$. It is known that π then occurs in the correspondence for any pair $(G'_n, O(V_m^\pm))$ with $m \geq m_\chi^\pm(\pi)$, and that, by Howe's stable range condition, $m_\chi^\pm(\pi) \leq m_0 + 2n$, where $m_0 = m_0(\pm, \chi)$ is the dimension of the smallest quadratic space in the tower $\{V_m^\pm\}$. The **conservation relation** can be stated as follows.

Conjecture A. *For any* $\pi \in \mathrm{Irr}(G'_n)$,

$$m_\chi^+(\pi) + m_\chi^-(\pi) = 4n + 4.$$

In the case of odd parity, Conjecture A implies the following **theta dichotomy** phenomenon:

Conjecture B. *Fix a quadratic character* χ. *Any* $\pi \in \mathrm{Irr}(G'_n)$ *occurs in the theta correspondence with precisely one of the spaces* V_{2n+1}^+ *and* V_{2n+1}^-.

In fact, this dichotomy is controlled by a local root number, cf. **[3]** in the unitary group case, and this phenomenon leads to interesting global arithmetic related to central values and derivatives of L-functions, **[8]**. We will not discuss these aspects further in the present paper, see, however, **[11]**.

Conjecture A asserts that if a representation π occurs 'early' in one tower, it must occur late in the other. The most extreme example of this is the case of the trivial representation $\mathbb{1}$ of $\mathrm{Sp}(W_n)$, which we view (formally) as the Weil representation of the zero dimensional space at the bottom of Witt tower of split quadratic space $\{V_m^+\}$ for the trivial character $\chi = \chi_0 = 1$. Thus, $m_{\chi_0}^+(\mathbb{1}) = 0$. On the other hand, we show:

Theorem 1. *The conservation relation holds for* $\pi = \mathbb{1}$ *and for any quadratic character* χ. *In fact,* $m_{\chi_0}^-(\mathbb{1}) = 4n + 4$, *and* $m_\chi^\pm(\mathbb{1}) = 2n + 2$, *for* $\chi \neq \chi_0$.

This result is actually an essential intermediate step in obtaining the following criterion in the general case. Let

$$\iota : G'_n \times G'_n \to G'_{2n} \tag{0.1}$$

be the doubling embedding, and let $I'_{2n}(s,\chi)$ be the degenerate principal series representation of G'_{2n} defined by the character χ and $s \in \mathbb{C}$.

Theorem 2. *Fix a quadratic character χ and a parity. Suppose that π is an irreducible admissible representation of G'_n such that*

$$\dim \operatorname{Hom}_{G'_n \times G'_n}(I_{2n}(s_0,\chi), \pi \otimes \pi^\vee) = 1,$$

for all s_0 in the set:

$$\mathrm{Crit} = \begin{cases} \{ -n-\frac{1}{2}, -n+\frac{1}{2}, \ldots, n-\frac{1}{2}, n+\frac{1}{2} \} & \text{for even parity and } \chi = 1, \\ \{ -n+\frac{1}{2}, \ldots, n-\frac{1}{2} \} & \text{for even parity and } \chi \neq 1, \\ \{ -n, -n+1, \ldots, n-1, n \} & \text{for odd parity.} \end{cases}$$

Then

$$m^+_\chi(\pi) + m^-_\chi(\pi) = 4n+4.$$

Of course, many representations $\pi \in \mathrm{Irr}(G'_n)$ satisfy the condition of this Theorem. For example:

Corollary 3. *The conservation relation of Conjecture A holds for any supercuspidal representation $\pi \in \mathrm{Irr}(G'_n)$. In particular, in the metaplectic case, such representations satisfy theta dichotomy.*

Analogous conservations relations should hold for other dual pairs.

For example, for an irreducible admissible representation $\tau \in \mathrm{Irr}(O(V))$ of the orthogonal group of some quadratic space V over F, let $n(\tau)$ be the smallest n for which τ occurs in the theta correspondence for the dual pair $(O(V), G'_n)$. Again, we know that τ then occurs in the correspondence for the pair $(O(V), G'_n)$ for any $n \geq n(\tau)$, and that $n(\tau) \leq m = \dim_F(V)$. Let $\nu : O(V) \to F^\times/F^{\times,2}$ be an extension to $O(V)$ of the spinor norm on $SO(V)$. For any quadratic character χ of $F^\times$, let $\tilde{\chi} = \chi \circ \nu$ be the corresponding character of $O(V)$ and let $\mathrm{sgn} \cdot \tilde{\chi}$ be its twist by the sgn character. The analogue of the conservation relation in this case is the following.

Conjecture C. *For a fixed quadratic character χ and for any $\pi \in \mathrm{Irr}(O(V))$,*

$$n(\tilde{\chi} \cdot \tau) + n(\mathrm{sgn} \cdot \tilde{\chi} \cdot \tau) = \dim_F(V).$$

The analogous conjectures in the case of unitary groups are made in [3]. The methods of the present paper can be applied in the orthogonal and unitary cases as well. We hope to discuss the results elsewhere.

Since this paper was first written, a number of people have established conservation relations and dichotomy in various cases. For example, for unitary groups over $\mathbb{R}$, quite precise information about the correspondence was obtained by A. Paul, [15] and the conservation relations were verified[2]. In the nonarchimedean case, there is work of Muiç, [14], who proves the conservation relation for discrete series representations $\pi \in \mathrm{Irr}_{ds}(G'_n)$ in the case of even parity, using the work of Moeglin and Tadic [12].

The authors were aware of elementary cases of dichotomy, a number of which are contained in [18]. This lead to some of the problems studied below. In particular, the conservation conjecture stated above and its analogue for unitary groups were found by the first author while visiting the Tata Institute in Bombay in January 1994. A serious effort to try to prove these conjectures was undertaken jointly soon thereafter, and the present paper, which utilizes in large part the methods of [18], was essentially written by May of 1995. The results were summarized in the lecture notes [7] in 1996. The present volume seemed like a good occasion to finally finish the task of polishing it for publication. The first author would like to express his heartfelt appreciation to the second for all of the mathematics and encouragment he has shared so generously over the years.

1. Quotients of degenerate principal series

In this section we recall some of the basic facts about certain degenerate principal series representations and their restrictions via doubling.

Let F be a non-archimedean local field of characteristic 0, and let $W, <\ ,\ >$ be a non-degenerate symplectic vector space of dimension $2n$ over F. We fix an identification $W = F^{2n}$ (row vectors) such that

$$< w_1, w_2 > = w_1 J\,{}^t w_2, \qquad J = \begin{pmatrix} & 1_n \\ -1_n & \end{pmatrix},$$

and we view the symplectic group $G = G_n = Sp(W)$ as acting on W on the right. We let $e_1, \dots, e_n, e'_1, \dots, e'_n$ be the fixed basis of W. Let P_n be the stabilizer of the maximal isotropic subspace of W spanned by the rows of the matrix $(0, 1_n) \in M_{n,2n}(F)$, i.e., by the vectors $e'_1, \dots, e'_n$, and let

$$M = \{\ m(a) = \begin{pmatrix} a & \\ & {}^t a^{-1} \end{pmatrix} \mid a \in \mathrm{GL}_n(F)\ \},$$

and

$$N = \{\ n(b) = \begin{pmatrix} 1 & b \\ & 1 \end{pmatrix} \mid b = {}^t b \in \mathrm{M}_n(F)\ \},$$

so that $P_n = MN$ is a Levi decomposition of the Siegel parbolic P_n. Let $K = Sp_n(\mathcal{O})$ be the maximal compact subgroup stabilizing the lattice $\mathcal{O}^{2n}$ in

[2] Here there are many Witt towers and one must group them suitably.

W. In the Iwasawa decomposition $G = PK = NMK$, write $g = nm(a)k$, and note that the quantity

$$|a(g)| = |\det a|$$

is well defined, independent of the choice of $a \in \mathrm{GL}_n(F)$. Similarly, for any r, $1 \leq r \leq n$, let P_r be the maximal parabolic of G_n which stabilizes the isotropic r-plane spanned by the first r rows of the matrix $(0, 1_n)$, i.e., by the vectors $e'_1, \ldots, e'_r$. Then there is a Levi decomposition $P_r = M_r N_r$ where $M_r \simeq \mathrm{GL}_r \times G_{n-r}$ is the subgroup of P_r which stabilizes the span of the first r rows of the matrix $(1_n, 0)$ and where

$$N_r = \{ \begin{pmatrix} 1_r & x & z & y \\ & 1_{n-r} & {}^t y & 0 \\ & & 1_r & \\ & & -{}^t x & 1_{n-r} \end{pmatrix} \}.$$

Now let

$$\mathbb{W} = W + W_-,$$

where $W_- = W$ with symplectic form $- < , >$. Let $H = Sp(\mathbb{W})$, and note that, for a suitable choice of basis, we have $H \simeq G_{2n}$, the natural map

$$\iota : G_n \times G_n \longrightarrow H = G_{2n}$$

is given by

$$\iota(g_1, g_2) = \iota_0(g_1, g_2^\vee),$$

with

$$\iota_0 \left(\begin{pmatrix} a_1 & b_1 \\ c_1 & d_1 \end{pmatrix}, \begin{pmatrix} a_2 & b_2 \\ c_2 & d_2 \end{pmatrix} \right) = \begin{pmatrix} a_1 & & b_1 & \\ & a_2 & & b_2 \\ c_1 & & d_1 & \\ & c_2 & & d_2 \end{pmatrix},$$

and

$$g^\vee = \begin{pmatrix} 1 & \\ & -1 \end{pmatrix} g \begin{pmatrix} 1 & \\ & -1 \end{pmatrix}^{-1}.$$

Let $P = P_{2n} \subset H$ be the Siegel parabolic in H. For $s \in \mathbb{C}$ and for a (unitary) character χ of $F^\times$, let

$$I_{2n}(s, \chi) = I_P^H(\chi| \ |^s)$$

be the degenerate principal series representation of H consisting of smooth complex valued functions $\Phi(s)$ on H (sections) satisfying

$$\Phi(nm(a)h, s) = \chi(a)|a|^{s+\rho_{2n}} \Phi(h, s),$$

where we write $\chi(a)$ for $\chi(\det(a))$ and where $\rho_{2n} = n + \frac{1}{2}$. Such a section is called standard for its restriction to K is independent of s. Information about the points of reducibility of these representations and their composition series can be found in **[2]**, **[9]**, **[20]**.

The proof of the following uniqueness result plays a fundamental role. Its proof will be give in the next section.

Theorem 1.1. *Let $\mathbb{1}$ be the trivial representation of $G \times G$. For all $s \in \mathbb{C}$ and for all characters χ,*

$$\dim \mathrm{Hom}_{G\times G}(I_{2n}(s,\chi), \mathbb{1}) = 1.$$

Here $I_{2n}(s,\chi)$ is viewed as a representation of $G \times G$ via ι.

In fact, we believe the following:

Conjecture 1.2. *For all irreducible admissible representations π of G, and for all s and χ,*

$$\dim \mathrm{Hom}_{G\times G}(I_{2n}(s,\chi), \pi \otimes \pi^{\vee}) = 1.$$

In the remainder of this section, we review some of the facts which are involved in the proofs.

The geometry of the double cosets for P and $G\times G$ in H, or, equivalently, of the $G \times G$ orbits in the symplectic Grassmannian $P\backslash H$ of isotropic $2n$-planes in $\mathbb{W}$ is well known **[17]**, **[7]**. Recall that there is a double coset decomposition

$$H = \coprod_{r=0}^{n} P\delta_r \iota(G \times G),$$

where the image of

$$\Omega_r := P\delta_r\iota(G \times G)$$

in $P\backslash H$ consists of those isotropic $2n$-planes U in $\mathbb{W}$ whose intersections $U \cap W$ and $U \cap W_-$ have dimension r. Moreover, the closure of an orbit is given by

$$\overline{\Omega}_r = \coprod_{j\geq r} \Omega_j.$$

We write

$$\mathrm{St}_r := \iota^{-1}\Big((\delta_r^{-1}P\delta_r) \cap \iota(G \times G)\Big),$$

for the stabilizer of the point $P \cdot \delta_r$ in $P\backslash H$, and we choose the orbit representatives δ_r as follows. First, for the open orbit Ω_0, let

$$\delta_0 = \delta_0^{(n)} = \begin{pmatrix} 0 & & 1 & \\ & 1 & & 0 \\ -1 & 1 & 0 & \\ & 0 & 1 & 1 \end{pmatrix},$$

where the blocks have size $n \times n$. For this choice of orbit representative,

$$\mathrm{St}_0 = \Delta(G),$$

the diagonal in $G \times G$. For the closed orbit $\Omega_n = P\delta_n\iota(G \times G)$, we take $\delta_n = 1$, so that

$$\mathrm{St}_n = P_n \times P_n \ \subset \ G \times G,$$

where $P_n \subset G_n = G$ is the Siegel parabolic of G. Finally, for any r, let $W = W' + W''$ where W' is the span of the vectors $e_1, \ldots, e_r, e_1', \ldots, e_r'$ and W'' is the span of the vectors $e_{r+1}, \ldots, e_n, e_{r+1}', \ldots, e_n'$. Then consider the decomposition

$$\mathbf{W} = \mathbf{W}' + \mathbf{W}'',$$

with $\mathbf{W}' = W' + W_-'$ and $\mathbf{W}'' = W'' + W_-''$. There is a corresponding imbedding

$$Sp(\mathbf{W}') \times Sp(\mathbf{W}'') \longrightarrow H,$$

and we let δ_r be the image $1_{\mathbf{W}'} \times \delta_{0,\mathbf{W}''}$. Explicitly, this is

$$\delta_r = \begin{pmatrix} 1 & & & & 0 & & & \\ & 0 & & & & 1 & & \\ & & 1 & & & & 0 & \\ & & & 1 & & & & 0 \\ 0 & & & & 1 & & & \\ & -1 & & 1 & & 0 & & \\ & & 0 & & & & 1 & \\ & & & 0 & & 1 & & 1 \end{pmatrix}.$$

Note that this choice corresponds to an isotropic $2n$-plane in $\mathbf{W}$ which intersects W' and W_-' in isotropic r-planes and which is transverse to W'' and W_-''. It is then not difficult to check that

$$\mathrm{St}_r = (\mathrm{GL}_r \times \mathrm{GL}_r \times \Delta(G_{n-r})) \ltimes (N_r \times N_r) \ \subset \ P_r \times P_r.$$

Later it will be useful to know that, if $\gamma = (a_1, a_2, \begin{pmatrix} a & b \\ c & d \end{pmatrix}) \in \mathrm{St}_r$, with a_1 and $a_2 \in \mathrm{GL}_r$ and $\begin{pmatrix} a & b \\ c & d \end{pmatrix} \in G_{n-r}$, then

$$(1.1) \qquad \delta_r \iota(\gamma) \delta_r^{-1} = \begin{pmatrix} a_1 & & & & & & & \\ & d & & c & & -c & & \\ & & a_2 & & & & & \\ & b & & a & & & & -b \\ & & & & {}^t a_1^{-1} & & & \\ & & & & & a & & -b \\ & & & & & & {}^t a_2^{-1} & \\ & & & & & -c & & d \end{pmatrix}.$$

The stratification of H by the orbits Ω_r gives rise to a filtration

$$I_{2n}(s,\chi) = I_{2n}(s,\chi)^{(n)} \supset \cdots \supset I_{2n}(s,\chi)^{(r)} \supset \cdots \supset I_{2n}(s,\chi)^{(0)},$$

where $I_{2n}(s,\chi)^{(r)}$ is the set of functions in $I_{2n}(s,\chi)$ whose restriction to the closure of Ω_{r+1} vanishes. The successive quotients are then induced representations of $G \times G$ given by

$$\begin{aligned} Q_r(s,\chi) &:= I_{2n}(s,\chi)^{(r)} / I_{2n}(s,\chi)^{(r+1)} \\ &\simeq I_{P_r \times P_r}^{G \times G} \Big(\chi|\ |^{s+\frac{r}{2}} \otimes \chi|\ |^{s+\frac{r}{2}} \otimes S(G_{n-r}) \Big), \end{aligned}$$

where the induction is *normalized.* Here we are inducing from the representation of $P_r \times P_r$ in which $N_r \times N_r$ acts trivially, the factor $\mathrm{GL}_r \times \mathrm{GL}_r$ acts by the character $\chi|\ |^{s+\frac{r}{2}} \otimes \chi|\ |^{s+\frac{r}{2}}$, and the factor $G_{n-r} \times G_{n-r}$ acts by left-right multiplication on the Schwartz space $S(G_{n-r})$. In particular,

$$Q_0(s,\chi) \simeq S(G),$$

is a subrepresentation of $I_{2n}(s,\chi)$, and

$$Q_n(s,\chi) \simeq I_n(s + \frac{n}{2}, \chi) \otimes I_n(s + \frac{n}{2}, \chi)$$

is a quotient of $I_{2n}(s,\chi)$.

Definition 1.3. For χ fixed, an irreducible admissible representation π of G is said to *occur on the boundary at the point* s if

$$\mathrm{Hom}_{G \times G}(Q_r(s,\chi), \pi \otimes \pi^\vee) \neq 0,$$

for some $r > 0$.

Since the matrix coefficients of any irreducible admissible representation π of G lie in $C^\infty(G) = Ind_1^G(\mathbb{1})$, the contragradient of $S(G) = ind_1^G(\mathbb{1})$, it follows that every representation occurs with multiplicity one on the open cell, i.e.,

Lemma 1.4. *For any fixed χ and s, and for any irreducible admissible representation π of G,*

$$\dim \mathrm{Hom}_{G\times G}(Q_0(s,\chi), \pi\otimes\pi^\vee) = \dim \mathrm{Hom}_{G\times G}(S(G), \pi\otimes\pi^\vee) = 1.$$

When $r > 0$, we apply Frobenius reciprocity to obtain:

$$\begin{aligned}
&\mathrm{Hom}_{G\times G}(Q_r(s,\chi), \pi\otimes\pi^\vee)\\
&\quad\simeq \mathrm{Hom}_{G\times G}\Big(\pi^\vee\otimes\pi,\ I^{G\times G}_{P_r\times P_r}(\chi^{-1}|\ |^{-s-\frac{r}{2}}\otimes\chi^{-1}|\ |^{-s-\frac{r}{2}}\otimes C^\infty(G_{n-r}))\Big)\\
&\quad\simeq \mathrm{Hom}_{M_r\times M_r}((\pi^\vee)_{N_r}\otimes\pi_{N_r},\ \chi^{-1}|\ |^{-s-\frac{r}{2}}\otimes\chi^{-1}|\ |^{-s-\frac{r}{2}}\otimes C^\infty(G_{n-r})).
\end{aligned}$$

Here $M_r \simeq \mathrm{GL}_r\times\mathrm{GL}_r\times G_{n-r}\times G_{n-r}$. Just as in the case of the open orbit, if π_{N_r} is non-zero, the restriction of $(\pi^\vee)_{N_r}\otimes\pi_{N_r}$ to $G_{n-r}\times G_{n-r}$ always has a non-zero intertwining map to $C^\infty(G_{n-r})$. Thus, occurrence on the boundary is controlled by the spaces $\mathrm{Hom}_{\mathrm{GL}_r}((\pi^\vee)_{N_r}, \chi^{-1}|\ |^{-s-\frac{r}{2}})$ and $\mathrm{Hom}_{\mathrm{GL}_r}(\pi_{N_r}, \chi^{-1}|\ |^{-s-\frac{r}{2}})$.

Lemma 1.5. *An irreducible admissible representation π of G occurs on the boundary at the point s if and only if, for some r, with $1\le r\le n$,*

$$\mathrm{Hom}_{\mathrm{GL}_r}((\pi^\vee)_{N_r}, \chi^{-1}|\ |^{-s-\frac{r}{2}}) \neq 0$$

and

$$\mathrm{Hom}_{\mathrm{GL}_r}(\pi_{N_r}, \chi^{-1}|\ |^{-s-\frac{r}{2}}) \neq 0.$$

Next, we recall the basic doubling integral **[16]** associated to an irreducible admissible representation π of G. Let $\phi(g) = <\pi^\vee(g)\xi^\vee, \xi>$, for $\xi\in\pi$ and $\xi^\vee\in\pi^\vee$, be a matrix coefficient of $\pi^\vee$. Then, for any section $\Phi(s)\in I_{2n}(s,\chi)$, let

$$Z(s,\chi,\phi,\Phi) := \int_G \phi(g)\,\Phi(\delta_0\,\iota(g,1),s)\,dg.$$

In **[16]** it is proved that this integral converges for $\mathrm{Re}(s)$ sufficiently large, and that, for fixed data ϕ and $\Phi(s)$, with $\Phi(s)$ standard, $Z(s,\chi,\phi,\Phi)$ has a meromorphic continuation to the whole s plane and is given by a rational function of q^{-s}. Moreover, the family of such functions obtained by varying ϕ and $\Phi(s)$ has a gcd of the form $f(q^{-s})^{-1}$ for some polynomial $f(x)$. Since the representation of $G\times G$ on $I_{2n}(s,\chi)$ is smooth, $Z(s_0,\chi,\cdot,\cdot)$ defines an element of

$$Z(s_0,\chi)\ \in\ \mathrm{Hom}_{G\times G}(I_{2n}(s_0,\chi), \pi\otimes\pi^\vee)$$

at any value s_0 for which $f(q^{-s_0}) \neq 0$. In general, the leading term of the Laurent expansion of $Z(s,\chi,\cdot,\cdot)$ at $s = s_0$ defines an element

$$Z^*(s_0,\chi) \in \mathrm{Hom}_{G\times G}(I_{2n}(s_0,\chi),\pi\otimes\pi^\vee).$$

Next suppose that χ is unramified and note that there is, then, a unique standard section $\Phi^0(s)$ determined by the condition $\Phi(k,s) = 1$ for all $k \in K_H$. Suppose that $\pi = \pi(\lambda)$ is the unramified representation of G with Satake parameter

$$\mathrm{diag}(q^{\lambda_1},q^{\lambda_2},\dots,q^{\lambda_n},1,q^{-\lambda_1},q^{-\lambda_2},\dots,q^{-\lambda_n})\times Fr \in \mathrm{SO}(2n+1,\mathbb{C})\times W_F = {}^LG.$$

Let

$$\phi^0(g) = <\pi^\vee(g)\xi_0^\vee,\xi_0>,$$

be the unramified matrix coefficient of $\pi^\vee$, with $\phi^0(e) = 1$. Here ξ_0 (resp. $\xi_0^\vee$) is a K-fixed vector in π (resp. $\pi^\vee$).

Proposition 1.6. (i) ([16], [17]) *For χ unramified,*

$$Z(s,\chi,\phi^0,\Phi^0) = \frac{1}{b_{2n}(s,\chi)}\cdot(1-\chi q^{-s-\frac{1}{2}})^{-1}\prod_{j=1}^{n}(1-\chi q^{\lambda_j-s-\frac{1}{2}})^{-1}(1-\chi q^{-\lambda_j-s-\frac{1}{2}})^{-1},$$

where $\chi = \chi(\varpi)$ and

$$b_{2n}(s,\chi) = L(s+n+\frac{1}{2},\chi)\cdot\prod_{j=1}^{n}L(2s-2j+2n+1,\chi^2),$$

for $L(s,\chi) = (1-\chi\cdot q^{-s})^{-1}$.
(ii) *For any matrix coefficient ϕ of $\pi^\vee$ and any standard section $\Phi(s)$, the product*

$$\prod_{j=1}^{n}(1-\chi q^{\lambda_j-s-\frac{1}{2}})(1-\chi q^{-\lambda_j-s-\frac{1}{2}})\cdot Z(s,\chi,\phi,\Phi)$$

is an entire function of s.

The proof of part (ii) of this result, which is rather long, will be given in section 5.

2. Proof of Theorem 1.1

We now turn to the particular case $\pi = \mathbb{1}$. Throughout our discussion, we may assume that either $\chi = 1$ or χ is ramified, since any unramified χ could be absorbed into s via a shift. To avoid confusion, we frequently write χ_0 rather than 1 for the trivial character.

The ramified case, in which $\chi \neq 1$, is easy to handle.

Proposition 2.1. *If χ is ramified, then $\mathbb{1}$ does not occur on the boundary for any value of s. Moreover, for any standard section $\Phi(s)$, the zeta integral $Z(s,\chi,1,\Phi)$ is entire and*

$$\mathrm{Hom}_{G\times G}(I_{2n}(s,\chi),\mathbb{1}) = \mathbb{C}\cdot Z(s,\chi,1,\cdot)$$

for all s.

Proof. Applying the criterion of Lemma 1.5 for $\pi = \mathbb{1}$, we find

$$\begin{aligned}
&\dim \mathrm{Hom}_{G\times G}(Q_r(s,\chi),\mathbb{1})\\
&\quad = \dim \mathrm{Hom}_{M_r\times M_r}\Big(\mathbb{1}, \chi^{-1}|\ |^{-s+n-r+\frac{1}{2}} \otimes \chi^{-1}|\ |^{-s+n-r+\frac{1}{2}} \otimes C^{\infty}(G_{n-r})\Big)\\
&\quad = \dim \mathrm{Hom}_{\mathrm{GL}_r\times \mathrm{GL}_r}(\mathbb{1}, \chi^{-1}|\ |^{-s+n-r+\frac{1}{2}} \otimes \chi^{-1}|\ |^{-s+n-r+\frac{1}{2}}).
\end{aligned}$$

This is zero for all s if χ is ramified. Note that, if $\chi = \chi_0$, then this dimension is 1 if and only if $s = n - r + \frac{1}{2}$ (modulo $\frac{2\pi i}{\log q}\mathbb{Z}$, of course). This fact will be used below.

Now consider the Laurent expansion of the zeta integral at a point $s = s_0$, on the space of standard sections:

$$Z(s,\chi,1,\Phi) = \frac{\tau_{-k}(\chi,\Phi)}{(s-s_0)^k} + \cdots + \tau_0(\chi,\Phi) + \ldots .$$

Note that, as mentioned above, the leading term defines a $G \times G$-invariant linear functional

$$\tau_{-k} : I_{2n}(s_0,\chi) \longrightarrow \mathbb{C}.$$

If $k > 0$, so that the doubling zeta integral actually has a pole at s_0, then this functional restricts to zero on the subspace $I_{2n}(s_0,\chi)^{(0)}$, since the zeta integral is entire for sections in this subspace. It must then give rise to a non-zero element of $\mathrm{Hom}_{G\times G}(Q_r(s_0,\chi),\mathbb{1})$, for some $r > 0$. This contradicts the vanishing of these spaces, so that no pole can occur. The same argument shows that any non-zero element of $\mathrm{Hom}_{G\times G}(I_{2n}(s_0,\chi),\mathbb{1})$ must have non-zero restriction to $I_{2n}(s_0,\chi)^{(0)}$. Since the restriction of $Z(s_0,\chi,1)$ to $I_{2n}(s_0,\chi)^{(0)}$ is non-zero, and is hence a basis for $\mathrm{Hom}_{G\times G}(I_{2n}(s_0,\chi)^{(0)},\mathbb{1})$, we obtain the claimed result. □

For the rest of this section, we consider the more difficult case $\chi = \chi_0 = 1$. We first record some useful facts about the zeta integral in this situation.

Proposition 2.2. *Suppose that $\chi = \chi_0$ is trivial.*
(i)

$$Z(s,\chi_0,1,\Phi^0) = \frac{1}{b_{2n}(s,\chi_0)} \cdot (1-q^{-s-\frac{1}{2}})^{-1} \prod_{j=1}^{n} (1-q^{j-s-\frac{1}{2}})^{-1}(1-q^{-j-s-\frac{1}{2}})^{-1}.$$

In particular, this function has only simple poles, and these occur at the points

$$s_0 \in X(\mathbb{1}) := \{ \frac{1}{2}, \frac{3}{2}, \dots, n-\frac{1}{2} \}.$$

(ii) *If $s \notin X(\mathbb{1})$, then $\mathbb{1}$ does not occur on the boundary at s, i.e.,*

$$\mathrm{Hom}_{G\times G}(Q_r(s,\chi_0),\mathbb{1}) = 0$$

for $1 \le r \le n$.
(iii) *Suppose that $s_0 = n - r_0 + \frac{1}{2} \in X(\mathbb{1})$. Then*

$$\dim \mathrm{Hom}_{G\times G}(Q_r(s_0,\chi_0),\mathbb{1}) = \begin{cases} 1 & \text{if } r = r_0 \\ 0 & \text{otherwise.} \end{cases}$$

In particular, $\mathbb{1}$ occurs on a unique boundary component when $s_0 \in X(\mathbb{1})$.
(iv) *For any standard section $\Phi(s) \in I_{2n}(s,\chi)$, the function $Z(s,\chi_0,1,\Phi)$ has at most simple poles, and these can only occur at the points $s_0 \in X(\mathbb{1})$.*
(v) *Finally, if $s \notin X(\mathbb{1})$, then*

$$\mathrm{Hom}_{G\times G}(I_{2n}(s,\chi),\mathbb{1}) = \mathbb{C}\cdot Z(s,\chi_0,1,\cdot) \neq 0.$$

Proof. Part (i) is just a specialization of Proposition 1.6 (i), while parts (ii) and (iii) were proved in the course of the proof of Proposition 2.1. The argument about leading terms in that proof also implies that $Z(s,\chi_0,1,\cdot)$ cannot have a pole at any point $s \notin X(\mathbb{1})$. This together with (ii) of Proposition 1.6 proves (iv). Finally, the leading term arguments from the proof of Proposition 2.1 imply that (v) holds for $s \notin X(\mathbb{1})$. □

To complete the proof of Theorem 1.1, we must consider the case $\chi = \chi_0$ and $s_0 = n - r_0 + \frac{1}{2} \in X(\mathbb{1})$, so that $\mathbb{1}$ occurs on a unique boundary component. The basic idea is to show that the $G \times G$–invariant functional on this boundary component extends to a functional on all of $I_{2n}(s_0)$, while the invariant functional defined on the space of functions $S(G)$ on the open cell *does not extend.* The first step is to compose invariant functionals with the projection from $S(H)$ to $I_{2n}(s,\chi)$. This allows us to view linear functionals on $I_{2n}(s,\chi)$ as distributions on $S(H)$ with certain invariance properties under $P \times G \times G$. Note that H can be stratified by $P \times G \times G$ orbits. The second step is to consider certain vector valued distributions whose (unipotent) transformation law under P arises from the Laurent expansion of the induced representation transformation law at s_0. The standard analysis of Jacquet functors allows us to show that, at a given critical point s_0, the space of such distributions is nonzero for a unique boundary component. The vector valued distributions constructed from the Laurent expansion of the pullback of the zeta integral give an explicit basis element for this space. Finally, by expressing the distribution associated to an arbitrary $G \times G$–invariant functional on a boundary component or on the open cell in terms of these explicitly constructed distributions, we obtain the claimed extension/non-extension properties.

We begin by setting up a little more machinery.

For any s, there is a surjective map

$$\mathrm{pr}(s,\chi) : S(H) \longrightarrow I_{2n}(s,\chi),$$

given by

$$\mathrm{pr}(s,\chi)(\varphi)(h) := \int_P \chi(p)^{-1}|p|^{-s+\rho_{2n}}\,\varphi(ph)\,d_lp,$$

where d_lp is a left Haar measure on P, [1]. The kernel of this map is

$$\ker(\mathrm{pr}(s,\chi)) = \big\langle\ \ell(p)\varphi - \chi(p)^{-1}|p|^{-s+\rho_{2n}}\cdot\varphi \ \ \big|\ \ \text{for } p\in P \text{ and } \varphi\in S(H)\big\rangle.$$

Here the right hand side denotes the subspace of $S(H)$ spanned by these functions; note that it depends on s. Also, we have

$$\mathrm{pr}(s,\chi)(\ell(p)\varphi) = \chi(p)^{-1}|p|^{-s+\rho_{2n}}\cdot \mathrm{pr}(s,\chi)(\varphi),$$

where $\ell(p)\varphi(h) = \varphi(p^{-1}h)$. Composing this map with the zeta integral associated to the trivial representation, we define, for $\mathrm{Re}(s)$ sufficiently large,

$$z(s,\chi,\varphi) := \int_G\int_P \chi(p)^{-1}|p|^{-s+\rho_{2n}}\,\varphi(p\delta_0 i(g,1))\,d_lp\,dg. \tag{2.1}$$

This integral defines a distribution on $S(H)$ such that, for $p\in P$

$$z(s,\chi)(\ell(p)\varphi) = \chi(p)^{-1}|p|^{-s+\rho_{2n}}\cdot z(s,\chi)(\varphi),$$

and, for g_1 and $g_2 \in G$,

$$z(s,\chi)(r(g_1,g_2)\varphi) = z(s,\chi)(\varphi),$$

where $r(g_1,g_2)\varphi(h) = \varphi(h\cdot i(g_1,g_2))$.

The results about the doubling integral reviewed above imply that $z(s,\chi)$ has a meromorphic analytic continuation with at most simple poles at the points $s_0 \in X(\mathbb{1})$, in the case $\chi = \chi_0$.

Suppose that $\chi=\chi_0$ and write

$$z(s,\chi_0)(\varphi) = \frac{\mu_{-1}(\varphi)}{(s-s_0)} + \mu_0(\varphi) + \dots,$$

for the Laurent expansion at s_0. For any $m \geq 1$, let

$$\tilde{\mu}^{(m)}(\varphi) = \begin{pmatrix} \mu_{m-2}(\varphi) \\ \mu_{m-3}(\varphi) \\ \vdots \\ \mu_0(\varphi) \\ \mu_{-1}(\varphi) \end{pmatrix}, \tag{2.2}$$

so that $\tilde{\mu}^{(m)}$ is a distribution on H valued in $\mathbb{C}^m$.

For $x \in F^\times$, let

$$\gamma_r(x) = \frac{(\log|x|)^r}{r!}.$$

By writing out the Laurent expansion, at s_0, of both sides of the identity

$$z(s,\chi)(\ell(p)\varphi) = \chi(p)^{-1}|p|^{-s_0+\rho_{2n}}\left(|p|^{-(s-s_0)} \cdot z(s,\chi)(\varphi)\right),$$

we obtain the following transformation law for $\tilde{\mu}^{(m)}$.

Lemma 2.3. *For $p \in P$,*

$$\tilde{\mu}^{(m)}(\ell(p)\varphi) = \chi(p)^{-1}|p|^{-s_0+\rho_{2n}} \cdot \begin{pmatrix} 1 & \gamma_1(p) & \gamma_2(p) & \dots & \gamma_{m-1}(p) \\ 0 & 1 & \gamma_1(p) & \dots & \gamma_{m-2}(p) \\ 0 & 0 & 1 & \dots & \gamma_{m-3}(p) \\ \vdots & \vdots & & \ddots & \vdots \\ 0 & 0 & 0 & \dots & 1 \end{pmatrix} \cdot \tilde{\mu}^{(m)}(\varphi),$$

and

$$\tilde{\mu}^{(m)}(r(g_1,g_2)\varphi) = \tilde{\mu}^{(m)}(\varphi).$$

For convenience, we write $\sigma_m(p)$ for the matrix on the right hand side of the first identity of Lemma 2.3. For any $s_0 \in \mathbb{C}$ and χ, let

$$D^{(m)}(s_0,\chi) = \mathrm{Hom}_{P\times G\times G}(S(H), \chi^{-1}|\ |^{-s_0+\rho_{2n}}\sigma_m \otimes \mathbb{1} \otimes \mathbb{1}),$$

be the space of $\mathbb{C}^m$–valued distributions on H which satisfy:

$$\eta(\ell(p)\varphi) = \chi(p)^{-1}|p|^{-s_0+\rho_{2n}}\,\sigma_m(p) \cdot \eta(\varphi),$$

and

$$\eta(r(g_1,g_2)\varphi) = \eta(\varphi),$$

for all $p \in P$ and g_1, $g_2 \in G$.

We will only need to consider the case $m = 2$ when $s_0 \in X(\mathbb{1})$ and $\chi = \chi_0$. We write

$$\zeta_1 = \begin{pmatrix} \mu_{-1} \\ 0 \end{pmatrix} \qquad \text{and} \qquad \zeta_2 = \begin{pmatrix} \mu_0 \\ \mu_{-1} \end{pmatrix} \in D^{(2)}(s_0,\chi_0),$$

for the two vector valued distributions which arise from the Laurent expansion of the zeta integral.

Again considering the stratification of H, we obtain a filtration

$$S(H) = S(H)^{(n)} \supset \cdots \supset S(H)^{(r)} \supset S(H)^{(0)} = S(\Omega_0),$$

with successive quotients

$$S(H)^{(r)}/S(H)^{(r-1)} \simeq S(\Omega_r).$$

Proposition 2.4. *(i) The restriction of μ_0 to $S(\Omega_0) \simeq S(P) \otimes S(G)$ gives a basis for the space*

$$\mathrm{Hom}_{P\times G\times G}(S(\Omega_0), \chi^{-1}|\ |^{-s_0+\rho_{2n}} \otimes \mathbb{1} \otimes \mathbb{1}) = \mathbb{C} \cdot \mu_0.$$

(ii) For $1 \le r \le n$, and for $s_0 = n - r_0 + \frac{1}{2}$,

$$\dim \mathrm{Hom}_{P\times G\times G}(S(\Omega_r), \chi^{-1}|\ |^{-s_0+\rho_{2n}} \sigma_2 \otimes \mathbb{1} \otimes \mathbb{1}) = \begin{cases} 1 & \text{if } r = r_0, \\ 0 & \text{otherwise.} \end{cases}$$

Proof. By (2.1) and (2.2), part (i) just amounts to the uniqueness of the Haar measures.

Next we prove (ii). If $1 \le r \le n$, and using *unnormalized* induction, we have

$$\begin{aligned}
&\mathrm{Hom}_{P\times G\times G}\big(S(\Omega_r),\ \chi^{-1}|\ |^{-s_0+\rho_{2n}} \sigma_2 \otimes \mathbb{1} \otimes \mathbb{1} \big) \\
&\simeq \mathrm{Hom}_{P\times G\times G}\Big(ind_{St_r}^{P\times G\times G}(\mathbb{1}),\ \chi^{-1}|\ |^{-s_0+\rho_{2n}} \sigma_2 \otimes \mathbb{1} \otimes \mathbb{1} \Big) \\
&\simeq \mathrm{Hom}_{P\times G\times G}\Big(\chi|\ |^{s_0-\rho_{2n}} \sigma_2^\vee \otimes \mathbb{1} \otimes \mathbb{1},\ Ind_{\mathrm{St}_r}^{P\times G\times G}(|\ |^{-r} \otimes |\ |^{-r} \otimes \mathbb{1}) \Big) \\
&\simeq \mathrm{Hom}_{\mathrm{St}_r}\Big(\chi|\ |^{s_0-\rho_{2n}} \sigma_2^\vee \otimes \mathbb{1} \otimes \mathbb{1},\ |\ |^{-r} \otimes |\ |^{-r} \otimes \mathbb{1} \Big) \\
&\simeq \mathrm{Hom}_{\mathrm{GL}_r\times \mathrm{GL}_r}\big(|\ |^{s_0-\rho_{2n}+r} \otimes |\ |^{s_0-\rho_{2n}+r},\ \sigma_2 \big).
\end{aligned}$$

Here, in the last line, σ_2 denotes the pullback of the two dimensional representation of P via the imbedding of $\mathrm{GL}_r \times \mathrm{GL}_r$ into that group given in (1.1) above. In particular, for this embedding, the quantity $|p|$ pulls back to $|a_1||a_2|$. Also note that, in taking a contragradient, we have used the fact that

$$\frac{\Delta_{\mathrm{St}_r}}{\Delta_{P\times G\times G}} = |\ |^{-r} \otimes |\ |^{-r} \otimes \mathbb{1}$$

on St_r. The last space is zero unless $s_0 = n - r + \frac{1}{2}$, and in that case has dimension 1, as claimed.

□

In general, if $\eta \in D^{(2)}(s_0, \chi_0)$ restricts to zero on $S(H)^{(r-1)}$, then it induces a homomorphism

$$\bar{\eta} \in \mathrm{Hom}_{P\times G\times G}(S(\Omega_r), \chi^{-1}|\ |^{-s_0+\rho_{2n}} \sigma_2 \otimes 1\!\!1 \otimes 1\!\!1)$$

on the quotient of the next step of the filtration. Part (ii) of Proposition 2.4 forces the $\bar{\eta}$'s to vanish when $r \neq r_0$. Note that μ_{-1} and hence ζ_1 vanish on $S(\Omega_0)$, since the zeta integral has no poles for φ in this space. On the other hand, ζ_1 is not identically zero, since a pole occurs. Thus, its restriction to $S(H)^{(r_0)}$ is non-zero, and the resulting $\bar{\zeta_1}$ gives a basis for the space in (ii) of Proposition 2.4. These considerations yield the following result.

Corollary 2.5. *Suppose that* $s_0 = n - r_0 + \frac{1}{2} \in X(1\!\!1)$.
(i) If $\eta \in D^{(2)}(s_0, \chi_0)$ *restricts to zero on* $S(\Omega_0)$, *then* η *is zero on* $S(H)^{(r_0-1)}$.
(ii) If $\eta \in D^{(2)}(s_0, \chi_0)$ *restricts to zero on* $S(H)^{(r_0)}$, *then* $\eta = 0$.
(iii) The distribution ζ_1 *restricts to zero on* $S(H)^{(r_0-1)}$, *and its restriction to* $S(H)^{(r_0)}$ *gives a basis for the space*

$$\mathrm{Hom}_{P\times G\times G}(S(\Omega_{r_0}), \chi^{-1}|\ |^{-s_0+\rho_{2n}} \sigma_2 \otimes 1\!\!1 \otimes 1\!\!1) = \mathbb{C} \cdot \bar{\zeta}_1.$$

We now return to the proof of Theorem 1.1.
Now suppose that $s_0 \in X(1\!\!1)$ and that $\lambda \in \mathrm{Hom}_{G\times G}(I_{2n}(s_0, \chi_0), 1\!\!1)$. Let

$$\lambda' = \lambda \circ \mathrm{pr}(s_0, \chi_0) \ \in \ \mathrm{Hom}_{P\times G\times G}(S(H), |\ |^{-s_0+\rho_{2n}} \otimes 1\!\!1 \otimes 1\!\!1)$$

be the pullback of λ via the projection map, and let

$$\lambda'' = \begin{pmatrix} \lambda' \\ 0 \end{pmatrix} \in D^{(2)}(s_0, \chi_0).$$

By (i) of Proposition 2.4, there is a constant c such that

$$\lambda'|_{S(\Omega_0)} = c \cdot \mu_0|_{S(\Omega_0)}.$$

Since the zeta integral has no pole for $\varphi \in S(\Omega_0)$, μ_{-1} restricts to zero on this space. Thus, passing to elements of $D^{(2)}(s_0, \chi_0)$, we have

$$(\lambda'' - c \cdot \zeta_2)|_{S(\Omega_0)} = 0.$$

Set $\eta = \lambda'' - c\cdot\zeta_2$. By (i) and (iii) of Corollary 2.5, there is a constant c' such that $\eta - c' \cdot \zeta_1$ restricts to zero on $S(H)^{(r_0)}$, and hence, by (i) of Corollary 2.5, $\eta = c' \cdot \zeta_1$. But then, the second component, $-c \cdot \mu_{-1}$, of η, must be identically zero. This implies that $c = 0$, and that, in fact,

$$\lambda = c' \cdot \mu_{-1}.$$

This completes the proof of Theorem 1.1 and yields some extra information.

Corollary 2.6. *(i) For all values of s_0 and χ, the one dimensional space*

$$\mathrm{Hom}_{G\times G}(I_{2n}(s_0,\chi),\mathbb{1})$$

is spanned by the leading term of the Laurent expansion at s_0 of the zeta integral $Z(s,\chi,1)$.
(ii) If $\chi=\chi_0$ and $s_0\in X(\mathbb{1})$, then the space

$$\mathrm{Hom}_{P\times G\times G}(S(H),\chi^{-1}|\ |^{-s_0+\rho_{2n}}\sigma_2\otimes\mathbb{1}\otimes\mathbb{1})$$

has dimension 2 and is spanned by the Laurent vectors ζ_1 and ζ_2 defined using the first two terms of the Laurent expansion of the zeta integral $z(s,\chi_0)$ at s_0.

3. The local theta correspondence

In this section, we prove some basic results about 'early' and 'late' occurrence in the theta correspondence, for dual pairs of type (Sp,O). We consider both the symplectic and metaplectic case. We fix, once and for all, a nontrivial additive character ψ of F which we assume to be trivial on $\mathcal{O}_F$ and non-trivial on $\mathcal{P}_F^{-1}$, where $\mathcal{P}_F=\varpi\mathcal{O}_F$ is the maximal ideal of $\mathcal{O}_F$.

Let V, $(\ ,\ )$ be a vector space of dimension m over F with a non-degenerate symmetric bilinear form. Let

$$G'=G'_n=\begin{cases}Mp(W) & \text{if } m \text{ is odd}\\ Sp(W) & \text{if } m \text{ is even,}\end{cases}$$

where $W=W_n$ is a symplectic vector space of dimension $2n$, as in section 1. Here $Mp(W)$ denotes the metaplectic extension of $Sp(W)$, which we take to be the extension

$$1\longrightarrow\mathbb{C}^1\longrightarrow Mp(W)\longrightarrow Sp(W)\longrightarrow 1,$$

obtained from the usual 2-fold covering via the inclusion of $\{\pm1\}$ into $\mathbb{C}^1$.

In the symplectic group $Sp(V\otimes_F W)$ of the vector space $V\otimes_F W$, of dimension $2mn$ over F with symplectic form $(\ ,\ )\otimes<\ ,\ >$, we then have the usual reductive dual pair $(O(V),Sp(W))$. The homomorphism

$$i:O(V)\times Sp(W)\longrightarrow Sp(V\otimes_F W)$$

has a lifting to a homomorphism

$$\tilde{i}:O(V)\times G'\longrightarrow\mathbb{G}',$$

where

$$\mathbb{G}'=\begin{cases}Mp(V\otimes_F W) & \text{if } m \text{ is odd}\\ Sp(V\otimes_F W) & \text{if } m \text{ is even.}\end{cases}$$

The pullback to $G' \times O(V)$ of the Weil representation ω_ψ of $\mathbb{G}'$, yields a representation of this product group on the space $S(V^n)$, in the standard Schrödinger model. The lifting $\tilde{i}$ is unique provided we require that $h \in O(V)$ acts by $\omega(h)\varphi(x) = \varphi(h^{-1}x)$, and that, when m is odd, the central $\mathbb{C}^1$ in G' acts in $S(V^n)$ by $\omega(z)\varphi = z \cdot \varphi$.

If π is an irreducible admissible representation of G', we let $S(\pi)$ be the maximal quotient of $S = S(V^n)$ on which G' acts by a multiple of π, and we write

$$S(\pi) \simeq \pi \otimes \Theta(\pi, V).$$

The Howe duality principle is the following.

Theorem 3.1. (Howe, [4], Waldspurger, [22]) *Assume that the residue characteristic of F is odd.*
(i) $\Theta(\pi, V)$ is a finitely generated representation of $O(V)$.
(ii) If $S(\pi) \neq 0$, then $\Theta(\pi, V)$ has a unique irreducible quotient $\theta(\pi, V)$.
(iii) The correspondence $\pi \mapsto \theta(\pi, V)$ defines a bijection between the subsets of $Irr(G')$ and $Irr(O(V))$ which occur in the correspondence.

This duality principle is also expected to hold when the residue characteristic of F is even. Note that the space $\Theta(\pi, V)$ is defined, in any case. As in [3], we will make use of only the following weak version of (i):

Proposition 3.2. *For any irreducible admissible representation π of G' the space $\Theta(\pi, V)$ is either 0, or, as a representation of $O(V)$, it has finite length and hence has an irreducible admissible quotient.*

Note that there is no restriction on the residue characteristic here. This result is proved in Chapitre 3, Theorem 4, 2) a), p. 69 of [13], via the methods in [18] and [5].

We now consider the behavior of the correspondence with respect to Witt towers. For any integer $r \geq 1$, let $V_{r,r} = F^{2r}$ with inner product given by the matrix $\begin{pmatrix} & 1_r \\ 1_r & \end{pmatrix}$. Two quadratic spaces V and V' are said to lie in the same Witt tower if, up to isomorphism, they differ by a $V_{r,r}$, e.g., if $V' \simeq V + V_{r,r}$. The following facts are well known.

Lemma 3.3. *(i) (persistence) For any irreducible admissible representation π of G', and for any $r \geq 1$,*

$$\Theta(\pi, V) \neq 0 \implies \Theta(\pi, V + V_{r,r}) \neq 0.$$

(ii) (stable range (Howe)*) If* $V = V_0 + V_{r,r}$ *with* $r \geq 2n$*, then, for all irreducible admissible representations* π *of* G'

$$\Theta(\pi, V) \neq 0,$$

provided, in the case $m = \dim V$ *is odd,* π *is assumed to be genuine.*

The isomorphism class of V is determined by the invariants $m = \dim_F V$, the quadratic character χ_V of $F^\times$ defined by

$$\chi_V(x) = (x, (-1)^{\frac{m(m-1)}{2}} \det V),$$

and the Hasse invariant $\epsilon(V) = \pm 1$. We normalize our Hasse invariant so that it is constant on Witt towers. Note that the character χ_V also has this property, so that a Witt tower is determined by a choice of $\chi = \chi_V$ and $\epsilon = \epsilon(V)$. More precisely, the Witt towers can then be described as follows, [18].

For m **even and** $\chi = 1$, we have the split tower, $\{V_m^+ = V_{r,r}\}$, with $m = 2r$. Here we view the first element of the tower to be the 0 dimensional space $V_{0,0}$ (!). Let B be the division quaternion algebra over F, and let $V_4^- = B$, with $(x, y) = \mathrm{tr}(xy^\iota)$, where ι is the main involution of B. Then the quaternionic tower is $\{V_m^-\}$, $m = 2r \geq 4$, where $V_m^- = V_4^- + V_{r-2,r-2}$.

For m **even and** $\chi \neq 1$, we let k be the quadratic extension of F corresponding to the character χ, and let $V_2^+ = k$ with $(x, y) = tr(xy^\iota)$, where ι is the non-trivial Galois automorphism of k over F. Let $V_2^- = k$ with $(x, y) = \kappa \cdot \mathrm{tr}(xy^\iota)$, where $\chi(\kappa) = -1$. Then let $V_{2r}^\pm = V_2^\pm + V_{r-1,r-1}$, so that $\{V_m^+\}$ and $\{V_m^-\}$, $m \geq 2$ are the two tower associated to χ.

For $m = 2r + 1$ **odd**, and for a fixed quadratic character $\chi(x) = (x, \alpha)_F$, where $(\ ,\ )_F$ is the quadratic Hilbert symbol for F, let $V_1^+ = F$ with $(x, y) = \alpha \cdot xy$. Let B be as before, and let $V_3^- = \{x \in B \mid \mathrm{tr}(x) = 0\}$. Here we set $(x, y) = \alpha' \cdot \mathrm{tr}(xy)$, where the scalar $\alpha' \in F^\times$ is chosen so that $\chi_{V_3^-} = \chi$. Then our Witt towers associated to χ are $\{V_m^+\}$, $m = 2r+1 \geq 1$, with $V_m^+ = V_1^+ + V_{r,r}$ and $\{V_m^-\}$, $m = 2r + 1 \geq 3$, with $V_m^- = V_3^- + V_{r-1,r-1}$.

Note that, in all cases, the Hasse invariant is given by $\epsilon(V_m^\pm) = \pm 1$.

Definition 3.4. Fix a quadratic character χ and a parity[3]. Then, for any irreducible admissible representation π of G', let $m_\chi^\pm(\pi)$ be the smallest integer m, of the fixed parity, such that $\Theta_\chi(\pi, V_m^\pm) \neq 0$. In the case of odd parity, π is assumed to be genuine.

Part (iii) of Lemma 3.1, i.e., the stable range condition, immediately yields the following.

[3] By this we mean that we consider quadratic spaces of dimension m where the parity of m is fixed.

Lemma 3.5. *(i) For $\chi = \chi_0 = 1$ and even parity,*

$$0 \le m^+_{\chi_0}(\pi) \le 4n, \qquad \text{and} \qquad 4 \le m^-_{\chi_0}(\pi) \le 4n+4.$$

(ii) For $\chi \neq 1$ and even parity, $2 \le m^\pm_\chi(\pi) \le 4n+2$.
(iii) For odd parity,

$$1 \le m^+_\chi(\pi) \le 4n+1, \qquad \text{and} \qquad 3 \le m^-_\chi(\pi) \le 4n+3.$$

In fact, these first occurrence indices should be linked by the following **conservation relation** (Conjecture A, of the introduction):

Conjecture 3.6. *Fix a parity and a quadratic character χ. Then, for all irreducible admissible representations π of G',*

$$m^+_\chi(\pi) + m^-_\chi(\pi) = 4n+4.$$

Here, in the case of odd parity, π is assumed to be genuine.

In effect, if a representation π occurs 'early' in one tower, it must occur 'late' in the other! For example, in the case of even parity, π occurs 'early' if $\Theta(\pi, V) \neq 0$ for $\dim V = 2n+2-2r \le 2n$. For V' in the other tower, the conjectured relation then requires that $\Theta(\pi, V') = 0$ if $\dim V' \le 2n+2r$. In the case of odd parity, the conjectured relation implies **theta dichotomy** (Conjecture B of the introduction).

Conjecture 3.7. *Fix a quadratic character χ. For any irreducible admissible genuine representation π of G', precisely one of the spaces $\Theta_\chi(\pi, V^+_{2n+1})$ and $\Theta_\chi(\pi, V^-_{2n+1})$ is non-zero.*

Here the subscript χ serves as a reminder that we are considering the pair of Witt towers associated to χ.

The analogous conjectures for orthogonal groups are mentioned in the introduction. In the case of unitary groups, the conjectures appear in section 7 of [3].

Let $H = Sp(\mathbb{W})$ be as in section 1, where $\mathbb{W} = W + W_-$, and let

$$H' = \begin{cases} Sp(\mathbb{W}) & \text{in the even parity case} \\ Mp(\mathbb{W}) & \text{in the odd parity case.} \end{cases}$$

The map $\iota : G \times G \longrightarrow H$ of section 1 yields a map

$$\iota' : G' \times G' \longrightarrow H',$$

which is unique if we require that, in the case of odd parity,

$$\iota' : \mathbb{C}^1 \times \mathbb{C}^1 \longrightarrow \mathbb{C}^1, \qquad \iota'(z_1, z_2) = z_1 z_2^{-1}.$$

In the case of even parity, we let $I'_{2n}(s,\chi)$ be the degenerate principal series representation $I_{2n}(s,\chi)$ of $H' = H$, as in section 1. In the case of odd parity, we note that the restriction of the extension H' to the Levi factor of the Siegel parabolic P is given by

$$M' \simeq M \times \mathbb{C}^1,$$

with multiplication

$$(m(a_1), z_1) \cdot (m(a_2), z_2) = \big(m(a_1 a_2), z_1 z_2 (\det a_1, \det a_2)_F\big).$$

Here we use the conventions of **[6]**. For a character χ of $F^\times$, we set

$$\chi^\psi(x) = \chi(x)\, \gamma_F(x, \frac{1}{2}\psi)^{-1},$$

where $\gamma_F(x, \frac{1}{2}\psi)$ is in **[19]**,**[6]**. We then define a (genuine) character χ^ψ of M' by

$$\chi^\psi\big((m(a), z)\big) = z \cdot \chi^\psi(\det a).$$

We extend this to $P' = M'N$, trivially on N, and let

$$I'_{2n}(s, \chi) = I_{P'}^{H'}(\chi^\psi |\ |^s),$$

be the (genuine) degenerate principal series of H'. In both cases, we restrict the representation $I'_{2n}(s, \chi)$ to $G' \times G'$ via ι'.

First, we extend Conjecture 1.2 to the metaplectic case:

Conjecture 1.2′ For all irreducible admissible (genuine) representations π of G', and for all s and χ,

$$\dim \operatorname{Hom}_{G' \times G'}(I'_{2n}(s, \chi), \pi \otimes \pi^\vee) = 1.$$

Our main results are the following:

Theorem 3.8. *Fix a quadratic character χ and a parity. Then for any irreducible admissible (genuine) representation π of G',*

$$m_\chi^+(\pi) + m_\chi^-(\pi) \geq 4n + 4.$$

Theorem 3.9. *Fix a quadratic character χ and a parity. Suppose that π is an irreducible admissible representation of G' such that*

$$\dim \operatorname{Hom}_{G_n \times G_n}(I_{2n}(s_0, \chi), \pi \otimes \pi^\vee) = 1,$$

for all s_0 in the set:

$$Crit = \begin{cases} \{-n - \frac{1}{2}, -n + \frac{1}{2}, \ldots, n - \frac{1}{2}, n + \frac{1}{2}\} & \textit{for even parity and } \chi = 1, \\ \{-n + \frac{1}{2}, \ldots, n - \frac{1}{2}\} & \textit{for even parity and } \chi \neq 1, \\ \{-n, -n + 1, \ldots, n - 1, n\} & \textit{for odd parity.} \end{cases}$$

Then

$$m_\chi^+(\pi) + m_\chi^-(\pi) = 4n + 4.$$

In particular, Conjectures 3.6 and 3.7 hold for any supercuspidal representation π.

Of course, Theorem 3.9 together with Conjectures 1.2 and 1.2′ imply Conjecture 3.6 and 3.7 for all π.

Corollary 3.10. *For the trivial representation* $\mathbb{1}$ *of* G', *and for even parity, if* $\chi = \chi_0 = 1$,

$$m^{+}_{\chi_0}(\mathbb{1}) = 0 \qquad \text{and} \qquad m^{-}_{\chi_0}(\mathbb{1}) = 4n+4,$$

while, if $\chi \neq 1$,

$$m^{+}_{\chi}(\mathbb{1}) = m^{-}_{\chi}(\mathbb{1}) = 2n+2.$$

In particular, $\Theta_{\chi_0}(\mathbb{1}, V^{-}_{4n+2}) = 0$.

Note that Conjecture 3.6 implies that $\mathbb{1}$ is the only irreducible admissible representation of G' for which $\Theta_{\chi_0}(\pi, V^{-}_{4n+2}) = 0$, and that any representation $\pi \neq \mathbb{1}$ for which the conservation principle holds must have $\Theta_{\chi_0}(\pi, V^{-}_{4n+2}) \neq 0$.

More precise information about first occurrence can be given in certain cases. In the case of even parity with $\chi \neq 1$, choose $\alpha \in F^{\times}$ such that $\chi(\alpha) = -1$. For a representation σ of $G' = G = Sp(W)$, let $\delta_\alpha(\sigma)(g) = \sigma(g^\alpha)$, where

$$g^\alpha = \begin{pmatrix} 1 & \\ & \alpha \end{pmatrix} g \begin{pmatrix} 1 & \\ & \alpha \end{pmatrix}^{-1}. \tag{3.1}$$

Note that, if $\sigma = \omega_{V^{\epsilon}_m}$ is the Weil representation of G determined by the space V^{ϵ}_m of character χ, then $\delta_\alpha(\sigma) \simeq \omega_{V^{-\epsilon}_m}$ is the Weil representation defined by $V^{-\epsilon}_m$. Thus, for any irreducible admissible representation π, we have

$$m^{\epsilon}_{\chi}(\pi) = m^{-\epsilon}_{\chi}(\delta_\alpha(\pi)).$$

Corollary 3.11. *In the case of even parity with* $\chi \neq 1$, *suppose that* $\delta_\alpha(\pi) \simeq \pi$. *Then* $m^{+}_{\chi}(\pi) = m^{-}_{\chi}(\pi) \geq 2n+2$. *Moreover, if* π *satisfies the hypothesis of Theorem 3.9, then*

$$m^{+}_{\chi}(\pi) = m^{-}_{\chi}(\pi) = 2n+2.$$

For example, $m^{\pm}_{\chi}(\mathbb{1}) = 2n+2$.

4. Proofs of Theorems 3.8 and 3.9

Before proving these results, we need a few more preliminaries.

For any quadratic space V, let

$$R_{2n}(V) = S(V^{2n})_{O(V)}$$

be the maximal quotient of $S(V^{2n})$ on which $O(V)$ acts trivially, and note that this is a representation of H'. Recall that there is a natural H' equivariant map

$$\lambda : S(V^{2n}) \longrightarrow I'_{2n}(s_0, \chi), \qquad \varphi \mapsto (h \mapsto (\omega(h)\varphi)(0)),$$

where $s_0 = \frac{m}{2} - \rho_{2n}$ and $\chi = \chi_V$. Moreover, this map induces an inclusion, [18],

$$R_{2n}(V) \hookrightarrow I'_{2n}(s_0, \chi).$$

A complete account of the reducibility of the degenerate principal series $I'_{2n}(s, \chi)$ in terms of these submodules can be found in [2], [9], [8], when $m = \dim V$ is even, and in [20] when m is odd. These results will be briefly reviewed below.

The following result is, by now, quite standard [4], [18], [3], and its proof is just like that of Proposition 3.1 of [3].

Proposition 4.1. *Let π be an irreducible admissible representation of G'. Then*

$$\Theta(\pi, V) \neq 0 \iff \mathrm{Hom}_{G' \times G'}(R_{2n}(V), \pi \otimes \pi^{\vee}) \neq 0.$$

We now begin the proof of Theorems 3.8 and 3.9 by proving the first part of Corollary 3.10:

Lemma 4.2. *For even parity and for $\chi = \chi_0 = 1$,*

$$\mathrm{Hom}_{G' \times G'}(R_{2n}(V_m^-), \mathbb{1} \otimes \mathbb{1}) = 0,$$

for $m \leq 4n + 2$, and

$$\dim \mathrm{Hom}_{G' \times G'}(R_{2n}(V_{4n+4}^-), \mathbb{1} \otimes \mathbb{1}) = 1.$$

Thus, the trivial representation does not occur in the quaternionic tower until the stable range, and

$$m_{\chi_0}^+(\mathbb{1}) + m_{\chi_0}^-(\mathbb{1}) = 4n + 4.$$

Proof. First observe that we view the trivial representation as the Weil representation associated to the space $V_{0,0}$ at the base of the split tower. Thus, $\Theta(\mathbb{1}, V_{0,0}) \neq 0$ (formally), and so, by (i) of Lemma 3.3 together with Lemma 4.1, we have

$$\mathrm{Hom}_{G' \times G'}(R_{2n}(V_m^+), \mathbb{1} \otimes \mathbb{1}) \neq 0,$$

for $m = 0, 2, \ldots$ in the split tower. Moreover, for $m \geq 4n+4$ even, we have, [9]

$$R_{2n}(V_m^+) = R_{2n}(V_m^-) = I_{2n}(s_0, \chi_0),$$

so that the combination of Theorem 1.1 and Proposition 4.1 yield the non-vanishing assertion. One could also invoke (ii) of Lemma 3.3. Also, by (i) of Lemma 3.3, to prove that $m_{\chi_0}^-(\mathbb{1}) = 4n+4$, it suffices to show that the space

$$\mathrm{Hom}_{G' \times G'}(R_{2n}(V_{4n+2}^-), \mathbb{1} \otimes \mathbb{1}) \tag{4.1}$$

is zero. But, as shown in [2] and [9], $R_{2n}(V_{4n+2}^-)$ is the unique irreducible submodule of $I'_{2n}(\rho_{2n}, \chi_0)$ and is precisely the image of the normalized intertwining operator

$$M_{2n}^*(-\rho_{2n}, \chi_0) : I_{2n}(-\rho_{2n}, \chi_0) \longrightarrow I_{2n}(\rho_{2n}, \chi_0). \tag{4.2}$$

If the space (4.1) were non-zero, we would obtain a non-zero element of the space

$$\mathrm{Hom}_{G' \times G'}(I_{2n}(-\rho_{2n}, \chi_0), \mathbb{1} \otimes \mathbb{1}) \tag{4.3}$$

by composing with the intertwining operator (4.2). This element would be trivial on the vector $\Phi^0(-\rho_{2n})$, which is, in fact, the constant function 1 and spans the kernel of $M_{2n}^*(-\rho_{2n}, \chi_0)$. But since $-\rho_{2n} \notin X(\mathbb{1})$, (v) and (i) of Proposition 2.2 imply that the zeta integral $Z(-\rho_{2n}, \chi_0, 1)$ spans the space (4.3) and does not vanish on $\Phi^0(-\rho_{2n})$. This proves that (4.1) must be zero, as required. □

Proof of Theorem 3.8. Fix χ and a parity, and suppose that π is an irreducible admissible representation of G', which is genuine in the case of odd parity. Suppose that $\Theta(\pi, V_a^+) \neq 0$ and $\Theta(\pi, V_b^-) \neq 0$, where V_a^+ and V_b^- are spaces of character χ, dimensions a and b, respectively, and Hasse invariants $+1$ and -1. There are then non-zero G'–intertwining maps

$$\lambda : S((V_a^+)^n) \longrightarrow \pi \qquad \text{and} \qquad \mu : S((V_b^-)^n) \longrightarrow \pi.$$

Recall that the MVW involution [13],[3] on the set of smooth representations of G' is defined by set $\tau^c(g) = \tau(g^c)$, where, in the case of even parity, $g^c = cgc^{-1}$, for $c = \begin{pmatrix} 1_n & 0 \\ 0 & -1_n \end{pmatrix}$. In the case of odd parity, we use the same notation to denote the unique lift of this automorphism to G' whose restriction to the central $\mathbb{C}^1$ is given by $z \mapsto z^{-1}$, [3]. Also recall that, if τ is irreducible, then $\tau^c \simeq \tau^\vee$, the contragredient of τ. We apply this involution to obtain an intertwining map

$$\mu^c : S((V_b^-)^n)^c \longrightarrow \pi^\vee.$$

On the other hand, the representation of G' on $S((V_b^-)^n)^c$ is isomorphic to the complex conjugate of its representation on $S((V_b^-)^n)$. Note that these representations are preunitary. We thus obtain a nonzero intertwining map

$$\lambda \otimes \mu^c \in \mathrm{Hom}_{G' \times G'}(S((V_a^+)^n) \otimes S((V_b^-)^n)^c, \pi \otimes \pi^\vee).$$

We next restrict these representations to the diagonal G' in $G' \times G'$. Since the central $\mathbb{C}^1$ now acts trivially, we obtain representations of $G = Sp(W)$. Moreover, the representation of G on $S((V_a^+)^n) \otimes S((V_b^-)^n)^c$ is isomorphic to the Weil representation of this group on $S((V_a^+ + (-V_b^-))^n)$, where $-V_b^-$ is the space V_b^- with the negative of its quadratic form. Thus, we obtain a nonzero intertwining map in

$$\mathrm{Hom}_G(S((V_a^+ + (-V_b^-))^n), \mathbb{1}),$$

by composing with the invariant pairing $\pi \otimes \pi^\vee \longrightarrow \mathbb{1}$. The following fact is easily checked:

Lemma 4.3. *The space $V_a^+ + (-V_b^-)$ has trivial character, i.e., square discriminant, and Hasse invariant -1. Specifically, it is the quaternionic space of dimension $a + b$.*

But now, we have $\Theta(\mathbb{1}, V_a^+ + (-V_b^-)) \neq 0$, so that, by Lemma 4.2, $a + b \geq 4n + 4$, as claimed. □

Proof of Theorem 3.9. We now must make use of the way in which the representations $R_{2n}(V)$ sit inside of $I'_{2n}(s_0, \chi)$, where $\chi = \chi_V$ and $s_0 = \frac{m}{2} - \frac{n+1}{2}$. Here χ will be fixed throughout our argument. This structure can be summarized as follows, [2], [9], [20], [21]. First of all, we recall that $I'_{2n}(s, \chi)$ is irreducible unless

(4.4)
$$s = s_0 \in \begin{cases} \{-n - \frac{1}{2}, -n + \frac{1}{2}, \dots, n - \frac{1}{2}, n + \frac{1}{2}\} & \text{for even parity and } \chi = \chi_0 = 1, \\ \{-n + \frac{1}{2}, \dots, n - \frac{1}{2}\} & \text{for even parity and } \chi \neq 1, \\ \{-n, -n + 1, \dots, n - 1, n\} & \text{for odd parity.} \end{cases}$$

Moreover, there are 3 constituents, except in the cases

$$s = s_0 = \begin{cases} \{-n - \frac{1}{2}, -n + \frac{1}{2}, n - \frac{1}{2}, n + \frac{1}{2}\} & \text{for even parity and } \chi = \chi_0 = 1, \\ \{-n, n\} & \text{for odd parity,} \end{cases}$$

when the composition series has length 2. The constituents can be described in terms of the $R_{2n}(V)$'s. We say that the spaces $V_{m_0}^+$ and V_m^+ (resp. $V_{m_0}^-$ and V_m^-), with $0 \leq m_0 \leq m$, are complementary if $m + m_0 = 4n + 2$. In this case, note that

$$s_0 = \frac{m}{2} - \frac{2n+1}{2} \qquad \text{and} \qquad -s_0 = \frac{m_0}{2} - \frac{2n+1}{2}.$$

Note that certain values of the parameters χ and m_0 are excluded. Specifically, the spaces V_0^- and V_1^- for any χ, V_2^- for $\chi = \chi_0$, and V_0^+ for $\chi \neq 1$ do not exist. In these cases, we formally take $R_{2n}(V_{m_0}^{\pm}) = 0$. Similarly, if $m_0 = 0$, we take $R_{2n}(V_0^+)$ to be the trivial representation, which is the unique irreducible submodule of $I_{2n}(-n - \frac{1}{2}, \chi_0)$ in the case of even parity. If $0 \leq m_0 \leq 2n + 1$, then $R_{2n}(V_{m_0}^{\pm})$ is irreducible, provided the space $V_{m_0}^{\pm}$ exists. As submodules of $I'_{2n}(-s_0, \chi)$,

$$R_{2n}(V_{m_0}^+) \cap R_{2n}(V_{m_0}^-) = 0. \tag{4.5}$$

When $m_0 = m = 2n + 1$, we have **[21]**

$$R_{2n}(V_{m_0}^+) \oplus R_{2n}(V_{m_0}^-) = I'_{2n}(0, \chi). \tag{4.6}$$

If $0 \leq m_0 < 2n + 1$,

$$R_{2n}(V_{m_0}^+) \oplus R_{2n}(V_{m_0}^-) = \mathrm{Im}(M_{2n}^*(s_0, \chi)) = \mathrm{Ker}(M_{2n}^*(-s_0, \chi)), \tag{4.7}$$

is the maximal proper submodule of $I'_{2n}(-s_0, \chi)$. Here $M_{2n}^*(s, \chi)$ is the normalized intertwining operator **[9]**, **[21]**:

$$M_{2n}^*(s, \chi) : I'_{2n}(s, \chi) \longrightarrow I'_{2n}(-s, \chi).$$

Moreover, the irreducible module

$$X(s_0) = \mathrm{Im}(M_{2n}^*(-s_0, \chi)) \simeq I'_{2n}(-s_0, \chi) / \big(R_{2n}(V_{m_0}^+) \oplus R_{2n}(V_{m_0}^-)\big), \tag{4.8}$$

is the maximal proper submodule of $I'_{2n}(s_0, \chi)$, and

$$X(s_0) = \mathrm{Ker}(M_{2n}^*(s_0, \chi)). \tag{4.9}$$

Also, in the range $2n + 1 < m \leq 4n + 2$, where $s_0 = \frac{m}{2} - \rho_{2n}$,

$$I'_{2n}(s_0, \chi) = R_{2n}(V_m^+) + R_{2n}(V_m^-), \tag{4.10}$$

$$X(s_0) = R_{2n}(V_m^+) \cap R_{2n}(V_m^-), \tag{4.11}$$

and

$$R_{2n}(V_m^{\pm}) / X(s_0) \simeq R_{2n}(V_{m_0}^{\pm}). \tag{4.12}$$

In the cases in which $V_{m_0}^-$ does not exist, this last relation means that $X(s_0) = R_{2n}(V_m^-)$.

Now, turning to the proof of Theorem 3.9, we note that $m_\chi^+(\pi) = 0$ implies that $\pi = \mathbb{1}$, so this case is covered by Lemma 4.2. Also note that, (4.10), together with Proposition 4.1 and our hypothesis on π, imply that

$$\min\{m_\chi^+(\pi), m_\chi^-(\pi)\} \leq \begin{cases} 2n+1 & \text{for odd parity} \\ 2n+2 & \text{for even parity.} \end{cases}$$

Similarly, the fact that $I'_{2n}(s_0,\chi)$ is irreducible in the cases (i) $s_0 = n+1$ and arbitrary χ for odd parity, (ii) $s_0 = n+\frac{1}{2}$ and $\chi \neq 1$ and (iii) $s_0 = n+\frac{3}{2}$ and $\chi = 1$, for even parity, implies that

$$\max\{m_\chi^+(\pi), m_\chi^-(\pi)\} \leq \begin{cases} 4n+3 & \text{for odd parity,} \\ 4n+2 & \text{for even parity with } \chi \neq 1, \text{ and,} \\ 4n+4 & \text{for even parity with } \chi = 1. \end{cases}$$

These bounds are a little weaker than the stable range condition of Lemma 3.3 (ii).

We suppose that $\epsilon = \pm 1$ is taken so that $m_\chi^\epsilon(\pi) = \min\{m_\chi^+(\pi), m_\chi^-(\pi)\}$.

In the cases $m_\chi^\epsilon(\pi) = m_0$ with $m_0 = 1$ and χ arbitrary, (resp. $m_0 = 2$ with $\chi \neq 1$), we must have $m_\chi^{-\epsilon}(\pi) \geq 4n+3$ (resp. $m_\chi^{-\epsilon}(\pi) \geq 4n+2$) by Theorem 3.8. The upper bounds just described then force $m_\chi^{-\epsilon}(\pi) = 4n+3$ (resp. $m_\chi^{-\epsilon}(\pi) = 4n+2$), as claimed.

Next, suppose that $2 \leq m_\chi^\epsilon(\pi) = m_0 \leq 2n+1$. By Theorem 3.8, this forces $m_\chi^{-\epsilon}(\pi) \geq 4n+4-m_0 \geq 2n+3$. With the conventions about complementary spaces just introduced, the minimality of $m_\chi^\epsilon(\pi)$ implies that

$$\dim \operatorname{Hom}_{G'\times G'}\big(R_{2n}(V^+_{m_0-2}) \oplus R_{2n}(V^-_{m_0-2}), \pi \otimes \pi^\vee\big) = 0. \tag{4.13}$$

Thus, every element of $\operatorname{Hom}_{G'\times G'}\big(I'_{2n}(-s_0-1,\chi), \pi\otimes\pi^\vee\big)$ must factor through $X(s_0+1)$, and

$$\dim \operatorname{Hom}_{G'\times G'}\big(X(s_0+1), \pi \otimes \pi^\vee\big) = 1.$$

On the other hand, let

$$\mu \in \operatorname{Hom}_{G'\times G'}\big(I'_{2n}(s_0+1,\chi), \pi \otimes \pi^\vee\big)$$

be a nonzero element. The restriction of μ to $X(s_0+1)$ must be nonzero, since, otherwise, it would factor through the quotient $R_{2n}(V^+_{m_0-2}) \oplus R_{2n}(V^-_{m_0-2})$, contradicting (4.13). But, in the range in question, $X(s_0+1) \subset R_{2n}(V^{-\epsilon}_{m+2})$ is a nonzero submodule, and hence

$$\mu\big|_{R_{2n}(V^{-\epsilon}_{m+2})} \neq 0.$$

Thus

$$m_\chi^{-\epsilon}(\pi) \leq m+2 = 4n+4-m_0.$$

This yields the equality $m_\chi^+(\pi) + m_\chi^-(\pi) = 4n+4$ in this case.

Only the case $m_\chi^\epsilon(\pi) = 2n+2$ remains. Let

$$\mu \in \operatorname{Hom}_{G'\times G'}(I'_{2n}(\frac{1}{2},\chi), \pi \otimes \pi^\vee)$$

be a nonzero element, and note that the restriction of μ to the submodule $X(\frac{1}{2})$ cannot be zero, since μ does not factor through the quotient

$$I_{2n}(\frac{1}{2},\chi)/X(\frac{1}{2}) \simeq R_{2n}(V_{2n}^{+}) \oplus R_{2n}(V_{2n}^{-}).$$

But then, the restriction of μ to $R_{2n}(V_{2n+2}^{-\epsilon}) \supset X(\frac{1}{2})$ is nonzero, as required. $\square$

Remark 4.4. An examination of our proof confirms that we have not made use of the stable range result Lemma 3.3 (ii), and that we have in fact reproved this fact.

5. Proof of (ii) of Proposition 1.6: simplicity of poles

In this section, we give the proof of the following result, which is just a restatement of part (ii) of Proposition 1.6. Note that we may absorb the unramified character χ into s by a suitable shift, so that it can be omitted in what follows. Also, for convenience, we will write Φ_s rather than $\Phi(s)$ for a given standard section.

Proposition 5.1. *Let $\mathfrak{z}_s \in \mathcal{H}(G//K) \otimes \mathbb{C}[q^s, q^{-s}]$ be the element defined by*

$$\mathfrak{z}_s = \prod_{i=1}^{n} (1 - q^{-s-\frac{1}{2}}\, t_i)\,(1 - q^{-s-\frac{1}{2}}\, t_i^{-1}).$$

For an unramified representation π of G, let $\pi(\mathfrak{z}_s)$ be the scalar by which $\mathfrak{z}_s$ acts on the unramified vector in π. Then for all matrix coefficients ϕ of $\pi^\vee$ and all standard sections $\Phi(s) \in I_{2n}(s)$, the function

$$\pi(\mathfrak{z}_s) \cdot Z(s, \chi_0, \phi, \Phi)$$

is an entire function of s.

Proof. The proof will be broken into several steps.

Step 1. By linearity, it is sufficient to consider matrix coefficients of the form

$$\phi(g) \;= <\pi^\vee(g)\pi^\vee(g_0)\xi_0^\vee\,,\,\pi(g_1)\xi_0>, \tag{5.1}$$

where $\xi_0^\vee$ and ξ_0 are spherical vectors in $\pi^\vee$ and π, as above and g_0 and $g_1 \in G$. Then

$$
\begin{aligned}
Z(s,\phi,\Phi) &= \int_G \Phi_s(\delta_0\, \iota(g,1)) \cdot <\pi^\vee(g)\pi^\vee(g_0)\xi_0^\vee\,,\, \pi(g_1)\xi_0> dg \\
&= \int_G \Phi_s(\delta_0\, \iota(g_1 g g_0^{-1},1)) \cdot <\pi^\vee(g)\xi_0^\vee\,,\, \xi_0> dg \qquad (5.2)\\
&= \int_G \Phi_s(\delta_0\, \iota(g,1)\iota(g_0^{-1},g_1^{-1})) \cdot \phi^0(g)\, dg.
\end{aligned}
$$

Since ϕ^0 is bi-K-invariant, the last integral here is equal to

$$
\begin{aligned}
(5.3) \qquad & \int_G \Phi_s(\delta_0\, \iota(k_1^{-1} g k_2,1)\iota(g_0^{-1},g_1^{-1})) \cdot \phi^0(g)\, dg \\
&= \int_G \Phi_s(\delta_0\, \iota(g,1)\iota(k_1,k_2)\iota(g_0^{-1},g_1^{-1})) \cdot \phi^0(g)\, dg.
\end{aligned}
$$

Thus we have

$$
(5.4) \qquad Z(s,\phi,\Phi) = \int_G \Psi_s(\delta_0\, \iota(g,1)) \cdot \phi^0(g)\, dg,
$$

where, for any $h \in H = G_{2n}$,

$$
(5.5) \qquad \Psi_s(h) := \int_{K\times K} \Phi_s(h\iota(k_1,k_2)\iota(g_0^{-1},g_1^{-1}))\, dk_1\, dk_2.
$$

Note that Ψ_s is again a section of $I_{2n}(s)$ which is right $K\times K$-invariant. It may not be standard, however!

Step 2. We next define an element $\mathfrak{z}$ of the algebra

$$
(5.6) \quad \mathcal{A} = \mathbb{C}[X,X^{-1}] \otimes \mathcal{H}(G//K) \simeq \mathbb{C}[X,X^{-1}] \otimes \mathbb{C}[t_1,t_1^{-1},\dots,t_n,t_n^{-1}]^{W_G},
$$

where $\mathcal{H}(G//K)$ is the K–spherical Hecke algebra of G, by

$$
(5.7) \qquad \mathfrak{z} = \prod_{i=1}^{n} (1 - q^{-\frac{1}{2}}\, X\, t_i)\,(1 - q^{-\frac{1}{2}}\, X\, t_i^{-1}).
$$

The algebra $\mathcal{A}$ acts on the space $I_{2n}(s)^{K\times 1}$ of $K\times 1$-invariant vectors in $I_{2n}(s)$ where the action of $\mathcal{H}(G//K)$ comes from the action of $G\times 1$, via ι, and where X acts by scalar multiplication by q^{-s}. Note that this action commutes with the action of $1\times G$. The operator $\mathfrak{z}$ is defined to kill support of sections.

Proposition 5.2. *For any standard section Φ_s with associated section Ψ_s defined by (5.5)*

$$
\Psi_s * \mathfrak{z} \in I_{2n}^{(0)}(s)^{K\times K}.
$$

Proof. To illustrate the argument, we first consider the restriction to the closed orbit,

$$(5.8)\qquad \mathrm{pr}_n : I_{2n}(s) = I_{2n}^{(n)}(s) \longrightarrow Q_n(s) \simeq I_n(s+\frac{n}{2}) \otimes I_n(s+\frac{n}{2}),$$

given by $\Phi_s \mapsto (\ (g_1, g_2) \mapsto \Phi_s(\iota(g_1, g_2))\ )$, the map obtained by restriction of sections to Ω_n. Then we have

$$(5.9)\qquad \mathrm{pr}_n(\Psi_s * \mathfrak{z}) = \mathrm{pr}_n(\Psi_s) * \mathfrak{z},$$

where $\mathfrak{z}$ is acting on the K–invariant vectors in the first factor of the tensor product on the right side of (5.8). Thus, it suffices to prove that $\mathfrak{z}$ acts by zero on the space $I_n(s+\frac{n}{2})^K$. But

$$(5.10)\qquad I_n(s+\frac{n}{2}) \subset I_B^G(\lambda)$$

where B is the standard Borel subgroup of G and $I_B^G(\lambda)$ is the unramified principal series representation of G with Satake parameter

$$(5.11)\qquad \lambda = (q^{-s-\frac{1}{2}}, q^{-s-\frac{3}{2}}, \dots, q^{-s-n+\frac{1}{2}}).$$

The operator $\mathfrak{z}$ then acts on the (common) K-fixed vector in these representations by the scalar

$$(5.12)\qquad \prod_{i=1}^{n}(1 - q^{-s-\frac{1}{2}}\, q^{-s-i+\frac{1}{2}})(1 - q^{-s-\frac{1}{2}}\, q^{s+i-\frac{1}{2}}) = 0.$$

Thus $\mathrm{pr}_n(\Psi_s * \mathfrak{z}) = 0$ so that $\Psi_s * \mathfrak{z} \in I_{2n}^{(n-1)}(s)$.

In general, restriction of sections to the orbit Ω_r yields a map

$$(5.13)\qquad \mathrm{pr}_r : I_{2n}(s) \longrightarrow I_{P_r\times P_r}^{G\times G}\left(|\ |^{s+\frac{r}{2}} \otimes |\ |^{s+\frac{r}{2}} \otimes C(G_{n-r})\right) =: B_r(s),$$

where $C(G_{n-r})$ is the space of all smooth functions on G_{n-r}. There is a nondegenerate pairing

$$(5.14)\qquad Q_r(s) \otimes B_r(-s-r) \longrightarrow \mathbb{C}$$

given by

$$(5.15)\qquad \langle f_1, f_2 \rangle = \int_{P_r\times P_r\backslash G\times G} \langle f_1(g_1, g_2), f_2(g_1, g_2) \rangle_{G_{n-r}}\, d\mu(g_1)\, d\mu(g_2)$$

where the integrand is the pairing on the values of function $f_1 \in B_r(-s-r)$ and $f_2 \in Q_r(s)$ given by integration over G_{n-r} and the outer 'integral' is the $G \times G$ invariant functional on the spaces of functions on $G \times G$ transforming on the left under $P_r \times P_r$ according to the square of the modulus character. A straightforward density argument shows that an element $\phi \in Q_r(s)$ is zero if and only if it pairs to zero against all elements of the subspace $Q_r(-s-r)$

of $B_r(-s-r)$. If $\phi \in Q_r(s)^{K\times K}$, then it suffices to pair with elements $f \in Q_r(-s-r)^{K\times K}$.

We consider the pairing

$$\langle \mathrm{pr}(\Psi_s * \mathfrak{z}), f_s \rangle = \langle \mathrm{pr}(\Psi_s), f_s * \mathfrak{z}_s^\vee \rangle \tag{5.16}$$

for any function $f_s \in Q_r(-s-r)^{K\times K}$, where $\mathfrak{z}_s^\vee$ is the adjoint of $\mathfrak{z}_s = \mathfrak{z}|_{X=q^{-s}}$.

Lemma 5.3. *For any function* $f_s \in Q_r(-s-r)^{K\times K}$,

$$f_s * \mathfrak{z}_s^\vee = 0.$$

Proof. Since a function in $f_s \in Q_r(-s-r)^{K\times K}$ is determined by its value at the identity element, we can consider $f_s(e) \in S(G_{n-r})$. Note that, for $x \in G_{n-r}$ and $k_1, k_2 \in K_{n-r} = K \cap G_{n-r}$, we have

$$f_s(e)(k_1^{-1}xk_2) = f_s(k_1,k_2)(x) = f_s(e)(x), \tag{5.17}$$

so that, in fact,

$$f_s(e) \in S(G_{n-r})^{K_{n-r}\times K_{n-r}}. \tag{5.18}$$

If τ is any irreducible admissible representation of G_{n-r}, the action of the space of functions $S(G_{n-r})$ on τ determines a $G_{n-r}\times G_{n-r}$-equivariant map

$$\mu_\tau : S(G_{n-r}) \longrightarrow \mathrm{Hom}^{\mathrm{smooth}}(\tau,\tau) \;\simeq\; \tau^\vee \otimes \tau. \tag{5.19}$$

Here, note that $G\times G$ acts on $\mathrm{Hom}(\tau,\tau)$ by pre and pos-multiplication, so that elements of $\mathrm{Hom}^{\mathrm{smooth}}(\tau,\tau)$, which by definition are fixed by some compact open subgroup $K\times K$, have finite dimensional image. A function $\phi \in S(G_{n-r})^{K_{n-r}\times K_{n-r}}$ is nonzero if and only if there is an unramified representation τ of G_{n-r} such that $\tau(\phi) \neq 0$, i.e., $\mu_\tau(\phi) \neq 0$.

Applying this to $f_s * \mathfrak{z}_s^\vee$, we suppose that there is an irreducible unramified representation τ of G_{n-r} such that

$$\mu_\tau(f_s * \mathfrak{z}_s^\vee(e)) \neq 0. \tag{5.20}$$

There is a map of induced representations

$$\begin{aligned} I(\mu_\tau) : I_{P_r\times P_r}^{G\times G}&\left(|\,|^{-s-\frac{r}{2}}\otimes|\,|^{-s-\frac{r}{2}}\otimes S(G_{n-r})\right)\\ &\longrightarrow I_{P_r\times P_r}^{G\times G}\left(|\,|^{-s-\frac{r}{2}}\otimes|\,|^{-s-\frac{r}{2}}\otimes\tau^\vee\otimes\tau\right), \end{aligned} \tag{5.21}$$

and

$$I(\mu_\tau)(f)(e) = \mu_\tau(f(e)). \tag{5.22}$$

We can embed the latter of these representations into a full induced representation

$$I_{P_r}^G(|\ |^{-s-\frac{r}{2}} \otimes \tau^\vee) \otimes I_{P_r}^G(|\ |^{-s-\frac{r}{2}} \otimes \tau) \quad \hookrightarrow \quad I_B^G(\lambda_1) \otimes I_B^G(\lambda_2), \tag{5.23}$$

where the Satake parameter λ_1 is given by

$$\lambda_1 = (q^{s+r-\frac{1}{2}}, q^{s+r-\frac{3}{2}}, \dots, q^{s+\frac{1}{2}}, q^{\nu_1}, \dots, q^{\nu_{n-r}}). \tag{5.24}$$

Note that $\mathfrak{z}_s^\vee = \mathfrak{z}_s$ acts on the unique K-invariant vector in this representation by the scalar

$$\prod_{i=1}^{r}(1 - q^{-s-\frac{1}{2}}\, q^{s+r+\frac{1}{2}-i})(1 - q^{-s-\frac{1}{2}}\, q^{-s-r-\frac{1}{2}+i}) \cdot (\star) = 0. \tag{5.25}$$

But then, we have

$$\begin{aligned} \mu_\tau(f_s * \mathfrak{z}_s^\vee(e)) &= I(\mu_\tau)(f_s * \mathfrak{z}_s^\vee)(e) \\ &= (I(\mu_\tau)(f_s) * \mathfrak{z}_s^\vee)(e) \\ &= 0, \end{aligned} \tag{5.26}$$

since $I(\mu_\tau)(f_s)$ is a multiple of the $K \times K$-invariant vector in $I_B^G(\lambda_1) \otimes I_B^G(\lambda_2)$. This finishes the proof of Lemma 5.3 and hence of Proposition 5.2. □

Step 3. Consider the isomorphism

$$\mathrm{pr}_0 : I_{2n}^{(0)}(s) \longrightarrow Q_0(s) \simeq S(G), \tag{5.27}$$

given by $\Phi_s \mapsto \big(\, (g,1) \mapsto \Phi_s(\delta_0 \iota(g,1))\, \big)$. Proposition 1.7 shows that, *for a fixed* s, the image of $\Psi_s * \mathfrak{z}$ under pr_0 lies in $S(G)^{K\times K}$. But the support could, a priori, vary with s. The following result shows that the support is, in fact, bounded *uniformly* in s.

Lemma 5.4.

$$\mathrm{pr}_0(\Psi_s * \mathfrak{z}) \in \mathbb{C}[q^s, q^{-s}] \otimes S(G)^{K\times K} = \mathbb{C}[q^s, q^{-s}] \otimes \mathcal{H}(G//K).$$

Proof. Using the Cartan decomposition, write

$$\mathrm{pr}_0(\Psi_s * \mathfrak{z}) = \sum_{\lambda \in \Lambda} c_\lambda(s)\, L_\lambda, \tag{5.28}$$

where L_λ is the characteristic function of the double coset $K g_\lambda K$ and Λ is the usual semigroup.

Lemma 5.5. *For each* $\lambda \in \Lambda$, *the function* $c_\lambda(s)$ *is an entire analytic function of* s. *In fact,* $c_\lambda(s) \in \mathbb{C}[q^s, q^{-s}]$.

Assuming this result for the moment, let Λ_1 be the set of λ's for which the function $c_\lambda \neq 0$, and let

$$D_\lambda = \{s \in \mathbb{C} \mid c_\lambda(s) \neq 0\}. \tag{5.29}$$

For $\lambda \in \Lambda_1$, D_λ is an open dense subset of $\mathbb{C}$. Thus, by the Baire category theorem, the set

$$\bigcap_{\lambda \in \Lambda_1} D_\lambda \tag{5.30}$$

is non-empty, and so, there is an $s_0 \in \mathbb{C}$ such that $c_\lambda(s_0) \neq 0$ for all $\lambda \in \Lambda_1$. Since

$$\mathrm{pr}_0(\Psi_{s_0} * \mathfrak{z}) = \sum_{\lambda \in \Lambda_1} c_\lambda(s_0)\, L_\lambda, \tag{5.31}$$

has compact support, it follows that the set Λ_1 is finite, as claimed! □

Proof of Lemma 5.5. We have the formula

$$c_\lambda(s) \cdot ||L_\lambda||^2 = \int_G (\Psi_s * \mathfrak{z})(\delta_0 \iota(g,1)) \cdot L_\lambda(g)\, dg. \tag{5.32}$$

The right side of (5.32) is a finite linear combination, with coefficients in $\mathbb{C}[q^s, q^{-s}]$, of integrals of the form

$$\begin{aligned} &\int_G \int_G \Psi_s(\delta_0 \iota(g,1)\, \iota(g_0,1)) \cdot L_\mu(g_0)\, dg_0 \cdot L_\lambda(g)\, dg \\ &= \int_G \int_G \Psi_s(\delta_0\, \iota(g_0,1)) \cdot L_\mu(g^{-1}g_0) \cdot L_\lambda(g)\, dg_0\, dg \\ &= \int_G \Psi_s(\delta_0\, \iota(g_0,1)) \cdot \varphi(g_0)\, dg_0 \end{aligned} \tag{5.33}$$

where $\varphi \in \mathcal{H}(G//K)$ is an element depending on λ and μ. Since φ is a finite linear combination of characteristic functions of cosets gK, the integral in the last line of (5.33) is a finite linear combination of terms of the form

$$\int_K \int_{K\times K} \Phi_s(\delta_0\, \iota(gk,1)\iota(k_1,k_2)\iota(g_1,g_2))\, dk_1\, dk_2\, dk. \tag{5.34}$$

But, noting that (g_1, g_2) is fixed and recalling that Φ_s is standard and hence is right invariant under some compact open subgroup of H, *uniformly in* s, this expression is just a finite linear combination of values of Φ_s at a set of points not depending on s. Such values have the form cq^{rs} for $r \in \mathbb{Z}$. This finishes the proof of Lemma 5.5. □

Step 4. Now returning to the zeta integral (5.4) and inserting the operator $\mathfrak{z}$, we find that

$$Z^*(s,\phi^0,\Psi) := \int_G (\Psi_s * \mathfrak{z})(\delta_0\, \iota(g,1))\, \phi^0(g)\, dg \tag{5.35}$$

is equal to the scalar by which the element $\mathrm{pr}_0(\Psi_s * \mathfrak{z})$ of $\mathcal{A}$ acts on the spherical vector $\xi_0^\vee$ in $\pi^\vee$. This scalar lies in $\mathbb{C}[q^s, q^{-s}]$, and, in particular, is an entire function of s. On the other hand, for Re(s) large,

$$\begin{aligned} Z^*(s,\phi^0,\Psi) &= \pi^\vee(\mathfrak{z}_s)\cdot \int_G \Psi_s(\delta_0\, \iota(g,1))\, \phi^0(g)\, dg \\ &= \pi^\vee(\mathfrak{z}_s)\cdot Z(s,\phi,\Phi) \end{aligned} \tag{5.36}$$

where $\pi^\vee(\mathfrak{z}_s)$ is the scalar by which $\mathfrak{z}_s := \mathfrak{z}|_{X=q^{-s}}$ acts on the spherical vector in $\pi^\vee$. Since $Z^*(s,\phi^0,\Psi)$ is an entire function of s, this completes the proof of the proposition.

□

Note that

$$\begin{aligned} \pi^\vee(\mathfrak{z}_s) &= \prod_{i=0}^{n} (1 - q^{-s-\frac{1}{2}}\pi^\vee(t_i))\, (1 - q^{-s-\frac{1}{2}}\pi^\vee(t_i^{-1})) \\ &= b_{2n}(s)^{-1}\, (1 - q^{-s-\frac{1}{2}})^{-1}\cdot Z(s,\phi^0,\Phi^0)^{-1}. \end{aligned} \tag{5.37}$$

Thus

$$Z(s,\phi^0,\Phi^0)^{-1}\cdot Z(s,\phi,\Phi) = b_{2n}(s)\, (1 - q^{-s-\frac{1}{2}})\, Z^*(s,\phi^0,\Psi_s). \tag{5.38}$$

This shows that the unramified zeta integral $Z(s,\phi^0,\Phi^0)$ can have some extra zeros, compared to the whole family of zeta integrals.

References

[1] F. Bruhat, *Distributions sur un groupe localement compact et applications à l'étude des representations des groupes p-adiques*, Bull. Soc. Math. France, **89** (1961), 43–75.

[2] R. Gustafson, *The degenerate principal series of Sp(2n)*, Memoirs of AMS. **248** (1981).

[3] M. Harris, S. Kudla, and W. Sweet, *Theta dichotomy for unitary groups*, Jour. of the AMS, **9** (1996), pp. 941–1004.

[4] R. Howe, *θ-series and invariant theory*, in Automorphic Forms, Representations and L-functions, Proc. Symp. Pure Math., **33** (1979), pp. 275–285.

[5] S. Kudla, *On the local theta correspondence*, Invent. math., **83** (1986), pp. 229–255.

[6] ______, *Splitting metaplectic covers of dual reductive pairs*, Israel J. Math., **87** (1994), pp. 361–401.

[7] ______, *Notes on the local theta correspondence*, Lecture notes from the European School on Group Theory, Schloß Hirschberg, Germany, 1996

[8] ______, *Central derivatives of Eisenstein series and height pairings*, Ann. of Math., **146** (1997), pp. 545–646.

[9] S. Kudla and S. Rallis, *Ramified degenerate principal series*, Israel Math. J., **78** (1992), pp. 209–256.

[10] S. Kudla and W. Sweet, *Degenerate principal series for unitary groups*, Israel J. Math. **98** (1997), 253–306.

[11] E. Lapid and S. Rallis, *On the local factors of classical groups*, this volume.

[12] C. Moeglin and M. Tadic, *Construction of discrete series for classical p-adic groups*, Jour. of the AMS, **15** (2002), 715–786.

[13] C. Moeglin, M.-F. Vigneras, and J.-L. Waldspurger, Correspondances de Howe sur un corps p-adique, Lecture Notes in Math., 1291, Springer-Verlag, New York, 1987.

[14] G. Muiç, *Howe correspondence for discrete series representations; the case of (Sp(n),O(V))*, to appear in Crelle Math. J.

[15] A. Paul, *First occurrence for the dual pairs (U(p,q),U(r,s))*, Canadian J. Math. **51** (1999), 636–657.

[16] I. I. Piatetski-Shapiro and S. Rallis, *ϵ-factors of representations of classical groups*, Proc. Nat. Acad. Sci., **83** (1986), pp. 4589–4593.

[17] I. I. Piatetski-Shapiro and S. Rallis, L-functions for classical groups, Lecture Notes in Math., 1254, Springer-Verlag, New York, 1987, pp. 1–52.

[18] S. Rallis, *On the Howe duality conjecture*, Compositio Math., **51** (1984), pp. 333–399.

[19] R. Rao, *On some explicit formulas in the theory of the Weil representation*, Pacific J. Math., **157** (1993), pp. 335–371.

[20] W. Sweet, *A computation of the gamma matrix of a family of p-adic zeta integrals*, J. of Number Theory, **55** (1995), 222–260.

[21] ______, *Functional equations of p-adic zeta integrals and representations of the metaplectic group*, preprint, 1995.

[22] J.-L. Waldspurger, *Demonstration d'une conjecture de duality de Howe dans le case p-adiques, $p \neq 2$*, Festschrift in Honor of Piatetski-Shapiro, Israel Math. Conf. Proc., **2**, 1990, pp. 267–324.

Stephen S. Kudla
Math Department
University of Maryland
College Park, MD 20742, USA
ssk@math.umd.edu

Stephen Rallis
Department of Mathematics
The Ohio State University
231W 18th Ave
Columbus OH 43210. USA
haar@math.ohio-state.edu

On the local factors of representations of classical groups

*Erez M. Lapid and Stephen Rallis**

Abstract. We elaborate on the local aspects of the doubling method of Piatetski-Shapiro and Rallis and define the local factors of representations of classical groups. In particular, we determine the root number of any irreducible representation of a symplectic or an even orthogonal group. This is an analogue of a result of Deligne ([**Del76**]).

1. Introduction

The main purpose of this note is to elaborate on the doubling method of Piatetski-Shapiro and Rallis developed in [**PSR86**], [**GPSR87, Part A**]. We will mainly focus on the local aspects. The setup consists of a classical group G over a number field F,[1] i.e. the group of isomorphisms of a pair $\mathcal{V} = (V, h)$ consisting of an n-dimensional vector space V over E and a symplectic, symmetric or hermitian form h on V. Here $E = F$ in the symplectic and symmetric cases and E is a quadratic extension of F in the hermitian case. For any cuspidal representation π of $G(\mathbb{A})$, where $\mathbb{A}$ is the ring of adèles of F, and a Hecke character ω of $\mathbb{I}_E$, the doubling method gives a certain zeta-integral of Rankin-Selberg type which yields (up to some normalizing factors) at least the partial L-function $L_S^{\mathcal{V}}(s, \pi \times \omega)$ corresponding to the transfer of π to an appropriate GL_N. The analytic properties of the L-functions are then deduced, at least in principle, from those of the Eisenstein series appearing in the integral.

The doubling method applies equally well in the local situation to define local L and ε factors. It is simpler to consider first the local gamma factors (depending on a choice of a non-trivial additive character of the field). They are essentially given as the proportionality constants for the local zeta-integrals, with respect to a suitably normalized intertwining operator. These factors are better behaved than the L and ε factors themselves, namely, they are compatible, in a strong way, with parabolic induction. The functional equation

*Second named author partially supported by NSF grant DMS 9970342.

[1] The restriction to characteristic zero is only for convenience.

of the L-function translates informally to "$\prod_v \gamma^{V_v}(s, \pi_v \otimes \omega_v, \psi_v)$"$= 1$ for a cuspidal $\pi = \otimes_v \pi_v$, a Hecke character $\omega = \otimes_v \omega_v$ of $E^* \backslash \mathbb{I}_E$ and a non-trivial character $\psi = \otimes \psi_v$ of $F \backslash \mathbb{A}_F$. In fact, these properties, together with some other simpler ones, completely characterize the gamma factors. Granted the gamma factors, one can define, following Shahidi ([**Sha90**]), the local L and ε-factors using the relation

$$\gamma^V(s, \pi \times \omega, \psi) = \varepsilon^V(s, \pi \times \omega, \psi) \frac{L^V(1-s, \tilde{\pi} \times \omega^{-1})}{L^V(s, \pi \times \omega)}.$$

Namely, (say in the p-adic case) $L^V(s, \pi \times \omega)$ is the numerator of $\gamma^V(s, \pi \times \omega, \psi)$ in the tempered case, and is defined using the Langlands classification in general. The payoff is a definition and an exact functional equation for the global L-function as predicted in [**Lan70a**]. One expects that

$$L^V(s, \pi \times \omega) = L(s, \mathrm{BC}(\pi) \otimes \omega), \quad \varepsilon^V(s, \pi \times \omega, \psi) = \varepsilon(s, \mathrm{BC}(\pi) \otimes \omega, \psi)$$

where the factors on the right-hand side are the usual ones defined by Godement and Jacquet ([**GJ72**]) and where $\mathrm{BC}(\pi)$ is the (hypothetical) transfer of π to $GL_N(E)$, where N is either $n+1$, n or $n-1$ (depending on the type of h).

It is of interest to define the L-factors more directly from the local zeta integrals as a "g.c.d." of "good sections". This may be important, for example, for the analysis of the poles of the global L-function. We will not discuss these aspects here, but refer the reader to [**PSR87**] and [**HKS96**] for further analysis. Perhaps the geometric interpretation of [**BK99**] will shed some more light on these issues.

Let now F be a local field and suppose that h is either symplectic or symmetric with n even. In the latter case, let D be $(-1)^{n/2}$ times the discriminant of h and let τ_D be the quadratic character $x \mapsto (x, D)$ of F^* where $(\cdot, \cdot)$ denotes the Hilbert symbol. As a by-product of our analysis we will obtain the following result.

Theorem 1. *Let π be an irreducible representation of $G(F)$ and suppose that $\omega^2 = 1$. Then*

$$\varepsilon^V(\frac{1}{2}, \pi \times \omega, \psi) = \pi(-1)\omega(-1)^{n/2} \cdot \begin{cases} \varepsilon(\frac{1}{2}, \omega, \psi) & \textit{symplectic case,} \\ \omega(D)\varepsilon(\frac{1}{2}, \tau_D, \psi) & \textit{orthogonal case.} \end{cases}$$

The factor $\varepsilon(\frac{1}{2}, \tau_D, \psi)$ is the ratio $\gamma_F(\psi_D)/\gamma_F(\psi)$ where $\psi_D(x) = \psi(Dx)$ and $\gamma_F(\eta)$ is the Weil index of the character of the second degree $\eta(x^2)$ ([**Wei64**]). It is also the Langlands constant $\lambda(F(\sqrt{D})/F, \psi)$ of ([**Lan70b**]).

Theorem 1 is the analogue, under the local Langlands conjecture, of a result of Deligne for orthogonal representations of the Weil–Deligne group ([**Del76**]). In an earlier paper ([**Lap04**]) the first named author proved a special case of Theorem 1 for generic representations in the split case using global methods

(Eisenstein series). While this method potentially carries over, with some difficulty, to the general case, it is more natural to prove Theorem 1 in the context of the doubling method.

From Theorem 1 we immediately deduce the following theorem.

Theorem 2. *Let π be a cuspidal representation of either a symplectic group or an even orthogonal group over a number field. Then $\varepsilon(\frac{1}{2}, \pi) = 1$.*

This theorem answers a conjecture of Prasad and Ramakrishnan (**[PR99]**). Their motivation came from its analogue to the Galois side, which had been proved by Fröhlich and Queyrut (**[FQ73]**), and which inspired Deligne's result as well. It also plays a role, at least implicitly, in the far-reaching multiplicity conjectures of Arthur (**[AG91]**).

It is worthwhile to explicate Theorem 1 in the case where $n = 2$ and h is the norm form of a quadratic extension E/F of discriminant D with Galois involution $x \mapsto \bar{x}$. Let χ be a character of E^1 and ω a quadratic character of F^*, and let χ' and $\tilde{\omega}$ be the characters of E^* defined by $\chi'(x) = \chi(x/\bar{x})$ and $\tilde{\omega}(x) = \omega(x\bar{x})$ respectively. Finally, let ψ_F be a non-trivial additive character of F and let $\psi_E = \psi_F \circ \mathrm{Tr}_{E/F}$. Then (cf. **[Del76, Theorem 3.2]**, **[FQ73]**)

$$\varepsilon(\frac{1}{2}, \chi'\tilde{\omega}, \psi_E) = \chi(-1)\omega(-D).$$

We also have a version of Theorems 1 and 2 in the hermitian case (see §10).

The case of odd orthogonal groups is perhaps more interesting, since the root number is related to the so-called "theta dichotomy" – the non-vanishing of certain theta-lifts (cf. **[HKS96]**, where the hermitian case is discussed). In fact, the main idea of Theorem 1 shows up in **[HKS96]**. The point is that at the cases at hand, the induced representation appearing in the local integrals is irreducible at the point of symmetry. Hence, the normalized intertwining operator acts as the identity, so that the proportionality constant, which is essentially the gamma factor, is 1.

There is certainly a lot of overlap between the material here and previous works, especially, **[PSR86]**, **[GPSR87]** **[PSR87]** and **[KR90b]**. The main focus here is on a complete description of the gamma factors and their characterization. We also strive to fill some gaps and streamline the discussion about multiplicativity of the zeta integrals, and the normalization of intertwining operators. We do our best to minimize the amount of computation by using inductive and global arguments, deviating at times from the point of view of **[PSR86]**. Another contribution is the determination of the gamma factors in the archimedean case, by applying results of Branson, Ólafsson and Ørsted (**[BÓØ96]**). We also treat the odd orthogonal case which had not been previously discussed. In principle, the unitary groups coming from division algebras, as well as metaplectic groups can also be included in the setup, but we do not do it here.

The doubling method has many important consequences, for example, to the application of the converse theorem (**[RS]**), as well as to other types of L-functions (**[GPSR97]**). As we have already indicated, it is also tightly connected, through the Siegel-Weil formula, to the theta correspondence (cf. **[Ral87]**, **[KR94]**). In fact, historically, it was the theta correspondence which led to the doubling method. We will discuss none of this here. Other works on the doubling method include **[Tak97]** and **[Kim00]**. There is also a great deal of literature, going back to Siegel, on the special Eisenstein series occurring in the doubling method, and its local properties (see **[Sie36]**, **[Sie37]**, **[Wei65]**, **[Shi99c]**, **[Shi99a]**, **[Shi99b]**, **[Shi82]**, **[Kat01]**, **[Kat99]**, **[Kit89]**, **[Kit88]**, **[Kit86]**, **[SH00]**, **[HS88]**, to mention a few).

We would like to thank Bill Casselman, Jim Cogdell, Jean-Pierre Labesse, Dragan Miličić, Goran Muić, Omer Offen, Gestur Ólafsson, Gopal Prasad, Jonathan Rogawski, David Soudry and Yakov Varshavsky for their help on some fine points in the arguments. We also thank the referee for his careful reading and useful suggestions.

The first named author would like to express his deep appreciation to the second named author on the occasion of his 60th birthday.

2. Preliminaries

Until §8 F will be a local field of characteristic 0. If F is non-archimedean then we denote by q_F, or simply by q, the number of elements in the residue field. For any algebraic group G defined over F, we will denote the F-points of G also by G. For any such group we will denote by δ_G the modulus function of G. More generally, if G acts by automorphisms on a group H we denote by $\delta_{G;H}$ the modulus function of G on H. Thus $\delta_G = \delta_{G;G} = \delta_{G;\mathrm{rad}(G)}$ where $\mathrm{rad}(G)$ denotes the unipotent radical of G with G acting by conjugation. Recall the following well-known lemma.

Lemma 1. *Let P and Q be parabolic subgroups of a connected reductive group G. Denote by M and L the Levi parts of P and Q respectively and by U and V the corresponding unipotent radicals. Let proj_M be the canonical projection from P onto M and set $S = P \cap Q$. Then*

1. *The image of S under proj_M is the parabolic subgroup P' of M whose unipotent radical is $\mathrm{proj}_M(P \cap V)$. Similarly, $Q' = \mathrm{proj}_L(S)$ is a parabolic subgroup of L.*
2. *The Levi parts of S, P' and Q' are canonically isomorphic.*
3. *We have*
$$\delta_Q^{\frac{1}{2}} \cdot \delta_{S;U\cap Q\backslash U} = (\delta_{P'}/\delta_{Q'})^{\frac{1}{2}} \cdot \delta_P^{\frac{1}{2}}. \tag{1}$$

Note that the kernel of proj_M on $P \cap V$ coincides with the kernel of proj_L on $Q \cap U$ (and they are both equal to $U \cap V$). It follows that $\delta_{P'}/\delta_{Q'} = \delta_{S;P\cap V}/\delta_{S;Q\cap U}$. Hence (1) reduces to the relation

$$\delta_Q \cdot \delta_P = \delta_{S;U\cap Q} \cdot \delta_{S;V\cap P} \text{ on } S$$

which is proved in **[BZ77, §6]**.

Let now E be a local field of characteristic 0 with an automorphism θ of order 1 or 2, and let F be the fixed field of θ. Denote by $|\cdot| = |\cdot|_E$ the normalized absolute value of E. Let $\mathcal{V}$ be a pair (V, h) consisting of an n-dimensional vector space V over E and a sesqui-linear form h on V such that

$$\theta(h(v, u)) = \epsilon h(u, v) \quad \forall u, v \in V$$

where $\epsilon\theta(\epsilon) = 1$. We will also assume that either $h = 0$ or h is non-degenerate. If h is non-degenerate then it is either symmetric or anti-symmetric (if $E = F$), or ϵ-hermitian (if $E \neq F$). There is a slight abuse of notation in the case where $V = 0$: it will be convenient to artificially distinguish the case where h is symplectic from $h = 0$.

Unless otherwise mentioned we will always denote by G the group of isometries $\mathrm{Isom}(\mathcal{V})$ of $\mathcal{V}$, considered as an algebraic group over F. Note that in the ϵ-hermitian case we can assume, by multiplying h by an appropriate constant, that $\epsilon = 1$ (Hilbert 90). For any h-totally isotropic subspace W of V we denote by S_W the stabilizer of W in G. This is a maximal parabolic subgroup of G and all maximal parabolic subgroups of G are obtained this way.[2]

The unipotent radical R_W of S_W consists of the elements in G which act as the identity on both W and $W^\perp/W$, where

$$W^\perp = \{v \in V : h(v, w) = 0 \text{ for all } w \in W\}.$$

Note that if h is non-degenerate then it induces an identification of $V/W^\perp$ with the θ-dual of W.

For convenience we may also consider formal finite sums $\mathcal{V} = \boxplus \mathcal{V}_i = \boxplus (V_i, h_i)$, $i = 1, \ldots, k$ where each h_i is either 0 or non-degenerate. We set $\mathrm{Isom}(\mathcal{V}) = \prod \mathrm{Isom}(\mathcal{V}_i)$. We say that $\mathcal{V}$ is compound if $k > 1$ and simple if $k = 1$.

For example, the Levi part M_W of S_W is canonically isomorphic to $\mathrm{Isom}(\mathcal{W})$ where $\mathcal{W} = (W, 0) \boxplus (W^\perp/W, h)$. More generally, if $0 = W_0 \subset W_1 \subset \cdots \subset W_k$ is an isotropic flag then its stabilizer is a parabolic subgroup whose Levi part is isomorphic to $\mathrm{Isom}(\mathcal{W})$ where

$$\mathcal{W} = \boxplus_{i=1}^k (W_i/W_{i-1}, 0) \boxplus (W^\perp/W_k, h). \tag{2}$$

All parabolic subgroups are obtained this way.

[2]For the definition of parabolic subgroups in the non-connected case see e.g. **[MW95, Appendix IV]**.

In the proofs below we will often implicitly assume that $\mathcal{V}$ is simple. The extension to the compound case is straightforward.

We will also need to consider the case where $E = F \oplus F$, and h is either 0 or non-degenerate (hermitian). The data $\mathcal{V}$ consists of $V = (V_1, V_2)$ where V_i, $i = 1, 2$ are finite dimensional vector spaces over F (which are in duality under h if h is non-degenerate). In this case G is either $GL(V_1) \times GL(V_2)$ if $h = 0$, or $G \simeq GL(V_1) \simeq GL(V_2)$ if h is non-degenerate. The totally isotropic subspaces W are just pairs $W = (W_1, W_2)$ of subspaces of the V_i's with $W_1 \subset W_2^{\perp}$. If $h = 0$ then S_W is the product of the parabolic subgroups $S_{W_1} \times S_{W_2}$ in $G = GL(V_1) \times GL(V_2)$. If h is non-degenerate then S_W is the parabolic subgroup of $GL(V_1)$ stabilizing W_1 and $W_2^{\perp}$. As before, we can still view, if we wish, the Levi part of S_W as $\mathrm{Isom}(\mathcal{W})$.

3. The doubling method

Let $\mathcal{V} \times \mathcal{V}$ be the space $V \times V$ equipped with the form $h^{\square} = h \oplus -h$ and let $G^{\square} = \mathrm{Isom}(\mathcal{V} \times \mathcal{V})$. We identify $G \times G$ with the subgroup of $G^{\square}$ preserving $V_1 = V \times \{0\}$ and $V_2 = \{0\} \times V$. If $h \neq 0$ then the subspace $V^{\diamond} = \{(v, v) : v \in V\}$ is a maximal totally isotropic subspace of $\mathcal{V} \times \mathcal{V}$. Hence, the group $G^{\square}$ is quasi-split. We will consider the maximal parabolic subgroup (or, a product of two maximal parabolic subgroups if $E = F \oplus F$ and $h = 0$) $S_{V^{\diamond}}$ of $G^{\square}$ stabilizing $V^{\diamond}$. If h is non-degenerate then $S_{V^{\diamond}}$ is of Siegel type and its Levi part is canonically isomorphic to $GL(V^{\diamond})$. If $h = 0$ then $S_{V^{\diamond}}$ is of type (n, n) and its Levi part is canonically isomorphic to $GL(V^{\diamond}) \times GL(V \times V/V^{\diamond})$. In any case we have

$$(G \times G) \cap S_{V^{\diamond}} = G^{\diamond} = \{(g, g) : g \in G\}. \qquad (3)$$

We denote by $\Delta_{V^{\diamond}}$ or simply by Δ the following character on $S_{V^{\diamond}}$:

$$\Delta_{V^{\diamond}}(x) = \begin{cases} \det{}_{V^{\diamond}}(x) & h \text{ non-degenerate,} \\ \frac{\det_{V^{\diamond}}(x)}{\theta(\det_{V \times V/V^{\diamond}}(x))} & h = 0. \end{cases}$$

We also define the character $\kappa = \kappa^{\mathcal{V}}$ on $G \times G$ by

$$\kappa(x) = \begin{cases} \det{}_{V_2}(x)^{-1} & h \text{ non-degenerate,} \\ \frac{\det_{V_2}(x)^{-1}}{\theta(\det_{V_2}(x))^{-1}} & h = 0. \end{cases}$$

Note that $\Delta_{V^{\diamond}} = \kappa^{-1}$ on $G^{\diamond}$. Let ω be a character of E^* and let $I(\omega, s) = I^{\mathcal{V}}(\omega, s)$ be the induced space (in the category of smooth representations) $\mathrm{Ind}_{S_{V^{\diamond}}}^{G^{\square}} \omega_s \circ \Delta_{V^{\diamond}}$ where $\omega_s = \omega \cdot |\cdot|^s$ (all inductions are normalized).

Choose a maximal compact subgroup K of $G^{\square}$ such that $S_{\mathcal{V}^{\diamond}} \cdot K = G^{\square}$. Extend $|\Delta|$ to a right-K-invariant function on $G^{\square}$, which we still denote by $|\Delta|$. We may then identify $I(\omega, 0)$ with $I(\omega, s)$, for any s, (as K-modules) by $f \mapsto f_s := f \cdot |\Delta|^s$.

Define the intertwining operator

$$M^{\mathcal{V}}(\omega, s) = M(\omega, s) = M(s) : I^{\mathcal{V}}(\omega, s) \to I^{\mathcal{V}}(\theta(\omega)^{-1}, -s)$$

by

$$M(s)f_s(g) = \int_{R_{\mathcal{V}^{\diamond}}} f_s(wug)\, du$$

where $w = (1, -1) \in G \times G \subset G^{\square}$. The integral defining $M(s)$ converges for $\Re(s) \gg 0$ and $M(s)$ admits a meromorphic continuation.

Let π be an admissible representation of G and let $\widetilde{\pi}$ be its admissible dual. (We will often not distinguish between π and its underlying space.) Denote by $\mathcal{P} = \mathcal{P}_\pi : \pi \otimes \widetilde{\pi} \to \mathbf{C}$ the standard pairing.

For any $\alpha \otimes \widetilde{\alpha} \in \pi \otimes \widetilde{\pi}$ and $f \in I(\omega, 0)$ define the zeta-integral

$$Z^{\mathcal{V}}(f_s, \alpha \otimes \widetilde{\alpha}) = \int_{G^{\diamond} \backslash G \times G} f_s(g)\omega(\kappa(g))\mathcal{P}((\pi \otimes \widetilde{\pi})(g)(\alpha \otimes \widetilde{\alpha}))\, dg.$$

The following basic theorem is stated in **[PSR86]**.

Theorem 3. *Suppose that π is irreducible. The integral defining $Z^{\mathcal{V}}(f_s, \alpha \otimes \widetilde{\alpha})$ converges for $\Re(s) \gg 0$. The zeta-integral extends to a meromorphic function in s which in the non-archimedean case is a rational function in q^{-s}. Moreover it satisfies the functional equation*

$$Z(M(s)f_s, \alpha \otimes \widetilde{\alpha}) = \Gamma^{\mathcal{V}}(s, \pi, \omega) Z(f_s, \alpha \otimes \widetilde{\alpha})$$

for an appropriate scalar-valued function $\Gamma^{\mathcal{V}}(s, \pi, \omega)$.

We will recall the proof of Theorem 3 in the next section. For now, we observe that if h is non-degenerate, E is a field, and π is supercuspidal[3] then the integrand is compactly supported, and hence the integral converges for all s. In the square-integrable case we have the following.

Lemma 2. *Suppose that the form h is non-degenerate, E is a field, π is square-integrable and ω is unitary. Then the integral defining $Z^{\mathcal{V}}(f_s, \alpha \otimes \widetilde{\alpha})$ is absolutely convergent for $\Re(s) \geq 0$.*

---

[3] i.e., having compactly supported matrix coefficients (which in particular means that $\mathcal{V}$ is anisotropic in the archimedean case)

Proof. If $s \in i\mathbb{R}$ then $|f_s(\cdot)|^2 \in \mathrm{Ind}_{S_{V^\diamond}}^{G^\square}(\delta_{S_{V^\diamond}}^{\frac{1}{2}})$. Hence, since $G^\diamond\backslash G \times G$ is open and dense in $S_{V^\diamond}\backslash G^\square$ (cf. **[GPSR87]**), we have

$$\int_{G^\diamond\backslash G\times G} |f_s(g)|^2\, dg = \int_{S_{V^\diamond}\backslash G^\square} |f_s(h)|^2\, dh = \int_K |f_s(k)|^2\, dk < \infty.$$

By Cauchy-Schwartz it follows that $Z(f_s, \alpha\otimes\widetilde{\alpha})$ converges for $s \in i\mathbb{R}$ since π is square-integrable. The same will be true for $\Re(s) > 0$ since $|\Delta|$ is bounded on $G \times G$. To wit, let $V^{\overline{\diamond}} = \{(v, -v) : v \in V\}$ and consider the parabolic subgroup $S_{V^{\overline{\diamond}}}$ opposite to $S_{V^\diamond}$. It is obtained by conjugating $S_{V^\diamond}$ by w. Write $g \in G \times G$ as muk with $m \in S_{V^\diamond} \cap S_{V^{\overline{\diamond}}}$, $u \in R_{V^\diamond}$ and $k \in K$. Then

$$muk = g = wgw^{-1} = wmw^{-1}wukw^{-1},$$

and hence,

$$|\Delta(g)| = |\Delta(m)| = |\Delta(wmw^{-1})| \cdot |\Delta(wukw^{-1})| = |\Delta(m)|^{-1} \cdot |\Delta(wukw^{-1})|.$$

Thus,

$$|\Delta(g)| = |\Delta(wukw^{-1})|^{\frac{1}{2}}$$

and the latter is known to be bounded (cf. **[Art78, p. 936]**). □

In the compound case where $\mathcal{V} = \boxplus\mathcal{V}_i$, the "doubled" group is $\mathrm{Isom}(\mathcal{V} \times \mathcal{V})$ where by definition $\mathcal{V} \times \mathcal{V} = \boxplus(\mathcal{V}_i \times \mathcal{V}_i)$. We also set $\mathcal{V}^\diamond = \boxplus{\mathcal{V}_i}^\diamond$.

4. Multiplicativity

Suppose that P is a parabolic subgroup of G with unipotent radical U and Levi part M. Let σ be an admissible irreducible representation of M (or its pull-back to P) and let $\pi = \mathrm{Ind}_P^G(\sigma)$. The space of π consists of the smooth functions $\alpha : G \to V_\sigma$ such that

$$\alpha(pg) = \delta_P(p)^{\frac{1}{2}}\sigma(p)\,[\alpha(g)] \text{ for all } p \in P, g \in G.$$

Choosing a semi-invariant measure on $P\backslash G$ we may identify $\widetilde{\pi}$ with $\mathrm{Ind}_P^G(\widetilde{\sigma})$. We then identify $\pi\otimes\widetilde{\pi}$ with $\mathrm{Ind}_{P\times P}^{G\times G}\sigma\otimes\widetilde{\sigma}$. Let $\alpha\otimes\widetilde{\alpha} \in \pi\otimes\widetilde{\pi}$. Thus,

$$(\alpha\otimes\widetilde{\alpha})(pg) = \delta_{P\times P}(p)^{\frac{1}{2}}(\sigma\otimes\widetilde{\sigma})(p)\,[(\alpha\otimes\widetilde{\alpha})(g)] \tag{4}$$

for all $p \in P \times P$, $g \in G \times G$. We also have

$$\mathcal{P}_\pi(\alpha\otimes\widetilde{\alpha}) = \int_{P^\diamond\backslash G^\diamond} \mathcal{P}_\sigma((\alpha\otimes\widetilde{\alpha})(x))\, dx.$$

We may compute $Z^{\mathcal{V}}(f_s, \alpha\otimes\tilde{\alpha})$ as

$$\int_{G^{\diamond}\backslash G\times G} f_s(g)\omega(\kappa(g)) \int_{P^{\diamond}\backslash G^{\diamond}} \mathcal{P}_\sigma((\alpha\otimes\tilde{\alpha})(xg))\, dx\, dg$$
$$= \int_{P^{\diamond}\backslash G\times G} f_s(g)\omega(\kappa(g))\mathcal{P}_\sigma((\alpha\otimes\tilde{\alpha})(g))\, dg.$$

Integrating first over $P^{\diamond}\backslash P\times P$ and using (4) we obtain

$$\int_{P\times P\backslash G\times G}\int_{P^{\diamond}\backslash P\times P} \delta_{P\times P}(p)^{-\frac{1}{2}}\omega(\kappa(p)\kappa(g))f_s(pg)$$
$$\mathcal{P}_\sigma((\sigma\otimes\tilde{\sigma})(p)(\alpha\otimes\tilde{\alpha})(g))\, dp\, dg.$$

Writing $dp = du\, dm$ we finally get

$$Z^{\mathcal{V}}(f_s,\alpha\otimes\tilde{\alpha}) = \int_{P\times P\backslash G\times G}\int_{M^{\diamond}\backslash M\times M}\left(\int_{U^{\diamond}\backslash U\times U} f_s(umg)\, du\right)$$
$$\delta_{P\times P}^{-\frac{1}{2}}(m)\omega(\kappa(m))\omega(\kappa(g))\mathcal{P}_\sigma((\sigma\otimes\tilde{\sigma})(m)(\alpha\otimes\tilde{\alpha})(g))\, dm\, dg. \quad (5)$$

Assume now that P is a maximal parabolic subgroup, i.e. it is the stabilizer of a totally isotropic subspace W of V. Recall that U consists of the isometries acting as the identity on W and $W^{\perp}/W$. For simplicity of notation we assume that E is a field. Set $m = \dim(W)$. The maximal parabolic subgroup $S_{W\times W}$ of $G^{\square}$ stabilizing $W\times W$ contains $P\times P$ and its Levi part $M_{W\times W}$ is canonically isomorphic to $\mathrm{Isom}(\mathcal{W}\times\mathcal{W})$ where $\mathcal{W} = (W,0)\boxplus(W^{\perp}/W,h)$.

Lemma 3. *1. Suppose that h is non-degenerate and identify the θ-dual of $W\times W$ with $V\times V/W^{\perp}\times W^{\perp}$ through $h^{\square}$. Then the annihilator of $W^{\diamond}$ in $V\times V/W^{\perp}\times W^{\perp}$ is the image $(V/W^{\perp})^{\diamond}$ of $V^{\diamond}$ under the canonical projection.*

2. The canonical injection $\iota : U^{\diamond}\backslash U\times U\to(R_{W\times W}\cap S_{V^{\diamond}})\backslash R_{W\times W}$ is an isomorphism.

Proof. The first part is clear. The injectivity of ι follows from (3). To continue, we identify the set $S_{V^{\diamond}}\backslash G^{\square}$ with the set $\mathfrak{X}$ of n-dimensional totally isotropic subspaces Y of $\mathcal{V}\times\mathcal{V}$ via $S_{V^{\diamond}}g\mapsto g^{-1}(V^{\diamond})$. We claim that under this identification the set $(R_{W\times W}\cap S_{V^{\diamond}})\backslash R_{W\times W}$ (a subset of $S_{V^{\diamond}}\backslash G^{\square}$) becomes

$$\mathfrak{Y} = \{Y\in\mathfrak{X} : Y\cap W\times W = W^{\diamond}, Y\cap W^{\perp}\times W^{\perp}\equiv W^{\perp\diamond}\pmod{W\times W}\}$$

and moreover, any $Y\in\mathfrak{Y}$ is of the form $g^{-1}(V^{\diamond})$ with $g\in U\times U$. This will clearly imply the lemma. To prove the claim we first note that if $g\in R_{W\times W}$ then $Y = g^{-1}(V^{\diamond})\in\mathfrak{Y}$ since g acts as the identity on both $W\times W$ and

$(W \times W)^{\perp}/W \times W = W^{\perp} \times W^{\perp}/W \times W$. On the other hand, if $Y \in \mathfrak{Y}$ then Y satisfies

$$Y \equiv V^{\diamond} \pmod{W^{\perp} \times W^{\perp}}$$

by part (1). It follows that Y can be parameterized as $\{(j_1(v), j_2(v)) : v \in V\}$ where $j_1, j_2 \in GL(V)$ both act as the identity on W, $W^{\perp}/W$ and $V/W^{\perp}$. Since Y is $h^{\square}$-totally isotropic we have $h(j_1(v), j_1(v')) = h(j_2(v), j_2(v'))$ for all $v, v' \in V$. It follows that $j = j_1 j_2^{-1} \in \mathrm{Isom}(\mathcal{V})$ and since j acts as the identity on W and $W^{\perp}/W$, we obtain $j \in U$. Since $Y = \{(j(v), v) : v \in V\}$ our claim is proved. □

Denote by $\Delta_{\mathcal{W};\mathcal{V}}$ the character on $S_{W \times W}$ defined by

$$\Delta_{\mathcal{W};\mathcal{V}} = \begin{cases} \det{}_{W \times W}(\cdot)^{\frac{m-n}{2}}, & \text{if } h \text{ is non-degenerate;} \\ \dfrac{\det{}_{W \times W}(\cdot)^{\frac{m-n}{2}}}{\theta(\det{}_{V \times V/W \times W}(\cdot))^{-\frac{m}{2}}}, & \text{if } h = 0. \end{cases}$$

Lemma 4. *Let $S = S_{\mathcal{V}^{\diamond}} \cap S_{W \times W}$.*

1. *The image of S under $\mathrm{proj}_{M_{W \times W}}$ is the parabolic subgroup $S_{\mathcal{W}^{\diamond}}$ of $\mathrm{Isom}(\mathcal{W} \times \mathcal{W})$. Its unipotent radical is the image of $R_{\mathcal{V}^{\diamond}} \cap S_{W \times W}$.*
2. $\Delta_{\mathcal{V}^{\diamond}}|_S = \Delta_{\mathcal{W}^{\diamond}} \circ \mathrm{proj}_{M_{W \times W}}$.
3. *The group $\mathrm{proj}_{M_{\mathcal{V}^{\diamond}}}(S_{\mathcal{V}^{\diamond}} \cap R_{W \times W})$ is the unipotent radical of the parabolic subgroup $Q' = \mathrm{proj}_{M_{\mathcal{V}^{\diamond}}}(S)$ of $M_{\mathcal{V}^{\diamond}}$. In the non-degenerate case Q' is the stabilizer of the flag $W^{\diamond} \subset W^{\perp\diamond}$. If $h = 0$ then Q' is the stabilizer of $W^{\diamond} \subset V^{\diamond}$ and $W \times W/W^{\diamond} \subset V \times V/V^{\diamond}$.*
4. *The restriction of $|\Delta_{\mathcal{W};\mathcal{V}}|$ to S is $\delta_{Q'}^{-\frac{1}{2}} \circ \mathrm{proj}_{M_{\mathcal{V}^{\diamond}}}$*

Proof. Denote by P' the image of S under $\mathrm{proj}_{M_{W \times W}}$. Clearly $P' \subset S_{\mathcal{W}^{\diamond}}$ since any $h \in S$ stabilizes $\mathcal{W}^{\diamond}$. By a similar reasoning the image U' of $R_{\mathcal{V}^{\diamond}} \cap S_{W \times W}$ is contained in $R_{\mathcal{W}^{\diamond}}$. By Lemma 1 P' is a parabolic subgroup of $\mathrm{Isom}(\mathcal{W} \times \mathcal{W})$ whose unipotent radical is U'. Hence from the previous inclusions we infer that $P' = S_{\mathcal{W}^{\diamond}}$. Part 3 is proved analogously. To show part 2, observe that for any $m \in S$ we have

$$\det{}_{V^{\diamond}}(m) = \det{}_{W^{\diamond}}(m) \cdot \det{}_{(W^{\perp}/W)^{\diamond}}(m) \cdot \det{}_{(V/W^{\perp})^{\diamond}}(m).$$

If h is non-degenerate then

$$\det{}_{(V/W^{\perp})^{\diamond}}(m) = \theta(\det{}_{W \times W/W^{\diamond}}(m))^{-1} \tag{6}$$

since the $h^{\square}$-dual of $W \times W/W^{\diamond}$ is $(V/W^{\perp})^{\diamond}$ by Lemma 3, part 1. The second part follows in that case. The case $h = 0$ is proved in a similar way.

Finally, to prove the last part, suppose first that h is non-degenerate. Then $\dim(W^{\diamond}) = \dim((V/W^{\perp})^{\diamond})$ and we have

$$\delta_{Q'} = |\det{}^{n-m}_{W^{\diamond}} \cdot \det{}^{m-n}_{(V/W^{\perp})^{\diamond}}| = |\det{}^{n-m}_{W^{\diamond}} \cdot \det{}^{n-m}_{W\times W/W^{\diamond}}| = |\det{}^{n-m}_{W\times W}|$$

by (6). Similarly, if $h = 0$ then

$$\delta_{Q'} = \frac{|\det_{W^{\diamond}}|^{n-m}}{|\det_{V^{\diamond}/W^{\diamond}}|^{m}} \cdot \frac{|\det_{(W\times W+V^{\diamond})/V^{\diamond}}|^{n-m}}{|\det_{V\times V/(W\times W+V^{\diamond})}|^{m}} = \frac{|\det_{W\times W}|^{n-m}}{|\det_{V\times V/W\times W}|^{m}}$$

as required. □

Proposition 1. *The integral*

$$f_s \mapsto [\Psi(s)f_s](g) : p \mapsto |\Delta_{W;V}(p)|^{-1}\delta_{S_{W\times W}}(p)^{-\frac{1}{2}} \times \int_{U^{\diamond}\backslash U\times U} f_s(upg)\, du, \quad p \in S_{W\times W}, g \in G^{\square}$$

converges for $\Re(s) \gg 0$, admits a meromorphic continuation and defines an intertwining map (in the category of smooth representations)

$$\Psi^{V,W}(\omega, s) = \Psi(s) : I^{V}(\omega, s) \to \operatorname{Ind}_{S_{W\times W}}^{G^{\square}}(I^{W}(\omega, s) \otimes |\Delta_{W;V}|).$$

Proof. Set

$$\Phi(g) = \int_{U^{\diamond}\backslash U\times U} f_s(ug)\, du$$

so that $[\Psi(s)f_s](g)$ is the map $p \mapsto |\Delta_{W;V}(p)|^{-1}\delta_{S_{W\times W}}(p)^{-\frac{1}{2}}\Phi(pg)$. It follows from Lemma 3 part (2), that

$$\Phi(g) = \int_{(R_{W\times W}\cap S_{V^{\diamond}})\backslash R_{W\times W}} f_s(ug)\, du. \tag{7}$$

Thus, $\Psi(s)$ is an intertwining operator of the standard type, whose analytic properties are well-known (**[Wal92, Chapter 10]**, [Sha81]). It also follows from (7) that Φ is left-$R_{W\times W}$-invariant and that

$$\Phi(mg) = \delta^{\frac{1}{2}}_{S_{V^{\diamond}}}(m)\omega_s(\Delta_{V^{\diamond}}(m))\delta_{S;(R_{W\times W}\cap S_{V^{\diamond}})\backslash R_{W\times W}}(m)\Phi(g)$$

for $m \in S = S_{V^{\diamond}} \cap S_{W\times W}$. Using (1) (with $P = S_{W\times W}$ and $Q = S_{V^{\diamond}}$) together with Lemma 4 we infer that

$$\Phi(mg) = \delta^{\frac{1}{2}}_{S_{W^{\diamond}}}\omega_s(\Delta_{W^{\diamond}}(\operatorname{proj}_{M_{W\times W}} m)) \cdot |\Delta_{W;V}(m)| \cdot \delta^{\frac{1}{2}}_{S_{W\times W}}(m)\Phi(g).$$

It remains to take into account part 1 of Lemma 4. □

Observe that the restriction of $|\Delta_{\mathcal{W};\mathcal{V}}| \cdot \delta_{S_{\mathcal{W}\times\mathcal{W}}}^{\frac{1}{2}}$ to $P \times P$ is $\delta_{P\times P}^{\frac{1}{2}}$. Indeed, if h is non-degenerate then $\delta_P = |\det_{\mathcal{W}}|^{n-m+e}$ where

$$e = \begin{cases} 1 & h \text{ anti-symmetric} \\ -1 & h \text{ symmetric} \\ 0 & h \text{ hermitian} \end{cases}$$

Thus, $\delta_{P\times P} = |\det_{\mathcal{W}\times\mathcal{W}}|^{n-m+e}$. Similarly, $\delta_{S_{\mathcal{W}\times\mathcal{W}}} = |\det_{\mathcal{W}\times\mathcal{W}}|^{2n-2m+e}$. The case $h = 0$ is also easy.

To summarize, we have proved the following proposition.

Proposition 2. *We have*

$$Z^{\mathcal{V}}(f_s, \alpha \otimes \tilde{\alpha}) = \int_{P\times P\backslash G\times G} \omega(\kappa(g)) Z^{\mathcal{W}}([\Psi^{\mathcal{V},\mathcal{W}}(s) f_s](g), (\alpha \otimes \tilde{\alpha})(g))\, dg.$$

Indeed, this follows from the relation (5) and Proposition 1 noting that the restriction of $\kappa^{\mathcal{V}}$ to $M \times M$ is $\kappa^{\mathcal{W}}$.

Let us now go back to the proof of Theorem 3. We first prove the absolute convergence. The proof of Proposition 2, in conjunction with the subrepresentation theorems of Jacquet (in the p-adic case) and Casselman (in the archimedean cases), reduces the statement to the case where π is supercuspidal modulo the center. We have already remarked that if h is non-degenerate, then the integrand is compactly supported. If $h = 0$ then we can majorize the integrand by replacing π and ω by unramified characters. By applying Proposition 2 once again we reduce to the case $n = 1$. The zeta-integral is now easy to compute (cf. §7 below).

The argument also reduces the meromorphic continuation to the supercuspidal case. To go further we need the following lemma.

Lemma 5. *The space*

$$\operatorname{Hom}(I^{\mathcal{V}}(\omega, s), \operatorname{Ind}_{S_{\mathcal{W}\times\mathcal{W}}}^{G^{\Box}}(I^{\mathcal{W}}(\omega, s) \otimes |\Delta_{\mathcal{W};\mathcal{V}}|)) \tag{8}$$

(in the category of smooth representations) is at most one-dimensional for a generic value of s. Thus,

$$\Psi(\theta(\omega)^{-1}, -s) \circ M^{\mathcal{V}}(\omega, s) = \alpha(\omega, s) \cdot \operatorname{Ind}_{S_{\mathcal{W}\times\mathcal{W}}}^{G^{\Box}} M^{\mathcal{W}}(\omega, s) \circ \Psi(\omega, s)$$

for some scalar $\alpha(\omega, s)$ where we use $\operatorname{Ind}_{S_{\mathcal{W}\times\mathcal{W}}}^{G^{\Box}}$ as a functor.

Proof. Both $\Psi(\omega, s)$ and $\operatorname{Ind}_{S_{\mathcal{W}\times\mathcal{W}}}^{G^{\Box}} M^{\mathcal{W}}(\omega, s)^{-1} \circ \Psi(\theta(\omega)^{-1}, -s) \circ M^{\mathcal{V}}(\omega, s)$ belong to (8). Thus, the second statement follows from the first. We also note that if $\omega = 1$ then the first statement holds because $\operatorname{Ind}_{S_{\mathcal{W}\times\mathcal{W}}}^{G^{\Box}}(I^{\mathcal{W}}(\omega, s) \otimes |\Delta_{\mathcal{W};\mathcal{V}}|)$ contains a unique vector (up to a constant) which is fixed under the hyper-special maximal compact. In the general case we use Frobenius reciprocity ([**BZ77**], [**Cas89, Lemma 4.2**]) to write (8) as

$$\mathrm{Hom}_{\mathrm{Isom}(\mathcal{W}\times\mathcal{W})}(J_{S_{W\times W}} I^{\mathcal{V}}(\omega,s), I^{\mathcal{W}}(\omega,s)\otimes|\Delta_{\mathcal{W};\mathcal{V}}|) \tag{9}$$

where $J_{S_{W\times W}}$ denotes the (normalized) Jacquet functor in the p-adic case and the co-invariants under the action of the Lie-algebra of $R_{W\times W}$ (again, normalized by $\delta_{S_{W\times W}}^{-\frac{1}{2}}$) in the archimedean case.

We argue in the non-degenerate case. (The case $h = 0$ is similar.) The double cosets $S_{W\times W}\backslash G^{\square}/S_{\mathcal{V}^{\diamond}}$ are indexed by the connected components of the Grassmanian of $W \times W$. For each $k = 0, \dots, 2m$ let $U_k \subset W \times W$ with $\dim U_k = k$. In the p-adic case it follows from the geometric Lemma of [**BZ77**] that the module $J_{S_{W\times W}} I^{\mathcal{V}}(\omega, s)$ is "glued from"

$$X_k = \mathrm{Ind}_{S_{U_k}}^{GL(W\times W)} \omega_{s+\frac{k-m}{2}}\left(\frac{\det_{U_k}}{\theta(\det_{W\times W/U_k})}\right)\otimes I^{(W^{\perp}/W,h)}(\omega, s+k-m)\otimes|\Delta_{\mathcal{W};\mathcal{V}}|$$

$k = 1, \dots, 2m$. In the archimedean case, the situation is more involved. However, it was proved in [**CHM00, Theorem 4.1**] that principal series representations admit a "Bruhat filtration". It is possible to extend this to $I^{\mathcal{V}}(\omega, s)$. The co-invariants of each term in the filtration are given by X_k. We claim that in both cases only $k = m$ can contribute to (9) for a generic s. Indeed, for each k the central character of X_k is $\omega_s^{2(k-m)}|\Delta_{\mathcal{W};\mathcal{V}}|$ and the functor $J_{S_{W\times W}}$ is right-exact. As $X_m = I^{\mathcal{W}}(\omega, s) \otimes |\Delta_{\mathcal{W};\mathcal{V}}|$ we obtain that (9), and therefore (8) is indeed at most one-dimensional for generic s. □

We can now finish the proof of Theorem 3. Using the lemma, the statement reduces the functional equation to the case where π is supercuspidal modulo the center. In the p-adic case, the argument in [**HKS96, §4**] shows that

$$\dim \mathrm{Hom}_{G\times G}(I(\omega, s), \widetilde{\pi} \times (\pi \otimes \omega)) \leq 1 \tag{10}$$

for a generic value of s (in fact for all s if h is non-degenerate). This implies the functional equation, as well as the rationally statement, by a well-known principle of Bernstein (cf. [**GPSR87, Part B,§12**]).

Consider now the archimedean case (and π supercuspidal modulo the center). If h is non-degenerate then $\mathcal{V}$ is anisotropic and it is easy to see that $K = G \times G$ is the maximal compact of $G^{\square}$. Thus, as a $G \times G$-module, $I(\omega, s)$ is

$$\mathrm{Ind}_{G^{\diamond}}^{G\times G}\omega = \oplus_{\pi\in\hat{G}}\widetilde{\pi} \times (\pi \otimes \omega),$$

from which (10), and the functional equation, follow. (The entirety of Z is clear in this case.) It remains to consider the case where $h = 0$ (and $n = 1$). Here $G = GL_1$ and $G^{\square} = GL_2$. As in [**CHM00**], we use the Bruhat decomposition to identify the space $I_s = I(\omega, s)$ with the smooth sections of a certain real analytic line bundle on $\mathbb{P}^1(E)$. We have an exact sequence of G-modules

$$0 \to I_s^0 \to I_s \to I_{s,1} \oplus I_{s,-1} \to 0$$

where I_s^0 is the subspace of sections vanishing to infinite order at ± 1 and $I_{s,\pm 1}$ is the formal sections (Taylor series) of the line bundle at ± 1. The space I_s^0 is isomorphic as a G-module to $\mathcal{S}(E^*)$, while $I_{s,\pm 1}$ is isomorphic to $\mathbb{C}[[E]]$, the completion of the polynomial algebra on E (considered as a vector space over $\mathbb{R}$), with G acting by multiplying by $\omega_s^{\pm 1}$ and translating E. Thus, for a given character χ there is a unique (up to a scalar) smooth G-module morphism from I_s^0 to χ (given by the Fourier transform at χ). On the other hand, for a generic value of s (outside an arithmetic progression) there is no smooth morphism from $I_{s,\pm 1}$ to χ. This shows (10). The meromorphicity of Z can be done by a direct computation. It amounts to the meromorphicity (in s) of the function

$$\int_0^\infty \left(\frac{x}{x^2+1}\right)^s x^\alpha g(\tan^{-1} x)dx$$

for any given α and a smooth function g on $[0, \pi/2]$ (see §9 below).

Remark 1. The results of this section also hold, with little modification, in the case where $E = F \oplus F$ and $V = (V_1, V_2)$. Namely, one has to replace the dimension n of V by (n_1, n_2) where $n_i = \dim V_i$, and similarly for m. (Of course, $n_1 = n_2 = n$ if $h \neq 0$.) Thus, for example, the character $|\Delta_{\mathcal{W};\mathcal{V}}|$ is now

$$|\det{}_{W_1\times W_1}(\cdot)^{\frac{m_1-n}{2}} \cdot \det{}_{W_2\times W_2}(\cdot)^{\frac{m_2-n}{2}}|$$

if h is non-degenerate and it is given by

$$\left|\frac{\det{}_{W_1\times W_1}(\cdot)^{\frac{m_1-n_1}{2}} \cdot \det{}_{W_2\times W_2}(\cdot)^{\frac{m_2-n_2}{2}}}{\det{}_{V_1\times V_1/W_1\times W_1}(\cdot)^{-\frac{m_1}{2}} \cdot \det{}_{V_2\times V_2/W_2\times W_2}(\cdot)^{-\frac{m_2}{2}}}\right|$$

if $h = 0$.

5. Normalization of intertwining operators

For the sake of simplicity we will implicitly assume that E is a field. The case $E = F \oplus F$ requires little modification.

Let $\mathfrak{g}$ be the Lie algebra of $G = \mathrm{Isom}(\mathcal{V})$, identified with

$$\{X \in \mathrm{End}(V) : h(Xu, v) + h(u, Xv) = 0 \text{ for all } u, v \in V\}.$$

Similarly for $\mathfrak{g}^\square$ – the Lie algebra of $G^\square$.

Let $V^{\overline{\Diamond}} = \{(v, -v) : v \in V\}$. Then $S_{V^{\overline{\Diamond}}}$ is a parabolic subgroup opposite to $S_{V^\Diamond}$. The map $u \mapsto u - I$ defines an isomorphism of the unipotent radical

$R_{V\overline{\diamond}}$ and its Lie algebra $\mathfrak{r}_{V\overline{\diamond}}$ which, as a subalgebra of $\mathfrak{g}^{\Box}$, is in turn identified with

$$\{X \in \mathfrak{g}^{\Box} : \mathrm{Ker}X \supset V^{\overline{\diamond}} \supset \mathrm{Im}X\}.$$

Similarly for $R_{V^{\diamond}}$. Thus, $\mathfrak{r}_{V^{\diamond}}$ will be identified with a linear subspace of $\mathrm{Hom}(V \times V/V^{\diamond}, V^{\diamond})$. The Lie algebra $S_{V^{\diamond}}$ will be denoted by $\mathfrak{s}_{V^{\diamond}}$.

Lemma 6. *Suppose that $W \subset V$ is totally isotropic. Then*

$$R_{W\times W} = (R_{V\overline{\diamond}} \cap R_{W\times W}) \cdot (S_{V^{\diamond}} \cap R_{W\times W}).$$

Proof. It is enough to check the analogous statement for the Lie algebra $\mathfrak{r}_{W\times W}$. This amounts to saying that the projections corresponding to the decomposition $\mathfrak{g}^{\Box} = \mathfrak{r}_{V\overline{\diamond}} \oplus \mathfrak{s}_{V^{\diamond}}$ preserve $\mathfrak{r}_{W\times W}$. This is easily checked. □

Let ψ_F be a non-trivial character of F and set $\psi = \psi_E = \psi_F \circ \mathrm{Tr}_{E/F}$. For any $A \in \mathfrak{r}_{V^{\diamond}}$ we denote by ψ_A the character on $\mathfrak{r}_{V\overline{\diamond}}$ (or on $R_{V\overline{\diamond}}$) given by $\psi_A(X) = \psi(\mathrm{tr}\, XA)$ for $X \in \mathfrak{r}_{V\overline{\diamond}}$. (The product takes place in $\mathrm{End}(V \times V)$.) Note that $\mathrm{tr}\, XA \in F$ if h is non-degenerate.

For the rest of the section we are going to exclude the odd orthogonal case. We can then choose $A \in \mathfrak{r}_{V^{\diamond}}$ of rank n. Define for $f \in I(\omega, 0)$

$$l_{\psi_A}(f_s) = l^V_{\psi_A}(f_s) = \int_{R_{V\overline{\diamond}}} f_s(u)\psi_A(u)\, du.$$

The integral defining $l_{\psi_A}(f_s)$ converges for $\Re(s) \gg 0$. In the p-adic case the integral stabilizes and hence, admits analytic continuation ([**Kar79**]). In the archimedean case the analytic continuation of l_{ψ_A} (to an entire function) was proved by Wallach in [**Wal88a**]. Moreover, in both cases l_{ψ_A} is the unique (continuous) $(R_{V\overline{\diamond}}, \psi_A)$-equivariant functional on $I(\omega, s)$. Note also that for any s we can choose f_s so that $l_{\psi_A}(f_s) \neq 0$.

Suppose now that $W \subset V$ is totally isotropic and that A preserves $W \times W$ (that is $A(W \times W) \subset W^{\diamond}$). Then A preserves $W^{\perp} \times W^{\perp}$ as well, and hence, it gives rise to $B \in \mathfrak{r}_{\mathcal{W}^{\diamond}}$ in the Lie algebra of $\mathrm{Isom}(\mathcal{W} \times \mathcal{W})$. Note that if h is non-degenerate then the action of A on $(V/W^{\perp}) \times (V/W^{\perp})$ is the negative of the θ-dual of the action of A on $W \times W$. Clearly ψ_A is trivial on $R_{V\overline{\diamond}} \cap R_{W\times W}$. Indeed, if $X \in \mathfrak{r}_{V\overline{\diamond}} \cap \mathfrak{r}_{W\times W}$ then X acts trivially on $W \times W$, $W^{\perp} \times W^{\perp}/W \times W$ and $V \times V/W^{\perp} \times W^{\perp}$. Thus XA is nilpotent and its trace is zero.

For the next two lemmas we will denote by U' the subgroup of $S_{V\overline{\diamond}} \cap R_{W\times W}$ consisting of the elements which act as the identity on $(V/W)^{\overline{\diamond}}$.

Lemma 7. *We have*

$$[R_{V\overline{\diamond}}, U'] \subset R_{V\overline{\diamond}} \cap S_{W\times W}$$

and

$$R_{V^{\bar{\diamond}}} \cap S_{W\times W} = \{u \in R_{V^{\bar{\diamond}}} : \psi_A([u,u']) = 1 \text{ for all } u' \in U'\}.$$

Proof. Let $g \in U'$ and $X \in \mathfrak{r}_{V^{\bar{\diamond}}}$. Then $[g, 1+X] = 1 + Ad(g)X - X$. The first statement easily follows. It remains to prove that

$$\mathfrak{r}_{V^{\bar{\diamond}}} \cap \mathfrak{s}_{W\times W} = \{X \in \mathfrak{r}_{V^{\bar{\diamond}}} : \mathrm{tr}\,((\mathrm{Ad}(g)X - X)A) = 0 \text{ for all } g \in U'\}. \quad (11)$$

If $X \in \mathfrak{r}_{V^{\bar{\diamond}}} \cap \mathfrak{s}_{W\times W}$ and $g \in U'$ (or even, if $g \in S_{V^{\bar{\diamond}}} \cap R_{W\times W}$) then $(\mathrm{Ad}(g)X - X)A$ acts trivially on $W \times W$, $W^\perp \times W^\perp / W \times W$ and on $V \times V / W^\perp \times W^\perp$ because X and A preserve these spaces and g acts as the identity there. Hence, $\mathrm{tr}\,((\mathrm{Ad}(g)X - X)A) = 0$.

For the converse inclusion assume for concreteness that the form h is non-degenerate. For any $x \in GL(V^{\bar{\diamond}})$ there exists a unique element $g \in S_{V^{\diamond}} \cap S_{V^{\bar{\diamond}}}$ which acts as x on $V^{\bar{\diamond}}$. The action on $V^{\diamond}$ is $(x^*)^{-1}$ where

$$h^{\square}(xu, v) = h^{\square}(u, x^* v)$$

for any $u \in V^{\bar{\diamond}}, v \in V^{\diamond}$.

Suppose that $X \in \mathfrak{r}_{V^{\bar{\diamond}}}$ and let $C \in \mathrm{Hom}((V/W)^{\bar{\diamond}}, W^{\bar{\diamond}}) \subset \mathrm{End}(V^{\bar{\diamond}})$. Consider the element $g \in S_{V^{\diamond}} \cap S_{V^{\bar{\diamond}}}$ inducing $I + C$ on $V^{\bar{\diamond}}$. Then $g \in U'$ and

$$\begin{aligned} \mathrm{tr}\,((\mathrm{Ad}(g)X - X)A) &= \mathrm{tr}_{V^{\bar{\diamond}}}((\mathrm{Ad}(g)X - X)A) \\ &= \mathrm{tr}_{V^{\bar{\diamond}}}(((I+C)X(I+C^*) - X)A) \\ &= \mathrm{tr}_{V^{\bar{\diamond}}}(CXA) + \mathrm{tr}_{V^{\bar{\diamond}}}(XC^*A) + \mathrm{tr}_{V^{\bar{\diamond}}}(CXC^*A). \end{aligned} \quad (12)$$

Now, CXC^*A is nilpotent on $V^{\bar{\diamond}}$ since $\mathrm{Im}(C) \subset W^{\bar{\diamond}} \subset \mathrm{Ker}(C^*A)$. Moreover, $\mathrm{tr}_{V^{\bar{\diamond}}}(XC^*A) = \mathrm{tr}_{V^{\diamond}}(AXC^*) = \theta(\mathrm{tr}_{V^{\bar{\diamond}}}(CXA))$ since $AXC^* = (CXA)^*$. Thus, if X belongs to the right hand side of (11) then $\mathrm{Tr}_{E/F}\,\mathrm{tr}\,(CXA) = 0$ (hence $\mathrm{tr}\,(CXA) = 0$) for all $C \in \mathrm{Hom}((V/W)^{\bar{\diamond}}, W^{\bar{\diamond}})$. Thus, the image of XA under the canonical projection $\mathrm{Hom}(V^{\bar{\diamond}}, V^{\bar{\diamond}}) \to \mathrm{Hom}(W^{\bar{\diamond}}, (V/W)^{\bar{\diamond}})$ is zero. In other words, $XA(W^{\bar{\diamond}}) \subset W^{\bar{\diamond}}$. Since A is non-degenerate, it follows that $X \in \mathfrak{s}_{W\times W}$. □

Lemma 8. *With the assumptions and notation as above we have*

$$l^{V}_{\psi_A}(f_s) = \int_{R_{V^{\bar{\diamond}}} \cap S_{W\times W} \backslash R_{V^{\bar{\diamond}}}} l^{W}_{\psi_B}([\Psi(s) f_s](u))\, \psi_A(u)\, du. \quad (13)$$

Moreover, the integrand is a Schwartz function on $R_{V^{\bar{\diamond}}} \cap S_{W\times W} \backslash R_{V^{\bar{\diamond}}}$.

Proof. Since ψ_A is trivial on $R_{V^{\triangledown}} \cap R_{W\times W}$ we may write

$$l_{\psi_A}(f_s) = \int_{R_{V^{\triangledown}}\cap S_{W\times W}\backslash R_{V^{\triangledown}}} \int_{R_{V^{\triangledown}}\cap R_{W\times W}\backslash R_{V^{\triangledown}}\cap S_{W\times W}} \int_{R_{V^{\triangledown}}\cap R_{W\times W}} f_s(u''u'u)\psi_A(u'u)\, du''\, du'\, du.$$

By Lemma 6 we may replace the inner integral by an integral over $S_{V^{\diamond}} \cap R_{W\times W}\backslash R_{W\times W}$. Hence, by (7), we obtain

$$\iint \{[\Psi(s)f_s](u)\}(u')\psi_A(u')\, du' \quad \psi_A(u)\, du$$

where the integrals are taken over $u \in R_{V^{\triangledown}} \cap S_{W\times W}\backslash R_{V^{\triangledown}}$, $u' \in R_{V^{\triangledown}} \cap R_{W\times W}\backslash R_{V^{\triangledown}}\cap S_{W\times W}$. By an obvious analogue of Lemma 4 part 1, the integral over u' becomes $\phi(u) = l^W_{\psi_B}([\Psi(s)f_s](u))$.

Observe that ϕ is left-invariant under $S_{V^{\triangledown}}\cap R_{W\times W}$ and moreover, it follows from Lemma 7 that

$$\phi(uu') = \psi_A([u,u'])\phi(u)$$

for all $u \in R_{V^{\triangledown}}$ and $u' \in U'$. Write

$$f_s = I(\varphi,\omega,s)f'_s = \int_{G^{\square}} \varphi(g)f'_s(\cdot g)\, dg$$

for some $f'_s \in I(\omega,s)$ and $\varphi \in C_c^{\infty}(G^{\square})$. (In the archimedean case this is a consequence of the Dixmier-Malliavin Theorem **[DM78]**.) Set $\phi'(g) = l^W_{\psi_B}([\Psi(s)f'_s](g), s)$. Then ϕ' is smooth and of moderate growth and

$$\phi(u) = \int_{G^{\square}} \varphi(g)\phi'(ug)\, dg = \int_{U'\backslash G^{\square}} \int_{U'} \varphi(u'g)\phi'(uu'g)\, du'\, dg$$
$$= \int_{U'\backslash G^{\square}} \int_{U'} \varphi(u'g)\psi_A([u,u'])\, du'\, \phi'(ug)\, dg.$$

By Lemma 7, the inner integral is a Schwartz function on the space $(R_{V^{\triangledown}} \cap S_{W\times W})\backslash R_{V^{\triangledown}}$ (depending continuously on g) since it is the Fourier transform of $\varphi(\cdot g)$. Hence the same is true for $\phi(u)\psi_A(u)$ since g is integrated over a compact set of $U'\backslash G^{\square}$. □

Example . Consider the case where $n = 2$, F is non-archimedean and h is symplectic, so that $G = SL(2)$ and $G^{\square} = Sp_4$. Choose a standard basis e_1, e_2 for V and consider the basis $v_1, \dots, v_4$ of $V^{\square}$ consisting of $v_1 = (e_1, e_1)$, $v_2 = (e_2, e_2)$, $v_3 = (e_1, -e_1)$, $v_4 = (e_2, -e_2)$. Thus $S_{V^{\diamond}}$ is the "usual" Siegel parabolic subgroup consisting of matrices of the form $\begin{pmatrix} * & * & * & * \\ * & * & * & * \\ 0 & 0 & * & * \\ 0 & 0 & * & * \end{pmatrix}$. For an appropriate A the function l_{ψ_A} is given by

$$l_{\psi_A}(f_s) = \iiint_{F^3} f_s(\begin{pmatrix} 1 & 0 & 0 & 0 \\ 0 & 1 & 0 & 0 \\ y & x & 1 & 0 \\ z & y & 0 & 1 \end{pmatrix})\psi(y)\ dx\ dy\ dz$$

The integral over x becomes an intertwining operator $\Psi(s)$ from $I(s)$ to

$$\mathrm{Ind}_{B'}^{Sp_4}(\begin{pmatrix} t_1 & * & * & * \\ 0 & t_2 & 0 & * \\ 0 & * & t_2^{-1} & * \\ 0 & 0 & 0 & t_1^{-1} \end{pmatrix} \mapsto |t_1|^{s-\frac{1}{2}}|t_2|^{s+\frac{1}{2}})$$

where $B' = \{\begin{pmatrix} * & * & * & * \\ 0 & * & 0 & * \\ 0 & * & * & * \\ 0 & 0 & 0 & * \end{pmatrix}\}$. The integral over y then becomes a Jacquet integral in the context of SL_2. The integral over z is compactly supported. Indeed if we take the right translate of f_s by $\begin{pmatrix} 1 & t & 0 & 0 \\ 0 & 1 & 0 & 0 \\ 0 & 0 & 1 & -t \\ 0 & 0 & 0 & 1 \end{pmatrix}$ (for $|t|$ small) then we get

$$\iint_{F^2} f_s(\begin{pmatrix} 1 & t & 0 & 0 \\ 0 & 1 & 0 & 0 \\ 0 & 0 & 1 & -t \\ 0 & 0 & 0 & 1 \end{pmatrix}\begin{pmatrix} 1 & 0 & 0 & 0 \\ 0 & 1 & 0 & 0 \\ y+tz & x+t(2y+tz) & 1 & 0 \\ z & tz+y & 0 & 1 \end{pmatrix})\psi(y)\ dx\ dy$$

which after a change of variables becomes

$$\psi(-tz)\iint_{F^2} f_s(\begin{pmatrix} 1 & 0 & 0 & 0 \\ 0 & 1 & 0 & 0 \\ y & x & 1 & 0 \\ z & y & 0 & 1 \end{pmatrix})\psi(y)\ dx\ dy.$$

Thus, it vanishes for z large.

By uniqueness, the functional l_{ψ_A} admits a functional equation

$$l_{\psi_A} \circ M(s) = c(s, \omega, A, \psi) \cdot l_{\psi_A}. \tag{14}$$

Note that $c(s, \omega, A, \psi)$ also depends on a measure which is used to define $M(s)$. However, $c(s, \omega, A, \psi)$ depends only on the similitude class of h.

We will normalize $M(\omega, s)$ by setting

$$M^*(\omega, A, s) = c(s, \omega, A, \psi)^{-1} \cdot M(\omega, s)$$

so that

$$l_{\psi_A} \circ M^*(\omega, A, s) = l_{\psi_A}. \tag{15}$$

Note that $M^*(\omega, A, s)$ depends on ψ but not on any measure (and only on the homothety class of h). It is also clear that

$$M^*(\theta(\omega)^{-1}, A, -s) M^*(\omega, A, s) = I \tag{16}$$

since $M(\theta(\omega)^{-1}, -s) M(\omega, s)$ is a scalar for generic s.

Lemma 9. *Under the assumptions and notation of Lemma 8 we have*

$$\Psi(\theta(\omega)^{-1}, -s) \circ M^*_{\mathcal{V}}(\omega, A, s) = \mathrm{Ind}_{S_{W \times W}}^{G^{\square}} M^*_{\mathcal{W}}(\omega, B, s) \circ \Psi(\omega, s). \tag{17}$$

Here, as in Lemma 5, $\mathrm{Ind}_{S_{W \times W}}^{G^{\square}}$ *has a functorial meaning.*

Proof. We already know by Lemma 5 that (17) holds up to a constant. To determine the constant consider the functional $\mathcal{L}$ on

$$\mathrm{Ind}_{S_{W \times W}}^{G^{\square}} (I^{\mathcal{W}}(s) \otimes \Delta_{W;\mathcal{V}})$$

defined by

$$\mathcal{L}(F) = \int_{R_{V\bar{\delta}} \cap S_{W \times W} \backslash R_{V\bar{\delta}}} l_{\psi_B}^{\mathcal{W}}(F(u))\ \psi_A(u)\ du.$$

By Lemma 8 we have

$$l_{\psi_A} = \mathcal{L} \circ \Psi(s). \tag{18}$$

Hence, applying $\mathcal{L}$ to the left-hand side of (17) we obtain $l_{\psi_A} \circ M^*_{\mathcal{V}}(\omega, A, s) = l_{\psi_A}$. Similarly, applying $\mathcal{L}$ to the right-hand side of (17) gives

$$\int_{R_{V\bar{\delta}} \cap S_{W \times W} \backslash R_{V\bar{\delta}}} l_{\psi_B}^{\mathcal{W}}(M^*_{\mathcal{W}}(\omega, B, s)\,([\Psi(s)(\cdot)](u)))\ \psi_A(u)\ du.$$

Using the defining property of $M^*_{\mathcal{W}}(\omega, B, s)$ we obtain

$$\int_{R_{V\bar{\delta}} \cap S_{W \times W} \backslash R_{V\bar{\delta}}} l_{\psi_B}^{\mathcal{W}}([\Psi(s)(\cdot)](u))\ \psi_A(u)\ du$$

which is again equal to $l_{\psi_A}(\cdot)$ by Lemma 8. □

The dependence of $c(s, \omega, A, \psi)$ on A and ψ is described by the following lemma.

Lemma 10. *For any $m \in S_{V^\diamond}$ we have*

$$c(s,\omega,\mathrm{Ad}(m^{-1})A,\psi) = \omega_s(\Delta_{V^\diamond}(m))\omega_s(\theta(\Delta_{V^\diamond}(m)))c(s,\omega,A,\psi).$$

Moreover, let $\lambda \in F^*$ and set $\psi_\lambda(\cdot) = \psi(\lambda\cdot)$. Then

$$c(s,\omega,A,\psi_\lambda) = c(s,\omega,\lambda A,\psi) = \omega_s(\lambda)^{-ln}c(s,\omega,A,\psi)$$

where $l = 1$ if h is non-degenerate, and $l = 2$ if $h = 0$.

Proof. We write $l_{\psi_{\mathrm{Ad}(m^{-1})A}}(f)$ as

$$\begin{aligned}
&\int_{\mathfrak{r}_{V^{\overline{\diamond}}}} f_s(I+X)\psi(\mathrm{tr}\,(m^{-1}AmX))\,dX \\
&= \int_{\mathfrak{r}_{V^{\overline{\diamond}}}} f_s(I+X)\psi(\mathrm{tr}\,(AmXm^{-1}))\,dX \\
&= \delta_{S_{V^{\overline{\diamond}}}}(m)^{-1} \int_{\mathfrak{r}_{V^{\overline{\diamond}}}} f_s(m^{-1}(I+X)m)\psi(\mathrm{tr}\,(AX))\,dX \\
&= \delta_{S_{V^\diamond}}(m)^{\frac{1}{2}}\omega_s(\Delta_{V^\diamond}(m^{-1})) \int_{\mathfrak{r}_{V^{\overline{\diamond}}}} f_s((I+X)m)\psi(\mathrm{tr}\,(AX))\,dX
\end{aligned}$$

so that

$$l_{\psi_{\mathrm{Ad}(m^{-1})A}} = \delta_{S_{V^\diamond}}(m)^{\frac{1}{2}}\omega_s(\Delta_{V^\diamond}(m^{-1}))\cdot l_{\psi_A}\circ I(m,\omega,s).$$

The first part follows.

The second part follows from the first if $h = 0$. If h is non-degenerate then we consider, instead of $G^\square$, the similitude group $G\,\mathrm{Isom}(\mathcal{V}\times\mathcal{V})$ of $\mathcal{V}\times\mathcal{V}$ and its corresponding Siegel parabolic which we still denote by $S_{V^\diamond}$. Its Levi part is isomorphic to $GL(V^\diamond)\times GL(1,F)$, with the isomorphism taking p to its restriction to $V^\diamond$ and to its similitude factor $\mu(p)$. Given a character χ of F^* we denote by $I(\omega,\chi,s)$ the representation of $G\,\mathrm{Isom}(\mathcal{V}\times\mathcal{V})$ induced from $\omega_s\circ\Delta\times\chi_{-sn/2}\circ\mu$ on $S_{V^\diamond}$. The intertwining operator takes $I(\omega,\chi,s)$ to $I(\theta(\omega)^{-1},\chi\omega^n|_{F^*},-s)$. The functionals l_{ψ_A} and the relation (14) remain unchanged, with the same proportionality constant, which is independent of χ. Let $B\in G\,\mathrm{Isom}(V\times V)$ act by 1 on $V^\diamond$ and by λ on $V^{\overline{\diamond}}$. Its similitude factor is λ and $\mathrm{Ad}(B)X = \lambda X$ for all $X\in\mathfrak{r}_{V^{\overline{\diamond}}}$. By the same reasoning as before

$$l_{\psi_{\lambda A}} = \delta_{S_{V^\diamond}}(B)^{\frac{1}{2}}\chi(\lambda)^{-1}|\lambda|^{sn/2}\cdot l_{\psi_A}\circ I(B)$$

on $I(\omega,\chi,s)$ while

$$l_{\psi_{\lambda A}} = \delta_{S_{V^\diamond}}(B)^{\frac{1}{2}}\chi(\lambda)^{-1}\omega(\lambda)^{-n}|\lambda|^{-sn/2}\cdot l_{\psi_A}\circ I(B)$$

on $I(\theta(\omega)^{-1},\chi\omega^n,-s)$. Together, we get the last part of the lemma. □

Going back to the zeta-integrals considered in the previous sections, we define the factors $\Gamma^{\mathcal{V}}(s,\pi,\omega,A,\psi)$ by

$$Z(M^*(\omega,A,s)f_s,\alpha\otimes\tilde{\alpha}) = \Gamma^{\mathcal{V}}(s,\pi,\omega,A,\psi)Z(f_s,\alpha\otimes\tilde{\alpha}). \qquad (19)$$

Thus,

$$\Gamma^{\mathcal{V}}(s,\pi,\omega,A,\psi) = \Gamma^{\mathcal{V}}(s,\pi,\omega)/c(s,\omega,A,\psi)$$

Recall that both $\Gamma^{\mathcal{V}}(s,\pi,\omega)$ and $c(s,\omega,A,\psi)$ depend on a choice of a measure on $R_{V^\diamond}$, but their quotient does not. Similarly, it depends on h only up to homothety. Finally, it does not depend on the choice of the maximal compact K.

We will say that A is *split* if the form $(u,v)\mapsto h^\square(Au,v)$ on $V\times V/V^\diamond$ is split, i.e. admits a totally isotropic subspace of dimension $\lfloor n/2\rfloor$. This is the case, for example, if A preserves $W\times W$ for some h-totally isotropic space W of dimension $\lfloor n/2\rfloor$.

6. Odd orthogonal case

The discussion of the previous section requires some modification in the case where h is symmetric and n is odd. We are going to consider this case in this section. Choose $A\in\mathfrak{r}_{V^\diamond}$ of maximal rank, i.e., of rank $n-1$ and set $K = \operatorname{Ker}A\cap\overline{V^\diamond}$. Note that $\operatorname{Im}(A) = V^\diamond\cap K^\perp$. The functional l_{ψ_A} on $I(\omega,s)$ is defined in the same way as in the previous section and admits the same analytic properties. Moreover, it is the unique (continuous) $(R_{\overline{V^\diamond}},\psi_A)$-equivariant functional on $I(\omega,s)$ which is zero for sections which vanish on the connected component of $G^\square$.

Choose an anisotropic line L in $V\times V$ whose projection onto $\overline{V^\diamond}$ (along $V^\diamond$) is K. Denote by $\sigma=\sigma_L\in G^\square$ the orthogonal reflection around L.

Lemma 11. *The space $R_{\overline{V^\diamond}}\cap\sigma S_{V^\diamond}\sigma$ does not depend on L and it is canonically isomorphic to $K\otimes\overline{V^\diamond}/K$. The choice of L gives rise to a pairing on this space.*

Proof. By passing to the Lie algebra we can identify $R_{\overline{V^\diamond}}\cap\sigma S_{V^\diamond}\sigma$ with

$$\{X\in\mathfrak{g}^\square : \operatorname{Ker}X\supset\overline{V^\diamond}\supset\operatorname{Im}X, X(K^\perp)\subset K\}.$$

Since any such X is determined by its restriction to $K^\perp\cap V^\diamond$, and the latter can be identified with the dual of $\overline{V^\diamond}/K$, the first part follows.

The restriction of σ to K defines an isomorphism between K and $K' = \operatorname{proj}_{V^\diamond}L$. On the other hand, we can identify K' with the dual of K^* via $h^\square$. Similarly, A gives rise to an identification of $\overline{V^\diamond}/K$ with its dual space (which is isomorphic to $K^{\perp} \cap V^{\diamond}$). All in all, we obtain an isomorphism of $K \otimes V^{\bar{\diamond}}/K$ with its dual space. □

Remark 2. If we choose another anisotropic line L' whose projection onto $V^{\bar{\diamond}}$ (along $V^{\diamond}$) is K then the pairing on $K \otimes V^{\bar{\diamond}}/K$ changes by the factor $\Delta_{V^{\diamond}}(\sigma_{L'}\sigma_L)$. If $v \in L$ and $v' \in L'$ are chosen so that $v \equiv v' \mod V^{\diamond}$ then

$$\Delta_{V^{\diamond}}(\sigma_{L'}\sigma_L) = \frac{h^{\square}(v',v')}{h^{\square}(v,v)}.$$

We take the Haar measure on $R_{V^{\bar{\diamond}}} \cap \sigma S_{V^{\diamond}} \sigma$ which is self-dual with respect to the pairing of Lemma 11 and the character ψ. We also fix a Haar measure on $R_{V^{\bar{\diamond}}}$. Set

$$l'_{\psi_A}(f_s) = l'^{V}_{\psi_A,L}(f_s) = \int_{(R_{V^{\bar{\diamond}}} \cap \sigma S_{V^{\diamond}} \sigma)\backslash R_{V^{\bar{\diamond}}}} f_s(\sigma u)\psi_A(u)\, du$$

where we take the quotient measure. Clearly, $l'_{\psi_A,L}$ is $(R_{V^{\bar{\diamond}}}, \psi_A)$-equivariant. By uniqueness we have

$$l'_{\psi_A,L} \circ M(s) = c(s,\omega,A,L,\psi) \cdot l_{\psi_A}.$$

(Note that $\det(w) = \det(\sigma_L) = -1$.) The factor $c(s,\omega,A,L,\psi)$ depends in addition on a measure which is used to define $M(s)$.

As before, we can normalize $M(\omega,s)$ by the scalars $c(s,\omega,A,L,\psi)$ and get relations analogous to (15) and (16). The normalized operator does not depend on the measures on $R_{V^{\diamond}}$ and $R_{V^{\bar{\diamond}}}$ used in the definition of $M(s)$, l_{ψ_A}, l'_{ψ_A}. The analogues of Lemma 8 (for both l and l') and Lemma 9 still hold, provided that we require in addition that $K, L \subset W^{\perp} \times W^{\perp}$.

Lemma 12. *For any $m \in S_{V^{\diamond}}$ we have*

$$c(s,\omega,\mathrm{Ad}(m^{-1})A, m^{-1}L,\psi) = \omega_s(\Delta_{V^{\diamond}}(m))^2 c(s,\omega,A,L,\psi),$$
$$c(s,\omega,A,L,\psi_\lambda) = c(s,\omega,\lambda A,L,\psi) = \omega_s(\lambda)^{-n+1} c(s,\omega,A,L,\psi).$$

Finally, if L' is another anisotropic line whose projection on $V^{\bar{\diamond}}$ is K then

$$c(s,\omega,A,L',\psi) = \omega_s(\Delta_{V^{\diamond}}(\sigma_{L'}\sigma_L))^{-1} c(s,\omega,A,L,\psi)$$

Proof. As in Lemma 10 we have

$$l_{\psi_{\mathrm{Ad}(m^{-1})A}} = \delta_{S_{V^{\diamond}}}(m)^{\frac{1}{2}} \omega_s(\Delta_{V^{\diamond}}(m^{-1})) \cdot l_{\psi_A} \circ I(m,\omega,s).$$

On the other hand, it is easy to see that the measure on $R_{V^{\bar{\diamond}}} \cap \sigma_{m^{-1}L} S_{V^{\diamond}} \sigma_{m^{-1}L}$ is the push-forward by $\mathrm{Ad}(m)^{-1}$ of the measure on $R_{V^{\bar{\diamond}}} \cap \sigma_L S_{V^{\diamond}} \sigma_L$. Thus

$l'_{\psi_{\mathrm{Ad}(m^{-1})A},m^{-1}L}(f_s)$ equals

$$\int_{(\mathfrak{r}_{V\bar{\diamond}}\cap\sigma_{m^{-1}L}\mathfrak{s}_{V\bar{\diamond}}\sigma_{m^{-1}L})\backslash\mathfrak{r}_{V\bar{\diamond}}} f_s(\sigma_{m^{-1}L}(I+X))\psi(\mathrm{tr}\,(m^{-1}AmX))\,dX$$

$$=\int_{(\mathfrak{r}_{V\bar{\diamond}}\cap\sigma_{m^{-1}L}\mathfrak{s}_{V\bar{\diamond}}\sigma_{m^{-1}L})\backslash\mathfrak{r}_{V\bar{\diamond}}} f_s(\sigma_{m^{-1}L}(I+X))\psi(\mathrm{tr}\,(AmXm^{-1}))\,dX$$

$$=\delta_{S_{V\bar{\diamond}}}(m)^{-1}\int_{(\mathfrak{r}_{V\bar{\diamond}}\cap\sigma_L\mathfrak{s}_{V\bar{\diamond}}\sigma_L)\backslash\mathfrak{r}_{V\bar{\diamond}}} f_s(\sigma_{m^{-1}L}m^{-1}(I+X)m)\psi(\mathrm{tr}\,(AX))\,dX.$$

Since $\sigma_{m^{-1}L}=m^{-1}\sigma_L m$ we obtain

$$l'_{\psi_{\mathrm{Ad}(m^{-1})A},m^{-1}L}=\delta_{S_{V^\diamond}}(m)^{\frac{1}{2}}\omega_s(\Delta_{V^\diamond}(m^{-1}))\cdot l'_{\psi_A,L}\circ I(m,\omega,s).$$

The first part follows. The second part follows from the first by taking m such that $\mathrm{Ad}(m^{-1})A=\lambda A$ and $m^{-1}L=L$. To show the last part note that $\sigma_{L'}\sigma_L\in S_{V^\diamond}$ because $L\equiv L'\mod V^\diamond$. The integrand in $l'_{\psi_A,L'}(f_s)$ differs from the integrand of $l'_{\psi_A,L}(f_s)$ by the factor $\delta^{\frac{1}{2}}_{S_{V^\diamond}}(\sigma_{L'}\sigma_L)\omega_s(\Delta_{V^\diamond}(\sigma_{L'}\sigma_L))$. The spaces $R_{V\bar{\diamond}}\cap\sigma_L S_{V^\diamond}\sigma_L$ and $R_{V\bar{\diamond}}\cap\sigma_{L'}S_{V^\diamond}\sigma_{L'}$ are equal, but by remark 2 the self-dual measure changes by $|\Delta_{V^\diamond}(\sigma_{L'}\sigma_L)|^{(n-1)/2}$. Thus, the measure on $(R_{V\bar{\diamond}}\cap\sigma S_{V^\diamond}\sigma)\backslash R_{V\bar{\diamond}}$ changes by $|\Delta_{V^\diamond}(\sigma_{L'}\sigma_L)|^{-(n-1)/2}$. This factor will cancel with $\delta^{\frac{1}{2}}_{S_{V^\diamond}}(\sigma_{L'}\sigma_L)$. The Lemma follows. □

For future reference we choose an isomorphism $\tilde{A}_L : V\times V/V^\diamond\to V^\diamond$ such that

1. $\tilde{A}_L(K)=K'$,
2. $\tilde{A}_L$ induces the isomorphism $K\to K'$ determined by σ,
3. $\tilde{A}_L$ induces the map A on $V\times V/(K+V^\diamond)\to V^\diamond\cap K^\perp\simeq V^\diamond/K'$.

(The ambiguity in the choice of $\tilde{A}_L$ will not play any role.) Note that given L and L' we can choose

$$\tilde{A}_{L'}=\sigma_{L'}\sigma_L\tilde{A}_L. \tag{20}$$

Also, given m and $A'=\mathrm{Ad}(m)^{-1}A$ we can choose

$$\widetilde{A'}_{m^{-1}L}=\mathrm{Ad}(m)^{-1}\tilde{A}_L. \tag{21}$$

7. Unramified computation

Suppose now that F is of odd residual characteristic and $\mathcal{V}$ is defined over $\mathcal{O}_F$. We say that $\mathcal{V}$ is unramified if either $h = 0$ or if E/F is unramified (or split) and the reduction of h modulo the maximal ideal of $\mathcal{O}_F$ is non-degenerate. We will assume that this is the case throughout this section. If h is non-degenerate then $G = \mathrm{Isom}(\mathcal{V})$ is defined over $\mathcal{O}_F$, quasi-split over F and splits over an unramified extension of F, and $K = G(\mathcal{O}_F)$ is a hyper-special maximal compact. If $h = 0$ then $G = GL_n(E)$ and we let $K = GL_n(\mathcal{O}_E)$.

Let π be an unramified representation of G. We consider the representation $\mathrm{BC}(\pi)$ on $GL_N(E)$ given by the following. If $h = 0$ then $N = 2n$ and $\mathrm{BC}(\pi)$ is (the unramified subquotient of) the representation induced from π and $\theta(\widetilde{\pi})$. If h is non-degenerate then the Satake parameter of $\mathrm{BC}(\pi)$ is the transfer of that of π (or of an irreducible constituent of the restriction of π to the connected component of G in the orthogonal case) under the "standard" embedding of the L-group of G in GL_N. Thus, $N = n+1$ if h is symplectic, $N = n-1$ if h is symmetric and n is odd, and $N = n$ in the hermitian and even orthogonal cases. For an unramified character ω of E^* we set $L^{\mathcal{V}}(s, \pi \times \omega) = L(s, \mathrm{BC}(\pi) \otimes \omega)$ where on the right-hand side we consider the usual GL_N L-factor. Put in other words, this is the L-function corresponding to $\pi \times \omega$ and the "standard" representation *std* of the dual group ${}^L(G \times \mathrm{Res}_{E/F} GL_1)$ over F.

Denote by $f^0 = f^{0,\mathcal{V}}$ the K-invariant section in $I(\omega, 0)$, normalized so that $f^0(e) = 1$.

Let W be a totally isotropic subspace of V defined over $\mathcal{O}_F$ such that $\mathcal{W}$ is unramified. We have

$$[\Psi^{\mathcal{V},\mathcal{W}}(\omega, s)(f_s^{0,\mathcal{V}})](e) = d_{\mathcal{V};\mathcal{W}}(\omega, s) \cdot f_s^{0,\mathcal{W}}.$$

for a function $d_{\mathcal{V};\mathcal{W}}(\omega, s)$ which is rational in q^s. We also write $M(s) f_s^0 = m(\omega, s) f_s^0$.

If $\mathcal{V}$ is anisotropic then we set

$$d^{\mathcal{V}}(\omega, s) = \begin{cases} L(2s+1, \omega\theta(\omega)) \\ L(s+\frac{1}{2}, \omega) \\ L(2s+1, \omega^2) \\ 1 \\ L(2s+1, \omega^0) \end{cases}$$

according to whether h is 0 (with $n = 1$), symplectic (with $n = 0$), symmetric (with $n = 2$), symmetric (with $n = 1$) or hermitian (with $n = 1$) respectively. In the last case, ω^0 is the restriction of ω to F^*. If $n = 0$ (except for the symplectic case) then we set $d^{\mathcal{V}}(\omega, s) = 1$. If $\mathcal{V}$ is not anisotropic then we set inductively

$$d^{\mathcal{V}}(\omega, s) = d^{\mathcal{W}}(\omega, s) / d_{\mathcal{V};\mathcal{W}}(\omega, s)$$

where W is a non-trivial totally isotropic subspace of $\mathcal{V}$ for which $\mathcal{W}$ is unramified. This is well-defined by the following proposition.

Proposition 3. *Let $\alpha_0 \in \pi$, $\widetilde{\alpha_0} \in \widetilde{\pi}$ be K-fixed vectors with $\mathcal{P}(\alpha_0, \widetilde{\alpha_0}) = 1$. Then*

$$Z^{\mathcal{V}}(f_s^0, \alpha_0 \otimes \widetilde{\alpha_0}) = L^{\mathcal{V}}(s + \frac{1}{2}, \pi \times \omega)/d^{\mathcal{V}}(\omega, s). \tag{22}$$

Proof. We will use induction on n. If $\mathcal{V}$ is anisotropic, $E \neq F \oplus F$ and h is non-degenerate then G is compact, and thus the left-hand side of (22) is 1. It is also easy to check in each case that the right-hand side of (22) is 1. Consider the case where $h = 0$ and $n = 1$. Then π is a character χ of E^*. The imbedding of $G \times G = GL_1 \times GL_1$ to $G^{\square} = GL_2$ is via $t_1, t_2 \mapsto w\left(\begin{smallmatrix} t_1 & \\ & t_2 \end{smallmatrix}\right)w^{-1}$ where $w = \left(\begin{smallmatrix} 1 & 1 \\ 1 & -1 \end{smallmatrix}\right)$. The zeta integral is

$$\int_{E^*} f_s^0(w\left(\begin{smallmatrix} t & \\ & 1 \end{smallmatrix}\right)w^{-1})\chi(t)\, d^*t$$

where f_s^0 is the $GL_2(\mathcal{O}_E)$-invariant section in the representation induced from the character $t_1, t_2 \mapsto \omega(t_1\theta(t_2)^{-1})$. In the case where E is a field a simple computation gives

$$\sum_{n \in \mathbf{Z}} \omega(\varpi_E)^{|n|} q_E^{-|n|(s+\frac{1}{2})} \chi(\varpi_E)^n.$$

If $E = F \oplus F$ then writing $\omega = (\omega_1, \omega_2)$ and $\chi = (\chi_1, \chi_2)$ we get

$$\sum_{n,m \in \mathbf{Z}} [\omega_1(\varpi_F)^{n_+}\omega_2(\varpi_F)^{n_-}\omega_2(\varpi_F)^{m_+}\omega_1(\varpi_F)^{m_-}$$
$$\times\, q_F^{-(|n|+|m|)(s+\frac{1}{2})} \chi_1(\varpi_F)^n \chi_2(\varpi_F)^m]$$

where $n_\pm = \max(0, \pm n)$ and similarly for $m_\pm$. Evaluating the geometric series we obtain

$$\frac{1 - q_E^{-(2s+1)}\omega(\varpi_E)^2}{(1 - \omega(\varpi_E)q_E^{-(s+\frac{1}{2})})(1 - \omega^{-1}(\varpi_E)q_E^{-(s+\frac{1}{2})})}$$

or

$$\frac{(1 - q_F^{-(2s+1)}\omega_1(\varpi_F)\omega_2(\varpi_F))}{(1 - \omega_1(\varpi_F)\chi_1(\varpi_F)q_F^{-(s+\frac{1}{2})})(1 - \omega_2(\varpi_F)\chi_1(\varpi_F)^{-1}q_F^{-(s+\frac{1}{2})})}$$
$$\times \frac{(1 - q_F^{-(2s+1)}\omega_1(\varpi_F)\omega_2(\varpi_F))}{(1 - \omega_2(\varpi_F)\chi_2(\varpi_F)q_F^{-(s+\frac{1}{2})})(1 - \omega_1(\varpi_F)\chi_2(\varpi_F)^{-1}q_F^{-(s+\frac{1}{2})})}$$

respectively. In both cases, this is the right-hand side of (22). The case where $E = F \oplus F$, h hermitian and $n = 1$ is similar, but easier.

Assume now that $\mathcal{V}$ is not anisotropic and write π as a quotient of $\mathrm{Ind}_P^G\sigma$ where σ is unramified. By conjugating P we can assume that $P = S_{\mathcal{W}}$ with $\mathcal{W}$ unramified. Applying Proposition 2 we obtain

$$Z^{\mathcal{V}}(f_s^0, \alpha_0 \otimes \widetilde{\alpha_0}) = Z^{\mathcal{W}}([\Psi(s)(f_s^0)](e), \beta_0 \otimes \widetilde{\beta_0})$$

where β_0 and $\widetilde{\beta_0}$ are unramified vectors for σ and $\widetilde{\sigma}$ respectively with pairing $\mathcal{P}(\beta_0, \widetilde{\beta_0}) = 1$. Using the induction hypothesis and the inductive definition of $d^{\mathcal{V}}(\omega, s)$ we obtain the result. □

Remark 3. The factor $d_{\mathcal{V};\mathcal{W}}(\omega, s)$ can be computed by the method of Gindikin-Karpelevic (cf. **[GPSR87, p. 38-39]**). This gives the following formulas for $d^{\mathcal{V}}(\omega, s)$.

$$d^{\mathcal{V}}(\omega, s) = \begin{cases} \prod_{j=1}^{n} L(2s + j, \omega\theta(\omega)) \\ L(s + \frac{1}{2}(n+1), \omega) \cdot \prod_{j=1}^{n/2} L(2s + 2j - 1, \omega^2) \\ \prod_{j=1}^{n/2} L(2s + 2j - 1, \omega^2) \\ \prod_{j=1}^{(n-1)/2} L(2s + 2j, \omega^2) \\ \prod_{j=1}^{n} L(2s + j, \omega^0 \eta_{E/F}^{n-j}) \end{cases}$$

according to whether h is 0, symplectic, symmetric (n even), symmetric (n odd) or hermitian respectively. In the last case $\eta_{E/F}$ is the quadratic character of F^* attached to E by class field theory.

Proposition 4. *Suppose that ψ is unramified, A is split, and both A and A^{-1} are defined over $\mathcal{O}_E$. Then*

$$l_{\psi_A}^{\mathcal{V}}(f_s^0) = \frac{\beta(\omega, s)}{d^{\mathcal{V}}(\omega, s)}$$

where

$$\beta(\omega, s) = \begin{cases} L(s + \frac{1}{2}, \omega) & h \text{ symplectic}, \\ 1 & \text{otherwise}. \end{cases}$$

Similarly for l' in the odd orthogonal case.

Proof. We use Lemma 8 and note that for the unramified vector the integrand in (13) vanishes outside the integers. Thus, using induction we reduce to the anisotropic case. The latter is either trivial or reduces to a well-known rank one computation. □

8. Global theory

Suppose now that F is a number field and let $\mathbb{A}$ be its ring of adèles. We consider $\mathcal{V}$ defined over F, and E/F quadratic (or $E = F$). Let $G = \mathrm{Isom}(\mathcal{V})$ which is defined over F. We also consider the "doubled" group $G^\square$ and the induced representation $I(\omega, s)$ of $G^\square(\mathbb{A})$ for a Hecke character ω of the idèles $\mathbb{I}_E$ of E. Fix a maximal compact $K = \prod_v K_v$ of $G^\square(\mathbb{A})$ such that $G(\mathbb{A}) = S_{\mathcal{V}^\diamond}(\mathbb{A}) \cdot K$ and extend $|\Delta_{\mathcal{V}^\diamond}|$ to $G^\square(\mathbb{A})$ to be right-K-invariant.

For any $f \in I(\omega, 0)$ let $E(g, f, s)$ be the Eisenstein series on $G^\square(F)\backslash G^\square(\mathbb{A})$ defined for $\Re(s) \gg 0$ by

$$\sum_{\gamma \in S_{\mathcal{V}^\diamond}(F)\backslash G^\square(F)} f_s(\gamma g).$$

At least if f is K_∞-finite, $E(g, f, s)$ admits a meromorphic continuation and a functional equation

$$E(\cdot, f, s) = E(\cdot, M(s)f, -s).$$

Let π be a cuspidal automorphic representation of $G(\mathbb{A})$ realized on V_π. The contragredient representation $\widetilde{\pi}$ is realized on the complex conjugate of V_π. For any $\varphi \in V_\pi$, $\widetilde{\varphi} \in V_{\widetilde{\pi}}$ and $f \in I(\omega, 0)$ consider the global zeta integral defined by

$$Z^{\mathcal{V}}(f_s, \varphi \otimes \widetilde{\varphi}) = \int_{[G\times G](F)\backslash[G\times G](\mathbb{A})^{(1)}} E(g, f, s)\omega(\kappa(g))[\varphi \otimes \widetilde{\varphi}](g)\, dg$$

where $G \times G(\mathbb{A})^{(1)} = \{(g_1, g_2) \in [G \times G](\mathbb{A}) : |\det(g_1 g_2)| = 1\}$. The analytic properties of the Eisenstein series give a meromorphic continuation of the zeta integral and a functional equation

$$Z^{\mathcal{V}}(M(s)f_s, \varphi \otimes \widetilde{\varphi}) = Z^{\mathcal{V}}(f_s, \varphi \otimes \widetilde{\varphi}).$$

On the other hand, unfolding the Eisenstein series (**[GPSR87, §1]**) we get

$$Z^{\mathcal{V}}(f_s, \varphi \otimes \widetilde{\varphi}) = \int_{G^\diamond(\mathbb{A})\backslash[G\times G](\mathbb{A})} f_s(g)\omega(\kappa(g))\mathcal{P}((\pi \otimes \widetilde{\pi})(g)(\varphi \otimes \widetilde{\varphi}))\, dg$$

for $\Re(s) \gg 0$ where

$$\mathcal{P}(\varphi \otimes \widetilde{\varphi}) = \int_{G(F)\backslash G(\mathbb{A})^1} \varphi(g)\widetilde{\varphi}(g)\, dg$$

where $G(\mathbb{A})^1 = \{g \in G(\mathbb{A}) : |\det g| = 1\}$. If $\pi = \otimes \pi_v$ then

$$Z^{\mathcal{V}}(\otimes_v f_{v,s}, (\otimes \varphi_v) \otimes (\otimes \widetilde{\varphi}_v)) = \prod_v Z_v^{\mathcal{V}}(f_{v,s}, \varphi_v \otimes \widetilde{\varphi}_v).$$

Write $f = \otimes_v f_v$, $\varphi = \otimes_v \varphi_v$ and $\widetilde{\varphi} = \otimes \widetilde{\varphi}_v$ and let S be a large enough set of places containing all the archimedean ones such that for all $v \notin S$ $\mathcal{V}$, π_v, ω_v

are unramified and φ_v, $\widetilde{\varphi}_v$ are the standard sections. Using Proposition 3 we obtain

$$Z^{\mathcal{V}}(f_s, \varphi \otimes \widetilde{\varphi}) = \frac{L_S^{\mathcal{V}}(s + \frac{1}{2}, \pi \times \omega)}{d_S^{\mathcal{V}}(\omega, s)} \cdot \prod_{v \in S} Z_v^{\mathcal{V}}(f_v, s, \varphi_v \otimes \widetilde{\varphi}_v). \tag{23}$$

Here $L_S^{\mathcal{V}}(s, \pi \times \omega) = \prod_{v \notin S} L^{\mathcal{V}}(s, \pi_v \times \omega_v)$ is the partial L-function. Similarly, $d_S^{\mathcal{V}}(\omega, s) = \prod_{v \notin S} d_v^{\mathcal{V}}(\omega_v, s)$. We will soon see that $d_S^{\mathcal{V}}(\omega, s)$ admits a meromorphic continuation. (This can be also seen from Remark 3.) The relation (23) now implies the meromorphic continuation of the partial L-function.

Fix $A \in \mathfrak{r}_{V^\diamond}(F)$ of maximal rank and a non-trivial character ψ of $F\backslash\mathbb{A}$. The character ψ_A is defined on $R_{V^{\overline{\diamond}}}(F)\backslash R_{V^{\overline{\diamond}}}(\mathbb{A})$. We consider the Fourier coefficient

$$l_{\psi_A}(f_s) = \int_{R_{V^{\overline{\diamond}}}(F)\backslash R_{V^{\overline{\diamond}}}(\mathbb{A})} E(u, f, s)\psi_A(u)\, du$$

for any $f \in I(\omega, 0)$. Again by unfolding we get for $\Re(s) \gg 0$

$$l_{\psi_A}(f_s) = \int_{R_{V^{\overline{\diamond}}}(\mathbb{A})} f_s(u)\psi_A(u)\, du,$$

or, in the odd orthogonal case $l_{\psi_A}(f_s)$ equals

$$\int_{R_{V^{\overline{\diamond}}}(\mathbb{A})} f_s(u)\psi_A(u)\, du + \int_{(R_{V^{\overline{\diamond}}}(\mathbb{A}) \cap \sigma_L S_{V^\diamond}(\mathbb{A})\sigma_L)\backslash R_{V^{\overline{\diamond}}}(\mathbb{A})} f_s(\sigma_L u)\psi_A(u)\, du.$$

Thus, writing in brackets the additional term pertaining only to the odd orthogonal case

$$l_{\psi_A}(\otimes_v f_{v,s}) = \prod_v l_{\psi_A, v}(f_{v,s}) \left(+ \prod_v l'_{\psi_A, v}(f_{v,s}) \right).$$

From Proposition 4 we infer the meromorphic continuation of $d_S^{\mathcal{V}}(\omega, s)$. The functional equation of the Eisenstein series yields

$$l_{\psi_A}(M(\omega, s)f_s) = l_{\psi_A}(f_s),$$

or, in the odd orthogonal case

$$l'_{\psi_A}(M(\omega, s)f_s) = l_{\psi_A}(f_s).$$

For S as above we get

$$m_S(\omega, s) \cdot \frac{\beta_S(\omega^{-1}, -s)}{d_S^{\mathcal{V}}(\theta(\omega)^{-1}, -s)} \cdot \prod_{v \in S} l_{\psi_A, v}(M_v(\omega_v, s)f_{v,s}) = \frac{\beta_S(\omega, s)}{d_S^{\mathcal{V}}(\omega, s)} \cdot \prod_{v \in S} l_{\psi_A, v}(f_{v,s})$$

or, (in the odd orthogonal case),

$$m_S(\omega,s)\cdot\frac{1}{d_S^{\mathcal{V}}(\omega^{-1},-s)}\cdot\prod_{v\in S}l'_{\psi_A,v}(M_v(\omega_v,s)f_{v,s})=\frac{1}{d_S^{\mathcal{V}}(\omega,s)}\cdot\prod_{v\in S}l_{\psi_A,v}(f_{v,s}).$$

Using the local functional equations (14) we obtain

$$m_S(\omega,s)\cdot\frac{\beta_S(\omega^{-1},-s)}{d_S^{\mathcal{V}}(\theta(\omega)^{-1},-s)}\cdot\prod_{v\in S}c_v(s,A,\psi)=\frac{\beta_S(\omega,s)}{d_S^{\mathcal{V}}(\omega,s)},\tag{24}$$

and a similar relation in the odd orthogonal case.

9. Gamma factors

We now go back to the local situation. Let $G=\mathrm{Isom}(\mathcal{V})$ as before. Choose $A\in\mathfrak{r}_{V^\diamond}$ split. (This is merely for convenience.) We identify V with $V^\diamond$ via $v\mapsto(v,v)$. Similarly we identify $V\times V/V^\diamond$ with V via $(x,y)\mapsto x-y$. With these identifications $\det_V A$ is well-defined. Set

$$\gamma^{\mathcal{V}}(s+\frac{1}{2},\pi\times\omega,\psi)=\Gamma(s,\pi,\omega,A,\psi)\cdot\pi(-1)\cdot\omega_s(\det{}_V A)^{-1}$$
$$\times\begin{cases}\theta(\det{}_V A))^{-1} & h=0,\\ \gamma(s+\frac{1}{2},\omega,\psi) & h\text{ symplectic,}\\ \varepsilon(\frac{1}{2},\tau_D,\psi) & h\text{ symmetric, }n\text{ even}\\ 1 & h\text{ hermitian.}\end{cases}\tag{25}$$

(Recall that in the even orthogonal case $\tau_D=(\cdot,D)$ where D is $(-1)^{n/2}$ times the discriminant and $(\cdot,\cdot)$ denotes the Hilbert symbol on F^*.) It follows from Lemma 10 that the γ factor is independent of the choice of A. In the odd orthogonal case let L, $\tilde{A}_L$ be as in §6. Set

$$\gamma^{\mathcal{V}}(s+\frac{1}{2},\pi\times\omega,\psi)=\mathfrak{M}\cdot\omega_s(\det{}_V\tilde{A}_L)^{-1}\cdot\pi(-1)\cdot\Gamma^{\mathcal{V}}(s,\pi,\omega)/c(s,\omega,A,L,\psi)$$

where $\mathfrak{M}$ is the modified Witt invariant of $\mathcal{V}$ (as in **[Sch85, p. 80–81]**); that is $\mathfrak{M}=1$ if (and, in the non-archimedean case, only if) $\mathcal{V}$ is split, i.e., if the Witt index is $(n-1)/2$, and $\mathfrak{M}$ satisfies Hilbert's reciprocity law. Once again, it follows from Lemma 12 that the γ factor is independent of the choices of A and L. In all cases it depends only on the homothety class of h.

If $\mathrm{BC}(\pi)$ is the (hypothetical) transfer of π to $GL_N(E)$ (with N as in §7) then we expect that

$$\gamma^{\mathcal{V}}(s,\pi\times\omega,\psi)=\gamma^{GJ}(s,\mathrm{BC}(\pi)\otimes\omega,\psi)\tag{26}$$

where γ^{GJ} are the factors defined by Godement-Jacquet ([**GJ72**]). Note that in the orthogonal case both sides of (26) depend only on the restriction of π to the special orthogonal group.

In the compound case we set for $\mathcal{V} = \boxplus \mathcal{V}_j$ and $\pi = \otimes \pi_j$

$$\gamma^{\mathcal{V}}(s, \pi \times \omega, \psi) = \prod_j \gamma^{\mathcal{V}_j}(s, \pi_j \times \omega, \psi). \tag{27}$$

The "Ten Commandments" for the gamma factors are summarized in the following theorem.

Theorem 4. *The factors $\gamma^{\mathcal{V}}(s, \pi \times \omega, \psi)$ satisfy the following properties.*

1. *(unramified twisting)*
$$\gamma^{\mathcal{V}}(s, \pi \times \omega_{s_0}, \psi) = \gamma^{\mathcal{V}}(s + s_0, \pi \times \omega, \psi).$$

2. *(multiplicativity) If π is a subrepresentation of $\mathrm{Ind}_{S_{\mathcal{W}}}^{G}(\sigma)$ then*
$$\gamma^{\mathcal{V}}(s, \pi \times \omega, \psi) = \gamma^{\mathcal{W}}(s, \sigma \times \omega, \psi). \tag{28}$$

3. *(unramified factors) If $\mathcal{V}$, π, ω and ψ are unramified then*
$$\gamma^{\mathcal{V}}(s, \pi \times \omega, \psi) = \frac{L^{\mathcal{V}}(1 - s, \widetilde{\pi} \times \omega^{-1})}{L^{\mathcal{V}}(s, \pi \times \omega)}.$$

4. *(functional equation)*
$$\gamma^{\mathcal{V}}(s, \pi \times \omega, \psi)\gamma^{\mathcal{V}}(1 - s, \widetilde{\pi} \times \omega^{-1}, \psi^{-1}) = 1. \tag{29}$$

5. *(self-duality)*
$$\gamma^{\mathcal{V}}(s, \widetilde{\pi} \times \omega, \psi) = \gamma^{\mathcal{V}}(s, \pi \times \theta(\omega), \psi). \tag{30}$$

6. *(dependence on ψ) If $\psi_a(x) = \psi(ax)$, $a \in F^*$ then*
$$\gamma^{\mathcal{V}}(s, \pi \times \omega, \psi_a) = \omega_{s-\frac{1}{2}}(a)^N \{\cdot \tau_D(a)\} \cdot \gamma^{\mathcal{V}}(s, \pi \times \omega, \psi) \tag{31}$$
where $\{\dots\}$ appears only in the even orthogonal case.

7. *(minimal cases) Suppose that $\mathcal{V}$ is anisotropic and G is of (absolute) semisimple rank $r \leq 1$. If $r = 1$ then assume in addition that $\pi = 1$. Let* $\mathrm{BC}(\pi)$ *be the transfer of π to $GL_N(E)$. Then*
$$\gamma^{\mathcal{V}}(s, \pi \times \omega, \psi) = \gamma^{GJ}(s, \mathrm{BC}(\pi) \otimes \omega, \psi). \tag{32}$$

8. *(GL_n-factors) If $h = 0$ then*
$$\gamma^{\mathcal{V}}(s, \pi \times \omega, \psi) = \gamma^{GJ}(s, \pi \times \omega, \psi)\gamma^{GJ}(s, \theta(\widetilde{\pi}) \times \omega, \psi).$$

9. *(archimedean property) If F is archimedean then $\gamma^{\mathcal{V}}(s, \pi \times \omega, \psi)$ agrees with Langlands' prescription (see* [**Bor79**]*).*

10. *(global property) Let π be a cuspidal representation of $G(\mathbb{A})$, ω a Hecke character of $F^*\backslash\mathbb{I}_F$, and $\psi_F = \otimes_v \psi_{F_v}$ a non-trivial character of $F\backslash\mathbb{A}_F$. Let S be a finite set of places containing all the archimedean ones and the places where either $\mathcal{V}$, π, ω or ψ is ramified. Then*

$$L_S^{\mathcal{V}}(s, \pi \times \omega) = \prod_{v \in S} \gamma_v(s, \pi_v \times \omega_v, \psi_v) \cdot L_S^{\mathcal{V}}(1-s, \tilde{\pi} \times \omega^{-1}). \tag{33}$$

Moreover, the properties (27), (28) (31), (32), and (33) determine $\gamma^{\mathcal{V}}(s, \pi \times \omega, \psi)$ uniquely.

The rest of this section is devoted to the proof of Theorem 4.

Uniqueness. We first prove the uniqueness part. By the subrepresentation theorems of Casselman and Jacquet and (28), $\gamma^{\mathcal{V}}(s, \pi \times \omega, \psi)$ is completely determined by its values on the quadruples $(\mathcal{V}, \pi, \omega, \psi)$ where π is supercuspidal modulo the center (which include the case where $\mathcal{V}$ is anisotropic). Let F_0, E_0, $\mathcal{V}_0$, π_0, ω_0 and ψ_0 be given with π_0 supercuspidal modulo the center. The case $F_0 = \mathbb{C}$ follows from (32).

Consider the case where F_0 is non-archimedean. By standard local-global principles (e.g. [**Sch85**]) we can find a number field F, a quadratic extension E of F (if applicable), a place v_0 of F and a space $\mathcal{V}$ defined over F so that $F_{v_0} \simeq F_0$, $E_{v_0} \simeq E_0$, $\mathcal{V}$ is isomorphic to $\mathcal{V}_0$ over F_{v_0}, and $\mathcal{V}$ is quasi-split over all real places. Moreover, except for the even hermitian case or odd orthogonal case we can actually assume that $\mathcal{V}$ is quasi-split over any $v \neq v_0$. Now, we embed π_0 as a local component of a cuspidal representation π of $G(\mathbb{A})$ such that for all finite $v \neq v_0$ π_v has a non-zero vector fixed under the Iwahori subgroup of $G(F_v)$ (e.g. using a Poincare series, cf. [**Hen84, Appendice 1**]). We also take a Hecke character ω of $\mathbb{I}_E$ such that $\omega_{v_0} = \omega_0$ and a non-trivial character ψ_F of $F\backslash\mathbb{A}_F$ and, as always, set $\psi = \psi_F \circ \mathrm{Tr}_{E/F}$. Since every representation which admit a non-zero vector fixed under the Iwahori subgroup is a subquotient of an unramified principal series ([**Bor76**]), the factors $\gamma^{\mathcal{V}_v}(s, \pi_v \times \omega_v, \psi_v)$ are determined for all $v \neq v_0$ by (28), (32), and Casselman's subrepresentation theorem (if v is archimedean). The global property (33) determines $\gamma^{\mathcal{V}_0}(s, \pi_0 \times \omega_0, \psi_{v_0})$, and hence $\gamma^{\mathcal{V}_0}(s, \pi_0 \times \omega_0, \psi_0)$ by (31).

It remains to consider the case $F_0 = \mathbb{R}$. We can take $F = \mathbb{Q}$, $E = \mathbb{Q}$ or $\mathbb{Q}(\sqrt{-1})$ and $\mathcal{V}$ over F such that $\mathcal{V}$ is isomorphic to $\mathcal{V}_0$ over $\mathbb{R}$. We take a cuspidal representation π of $G(\mathbb{A})$ such that $\pi_\infty = \pi_0$ and ω as before. (This is always possible.) By (33) $\gamma^{\mathcal{V}_0}(s, \pi_0 \times \omega_0, \psi_\infty)$ is determined by the p-adic gamma factors, which are already known to be determined.

We now turn to the proof of the other properties of the γ-factors.

Unramified twisting. This follows directly from the definition.

Multiplicativity. The multiplicativity relation (28) follows from Proposition 2 and Lemma 9. Observe that since A is split, we can always assume, by changing A to a congruent form, that A preserves any given totally isotropic subspace of $\mathcal{V}$. Note also that the factor in (25) pertaining to A is also multiplicative, in the obvious sense.

Dependence on ψ. This follows from Lemma 10.

Unramified factors. This follows immediately from (28) and §7.

Functional equation. Follows from (16).

Global relation. We use (23) for $M(s)f_s$ and $\theta(\omega)^{-1}$ to obtain

$$Z^{\mathcal{V}}(M(s)f_s, \varphi \otimes \widetilde{\varphi})$$
$$= m_S(\omega, s)\frac{L_S^{\mathcal{V}}(\frac{1}{2} - s, \pi \times \theta(\omega)^{-1})}{d_S^{\mathcal{V}}(\theta(\omega)^{-1}, -s)} \prod_{v \in S} Z_v^{\mathcal{V}}(M_v(s)f_v, -s, \varphi_v \otimes \widetilde{\varphi}_v).$$

Comparing this with (23), and using the definition of $\gamma(s, \pi, \psi)$ we conclude that

$$\frac{L_S^{\mathcal{V}}(s + \frac{1}{2}, \pi \times \omega)\beta_S(\omega, s)}{d_S^{\mathcal{V}}(\omega, s)} = \prod_{v \in S} c_v(s, A, \psi) \prod_{v \in S} \gamma_v(s + \frac{1}{2}, A, \psi)$$
$$\times\, m_S(\omega, s)\frac{L_S^{\mathcal{V}}(\frac{1}{2} - s, \pi \times \theta(\omega)^{-1})\beta_S(\omega^{-1}, -s)}{d_S^{\mathcal{V}}(\theta(\omega)^{-1}, -s)}.$$

We note that

$$L_S^{\mathcal{V}}(s, \pi \times \theta(\omega)^{-1}) = L_S^{\mathcal{V}}(s, \widetilde{\pi} \times \omega^{-1}).$$

We now infer (33) from (24).

Minimal cases. We will work case by case.

9.1. The case $n = 1$, $h = 0$, $F = \mathbb{R}$. Here $G = GL_1$ and the imbedding of $G \times G$ to $G^{\square} = GL_2$ is via $t_1, t_2 \mapsto w\begin{pmatrix} t_1 & \\ & t_2 \end{pmatrix}w^{-1}$ where $w = \frac{1}{\sqrt{2}}\begin{pmatrix} 1 & 1 \\ 1 & -1 \end{pmatrix}$. In this identification the parabolic $S_{V^\diamond}$ is the Borel subgroup of upper triangular matrices. Set

$$E^1 = \{z \in E^* : |z| = 1\} = \{z \in E^* : z\theta(z) = 1\}.$$

Thus $E^* = E^1 \times \mathbb{R}_+$. By twisting ω by an unramified character we may assume that $\omega\theta(\omega) = 1$, that is ω is trivial on $\mathbb{R}_+$. Let $\alpha \in \mathbb{C}$ be such that $\chi(r) = |r|^\alpha$ for $r \in \mathbb{R}_+$ and let $m \in \mathbb{Z}$ be such that $\omega(z) = z^m$ for $z \in E^1$.

Let T be the diagonal torus in $G^\square$ and let K be the maximal compact of G obtained as the fixed point subgroup of the Cartan involution $g \mapsto \theta({}^t g^{-1})$ (i.e. either an orthogonal or a unitary group). Note that $w \in K$. We have

$$T \cap K = \{\left(\begin{smallmatrix} t_1 & 0 \\ 0 & t_2 \end{smallmatrix}\right) : t_1, t_2 \in E^1\}.$$

As a K-module, $I(\omega, s)$ is equivalent to $\mathrm{Ind}_{T\cap K}^K \omega \circ \det$. Thus, it decomposes as the direct sum of σ_k, $k = 0, 1, \ldots$ where in the notation of **[Wal79, §7]** $\sigma_k = \gamma_{2k}$ ($\sigma_0 = \gamma_0^m$) if $E = \mathbb{R}$ and $\sigma_k = \gamma_{2m,2k}$ if $E = \mathbb{C}$.

Consider the section $f_s^k \in I(\omega, s)$ whose restriction to K is given by

$$f_s^k(\left(\begin{smallmatrix} a & b \\ c & d \end{smallmatrix}\right)) = \omega(\det\left(\begin{smallmatrix} a & b \\ c & d \end{smallmatrix}\right)) \times \begin{cases} \det\left(\begin{smallmatrix} a & b \\ c & d \end{smallmatrix}\right)^{-k} \cdot (-bd)^k & k \geq 0, \\ \det\left(\begin{smallmatrix} a & b \\ c & d \end{smallmatrix}\right)^{k} \cdot (ac)^{-k} & k < 0. \end{cases}$$

It belongs to the K-type of $\sigma_{|k|}$ and satisfies

$$f_s(g\left(\begin{smallmatrix} t_1 & 0 \\ 0 & t_2 \end{smallmatrix}\right)) = \omega(t_1 t_2)\left(\frac{t_2}{t_1}\right)^k f_s(g) \quad \forall t_1, t_2 \in E^1,\ g \in G.$$

In particular, if we take k such that $\omega\chi(z) = z^k$ for all $z \in E^1$ then the zeta integral of the translate of f_s^k by w is

$$\int_{E^*} f_s(w\left(\begin{smallmatrix} t & \\ & 1 \end{smallmatrix}\right))\chi(t)\, d^*t = \int_{\mathbb{R}_+} f_s(w\left(\begin{smallmatrix} r & \\ & 1 \end{smallmatrix}\right))|r|^\alpha\, d^*r$$

since the integrand is invariant under E^1. The Iwasawa decomposition for $w\left(\begin{smallmatrix} r & \\ & 1 \end{smallmatrix}\right)$ in $GL_2(\mathbb{R})$ is

$$w\left(\begin{smallmatrix} r & \\ & 1 \end{smallmatrix}\right) = \begin{pmatrix} \frac{r}{\sqrt{\frac{r^2+1}{2}}} & * \\ 0 & \sqrt{\frac{r^2+1}{2}} \end{pmatrix} \left(\begin{smallmatrix} \sin\theta & \cos\theta \\ \cos\theta & -\sin\theta \end{smallmatrix}\right)$$

where $0 < \theta < \frac{\pi}{2}$ and $\tan(\theta) = 1/r$. We obtain

$$\chi(-1)\int_{\mathbb{R}_+} |\frac{2r}{r^2+1}|^{s+\frac{1}{2}} |r|^\alpha (\sin\theta\cos\theta)^{|k|}\, d^*r.$$

Changing the variable to θ the integral becomes

$$|2|^{s+\frac{1}{2}} \int_0^{\pi/2} |\sin(\theta)|^{s-\alpha+\frac{1}{2}} |\cos(\theta)|^{s+\alpha+\frac{1}{2}} (\sin\theta\cos\theta)^{|k|} \frac{d\theta}{\sin\theta\cos\theta}.$$

Letting $d = [E : F]$ and $d' = \frac{2}{d}$ this is equal to $|2|^{s+\frac{1}{2}}$ times

$$B(\frac{d}{2}(s+\alpha+\frac{1}{2}) + \frac{|k|}{2}, \frac{d}{2}(s-\alpha+\frac{1}{2}) + \frac{|k|}{2}) = \frac{L(s+\frac{1}{2}, \chi\omega)L(s+\frac{1}{2}, \theta(\chi)^{-1}\omega)}{d \cdot L(2s+d'|k|+1, \mathbf{1}_{E^*})}$$

where $B(\cdot,\cdot)$ denotes the Beta function and where $L(s,\chi)$ is the usual factor à la Tate.

The intertwining operator $M(s) : I(\omega, s) \to I(\omega, -s)$ is given by

$$M(s)f_s(g) = \int_E f_s((\begin{smallmatrix} 0 & 1 \\ 1 & 0 \end{smallmatrix})(\begin{smallmatrix} 1 & x \\ 0 & 1 \end{smallmatrix})g)\, dx.$$

On each σ_k, $M(s)$ acts by the scalar $\mu_k(s)$ which, in the notation of Lemma 7.17 of **[Wal79]** (resp. Lemma 7.23 of **[Wal79]**) is given by $\omega(-1)(-1)^k a_k(2s)$, (resp. $\omega(-1)(-1)^k b_{2k,k}(4s)$) if $E = \mathbb{R}$ (resp. $E = \mathbb{C}$). Note that the intertwining operator in **[Wal79]** differs from ours by composing it with left translation by $(\begin{smallmatrix} 0 & 1 \\ 1 & 0 \end{smallmatrix})$; hence the factor $\omega(-1)(-1)^k$. At any rate, $\mu_0(s) = \omega(-1)\frac{L(2s,\mathbf{1}_{E^*})}{L(2s+1,\mathbf{1}_{E^*})}$ and $\mu_{k+1}(s) = \frac{d'k+1-2s}{d'k+1+2s} \cdot \mu_k(s)$. (The latter can also be inferred from **[BÓØ96]**.)

We take $\psi(\cdot) = e^{2\pi i \cdot}$ and $A = \frac{1}{2}$, so that ψ_A is the character $(\begin{smallmatrix} 1 & 0 \\ x & 1 \end{smallmatrix}) \mapsto \psi(x)$. By **[Sha85]** $c(s,\omega,A,\psi) = \omega(-1)\frac{L(2s,\mathbf{1}_{E^*})}{L(1-2s,\mathbf{1}_{E^*})}$. The extra factor $\omega(-1)$ comes from the fact that for the intertwining operator we used the Weyl element $(\begin{smallmatrix} 0 & 1 \\ 1 & 0 \end{smallmatrix})$ while Shahidi uses $(\begin{smallmatrix} 0 & 1 \\ -1 & 0 \end{smallmatrix})$. By induction on k we obtain the relation

$$\mu_k(s) = c(s,\omega,A,\psi) \cdot \frac{L(-2s+d'k+1, \mathbf{1}_{E^*})}{L(2s+d'k+1, \mathbf{1}_{E^*})}$$

for $k = 0, 1, \ldots$. The factor $\frac{|2|^{-s+\frac{1}{2}}}{|2|^{s+\frac{1}{2}}}$ also cancels with $\omega_s(\det A)^{-2}$ and thus, taking k as before, $\gamma^{\mathcal{V}}(s+\frac{1}{2}, \chi \times \omega, \psi)$ is equal to

$$(-1)^k \frac{L(-s+\frac{1}{2}, \chi\omega)L(-s+\frac{1}{2}, \theta(\chi)^{-1}\omega)}{L(s+\frac{1}{2}, \chi\omega)L(s+\frac{1}{2}, \theta(\chi)^{-1}\omega)}.$$

Since $\varepsilon^{Tate}(s, \chi\omega, \psi) = \varepsilon^{Tate}(s, \theta(\chi)^{-1}\omega, \psi) = i^{|k|}$ and the L-factor is invariant under θ we finally obtain

$$\gamma^{\mathcal{V}}(s+\frac{1}{2}, \chi \times \omega, \psi) = \gamma^{Tate}(s+\frac{1}{2}, \chi\omega, \psi) \cdot \gamma^{Tate}(s+\frac{1}{2}, \theta(\chi)^{-1}\omega, \psi).$$

9.2. $n = 1$, $h = 0$, **general case.** We need to prove the relation

$$\gamma^{\mathcal{V}}(s, \chi_0 \times \omega_0, \psi) = \gamma^{Tate}(s, \chi_0\omega_0, \psi) \cdot \gamma^{Tate}(s, \theta(\chi_0)^{-1}\omega_0, \psi) \qquad (34)$$

for any characters χ_0, ω_0 of E_0^*. To simplify, we say that (34) holds for E_0/F_0 if it holds for any characters χ_0, ω_0 of E_0^*. Note that it is enough to check this for a dense subset of characters. By the previous part, we know that (34) holds for $F_0 = \mathbb{R}$ and E_0 either $\mathbb{C}$ or $\mathbb{R}$. We also know (34) if F_0 is non-archimedean of odd residual characteristic and χ_0 and ω_0 are unramified.

We will repeatedly use the following reduction principle.

Scholium . *Let E/F be a quadratic (or trivial) extension of number fields and let v_0, v_1 be two distinct places of F. Suppose that we know that (34) holds for E_{v_1}/F_{v_1} and for all $v \neq v_0, v_1$ (34) holds for E_v/F_v for unramified characters of E_v^*. Then (34) holds for E_{v_0}/F_{v_0}.*

Indeed, for any character χ_0 of $E_{v_0}^*$ (or at least, for a dense set of characters if v_0 splits in E) there exists a Hecke character χ of $\mathbb{I}_E$, ramified only at the places above v_0, v_1, such that $\chi_{v_0} = \chi_0$. Similarly for ω_0. The global property (33), the similar property for the usual gamma factors and the validity of (34) for all $v \neq v_0$ imply (34) for v_0, χ_0, ω_0.

Applying this principle to $F = \mathbb{Q}$, E either $\mathbb{Q}$ or a quadratic imaginary field, $v_0 = 2$, $v_1 = \infty$, we infer that (34) holds for $F_0 = \mathbb{Q}_2$ and any E_0. Then, we apply it with $F = \mathbb{Q}$, $E = \mathbb{Q}(\sqrt{2})$, $v_0 = \infty$, $v_1 = 2$ to conclude that (34) holds for $\mathbb{R} \oplus \mathbb{R}/\mathbb{R}$. Next, consider $F = \mathbb{Q}(r)$ where r is the real root of $r^3 - r + 8 = 0$, which has one real place $v_{\mathbb{R}}$, one complex place $v_{\mathbb{C}}$ and splits completely over 2. Let E be either F or its normal closure. Applying the same reasoning for $v_0 = v_{\mathbb{C}}$, $v_1 = v_{\mathbb{R}}$, we infer that (34) holds for $F = \mathbb{C}$. Consider now an extension of 2-adic fields E_0/F_0. We can find a quadratic (or trivial) extension E/F of number fields such that $[F : \mathbb{Q}] = [F_0 : \mathbb{Q}_2]$, $F \otimes \mathbb{Q}_2 = F_0$ and $E \otimes \mathbb{Q}_2 = E_0$. Again, applying the same argument with v_0 - even place, v_1 - any archimedean place, we infer that (34) holds for F_0. We can now prove that (34) holds for any extension E_0/F_0 of non-archimedean fields by choosing a quadratic (or trivial) extension E/F of number field and a place v_0 of F such that $F_{v_0} \simeq F_0$ and $E_{v_0} \simeq E_0$ and applying the reasoning above (with v_1 archimedean, or even).

9.3. The case $n = 1$, h hermitian. Consider first the case where E is a field. In the non-archimedean case, this is treated explicitly in **[HKS96, §8]**. Suppose that h is given by $h(x,y) = \delta x \theta(y)$ with $\theta(\delta) = -\delta$. Because of the different normalization, the operator $M^*(s,\omega)$ of [loc. cit.] is our $M^*(s,A,\omega)$ with $A = -\delta$. (This A will make our l_{ψ_A} consistent with $W_T(s)$ of [loc. cit.] for $T = 1$; the operator $M(s,\omega)$ of [loc. cit.] differs from ours by a scalar, but this is immaterial for $M^*(s,\omega)$.) Thus, we apply Corollary 8.3 (ii) of **[HKS96]** with $\eta = \widetilde{\pi}$.[4]

Corollary 8.3 of **[HKS96]** remains true for $E = F \oplus F$ and $F = \mathbb{R}$. For the split case, this is easy to check. For $F = \mathbb{R}$ this boils down to the computation of the previous paragraph, namely, to the eigenvalues of the intertwining operator (for $SL_2(\mathbb{R})$) on its K-types, i.e. to **[Wal79, Lemma 7.17]**. Alternatively,

[4] Note the following typo: $\rho(2s, \varepsilon_{E/F}, \psi)$ should be replaced by $\rho(2s, \chi^0, \psi)$.

we could give a global argument. Conversely, we could derive Corollary 8.3 of **[HKS96]** using the archimedean computation and a global argument.

9.4. $n = 2$, **h non-split symmetric, $\omega = 1$.** Note that $SO(2) \simeq U(1)$ with respect to $E = F(\sqrt{D})$. The doubled group is $O(4)$ whose connected component is isogenous to $SL_2 \times SL_2$. For $\omega = 1$ the intertwining operator, the Karel functional, and the integrand of the zeta-integral all factor through $SL_2/\{\pm 1\}$. Thus, the computation gives the same result as in the case $n = 1$, h hermitian considered before. The only difference is that because we compare it with the γ-factor with respect to F, rather than E, we have to multiply by Langlands' constant $\lambda(E/F, \psi)$, or what is the same, by $\varepsilon(\frac{1}{2}, \tau_D, \psi)$.

9.5. $n = 2$, **h non-split symmetric, general case.** We will use a global argument. Let F_0 and $\mathcal{V}_0$ be given and choose a number field F, a space $\mathcal{V}$ and a place v_0 of F such that F_{v_0} and $\mathcal{V}_{v_0}$ are our given F_0 and $\mathcal{V}_0$. Given a character ω_0 we can find a Hecke character ω of $\mathbb{I}_F$ such that $\omega_{v_0} = \omega_0$ and ω is unramified at all other places where $\mathcal{V}_v$ does not split (e.g. by choosing it to be ramified at only one place outside v_0). By the previous case and by (28) (in the split case), the relation (32) holds for all $v \neq v_0$. Hence it holds for v_0 as well by (33).

9.6. $n = 3$, **h non-split symmetric, $F = \mathbb{R}$, $\pi = 1$, ω unramified.** The doubled group $G^{\square}$ is $O(3,3)$ whose connected component is isomorphic to $SL_4/\pm 1$ and its maximal compact subgroup is $K = G \times G$. By unramified twisting we can assume that $\omega = 1$. Let f_s be the normalized section of $I(\omega, s)$. Its zeta integral is 1, so we have to compute the eigenvalue of $M(\omega, s)$ on f_s. By the Gindikin-Karpelevic formula (cf. **[Hel94]**) this eigenvalue is

$$\frac{\Gamma_{\mathbb{R}}(2s-1)}{\Gamma_{\mathbb{R}}(2s+2)}.$$

On the other hand for A, L, such that $\tilde{A}_L = \begin{pmatrix} 0 & \frac{1}{2} & 0 \\ -\frac{1}{2} & 0 & 0 \\ 0 & 0 & 1 \end{pmatrix}$ in an appropriate basis, we have

$$c(s, \omega, A, L, \psi) = \frac{\Gamma_{\mathbb{R}}(2s-1)}{\Gamma_{\mathbb{R}}(-2s+2)}.$$

Thus,

$$\Gamma^{\mathcal{V}}(s, \pi, \omega, A, \psi) = \frac{\Gamma_{\mathbb{R}}(-2s+2)}{\Gamma_{\mathbb{R}}(2s+2)}.$$

On the other hand $\mathrm{BC}(\pi)$ is the identity representation of $GL_2(\mathbb{R})$ and hence,

$$\gamma(s+\frac{1}{2},\mathrm{BC}(\pi),\psi)=\frac{L(-s,\mathbf{1}_{\mathbf{R}^*})L(1-s,\mathbf{1}_{\mathbf{R}^*})}{L(s,\mathbf{1}_{\mathbf{R}^*})L(s+1,\mathbf{1}_{\mathbf{R}^*})}=\frac{\Gamma_{\mathbf{R}}(-s)\Gamma_{\mathbf{R}}(1-s)}{\Gamma_{\mathbf{R}}(s)\Gamma_{\mathbf{R}}(s+1)}=\frac{\Gamma_{\mathbf{C}}(-s)}{\Gamma_{\mathbf{C}}(s)}.$$

It is now easy to see that (32) holds in this case (here $\mathcal{M}=-1$).

9.7. $n=3$, h **non-split symmetric,** $\pi=\mathbf{1}$**, general case.** We note that in the non-archimedean case we cannot use the usual analogue of the Gindikin-Karpelevic formula, even if $\omega=1$, since $G\times G$ is not contained inside a hyperspecial maximal compact of $\mathrm{Isom}(\mathcal{V}^{\square})$ (The group $G^{\square}$ is not simply connected!). Instead, we use a global argument. Given F_0 and ω_0 we choose a number fields F with a real place v_1 and a place v_0 such that $F_{v_0}\simeq F_0$. We take $\mathcal{V}$ over F which is non-split exactly at v_0 and v_1, and $\pi=\mathbf{1}$. We also choose a Hecke character ω of $\mathbb{I}_E$ such that $\omega_{v_0}=\omega_0$ and ω_{v_1} is unramified. Using (33) and (28) and the previous computation we conclude (32) for v_0.

9.8. $n=2$, h **non-split hermitian,** $F=\mathbb{R}$, $\pi=\mathbf{1}$, ω **unramified.** The doubled group $G^{\square}$ is $U(2,2)$ and its maximal compact subgroup is $K=G\times G$. By unramified twisting we can assume that $\omega=1$. Let f_s be the normalized section of $I(\omega,s)$. Its zeta integral is 1, so we have to compute the eigenvalue of $M(\omega,s)$ on f_s. By the Gindikin-Karpelevic formula (cf. **[Hel94]**) this eigenvalue is

$$\frac{\Gamma_{\mathbf{R}}(2s-1)\Gamma_{\mathbf{R}}(2s+1)}{\Gamma_{\mathbf{R}}(2s+2)^2}.$$

On the other hand we may apply Proposition 6.3 of **[HKS96]** with $T=\begin{pmatrix}1&0\\0&-1\end{pmatrix}$ and $\psi(\cdot)=e^{2\pi i\cdot}$ to conclude that

$$c(s,\omega,A,\psi)=-\frac{\Gamma_{\mathbf{R}}(2s-1)\Gamma_{\mathbf{R}}(2s+1)}{\Gamma_{\mathbf{R}}(-2s+2)^2}.$$

where $A\in GL(V)$ is $\frac{i}{2}\cdot T$ in an appropriate basis. Note that Proposition 6.3 of **[HKS96]** applies also to $F=\mathbb{R}$ by a global consideration (i.e. by applying (24) with for $\mathbb{Q}$ with $S=\{\infty\}$). Also, the sign comes from the fact that $\det T=-1$, and taking into account the fact that the Weil index and the root number of the signum character, are both equal to i, and hence cancel each other (cf. **[RR93, appendix]**). Thus,

$$\Gamma^{\mathcal{V}}(s,\pi,\omega,A,\psi)=-\left(\frac{\Gamma_{\mathbf{R}}(-2s+2)}{\Gamma_{\mathbf{R}}(2s+2)}\right)^2.$$

On the other hand $\mathrm{BC}(\pi)$ is the identity representation of $GL_2(E)$ and hence,

$$\gamma(s+\frac{1}{2},\mathrm{BC}(\pi),\psi)=\frac{L(-s,\mathbf{1}_{\mathbb{C}^*})L(1-s,\mathbf{1}_{\mathbb{C}^*})}{L(s,\mathbf{1}_{\mathbb{C}^*})L(s+1,\mathbf{1}_{\mathbb{C}^*})}=\frac{\Gamma_{\mathbb{C}}(-s)\Gamma_{\mathbb{C}}(1-s)}{\Gamma_{\mathbb{C}}(s)\Gamma_{\mathbb{C}}(s+1)}.$$

It is now easy to see that (32) holds in this case.

9.9. $n=2$, h **– non-split hermitian,** $\pi=\mathbf{1}$**, general case.** Once again, in the non-archimedean case we cannot use the usual analogue of the Gindikin-Karpelevic formula, even if $\omega=1$ and E/F is unramified, since $G\times G$ is not contained inside a hyperspecial maximal compact of $\mathrm{Isom}(\mathcal{V}^\square)$ (!). Instead, we use a global argument. Suppose that E_0/F_0 and ω_0 are given. We can choose a quadratic extension of number fields E/F and two places v_0, v_1 of F such that $E_{v_1}/F_{v_1}\simeq\mathbb{C}/\mathbb{R}$ and $E_{v_0}/F_{v_0}\simeq E_0/F_0$. We take $\mathcal{V}$ over F which is non-split exactly at v_0 and v_1, and $\pi=\mathbf{1}$. We also choose a Hecke character ω of $\mathbb{I}_E$ such that $\omega_{v_0}=\omega_0$ and ω_{v_1} is unramified. Using (33) and (28) and the previous computation we conclude (32) for v_0.

This finishes the list of minimal cases needed to be considered.

The case $h=0$**.** By multiplicativity we reduce to the case where π is supercuspidal (modulo the center). Let E_0, F_0, π_0, ω_0 be given. Let E/F be an extension of number fields, v_0 a place of F such that $E_{v_0}/F_{v_0}\simeq E_0/F_0$. We can find a cuspidal representation π of $GL_n(\mathbb{A}_E)$ such that $\pi_{v_0}\simeq\pi_0$ and π_v is a principal series representation for all $v\neq v_0$. Similarly there exists a Hecke character ω of $\mathbb{I}_E$ such that $\omega_{v_0}=\omega_0$. Thus, for all $v\neq v_0$

$$\gamma^{\mathcal{V}}(s,\pi_v\times\omega_v,\psi_v)=\gamma^{GJ}(s,\pi_v\times\omega_v,\psi_v)\gamma^{GJ}(s,\theta(\widetilde{\pi}_v)\times\omega_v,\psi_v) \qquad (35)$$

by (28) and (32). By (33) and the analogous property of the Godement-Jacquet local factors, we must have (35) for $v=v_0$ as well.

Self-duality. At this stage, we can infer the property (30) from the uniqueness of the γ-factors. We can also argue directly as follows. Let ι be the involution on $V^\square$ defined by $\iota(v_1,v_2)=(v_2,v_1)$. Then ι takes $h^\square$ to $-h^\square$, and hence it acts, by conjugation, on $G^\square$. Denote by ι^* the ensuing involution of $G^\square$. Clearly, ι^* stabilizes $G\times G$ and acts by interchanging the factors. It follows that

$$\omega(\kappa(\iota^*(g)))=\omega'(\det{}_{V\times V}(g))\cdot(\theta(\omega))(\kappa(g)) \qquad (36)$$

on $G\times G$ where $\omega'=\theta(\omega)$ if h is non-degenerate and $\omega'=\theta(\omega)/\omega$ if $h=0$. Also, ι^* stabilizes $S_{V^\diamond}$ and $S_{V^\triangledown}$ and acts trivially on their intersection. Note that $I(\theta(\omega),s)=I(\omega,s)\otimes\omega'\circ\det{}_{V\times V}$. Upon performing a change of variable

$(g_1, g_2) \mapsto (g_2, g_1)$ in the integral defining the zeta integral and using the relation (36) we will obtain

$$Z^{\mathcal{V}}(f_s, \widetilde{\alpha} \otimes \alpha) = Z^{\mathcal{V}}(\iota^*(f_s) \cdot \omega', \alpha \otimes \widetilde{\alpha})$$

where the zeta-integral on the left-hand side is with respect to ω and $\widetilde{\pi}$, and on the right-hand side – with respect to $\theta(\omega)$ and π. Clearly, $M(s)$ commutes with ι^*. Also, twisting by ω' plays no role on the integral defining l_{ψ_A}. Hence, $M^*(\theta(\omega), A, s) = M^*(\omega, A, s) \otimes \omega'$. The relation (30) follows.

Note that we didn't use the explicit description of $\widetilde{\pi}$ as a twist of π by an outer involution of G (cf. [**MVW87**]).

Archimedean factors. Consider now the archimedean case. Twisting ω by an unramified character we may assume that $\omega = \theta(\omega)^{-1}$. By multiplicativity we reduce to the case where $\mathcal{V}$ is anisotropic, i.e. $F = \mathbb{R}$ and G is either $O(2n)$, $O(2n+1)$ or $U(n)$. The doubled group $G^{\square}$ is $O(2n, 2n)$, $O(2n+1, 2n+1)$ or $U(n, n)$ respectively. The subgroup $K = G \times G$ is a maximal compact of $G^{\square}$ and $G^{\square} = S_{\mathcal{V}^{\diamond}} \cdot (G \times G)$ is the usual Iwasawa decomposition. As a K-module $I(\omega, s) = \mathrm{Ind}_{K \cap M}^{K} \omega$ is independent of s and it decomposes as $\oplus_{\pi \in \hat{G}} V_\pi$ where $V_\pi = \widetilde{\pi} \times (\pi \otimes \omega)$. In particular, it is multiplicity free. Moreover, V_π is realized by the matrix coefficients of π, twisted by ω (in the second variable). By our definition of the zeta integral the proportionality constants $\Gamma(s, \pi, \omega)$ will simply be the eigenvalue of $M(\omega, s)$ on $\widetilde{\pi} \times (\pi \otimes \omega)$. To compute these we use the procedure of Branson, Ólafsson and Ørsted ([**BÓØ96**]). Let $\lambda_\pi = (\lambda_1, \ldots, \lambda_n)$ be the highest weight of π, so that the λ_i's are integers, $\lambda_1 \geq \cdots \geq \lambda_n$, and in addition $\lambda_n \geq 0$ in the orthogonal case. (Note that in the latter case π is determined by its highest weight only up to a twist by the determinant.) In the unitary case let $k \in \mathbb{Z}$ be such that $\omega(z) = z^k$ on E^1. In the orthogonal case set $k = 0$ (even if ω is the signum character). Observe that $\lambda_{\widetilde{\pi}} = \lambda_\pi$ in the orthogonal case, while in the unitary case $\lambda_{\widetilde{\pi}} = (-\lambda_n, \ldots, -\lambda_1)$. The center of the Levi part of $\mathfrak{s}_{\mathcal{V}^{\diamond}}$ has a unique root in $\mathfrak{t}_{\mathcal{V}^{\diamond}}$. Thus, the *spectrum generating operator* in the language of [**BÓØ96**] is simply the Casimir operator with respect to K. On $\pi_1 \otimes \pi_2$ it acts by

$$\sum_{i=1,2} (\|\lambda_{\pi_i} + \rho\|^2 - \|\rho\|^2)$$

where ρ is the half-sum of positive roots of G given by

$$\rho = \begin{cases} (n-1, \ldots, 0) & \text{even orthogonal case} \\ (n - \frac{1}{2}, \ldots, \frac{1}{2}) & \text{odd orthogonal case} \\ (\frac{n-1}{2}, \ldots, -\frac{n-1}{2}) & \text{unitary case} \end{cases}$$

and

$$\|(\mu_1,\ldots,\mu_n)\|^2=\sum_{j=1}^n \mu_j^2.$$

Consider the K-module $\mathfrak{s}$ which is the orthogonal complement of $\mathfrak{k}$ in $\mathfrak{g}^{\square}$ with respect to the Killing form. We will identify $\mathfrak{s}$ with a subspace of $\mathrm{Ind}_{K\cap M}^K 1$ (cf. **[BÓØ96, 1.f]**, where this identification is denoted by the "cocycle" ω). The multiplication of functions gives a K-equivariant map

$$m:\mathfrak{s}\otimes \mathrm{Ind}_{K\cap M}^K\omega\to \mathrm{Ind}_{K\cap M}^K\omega.$$

We say that π and π' are (strongly) adjacent if the restriction of m to $\mathfrak{s}\otimes(\widetilde{\pi}\times(\pi\otimes\omega))$ has a non-trivial projection to $\widetilde{\pi'}\times(\pi'\otimes\omega)$. This is a symmetric condition, denoted by $\pi\overset{\omega}{\leftrightarrow}\pi'$ in **[BÓØ96, §4]**. Clearly, if π and π' are adjacent then

$$\mathrm{Hom}_K(\mathfrak{s}\otimes(\widetilde{\pi}\times(\pi\otimes\omega)),\widetilde{\pi'}\times(\pi'\otimes\omega))\neq 0.$$

If the latter condition is satisfied we say that π and π' are weakly adjacent. This relation is also symmetric and is denoted by $\pi\leftrightarrow\pi'$ in **[BÓØ96, §2]**.

Concretely, $\mathfrak{s}$ is either $St\otimes\widetilde{St}\oplus\widetilde{St}\otimes St$ (in the unitary case) or $St\otimes St$ (in the orthogonal case) where St is the "standard" n-dimensional (resp. $2n$-dimensional) representation of G. Thus, the weak adjacency condition is that

either $\mathrm{Hom}(St\otimes\pi,\pi')$ or $\mathrm{Hom}(\pi,St\otimes\pi')$ is non-zero.

This implies strong adjacency, because of the obvious fact that the matrix coefficients of $St\otimes\pi$ are the linear combinations of the products of matrix coefficients of St and π. Moreover, taking into account the classical branching rule for the decomposition of $St\otimes\pi$ (see e.g. **[KT87]**), the adjacency condition turns out to mean that $\lambda_\pi=(\lambda_1,\ldots,\lambda_n)$ and $\lambda_{\pi'}=(\lambda_1',\ldots,\lambda_n')$ differ in exactly one coordinate, in which they differ by 1.

By **[BÓØ96, (2.14)]** (with $r'=-r=2s$) we have the recursion relations

$$\begin{aligned}&(\|\lambda_{\widetilde{\pi}}+\rho\|^2+\|\lambda_{\pi\otimes\omega}+\rho\|^2-\|\lambda_{\widetilde{\pi'}}+\rho\|^2-\|\lambda_{\pi'\otimes\omega}+\rho\|^2+4s)\Gamma(s,\pi,\omega)\\ =&(\|\lambda_{\widetilde{\pi}}+\rho\|^2+\|\lambda_{\pi\otimes\omega}+\rho\|^2-\|\lambda_{\widetilde{\pi'}}+\rho\|^2-\|\lambda_{\pi'\otimes\omega}+\rho\|^2-4s)\Gamma(s,\pi',\omega)\end{aligned}$$

whenever π and π' are adjacent. Concretely, if, say, $\lambda_l'=\lambda_l+1$ then the recursion relation becomes

$$(\lambda_l+\rho_l+k/2+\frac{1}{2}-s)\Gamma(s,\pi,\omega)=(\lambda_l+\rho_l+k/2+\frac{1}{2}+s)\Gamma(s,\pi',\omega). \quad (37)$$

On the other hand if $\sigma: W_{\mathbf{R}}\to {}^LG$ is the parameter of π and std is as in §7 then

$$std\circ(\sigma\otimes\omega)=\oplus_{j=1}^n\mathrm{Ind}_{W_{\mathbf{C}}}^{W_{\mathbf{R}}}\left(z\mapsto\left(\frac{z}{\bar{z}}\right)^{\lambda_j+\rho_j}\{\omega(z)\}\right)$$

where we identify $W_{\mathbb{C}}$ with $\mathbb{C}^*$, and where on the right-hand side, $\omega(z)$ appears only in the unitary case. (In the orthogonal case, the representation obtained is stable under twisting by ω.) Hence,

$$\begin{aligned} L^{arch}(s,\widetilde{\pi}\times\omega^{-1},std) &= L^{arch}(s,\pi\times\omega,std) \\ &= \prod_{j=1}^{n}\Gamma_{\mathbb{C}}(s+|\lambda_j+\rho_j+k/2|), \\ \varepsilon^{arch}(s,\pi\times\omega,std,\psi) &= \prod_{j=1}^{n}\left[(-1)^{\lambda_j+\rho_j}i^k\lambda(E/F,\psi)\right]. \end{aligned}$$

The first equality holds since $std\circ\sigma\otimes\omega$ is self-dual. Thus,

$$\pi(-1)\gamma^{arch}(s+\frac{1}{2},\pi\otimes\omega,std,\psi) = c\prod_{j=1}^{n}\frac{\Gamma_{\mathbb{C}}(|\lambda_j+\rho_j+k/2|+\frac{1}{2}-s)}{\Gamma_{\mathbb{C}}(|\lambda_j+\rho_j+k/2|+\frac{1}{2}+s)} \tag{38}$$

where c is independent of π. It follows from (37) and (38) and the functional equation $\Gamma(s+1)=s\Gamma(s)$ that the quotient

$$\Gamma(s,\pi,\omega)/(\pi(-1)\gamma^{arch}(s+\frac{1}{2},\pi\otimes\omega,std,\psi))$$

is invariant under adjacency. Hence, it is independent of π. It remains to prove the required equality for $\pi=\mathbf{1}$. For this case we consider $F=\mathbb{Q}$, a quadratic imaginary field E (in the unitary case), $\pi=\mathbf{1}$ (globally) and a Hecke character of E^* which localizes to ω at the archimedean place. The gamma factors of π are compatible with those of the corresponding Weil group representation at all finite primes by (28) and (32). By (33) we conclude the same for the archimedean place.

This completes the proof of Theorem 4.

10. L and ε factors

To define the L and ε factors we proceed as follows. In the archimedean case we define $L^{\mathcal{V}}(s,\pi\times\omega)$ and $\varepsilon^{\mathcal{V}}(s,\pi\times\omega,\psi)$ as in **[Bor79]**. In the p-adic case we follow the procedure of **[Sha90]**. Namely, in the tempered case we define $L(s,\pi\times\omega)$ by the property that $L(s,\pi\times\omega)^{-1}$ is the numerator of $\gamma(s,\pi\times\omega,\psi)$ (a monic polynomial in q^{-s}). Then $\varepsilon(s,\pi\times\omega,\psi)$ is defined by

$$\varepsilon^{\mathcal{V}}(s,\pi\times\omega,\psi) = \gamma^{\mathcal{V}}(s,\pi\times\omega,\psi)\frac{L^{\mathcal{V}}(s,\pi\times\omega)}{L^{\mathcal{V}}(1-s,\widetilde{\pi}\times\omega^{-1})}.$$

It is holomorphic and by (29) it satisfies the functional equation

$$\varepsilon^{\mathcal{V}}(s,\pi\times\omega,\psi)\varepsilon^{\mathcal{V}}(1-s,\widetilde{\pi}\times\omega^{-1},\psi^{-1}) = 1.$$

Hence it is a monomial in q^s. If $\mathcal{V}$, π and ψ are unramified we retrieve the usual L-factor (and $\varepsilon = 1$). This follows from the fact that for tempered unramified representations the "Satake parameters" are unitary. (The latter follows from Macdonland's formula for the spherical function (**[Mac71]**); cf. **[Lab90, Proposition 6]** for a slightly weaker statement. The argument can be easily modified to deal with the general case.) More generally, if π contains a non-zero vector fixed under the Iwahori subgroup then $L^{\mathcal{V}}(s,\pi)$ is compatible with the L-factor of the corresponding representation of the Weil–Deligne group. This is because the latter is holomorphic for $\Re(s) > 0$ and the gamma factors agree.

In the case where $h = 0$ we have

$$L^{\mathcal{V}}(s,\pi\times\omega) = L^{GJ}(s,\pi\times\omega)L^{GJ}(s,\theta(\widetilde{\pi})\times\omega), \tag{39}$$
$$\varepsilon^{\mathcal{V}}(s,\pi\times\omega,\psi) = \varepsilon^{GJ}(s,\pi\times\omega,\psi)\varepsilon^{GJ}(s,\theta(\widetilde{\pi})\times\omega,\psi). \tag{40}$$

Thus, using this formula we can extend the definition of the L and ε-factors to quasi-tempered representations.

For $\mathcal{V} = \boxplus\mathcal{V}_j$ we set

$$L^{\mathcal{V}}(s,\otimes_j\pi_j\times\omega,\psi) = \prod_j L^{\mathcal{V}_j}(s,\pi_j\times\omega,\psi)$$
$$\varepsilon^{\mathcal{V}}(s,\otimes_j\pi_j\times\omega,\psi) = \prod_j \varepsilon^{\mathcal{V}_j}(s,\pi_j\times\omega,\psi).$$

Finally, a general irreducible representation π can be written essentially uniquely as a Langlands quotient of $\mathrm{Ind}_P^G\sigma$ where $P = MU$ is a parabolic subgroup of G (cf. **[Lan89]**, **[BW00]**, and also **[BJ01]** for the non-connected orthogonal case). If $P = S_{\mathcal{W}}$ for an isotropic flag $W_1 \subset \cdots \subset W_k$ where $\mathcal{W}$ is as in (2) then we define

$$L^{\mathcal{V}}(s,\pi\times\omega) = L^{\mathcal{W}}(s,\sigma\times\omega),$$
$$\varepsilon^{\mathcal{V}}(s,\pi\times\omega,\psi) = \varepsilon^{\mathcal{W}}(s,\sigma\times\omega,\psi).$$

The relation

$$\gamma^{\mathcal{V}}(s,\pi\times\omega,\psi) = \varepsilon^{\mathcal{V}}(s,\pi\times\omega,\psi)\frac{L^{\mathcal{V}}(1-s,\widetilde{\pi}\times\omega^{-1})}{L^{\mathcal{V}}(s,\pi\times\omega)} \tag{41}$$

holds for any π. If π contains a non-zero vector fixed under the Iwahori subgroup (in particular, if π is unramified) then $L^{\mathcal{V}}(s,\pi)$ is compatible with the L-factor of the corresponding representation of the Weil–Deligne group. This can be checked as in **[Sha90, §3]**.

In the case $h = 0$ the formula (39) still holds by (**[Jac79]**). Moreover, from the property (30) and the fact that $\widetilde{\pi}$ is the Langlands quotient of $\mathrm{Ind}_P^G\widetilde{\sigma}$

where $\bar{P}$ is an opposite parabolic to P, it follows that

$$L^{\mathcal{V}}(s,\widetilde{\pi}\times\omega)=L^{\mathcal{V}}(s,\pi\times\theta(\omega)),\quad \varepsilon^{\mathcal{V}}(s,\widetilde{\pi}\times\omega,\psi)=\varepsilon^{\mathcal{V}}(s,\pi\times\theta(\omega),\psi). \tag{42}$$

We remark that in the case where π is generic then $L^{\mathcal{V}}(s,\pi\times\omega)$ and $\varepsilon^{\mathcal{V}}(s,\pi\times\omega,\psi)$ agree with the factors defined by Shahidi in **[Sha90]**. This follows from the uniqueness part of **[Sha90, Theorem 3.5]**.

Suppose now that F is a number field and $\mathcal{V}$ is defined over F. Let π be a cuspidal representation of $G(\mathbb{A})$ and ω a Hecke character of $\mathbb{I}_E$. Define

$$L^{\mathcal{V}}(s,\pi\times\omega)=\prod_v L^{\mathcal{V}}(s,\pi_v\times\omega_v),\qquad \varepsilon^{\mathcal{V}}(s,\pi\times\omega)=\prod_v \varepsilon^{\mathcal{V}}(s,\pi_v\times\omega_v,\psi_v)$$

for some non-trivial character ψ_F of $F\backslash\mathbb{A}_F$. (As the notation suggests, $\varepsilon^{\mathcal{V}}(s,\pi\times\omega)$ does not depend on the choice of ψ_F.) Then $L(s,\pi\times\omega)$ extends to a meromorphic function and by (33) and (41) we have the functional equation

$$L^{\mathcal{V}}(s,\pi\times\omega)=\varepsilon^{\mathcal{V}}(s,\pi\times\omega)L^{\mathcal{V}}(1-s,\widetilde{\pi}\times\omega^{-1}).$$

We go back to the local situation.

Proposition 5. *Suppose that π is tempered ω is unitary. Then $L^{\mathcal{V}}(s,\pi\times\omega)$ is holomorphic for $\Re(s)\geq\frac{1}{2}$.*

Proof. In the archimedean case, this follows from the description of discrete series representations and the local L-factors (see **[Bor79]**).

Consider the non-archimedean case. By definition, the poles of $L^{\mathcal{V}}(s,\pi\times\omega)$ are the zeros of $\gamma^{\mathcal{V}}(s,\pi\times\omega)$, or of $\Gamma^{\mathcal{V}}(s-\frac{1}{2},\pi,\omega,A,\psi)$. By multiplicativity, it suffices to consider the case where π is square integrable. The case $h=0$ follows from (39). Assume that h is non-degenerate and consider the functional equation

$$Z(f_{-s},\alpha\otimes\widetilde{\alpha})=\Gamma^{\mathcal{V}}(s,\pi,\omega,A,\psi)Z(M^{*}(\theta(\omega)^{-1},A,-s)f_{-s},\alpha\otimes\widetilde{\alpha}).$$

The left-hand side can be taken to be non-zero by choosing f_s to be supported in a small neighborhood inside $G^{\diamond}\backslash G\times G$. On the other hand by Lemma 2 the zeta-integral is convergent, hence holomorphic, for $\Re(s)\geq 0$. Moreover, the operator $M^{*}(\theta(\omega)^{-1},A,s)$ is known to be holomorphic for $\Re(s)\leq 0$ (see **[HKS96, Lemma 6.7]** for the unitary case; **[KR92]** for the symplectic case; the orthogonal case is similar). The proposition follows. □

From our definition of the local factors we immediately get the following Corollary.

Corollary 1. *Suppose that π is a (tempered) constituent of* $\mathrm{Ind}_P^G\sigma$ *where σ is square-integrable. Then*

$$L^{V}(s,\pi\times\omega)=L^{W}(s,\sigma\times\omega),$$
$$\varepsilon^{V}(s,\pi\times\omega,\psi)=\varepsilon^{W}(s,\sigma\times\omega,\psi).$$

In particular, using (42) we obtain that if π is tempered and $\omega\theta(\omega)=1$ then

$$\varepsilon^{V}(\frac{1}{2},\pi\times\omega,\psi)=\gamma^{V}(\frac{1}{2},\pi\times\omega,\psi). \tag{43}$$

Note that for this fact we only used that $M^*(\omega,A,s)$ is holomorphic at $s=0$. This follows from its unitarity, which in turn follows from (16) and the simple fact that $M^*(\omega,A,s)$ is self-adjoint for $s\in\mathbb{R}$.

Let us now prove Theorem 1. The crucial point is the following irreducibility lemma.

Lemma 13. *Suppose that h is either symplectic or symmetric, n is even and $\omega^2=1$. Then the representation $I(\omega,0)$ is irreducible.*

Proof. In this p-adic case this is proved in [**Tad98**] and [**BJ03**]. For the archimedean case we make the following remark, which serves as a complement to Lemma 4.8 of [**BÓØ96**]. Consider the equivalence relation $\overset{\omega}{\rightleftharpoons}$ on the K-types of $\mathrm{Ind}_{K\cap M}^{K}\omega$ generated by $\overset{\omega}{\leftrightarrow}$ (see [**BÓØ96, §4**] for notation). We claim that this equivalence relation is in fact trivial. For otherwise, the K-module spanned by an equivalence class of K-types will be $\mathfrak{g}$-invariant for the induced action $I(\omega,s)$ for any s. This will contradict the generic irreducibility result of Bruhat ([**Bru56**]). Now, as in Lemma 4.8 of [**BÓØ96**], a sufficient condition for irreducibility will be that $\pi_\alpha-\pi_\beta\pm s\neq 0$ whenever $\alpha\overset{\omega}{\leftrightarrow}\beta$. In particular for $s=0$ we require that $\pi_\alpha\neq\pi_\beta$. This will certainly hold if $\pi_\alpha\neq\pi_\beta$ whenever $\alpha\leftrightarrow\beta$. We already analyzed the even orthogonal case for $F=\mathbb{R}$. (The linear factors in (37) are non-zero at $s=0$.) The symplectic case (with $\omega=1$) follows from [**BÓØ96, §3.e**]. The general symplectic case, and the cases $F=\mathbb{C}$ are similar. □

Remark 4. The subject of reducibility of degenerate principal series is vast. Some relevant works are [**Jan96b**], [**Jan96a**], [**Gus81**] for the p-adic case and [**Joh92**], [**HL99**], [**LL03**], [**Sah95**], [**Sah93**],[**KR90a**] for the archimedean case. Some of these references contain many (if not all) cases of the lemma above. However, we feel that in the archimedean case, the method of [**BÓØ96**] is arguably the simplest to apply.

We infer that $M^*(\omega,A,0)$ acts as a scalar, which is necessarily 1 by the relation (15). If π is tempered then the relation (19) yields $\Gamma^{V}(0,\pi,A,\psi)=1$.

Since $\det(A) \equiv (-1)^{n/2} \mod (F^*)^2$ in the symplectic case and $\det(A) \equiv (-1)^{n/2}D \mod (F^*)^2$ in the even orthogonal case, we obtain

$$\gamma^V(\frac{1}{2}, \pi \times \omega, \psi) = \pi(-1)\omega(-1)^{n/2} \cdot \begin{cases} \varepsilon(\frac{1}{2}, \omega, \psi) & \text{symplectic case,} \\ \omega(D)\varepsilon(\frac{1}{2}, \tau_D, \psi) & \text{orthogonal case.} \end{cases}$$

The same will be true for $\varepsilon^V(\frac{1}{2}, \pi \times \omega, \psi)$ by (43). In the general case, we use the Langlands classification and our prescription for $\varepsilon^V(s, \pi \times \omega, \psi)$. Thus, we need to consider the case $h = 0$ as well. However, this case follows from (39) since

$$\varepsilon^V(\frac{1}{2}, \pi \times \omega, \psi) = \varepsilon^{GJ}(\frac{1}{2}, \pi \times \omega, \psi)\varepsilon^{GJ}(\frac{1}{2}, \theta(\widetilde{\pi}) \times \omega, \psi)$$
$$= \varepsilon^{GJ}(\frac{1}{2}, \pi \times \omega, \psi)\varepsilon^{GJ}(\frac{1}{2}, \widetilde{\pi} \times \theta(\omega), \psi) = \varepsilon^{GJ}(\frac{1}{2}, \pi \times \omega, \psi)\varepsilon^{GJ}(\frac{1}{2}, \widetilde{\pi} \times \omega^{-1}, \psi)$$
$$= \varepsilon^{GJ}(\frac{1}{2}, \pi \times \omega, \psi)\varepsilon^{GJ}(\frac{1}{2}, \widetilde{\pi} \times \omega^{-1}, \psi^{-1})\pi(-1)\omega(-1)^n = \pi(-1)\omega(-1)^n.$$

(This works equally well if $\theta \neq \mathbf{1}$.)

There is an analogue of Theorem 1 for the unitary case (with the same proof). Namely, suppose that $\omega|_{F^*} = \eta_{E/F}^{n+1}$. Equivalently, $\omega\theta(\omega) = 1$ and $I(\omega, 0)$ is irreducible – see **[KS97]** for the p-adic case; the archimedean case follows from Lemma 4.8 of **[BÓØ96]** as in the orthogonal case. If h is ϵ-hermitian and n is even, let $\mathfrak{d}(h)$ be the class of $\epsilon^{n/2}\det(h(e_i, e_j))$ in F^*/NmE^* for any basis $\{e_i\}$ of V. If n is odd let $\delta \in E^*$ be such that $\theta(\delta) = -\delta$. The upshot of the previous argument is the equality

$$\varepsilon^V(\frac{1}{2}, \pi \times \omega, \psi) = \pi(-1) \cdot \begin{cases} \omega(\mathfrak{d}(h)) & n \text{ even,} \\ \omega(\delta) & n \text{ odd.} \end{cases}$$

(The argument of ω is $\det_V(A)^{-1}$ for an appropriate A.)

Note that if $\omega|_{F^*} = \eta_{E/F}^n$ then $\varepsilon^V(\frac{1}{2}, \pi \times \omega, \psi)$ is a much more subtle invariant – it is responsible for the "theta dichotomy" studied in **[HKS96]**.

References

[AG91] James Arthur and Stephen Gelbart. Lectures on automorphic L-functions. In *L-functions and arithmetic (Durham, 1989)*, volume 153 of *London Math. Soc. Lecture Note Ser.*, pages 1–59. Cambridge Univ. Press, Cambridge, 1991.

[Art78] James G. Arthur. A trace formula for reductive groups. I. Terms associated to classes in $G(\mathbf{Q})$. *Duke Math. J.*, 45(4):911–952, 1978.

[BJ01] Dubravka Ban and Chris Jantzen. The Langlands classification for non-connected p-adic groups. *Israel J. Math.*, 126:239–261, 2001.

[BJ03] Dubravka Ban and Chris Jantzen. Degenerate principal series for even-orthogonal groups. *Represent. Theory*, 7:440–480 (electronic), 2003.

[BK99] Alexander Braverman and David Kazhdan. On the Schwartz space of the basic affine space. *Selecta Math. (N.S.)*, 5(1):1–28, 1999.

[BÓØ96] Thomas Branson, Gestur Ólafsson, and Bent Ørsted. Spectrum generating operators and intertwining operators for representations induced from a maximal parabolic subgroup. *J. Funct. Anal.*, 135(1):163–205, 1996.

[Bor76] Armand Borel. Admissible representations of a semi-simple group over a local field with vectors fixed under an Iwahori subgroup. *Invent. Math.*, 35:233–259, 1976.

[Bor79] A. Borel. Automorphic L-functions. In *Automorphic forms, representations and L-functions (Proc. Sympos. Pure Math., Oregon State Univ., Corvallis, Ore., 1977), Part 2*, pages 27–61. Amer. Math. Soc., Providence, R.I., 1979.

[Bru56] François Bruhat. Sur les représentations induites des groupes de Lie. *Bull. Soc. Math. France*, 84:97–205, 1956.

[BW00] A. Borel and N. Wallach. *Continuous cohomology, discrete subgroups, and representations of reductive groups*. American Mathematical Society, Providence, RI, second edition, 2000.

[BZ77] I. N. Bernstein and A. V. Zelevinsky. Induced representations of reductive $\mathfrak{p}$-adic groups. I. *Ann. Sci. École Norm. Sup. (4)*, 10(4):441–472, 1977.

[Cas89] W. Casselman. Canonical extensions of Harish-Chandra modules to representations of G. *Canad. J. Math.*, 41(3):385–438, 1989.

[CHM00] William Casselman, Henryk Hecht, and Dragan Miličić. Bruhat filtrations and Whittaker vectors for real groups. In *The mathematical legacy of Harish-Chandra (Baltimore, MD, 1998)*, pages 151–190. Amer. Math. Soc., Providence, RI, 2000.

[Del76] Pierre Deligne. Les constantes locales de l'équation fonctionnelle de la fonction L d'Artin d'une représentation orthogonale. *Invent. Math.*, 35:299–316, 1976.

[DM78] Jacques Dixmier and Paul Malliavin. Factorisations de fonctions et de vecteurs indéfiniment différentiables. *Bull. Sci. Math. (2)*, 102(4):307–330, 1978.

[FQ73] A. Fröhlich and J. Queyrut. On the functional equation of the Artin L-function for characters of real representations. *Invent. Math.*, 20:125–138, 1973.

[GJ72] Roger Godement and Hervé Jacquet. *Zeta functions of simple algebras*. Springer-Verlag, Berlin, 1972. Lecture Notes in Mathematics, Vol. 260.

[GPSR87] Stephen Gelbart, Ilya Piatetski-Shapiro, and Stephen Rallis. *Explicit constructions of automorphic L-functions*, volume 1254 of *Lecture Notes in Mathematics*. Springer-Verlag, Berlin, 1987.

[GPSR97] D. Ginzburg, I. Piatetski-Shapiro, and S. Rallis. L functions for the orthogonal group. *Mem. Amer. Math. Soc.*, 128(611):viii+218, 1997.

[Gus81] Robert Gustafson. The degenerate principal series for $\mathrm{Sp}(2n)$. *Mem. Amer. Math. Soc.*, 33(248):vi+81, 1981.

[Hel94] Sigurdur Helgason. *Geometric analysis on symmetric spaces*, volume 39 of *Mathematical Surveys and Monographs*. American Mathematical Society, Providence, RI, 1994.

[Hen84] Guy Henniart. La conjecture de Langlands locale pour GL(3). *Mém. Soc. Math. France (N.S.)*, (11-12):186, 1984.

[HKS96] Michael Harris, Stephen S. Kudla, and William J. Sweet. Theta dichotomy for unitary groups. *J. Amer. Math. Soc.*, 9(4):941–1004, 1996.

[HL99] Roger Howe and Soo Teck Lee. Degenerate principal series representations of $\mathrm{GL}_n(\mathbf{C})$ and $\mathrm{GL}_n(\mathbf{R})$. *J. Funct. Anal.*, 166(2):244–309, 1999.

[HS88] Yumiko Hironaka and Fumihiro Satō. Spherical functions and local densities of alternating forms. *Amer. J. Math.*, 110(3):473–512, 1988.

[Jac79] Hervé Jacquet. Principal L-functions of the linear group. In *Automorphic forms, representations and L-functions (Proc. Sympos. Pure Math., Oregon State Univ., Corvallis, Ore., 1977), Part 2*, pages 63–86. Amer. Math. Soc., Providence, R.I., 1979.

[Jan96a] Chris Jantzen. Degenerate principal series for symplectic and odd-orthogonal groups. *Mem. Amer. Math. Soc.*, 124(590):viii+100, 1996.

[Jan96b] Chris Jantzen. Reducibility of certain representations for symplectic and odd-orthogonal groups. *Compositio Math.*, 104(1):55–63, 1996.

[Joh92] Kenneth D. Johnson. Degenerate principal series on tube type domains. In *Hypergeometric functions on domains of positivity, Jack polynomials, and applications (Tampa, FL, 1991)*, volume 138 of *Contemp. Math.*, pages 175–187. Amer. Math. Soc., Providence, RI, 1992.

[Kar79] Martin L. Karel. Functional equations of Whittaker functions on p-adic groups. *Amer. J. Math.*, 101(6):1303–1325, 1979.

[Kat99] Hidenori Katsurada. An explicit formula for Siegel series. *Amer. J. Math.*, 121(2):415–452, 1999.

[Kat01] H. Katsurada. Euler factor of a certain Dirichlet series attached to Siegel Eisenstein series. *Abh. Math. Sem. Univ. Hamburg*, 71:81–90, 2001.

[Kim00] Jahwan Kim. Gamma factors of certain supercuspidal representations. *Math. Ann.*, 317(4):751–781, 2000.

[Kit86] Yoshiyuki Kitaoka. Local densities of quadratic forms and Fourier coefficients of Eisenstein series. *Nagoya Math. J.*, 103:149–160, 1986.

[Kit88] Yoshiyuki Kitaoka. Local densities of quadratic forms. In *Investigations in number theory*, volume 13 of *Adv. Stud. Pure Math.*, pages 433–460. Academic Press, Boston, MA, 1988.

[Kit89] Yoshiyuki Kitaoka. Fourier coefficients of certain Eisenstein series. *Proc. Japan Acad. Ser. A Math. Sci.*, 65(7):253–255, 1989.

[KR90a] Stephen S. Kudla and Stephen Rallis. Degenerate principal series and invariant distributions. *Israel J. Math.*, 69(1):25–45, 1990.

[KR90b] Stephen S. Kudla and Stephen Rallis. Poles of Eisenstein series and L-functions. In *Festschrift in honor of I. I. Piatetski-Shapiro on the occasion of his sixtieth birthday, Part II (Ramat Aviv, 1989)*, pages 81–110. Weizmann, Jerusalem, 1990.

[KR92] Stephen S. Kudla and Stephen Rallis. Ramified degenerate principal series representations for $\mathrm{Sp}(n)$. *Israel J. Math.*, 78(2-3):209–256, 1992.

[KR94] Stephen S. Kudla and Stephen Rallis. A regularized Siegel-Weil formula: the first term identity. *Ann. of Math. (2)*, 140(1):1–80, 1994.

[KS97] Stephen S. Kudla and W. Jay Sweet, Jr. Degenerate principal series representations for $\mathrm{U}(n,n)$. *Israel J. Math.*, 98:253–306, 1997.

[KT87] Kazuhiko Koike and Itaru Terada. Young-diagrammatic methods for the representation theory of the classical groups of type B_n, C_n, D_n. *J. Algebra*, 107(2):466–511, 1987.

[Lab90] J.-P. Labesse. Fonctions élémentaires et lemme fondamental pour le changement de base stable. *Duke Math. J.*, 61(2):519–530, 1990.

[Lan70a] R. P. Langlands. Problems in the theory of automorphic forms. In *Lectures in modern analysis and applications, III*, pages 18–61. Lecture Notes in Math., Vol. 170. Springer, Berlin, 1970.

[Lan70b] Robert P. Langlands. On Artin's L-functions. *Rice University Studies*, 65:23–28, 1970.

[Lan89] R. P. Langlands. On the classification of irreducible representations of real algebraic groups. In *Representation theory and harmonic analysis on semisimple Lie groups*, pages 101–170. Amer. Math. Soc., Providence, RI, 1989.

[Lap04] Erez M. Lapid. On the root number of representations of orthogonal type. *Compos. Math.*, 140(2):274–286, 2004.

[LL03] Soo Teck Lee and Hung Yean Loke. Degenerate principal series representations of $\mathrm{Sp}(p,q)$. *Israel J. Math.*, 137:355–379, 2003.

[Mac71] I. G. Macdonald. *Spherical functions on a group of p-adic type*. Ramanujan Institute, Centre for Advanced Study in Mathematics,University of Madras, Madras, 1971. Publications of the Ramanujan Institute, No. 2.

[MVW87] Colette Mœglin, Marie-France Vignéras, and Jean-Loup Waldspurger. *Correspondances de Howe sur un corps p-adique*. Springer-Verlag, Berlin, 1987.

[MW95] C. Mœglin and J.-L. Waldspurger. *Spectral decomposition and Eisenstein series*. Cambridge University Press, Cambridge, 1995. Une paraphrase de l'Écriture [A paraphrase of Scripture].

[PR99] Dipendra Prasad and Dinakar Ramakrishnan. On the global root numbers of $\mathrm{GL}(n) \times \mathrm{GL}(m)$. In *Automorphic forms, automorphic representations, and arithmetic (Fort Worth, TX, 1996)*, pages 311–330. Amer. Math. Soc., Providence, RI, 1999.

[PSR86] I. Piatetski-Shapiro and S. Rallis. ϵ factor of representations of classical groups. *Proc. Nat. Acad. Sci. U.S.A.*, 83(13):4589–4593, 1986.

[PSR87] I. Piatetski-Shapiro and Stephen Rallis. Rankin triple L functions. *Compositio Math.*, 64(1):31–115, 1987.

[Ral87] Stephen Rallis. *L-functions and the oscillator representation*. Springer-Verlag, Berlin, 1987.

[RR93] R. Ranga Rao. On some explicit formulas in the theory of Weil representation. *Pacific J. Math.*, 157(2):335–371, 1993.

[RS] S. Rallis and D. Soudry. Stability of the local gamma factor arising from the doubling method. *Math. Ann., to appear.*

[Sah93] Siddhartha Sahi. Unitary representations on the Shilov boundary of a symmetric tube domain. In *Representation theory of groups and algebras*, volume 145 of *Contemp. Math.*, pages 275–286. Amer. Math. Soc., Providence, RI, 1993.

[Sah95] Siddhartha Sahi. Jordan algebras and degenerate principal series. *J. Reine Angew. Math.*, 462:1–18, 1995.

[Sch85] Winfried Scharlau. *Quadratic and Hermitian forms*, volume 270 of *Grundlehren der Mathematischen Wissenschaften [Fundamental Principles of Mathematical Sciences]*. Springer-Verlag, Berlin, 1985.

[SH00] Fumihiro Sato and Yumiko Hironaka. Local densities of representations of quadratic forms over p-adic integers (the non-dyadic case). *J. Number Theory*, 83(1):106–136, 2000.

[Sha81] Freydoon Shahidi. On certain L-functions. *Amer. J. Math.*, 103(2):297–355, 1981.

[Sha85] Freydoon Shahidi. Local coefficients as Artin factors for real groups. *Duke Math. J.*, 52(4):973–1007, 1985.

[Sha90] Freydoon Shahidi. A proof of Langlands' conjecture on Plancherel measures; complementary series for p-adic groups. *Ann. of Math. (2)*, 132(2):273–330, 1990.

[Shi82] Goro Shimura. Confluent hypergeometric functions on tube domains. *Math. Ann.*, 260(3):269–302, 1982.

[Shi99a] Goro Shimura. An exact mass formula for orthogonal groups. *Duke Math. J.*, 97(1):1–66, 1999.

[Shi99b] Goro Shimura. Generalized Bessel functions on symmetric spaces. *J. Reine Angew. Math.*, 509:35–66, 1999.

[Shi99c] Goro Shimura. Some exact formulas on quaternion unitary groups. *J. Reine Angew. Math.*, 509:67–102, 1999.

[Sie36] Carl Ludwig Siegel. Über die analytische Theorie der quadratischen Formen. II. *Ann. of Math. (2)*, 37(1):230–263, 1936.

[Sie37] Carl Ludwig Siegel. Über die analytische Theorie der quadratischen Formen. III. *Ann. of Math. (2)*, 38(1):212–291, 1937.

[Tad98] Marko Tadić. On reducibility of parabolic induction. *Israel J. Math.*, 107:29–91, 1998.

[Tak97] Keiji Takano. On standard *L*-functions for unitary groups. *Proc. Japan Acad. Ser. A Math. Sci.*, 73(1):5–9, 1997.

[Wal79] Nolan R. Wallach. Representations of reductive Lie groups. In *Automorphic forms, representations and L-functions (Proc. Sympos. Pure Math., Oregon State Univ., Corvallis, Ore., 1977), Part 1*, Proc. Sympos. Pure Math., XXXIII, pages 71–86. Amer. Math. Soc., Providence, R.I., 1979.

[Wal88a] Nolan R. Wallach. Lie algebra cohomology and holomorphic continuation of generalized Jacquet integrals. In *Representations of Lie groups, Kyoto, Hiroshima, 1986*, volume 14 of *Adv. Stud. Pure Math.*, pages 123–151. Academic Press, Boston, MA, 1988.

[Wal88b] Nolan R. Wallach. *Real reductive groups. I*, volume 132 of *Pure and Applied Mathematics*. Academic Press Inc., Boston, MA, 1988.

[Wal92] Nolan R. Wallach. *Real reductive groups. II*, volume 132 of *Pure and Applied Mathematics*. Academic Press Inc., Boston, MA, 1992.

[Wei64] André Weil. Sur certains groupes d'opérateurs unitaires. *Acta Math.*, 111:143–211, 1964.

[Wei65] André Weil. Sur la formule de Siegel dans la théorie des groupes classiques. *Acta Math.*, 113:1–87, 1965.

Erez M. Lapid
Einstein Institute of Mathematics
The Hebrew University of Jerusalem
Jerusalem 91904, Israel
erezla@math.huji.ac.il

Stephen Rallis
Department of Mathematics
The Ohio State University
231W 18th Ave
Columbus OH 43210, USA
haar@math.ohio-state.edu

Stabilité pour les représentations elliptiques de réduction unipotente; le cas des groupes unitaires.

Colette Mœglin

Résumé. This paper has two goals. The first one, which is of a technical nature, is to compute the restriction of a representation of a unitary group to its space of invariants under the pro-unipotent radical of a maximal parahoric subgroup. Of course, we assume that this restriction is not trivial and that it decomposes into Lusztig's unipotent representations. In the case of discrete series, we know (**[MW]**) that the knowledge of this restriction allows us to compute the orbital integrals of pseudo-coefficients of this discrete series on elliptic conjugacy classes. Moreover, by a general result essentially due to Arthur, we know that if such integrals are constant over stable conjugacy classes, then the discrete series (a linear combination of irreducible ones) is stable. Thus, a corollary of our technical result – and this is the second goal of the paper – is the determination of the stable packets of discrete series with unipotent reduction for unitary groups.

The computation of the restriction is certainly a technical tool. But its general form is like the one already obtained for orthogonal groups using a certain kind of Fourier transform on the parameters. This probably has a deeper explanation.

1. Introduction

Soit F un corps local p-adique et soit E/F une extension non ramifiée de degré 2. Soit encore V un E-espace vectoriel de dimension n. Il y a 2 classes de formes hermitiennes sur un tel espace qui sont distinguées par la parité de la valuation du déterminant de la forme ; si cette valuation est paire, on note $<,>_{iso}$ une forme de cette classe et si cette valuation est impaire on la note $<,>_{an}$. On note $G_\sharp$ le groupe d'une telle forme, où $\sharp = iso$ ou an suivant le contexte. On vérifie que la multiplication de la forme par une uniformisante entraîne que G_{iso} est isomorphe à G_{an} dans le cas où n est impaire ; par contre ces groupes sont formes intérieures l'un de l'autre si n est paire, G_{iso} est quasi-déployé. On note suivant la coutume $\mathcal{O}_E$ l'anneau des entiers de E et ϖ_E une uniformisante de E. Soit L un réseau de V ; pour $\natural = iso$ ou an, on note $L_\natural^*$ le dual de L pour la forme $<,>_\natural$. On suppose que L est presque autodual c'est-à-dire que $\varpi_E L_\natural^* \subset L \subset L_\natural^*$ et on pose :

$$\ell' := L/\varpi_E L_\natural^*; d' := dim_{\mathbb{F}_{q^2}} \ell', \qquad \ell'' := L_\natural^*/L; d'' := dim_{\mathbb{F}_{q^2}} \ell''.$$

Alors si $\natural = iso$ d'' est pair tandis que si $\natural = an$, d'' est impair. C'est le moyen que nous prenons de distinguer les 2 formes.

Une représentation de niveau 0 est une représentation ayant des invariants pour l'action du radical pro-p-unipotent d'un sous-groupe parahorique stabilisant un réseau presque autodual au sens ci-dessus. Comme en [**MW**], on dit qu'une représentation de niveau 0 est de réduction unipotente si quand on regarde l'action du groupe parahorique dans l'espace des invariants évoqué ci-dessus, cette action se fait par des représentations unipotentes (au sens de Lusztig) (c'est un produit de 2 groupes unitaires sur un corps fini qui opère).

Dans cet article, on s'intéresse essentiellement aux séries discrètes de niveau 0 et de réduction unipotente pour les groupes unitaires. Le résultat que nous avons en vue est plutôt technique ; il s'agit de calculer en termes de faisceaux caractères la représentation des sous-groupes parahoriques dans l'espace des invariants pour leur radical pro-p-unipotent. On a besoin de ce résultat pour compléter [**M**] qui décrit les distributions tempérées stables de niveau 0 pour les groupes orthogonaux impairs ; toutefois ceci n'est pas suffisant pour compléter [**M**]. En effet, suite aux travaux de Lusztig et de Aubert, Kutzko et Morris (travail en cours) on sait que le niveau zéro se ramène, par induction, au cas de réduction unipotente pour des groupes soient orthogonaux ET unitaires. Le problème est que le mot "ramener" sous-entend des calculs fins en particulier de signes qui devront être mis avec les signes qui apparaissent ici, ce qui n'est pas fait, d'autant que le travail d'AKM n'est pas disponible à l'heure où j'écris ces lignes.

Le résultat démontré ici se trouve ci-dessous en 7 ; il est amusant car il fait intervenir une sorte de transformation de Fourier (cf 6) qui intervenait naturellement dans [**MW**] comme un analogue p-adique de la transformation, définie par Lusztig sur les groupes finis et permettant de calculer les faisceaux caractères en terme de représentations irréductibles. Pour les groupes unitaires et sur les groupes finis, le faisceau caractère associé à une représentation irréductible du groupe symétrique est déjà (au signe près) le caractère d'une représentation irréductible du groupe unitaire sur le corps fini. Il n'était donc pas clair qu'une telle transformation ait un rôle à jouer, bien que pour les groupes p-adiques, il n'y ait pas beaucoup de différence dans la description des paramètres des représentations que ce soit pour un groupe unitaire, un groupe symplectique ou un groupe orthogonal. Il est donc rassurant de voir qu'une telle transformation intervient aussi. On aimerait évidemment en avoir d'autres applications qu'uniquement la formule technique de 7 puisque cette application est moralement un passage de la géométrie à la théorie des représentations.

Une fois le résultat technique démontré, il est facile d'obtenir la description des combinaisons linéaires stables de séries discrètes de niveau 0 ainsi que leur transfert (pour n pair) entre la forme anisotrope et la forme isotrope ; toutefois ici on utilise des résultats d'analyse harmonique qui supposent que p est grand, hypothèse que nous faisons donc. La description des distributions stables à support les représentations tempérées de réduction unipotente est faite en 8.1 mais la démonstration est, en plus simple, celle de **[MW]**. Le résultat est aussi simple qu'en **[MW]**, c'est-à-dire que c'est la somme des représentations discrètes dans un paquet qui engendre l'espace des distritutions stables combinaisons linéaires de caractères de représentations elliptiques.

Toutefois, il apparait ici un phénomène qui n'apparaissait pas en **[MW]** mais qui était prévu par Arthur ; considérons l'ensemble des représentations duales des représentations discrètes de niveau zéro, duales par la dualité qui généralise celle d'Iwahori et Matsumoto. Cette dualité a été définie entre autre par Aubert ([**Au**]) et Schneider-Stuhler ([**SS**]) ; par exemple dans [**Au**] elle est calculée complètement explicitement en terme d'induction restriction, il est donc clair qu'une telle transformation conserve la propriété de stabilité pour les caractères des représentations, mais elle introduit un signe. Précisément, à tout couple formé d'un morphisme $\psi : W_F \times SL(2,\mathbb{C})$ dans le L-groupe $GL(n,\mathbb{C}) \rtimes W_{E/F}$ de $U(n,E/F)$ qui ne se factorise par aucun Levi et d'un caractère, ϵ, du centralisateur de ce morphisme, Lusztig associe une représentation de $U(n,E/F)$ (la forme du groupe est déterminée par ϵ) qui est la duale (au sens de la dualité ci-dessus) d'une série discrète ; la série discrète, on la note $\pi(\psi,\epsilon)$ et sa duale $\pi_{unip}(\psi,\epsilon)$. Encore plus précisément, notons s_Z l'élément non trivial du centre du L-groupe de $U(n,E/F)$; si $\epsilon(s_Z) = +1$, alors $\pi(\psi,\epsilon)$ et $\pi_{unip}(\psi,\epsilon)$ sont des représentations de $U(n,E/F)_{iso}$ et sinon sont des représentations de $U(n,E/F)_{an}$. On écrit $\epsilon(s_Z) = \sharp$ si $\epsilon(s_Z) = +1$ quand $\sharp = iso$ et -1 sinon.

Avec ces notations et en fixant le morphisme ψ, on montre donc ici qu'il existe un signe (que l'on interprète ci-dessous) $\delta(\psi,\epsilon)$, tel que les combinaisons linéaires, pour $\sharp = iso$ ou an :

$\Pi(\sharp,\psi) := \sum_{\epsilon;\epsilon(s_Z)=\sharp} \pi(\psi,\epsilon)$ et $\Pi_{unip}(\sharp,\psi) \sum_{\epsilon;\epsilon(s_Z)=\sharp} \delta(\psi,\epsilon)\pi_{unip}(\psi,\epsilon)$ sont stables. Si n est pair, le transfert de $\Pi(iso,\psi)$ à $U(n,E/F)_{an}$ est $-\Pi(an,\psi)$ (idem pour Π_{unip}). De plus l'ensemble des distributions $\Pi(\sharp,\psi)$ quand ψ varie, engendre l'espace des distributions stables combinaisons linéaires de caractères de représentations elliptiques (au sens d'Arthur).

Expliquons le signe $\delta(\psi,\epsilon)$. Le groupe dual de $U(n,E/F)$ n'est pas linéaire mais il se plonge naturellement dans $GL(2n,\mathbb{C})$; En composant ψ avec ce plongement, on construit une représentation de $W_F \times SL(2,\mathbb{C})$ qui est symplectique si n est pair et orthogonale si n est impair. On peut facilement décomposer le morphisme $\psi_{|W_F}$ en la somme d'une représentation orthogonale et d'une représentation symplectique (ici on utilise le fait que le morphisme est discret) ; notons m_{orth} et m_{symp} la dimension de l'espace de ces représentations, d'où $2n = m_{ortho} + m_{symp}$. Alors ψ (composé avec l'inclusion) se factorise en un morphisme de $W_F \times SL(2,\mathbb{C})$ dans $GL(m_{ortho},\mathbb{C}) \times GL(m_{symp},\mathbb{C})$. Cela induit une décomposition de ψ en $\psi_{ortho} \oplus \psi_{symp}$; attention, la notation prête à confusion car ψ_{symp} n'est pas orthogonale c'est seulement sa restriction à W_F qui l'est, avec la même remarque pour ψ_{ortho}. La représentation ψ_{symp} restreinte à $SL(2,\mathbb{C})$ est orthogonale si n est pair et symplectique si n est impair. On note s_{symp} l'élément non trivial du centre de $GL(n_{symp},\mathbb{C})$ invariant sous W_F. On peut évaluer $\epsilon(s_{symp})$ et c'est le signe cherché. C'est-à-dire que $\delta(\psi,\epsilon) = \epsilon(s_{symp})$.

Pour terminer cette introduction, je voudrais remercier Anne-Marie Aubert et Jean-Loup Waldspurger pour les discussions que nous avons eues autour de ce travail.

Convention *Soit $d \in \mathbb{N}$, la notation $\sharp = (-1)^d$ signifie par convention $\sharp = iso$ si d est pair et $\sharp = an$ si d est impair et réciproquement, si $\sharp$ est connu c'est la parité de d qui est fixée par cette notation et $\sharp^d = 1$ si $\sharp = iso$ et $(-1)^d$ si $\sharp = an$.*

2. Paramètres discrets unipotents

2.1. Définition Par définition, un paramètre discret unipotent est la donnée d'un morphisme

$$\psi : W_F/I_F \times SL_2(\mathbb{C}) \to GL(n,\mathbb{C}) \rtimes W_F/I_F,$$

où W_F est le groupe de Weil de F et I_F son groupe de ramification et d'un caractère $\epsilon : Cent_{GL(n,\mathbb{C})}\psi \mapsto \{\pm 1\}$, soumis à un nombre certain de propriétés. On impose d'abord les propriétés habituelles à savoir que le diagramme complété par les projections sur W_F/I_F est commutatif, que la restriction de ψ à $SL(2,\mathbb{C})$ est un morphisme algébrique (nécessairement à valeurs dans $GL(n,\mathbb{C})$) et que la restriction de ψ à W_F/I_F composée avec la projection sur $GL(n,\mathbb{C})$ se factorise par un quotient fini. On revient sur la définition du produit semi-direct ci-dessous; pour cela on note $Fr_{E/F}$ un Frobenius pour l'extension E de F et on note w l'élément de $GL(n,\mathbb{C})$ défini par :

$$w = \begin{pmatrix} 0 & \cdots & \cdots & 1 \\ 0 & \cdots & -1 & 0 \\ \vdots & \vdots & \vdots & \vdots \\ (-1)^{n+1} & \cdots & 0 & 0 \end{pmatrix}.$$

Alors l'action de W_F/I_F sur $GL(n,\mathbb{C})$ se factorise par le quotient $Gal(E/F)$ qui est un groupe à 2 éléments, l'élément non trivial s'identifiant à $Fr_{E/F}$ et on a :

$$\forall g \in GL(n,\mathbb{C}), Fr_{E/F}.g = w^{-1}\,{}^t g^{-1}\, w.$$

Le fait que l'on ait mis discret dans la définition se traduit par le fait que l'image de ψ ne doit pas être incluse dans le L-groupe d'un sous groupe de Levi de $U(n,E/F)_\sharp$ (cela ne dépend pas de $\sharp$) et on impose en plus que la restriction de ϵ à l'élément diagonal de $GL(n,\mathbb{C})$ formé uniquement de (-1) vaille $+1$ si $\sharp = iso$ et -1 si $\sharp = an$. Quand cette condition est réalisée, on dit que $\epsilon_{|Z} = \sharp$.

Pour pouvoir classifier aisément de tels paramètres le mieux est de définir l'application

$$inc : GL(n,\mathbb{C}) \rtimes W_F/W_E \mapsto GL(2n,\mathbb{C})$$

en posant

$$\forall g \in GL(n,\mathbb{C}), inc(g) = \begin{pmatrix} g & 0 \\ 0 & {}^t g^{-1} \end{pmatrix}, inc(Fr_{E/F}) = \begin{pmatrix} 0 & w \\ w^{-1} & 0 \end{pmatrix}.$$

On remarque alors que l'on peut voir ψ comme un morphisme de $W_F/W_E \times SL(2,\mathbb{C})$ dans $GL(2n,\mathbb{C})$ qui se factorise par l'application *inc*.

2.2. Classification Soit O une orbite unipotente de $GL(n,\mathbb{C})$; on note $Jord(O)$ l'ensemble de ses blocs de Jordan, ensemble considéré avec multiplicités, éventuellement. Une application de $Jord(O)$ dans $\{\pm 1\}$ est une application qui à $\alpha \in Jord(O)$ associe soit $+1$ soit -1 indépendamment de la multiplicité de α dans $Jord(O)$; on dit évidemment que $Jord(O)$ est sans multiplicité si toutes les multiplicités sont 1.

Proposition 2.1. *L'ensemble des paramètres discrets unipotents pour le groupe $U(n,E/F)_\sharp$ vu à conjugaison près, est en bijection avec l'ensemble des couples (O,ϵ) où O est une orbite unipotente de $GL(n,\mathbb{C})$ telle que $Jord(O)$ soit sans multiplicité et ϵ est une application de $Jord(O)$ dans $\{\pm 1\}$ telle que $\prod_{\alpha \in Jord(O)} \epsilon(\alpha) = \sharp$.*

Soit (ψ,ϵ) un paramètre discret unipotent. On commence par considérer la restriction de ψ à $SL(2,\mathbb{C})$; comme on travaille à conjugaison près, c'est la même chose que de se donner une orbite unipotente O de $GL(n,\mathbb{C})$. On vérifie que si $Jord(O)$ n'est pas sans multiplicité, le paramètre n'est pas discret. Supposons donc que $Jord(O)$ soit sans multiplicité. On factorise $inc \circ$

$\psi_{|SL(2,\mathbb{C})}$ par un produit de groupes $\times_{\alpha\in Jord(O)}GL(2\alpha,\mathbb{C})$ (ici, on voit $\alpha \in Jord(O)$ comme un nombre) ; dans une telle factorisation, quitte à conjuguer, on peut s'arranger pour que la matrice $\begin{pmatrix} 0 & w \\ w^{-1} & 0 \end{pmatrix}$ provienne du produit $\times_{\alpha\in Jord(O)}\tilde{J}_\alpha$ où $\tilde{J}_\alpha = \begin{pmatrix} 0 & J_\alpha \\ J_\alpha^{-1} & 0 \end{pmatrix}$ où J_α est une matrice symétrique si α est impair et antisymétrique sinon, non dégénérée. Et à conjugaison près on peut prendre n'importe quelle matrice vérifiant ces conditions ; la conjugaison dont il est ici question est celle qui se fait sous $GL(n,\mathbb{C})$ plongé dans $GL(2n,\mathbb{C})$ comme expliqué ci-dessus, c'est-à-dire est $g \bullet {}^tg$. On utilise maintenant le fait qu'une représentation irréductible de $SL(2,\mathbb{C})$ de dimension paire est symplectique et qu'une représentation irréductible de dimension impaire est orthogonale. On peut donc supposer comme nous le ferons que l'image de $\psi(SL(2,\mathbb{C}))$ commute avec $\prod_{\alpha\in Jord(O)}\tilde{J}_\alpha$ et on peut écrire $\psi(Fr_{E/F})$ sous la forme $\times_{\alpha\in Jord(O)}\begin{pmatrix} \gamma_\alpha & 0 \\ 0 & {}^t\gamma_\alpha^{-1} \end{pmatrix}\tilde{J}_\alpha$. On vérifie que la condition de commutation avec l'image de $SL(2,\mathbb{C})$ est équivalent à ce que chaque γ_α soit scalaire.

Pour terminer la preuve de la proposition, il faut encore calculer le commutant de l'image de ψ dans $GL(n,\mathbb{C})$. On calcule d'abord le commutant de l'image de $SL(2,\mathbb{C})$ par ψ dans $GL(n,\mathbb{C})$ et on trouve que c'est un produit de $\mathbb{C}^*$ indexé par chaque élément de $Jord(O)$ interprété comme le centre de chaque $GL(\alpha,\mathbb{C})$. Quand on plonge dans $\times_\alpha GL(2\alpha,\mathbb{C})$, ce commutant s'écrit sous la forme d'un élément diagonal, $(t_\alpha, t_\alpha^{-1})_{\alpha\in Jord(O)}$, où t_α est une matrice scalaire de taille α. Quand on impose la commutation à la matrice $\times_{\alpha\in Jord(O)}\tilde{J}_\alpha$ comme ci-dessus on trouve la condition pour tout α, $t_\alpha {}^tt_\alpha = 1$ c'est-à-dire $t_\alpha = \pm 1$. Ainsi le commutant l'image de ψ dans $GL(n,\mathbb{C})$ est un produit de groupes $\{\pm 1\}$ indexés par les éléments de $Jord(O)$. Cela termine la preuve.

2.3. Le cas tempéré, versus elliptique On aura parfois besoin de travailler avec des paramètres unipotents tempérés et pas seulement discrets. On enlève la condition que l'image n'est pas incluse dans un Levi mais on garde le fait que ψ vu comme morphisme de $W_F/I_F \times SL(2,\mathbb{C})$ dans $GL(n,\mathbb{C}) \rtimes Gal(E/F)$ se factorise par le quotient $Gal(E/F)\times SL(2,\mathbb{C})$. Toutefois, ainsi on a trop de paramètres car les seules représentations qui nous intéressent sont les représentations elliptiques au sens d'Arthur ; ce sont des représentations virtuelles basées sur des représentations tempérées particulières. Sur les paramètres, la condition se traduit par le fait que la représentation $inc \circ \psi$ de $Gal(E/F) \times SL(2,\mathbb{C})$ est une somme de représentations irréductibles toutes orthogonales. On appelle ces paramètres des paramètres tempérés unipotents, versus elliptiques. On a dans ce cadre une généralisation immédiate de la proposition précédente :

Remarque 2.2. L'ensemble des paramètres tempérés unipotents, versus elliptiques, est en bijection avec l'ensemble des couples (O, ϵ) où O est une orbite unipotente de $GL(n, \mathbb{C})$ et où ϵ est une application de $Jord(O)$ dans $\{\pm 1\}$.

On reprend la démonstration précédente; la restriction de ψ à $SL(2, \mathbb{C})$ est, à conjugaison près, la donnée d'une orbite unipotente O de $GL(n, \mathbb{C})$. Quand on calcule le commutant de $\psi(SL(2, \mathbb{C}))$ on considère J_α comme dans cette démonstration mais il faut tenir compte de la multiplicité de α comme bloc de Jordan, c'est-à-dire que J_α est dans $GL(mult_\alpha(Jord(O))\alpha, \mathbb{C})$. On peut encore supposer (et nous le ferons) que $J_\alpha = (-1)^{\alpha+1\,t} J_\alpha$. On définit ensuite $\tilde{J}_\alpha := \begin{pmatrix} 0 & J_\alpha \\ J_\alpha^{-1} & 0 \end{pmatrix}$. Le commutant de $\psi(SL(2, \mathbb{C})$ dans $inc(GL(n, \mathbb{C})$ est isomorphe au produit semi-direct $GL(mult_\alpha(Jord(O)), \mathbb{C}) \ltimes \{1, \tilde{J}_\alpha\}$; pour montrer que l'orbite de $\psi(Fr_{E/F})$ est uniquement déterminée dans ce commutant, il faut utiliser l'hypothèse qu'elle a un représentant de la forme $\gamma_\alpha J_\alpha$ avec γ_α symétrique. L'orbite de γ_α par $h \bullet {}^t h$ est alors uniquement déterminée puisque γ_α est de rang maximum. Pour ce qui suit, on suppose, comme on en a le droit que $\gamma_\alpha = 1$ pour tout α.
Pour interpréter le caractère ϵ, il faut calculer le commutant dans $GL(n, \mathbb{C})$ de l'image de ψ. Un élément du commutant est nécessairement de la forme $\times_{\alpha \in Jord(O)} g_\alpha$ avec pour tout $\alpha \in Jord(O)$, $g_\alpha \in GL(mult_\alpha(Jord(O)), \mathbb{C})$ et la commutation avec $\psi(Fr_{E/F})$ se traduit par $g_\alpha {}^t g_\alpha = 1$. Ceci indique que g_α décrit un groupe orthogonal, le groupe des composantes est donc encore $\{\pm 1\}$. Ainsi ϵ s'interprète encore comme une application de $Jord(O)$ dans $\{\pm 1\}$. La condition de restriction au centre est maintenant

$$\times_{\alpha \in Jord(O)} \epsilon(\alpha)^{mult_\alpha(Jord(O))} = \sharp.$$

Remarque 2.3. cette notation, "versus elliptique" n'a pas vocation à entrer dans les annales. En fait, elle n'est pas bonne, car dans les constructions de représentations elliptiques telles que faites par Arthur, les représentations tempérées qui interviennent sont celles qui sont versus elliptique comme ci-dessus mais où *en plus* les multiplicités sont inférieures ou égales à 2.

2.4. Le cas tempéré On ne va quand même pas laisser en suspens la classification des paramètres de Langlands tempérés de réduction unipotente. D'abord, il faut regarder la restriction à W_F ; elle fait intervenir des caractères non ramifiés. On sépare les 2 caractères quadratiques des autres. Les "autres" font intervenir une induction irréductible sans mystère donc on s'intéresse aux caractères quadratiques. On appelle ces paramètres, autoduaux pour des raisons évidentes.

Lemme 2.4. *L'ensemble des paramètres de Langlands tempérés de réduction unipotente autoduaux, est en bijection avec les données,* $n_{symp}, n_{orth} \in \mathbb{N}$, *avec $n = n_{symp} + n_{orth}$, O_{symp} une orbite unipotente de $Sp(n_{symp}, \mathbb{C})$ et O_{orth} une orbite unipotente de $O(n_{orth}, \mathbb{C})$, et des applications $\epsilon_{symp}, \epsilon_{orth}$ de $Jord(O_{symp})$ et $Jord(O_{orth})$ dans $\{\pm 1\}$ soumises à la condition que $\epsilon_\delta(\alpha) = 1$ si $\delta = symp$ et α est impair ou si $\delta = orth$ et α est pair. Ceci se traduit plus simplement en disant que ϵ_δ pour $\delta = symp$ et $orth$ est un morphisme du groupe des composantes du centralisateur d'un point de O_δ dans $\{\pm 1\}$.*

Ici on retrouve la classification générale pour les groupes classiques, c'est-à-dire la donnée d'un ensemble de couples $\{(O_\rho, \epsilon_\rho)\}$, où ρ parcourt un ensemble des représentations irréductibles autoduales de W_F, O_ρ est une orbite unipotente soit d'un groupe symplectique soit d'un groupe orthogonal (ceci étant déterminé par ρ) et ϵ_ρ est un morphisme du groupe des composantes du centralisateur d'un point de l'orbite dans $\{\pm 1\}$. Avec un groupe unitaire, la notion de représentation de W_F est un peu plus compliquée car ce que l'on a est un morphisme de W_F dans $GL(n, \mathbb{C}) \ltimes W_F$ et que c'est lui qui par décomposition donne les représentations ρ. C'est l'intérêt de passer à $GL(2n, \mathbb{C})$ pour faire cette décomposition comme expliqué dans les preuves précédentes. Ce qui diffère ici par rapport à la preuve de 2.3 est, avec les notations de cette preuve, que l'on n'a pas que γ_α est symétrique. Il faut donc décomposer γ_α en la somme d'une matrice symétrique et d'une matrice antisymétrique ; c'est le rang de chacune des matrices de cette décomposition qui est l'invariant suppémentaire pour la classe de conjugaison. Ensuite la démonstration se termine de la même façon.

3. Construction de fonctions

3.1. Faisceaux caractères On note $D(n) := \{(d', d''); n = d' + d''\}$. Fixons $(d', d'') \in D(n)$ et pour $\sharp = (-1)^{d''}$ (cf la convention de l'introduction) on fixe un réseau presque autodual $L := L_{d',d''}$ de $U(n, E/F)_\sharp$ tel que $\ell' := L/\varpi_E L_\sharp^*$ soit de dimension (sur $\mathbb{F}_{q^2}$) d' et $\ell'' := L_\sharp^*/L$ soit de dimension d''. Les espaces ℓ' et ℓ'' héritent d'une forme unitaire, et on a donc des groupes unitaires, $U(\ell', \mathbb{F}_q)$ et $U(\ell'', \mathbb{F}_q)$ sur le corps fini $\mathbb{F}_q$ relatifs à l'extension $\mathbb{F}_{q^2}/\mathbb{F}_q$.

On note $\mathfrak{S}_{d'}$ et $\mathfrak{S}_{d''}$ les groupes symétriques sur d' et d'' éléments respectivement. On note $\hat{\mathfrak{S}}_{d'}$ l'ensemble des classes d'isomorphie de représentations irréductibles de $\mathfrak{S}_{d'}$ et cette définition sera généralisé à tous les groupe symétriques $\mathfrak{S}$ en $\hat{\mathfrak{S}}$. Grâce à Lusztig, on sait associer à une représentation de, disons $\mathfrak{S}_{d'}$, une fonction invariante par conjugaison sur $U(\ell', \mathbb{F}_q)$ (cf. par exemple [**DM**]). La propriété clé des faisceaux caractères pour les groupes unitaires est qu'à une représentation irréductible de $\mathfrak{S}_{d'}$ correspond une fonction invariante sur $U(\ell', \mathbb{F}_q)$ qui est la trace, à un signe près que l'on explicitera, d'une représentation irréductible de ce dernier groupe. D'où, en appliquant les mêmes constructions à ″, une application :

$$\mathbb{C}[\hat{\mathfrak{S}}_{d'}] \otimes \mathbb{C}[\hat{\mathfrak{S}}_{d''}] \to \mathbb{C}[U(\ell', \mathbb{F}_q)] \otimes \mathbb{C}[U(\ell'', \mathbb{F}_q)].$$

On note $K_{d',d''}$ le sous-groupe de $U(n, E/F)_\sharp$ qui stabilise le réseau $L_{d',d''}$; c'est un sous-groupe parahorique dont la réduction modulo le radical pro-p-unipotent est précisément $U(\ell', \mathbb{F}_q) \times U(\ell'', \mathbb{F}_q)$. On peut donc remonter les fonctions sur ce dernier groupe en des fonctions sur $K_{d',d''}$ de façon invariante par le radical pro-p-unipotent. On obtient ainsi une application de $\mathbb{C}[\hat{\mathfrak{S}}_{d'}] \otimes \mathbb{C}[\hat{\mathfrak{S}}_{d''}]$ dans l'ensemble des fonctions sur $K_{d',d''}$, invariantes par conjugaison sous le groupe et par translation sous son radical pro-p-unipotent. On identifie ces fonctions, via la trace, à des représentations semi-simples de $K_{d',d''}$ triviales sur le radical pro-p-unipotent.

Notation *On pose* $\mathbb{C}[\hat{\mathfrak{S}}_{D(n)}] := \otimes_{(d',d'') \in D(n)} \mathbb{C}[\hat{\mathfrak{S}}_{d'}] \otimes \mathbb{C}[\hat{\mathfrak{S}}_{d''}]$. *En sommant les constructions précédentes, on vient donc de définir une application notée* k *de* $\mathbb{C}[\hat{\mathfrak{S}}_{D(n)}]$ *dans la somme sur les couples* $(d', d'') \in D(n)$ *des ensembles de fonctions sur* $K_{d',d''}$ *définies ci-dessus.*

3.2. Formules explicites Les formules ci-dessous ont été tirées de [**AMR**]. Soit $d \in \mathbb{N}$ et $U(d, \mathbb{F}_q)$ un groupe unitaire pour l'extension de degré 2 du corps $\mathbb{F}_q$. Les représentations irréductibles unipotentes d'un tel groupe sont en bijection avec la donnée d'un entier k tel que $d - k(k+1)/2 \in 2\mathbb{N}$ et d'une représentation, ρ, irréductible, du groupe de Weyl de type C de rang $1/2(d - k(k+1)/2)$, noté $W_{1/2(d-k(k+1)/2)}$; en fait à l'entier k, on associe une représentation cuspidale du groupe $U(k(k+1)/2, \mathbb{F}_q)$ et l'algèbre d'entrelacement de l'induite de cette représentation cuspidale et du caractère trivial d'un tore convenable est l'algèbre de Hecke de $W_{1/2(d-k(k+1)/2)}$. Une représentation ρ comme ci-dessus est donnée par un couple de partition $(\alpha_1 \geq \cdots \geq \alpha_t \geq 0; \beta_1 \geq \cdots \geq \beta_t \geq 0)$ (t est ici un entier arbitrairement grand) vérifiant $\sum_{i \in [1,t]} \alpha_i + \beta_i = 1/2(d - k(k+1)/2)$. La bijection entre les couples (k, ρ) comme ci-dessus et les représentations irréductibles de $\mathfrak{S}_d$, c'est-à-dire les partitions de d se fait ainsi :

On fixe une partition $\mu := (a_1 \geq \cdots \geq a_\ell \geq 0)$ de d; on suppose que ℓ est à la fois grand et impair et on note $\mathcal{P} := \{a_i + \ell - i; (-1)^{a_i - i} = -1\}$ et $\mathcal{I} := \{a_i + \ell - i; (-1)^{a_i - i} = +1\}$. On pose encore $\tilde{k} := |\sharp\mathcal{P} - \sharp\mathcal{I}|$; ce nombre est impair. On distingue encore (ce qui définit l'entier k) $\tilde{k} = k - 1$ si $\sharp\mathcal{P} - \sharp\mathcal{I} > 0$ et $\tilde{k} = k$ sinon. Maintenant que l'on a défini k à partir de μ, il reste à définir ρ, c'est à dire les partitions $(\alpha_i; i \in [1, t'])$ et $(\beta_i; i \in [1, t'])$ ce qui est donné par :

$$\{\alpha_i\} = \{(a_j - j)/2 + \tilde{j} + (-1)^{k+1}/2\tilde{k}; a_j + \ell - j \in \mathcal{P}\};$$
$$\{\beta_i\} = \{(a_j - j - 1)/2 + \tilde{j} - (-1)^{k+1}/2\tilde{k}; a_j + \ell - j \in \mathcal{I}\},$$

où dans les 2 cas $\tilde{j}$ est le cardinal de $\{j' \leq j; (-1)^{a_{j'}-j'} = (-1)^{a_j - j}\}$.

On a expliqué que la correspondance entre faisceaux caractères et représentations introduisait un signe; pour μ comme ci-dessus, on le note σ_μ et d'après [DM] 9.6, $\sigma_\mu = (-1)^{n(n-1)/2+\sum_{i\in[1,\ell]} a_i(a_i-1)/2}$.

3.3. Filtration Soit $(d', d'') \in D(n)$ comme en 3.1 et soit ρ', ρ'' des représentations irréductibles de $\mathfrak{S}_{d'}$ et $\mathfrak{S}_{d''}$ correspondant à des partitions μ' et μ'' de d' et d'' respectivement. On définit la partition $\mu := \mu' + \mu''$ de n, en faisant la somme terme à terme, c'est-à-dire $\mu = (a_1' + a_1''), \cdots, (a_t' + a_t'')$ en ayant choisi t suffisamment grand et en écrivant les partitions comme un ensemble de nombres décroissants. A une telle partition correspond une orbite unipotente de $GL(n, \mathbb{C})$ que l'on note $O_{\mu',\mu''}$.

Pour une orbite unipotente O de $GL(n, \mathbb{C})$ et pour $(d', d'') \in D(n)$, on définit $\left(\mathbb{C}[\hat{\mathfrak{S}}_{d'}] \otimes \mathbb{C}[\hat{\mathfrak{S}}_{d''}]\right)_{\geq O}$ comme l'ensemble des combinaisons linéaires de représentations irréductibles correspondant à des couples de partitions (μ', μ'') (comme ci-dessus) avec $O \subset \overline{O}_{\mu',\mu''}$. Les termes correspondants à O seront alors dits les termes de plus bas degré.

On reprend la notation $\mathbb{C}[\hat{\mathfrak{S}}_{D(n)}]$ de 3.1 et pour O une orbite unipotente de $GL(n, \mathbb{C})$, on peut encore définir $\mathbb{C}[\hat{\mathfrak{S}}_{D(n)}]_{\geq O}$ en faisant la somme sur les éléments de $D(n)$ des espaces définis ci-dessus.

3.4. Avec en plus un caractère On revient à la situation de 3.3, où $(d', d'') \in D(n)$ et les partitions μ', μ'' sont fixées. On a déjà associé dans ce paragraphe à μ' et μ'' une orbite unipotente $O_{\mu',\mu''}$ de $GL(n, \mathbb{C})$ et on lui associe maintenant en plus une application $\epsilon_{\mu',\mu''} : Jord(O_{\mu',\mu''}) \to \{\pm 1\}$ en posant pour tout $i \in [1, t]$,

$$\epsilon_{\mu',\mu''}(a_i' + a_i'') = \begin{cases} (-1)^{a_i''} \breve{\imath}_{<,opp}, & \text{si } a_i' + a_i'' \text{ est pair}, \\ (-1)^{a_i''} \breve{\imath}_{>,opp}, & \text{si } a_i' + a_i'' \text{ est impair}, \end{cases}$$

où $\breve{\imath}_{<,opp}$ (resp. $_{>,opp}$) $:= |\{j; 0 < a_j' + a_j'' <$ (resp. $>$) $a_i' + a_i''$ et $(-1)^{a_i'+a_i''} \neq (-1)^{a_j'+a_j''}\}|$. Le lecteur est en droit de se demander pourquoi une formule si compliquée; de toute façon, il y a une part d'arbitraire d'autres choix sont sans doute possibles mais il y a 2 contraintes pour nos choix. On identifie $\sharp$ à ± 1, en posant $iso = +1$ et $an = -1$.

La première vient de ce que μ'' est une partition de d'' d'où $(-1)^{\sum_i \mu''_i} = \sharp$ et que l'on veut que

$$\prod_{\alpha \in Jord(O_{\mu',\mu''})} \epsilon_{\mu',\mu''}(\alpha) = \sharp.$$

Vérifions que cela est satisfait avec nos choix, plus exactement, nous allons vérifier que

$$(-1)^{\sum_i a''_i} = \prod_{\alpha \in Jord(O_{\mu',\mu''})} \epsilon_{\mu',\mu''}(\alpha).$$

En effet, on a

$$\prod_{\alpha \in Jord(O_{\mu',\mu''})} \epsilon_{\mu',\mu''}(\alpha) = \prod_i (-1)^{a''_i} \prod_{i | a'_i + a''_i \equiv 0} (-1)^{\mathrm{i}_{<,opp}} \prod_{i | a'_i + a''_i \equiv 1} (-1)^{\mathrm{i}_{>,opp}}.$$

Mais on remarque que $\prod_{i|a'_i+a''_i\equiv 1}(-1)^{\mathrm{i}_{>,opp}} = \prod_{i|a'_i+a''_i\equiv 0}(-1)^{\mathrm{i}_{<,opp}}$ d'où l'assertion cherchée.

La deuxième contrainte concerne le cas cuspidal que l'on va détailler ci-dessous ; c'est le cas où μ', μ'' correspondent à des représentations cuspidales des groupes unitaires, parce que dans ce cas là, on sait ce que l'on veut : μ', μ'' correspond (à un signe près) à un "type" et $(O_{\mu',\mu''}, \epsilon_{\mu',\mu''})$ doit être le paramétrage de la représentation cuspidale de $U(n, E/F)$ induite par ce "type".

Notation : *Soit O une orbite unipotente de $GL(n, \mathbb{C})$ dont on écrit $Jord(O)$ l'ensemble des blocs de Jordan. Pour $\alpha \in Jord(O)$, on pose $Jord_{<\alpha,opp}(O) := |\{\beta \in Jord(O); 0 < \beta < \alpha$ et $(-1)^\beta \neq (-1)^\alpha\}|$ et on définit de même $Jord_{>\alpha,opp}$; cela remplacera la notation peu explicite ci-dessus, $\mathrm{i}_{<,opp}$ et $\mathrm{i}_{>,opp}$. On utilisera aussi la notation* $Jord_{?\alpha,opp}(O) \begin{cases} = Jord_{<\alpha,opp} & \text{si } \alpha \text{ est pair} \\ = Jord_{>\alpha,opp} & \text{si } \alpha \text{ est impair.} \end{cases}$

Soient $(d', d'') \in D(n)$ et μ', μ'' des partitions de d' et d'' telles que $O = O_{\mu',\mu''}$. La notation la plus commode sera en générale d'écrire $\mu' := \{\mu'(\alpha); \alpha \in Jord(O)\}$ et $\mu'' := \{\mu''(\alpha); \alpha \in Jord(O)\}$ avec pour tout $\alpha \in Jord(O)$, $\mu'(\alpha) + \mu''(\alpha) = \alpha$ et alors $\epsilon_{\mu',\mu''}(\alpha) = (-1)^{\mu''(\alpha)}(-1)^{|Jord_{?\alpha,opp}(O)|}$.

Dans la notation ci-dessus, ce que l'on a utilisé implicitement est que pour $\alpha \in Jord(O)$, le couple $\mu'(\alpha), \mu''(\alpha)$ est bien déterminé ; il intervient dans le produit μ', μ'' avec la même multiplicité que α dans $Jord(O)$. Cela vient du fait que les partitions sont un ensemble de nombres rangés par ordre, ici décroissant.

3.5. Exemple, le cas cuspidal Soient donc $(d', d'') \in D(n)$ et μ', μ'' des partitions correspondant à des représentations cuspidales de $U(d', \mathbb{F}_q)$ et $U(d'', \mathbb{F}_q)$ (Rappelons que le passage d'une partition à une représentation introduit un signe mais il n'intervient pas ici). Cela entraîne que d' et d'' sont des nombres triangulaires c'est-à-dire qu'il existe k' et k'' des nombres entiers (éventuellement nuls) tel que $d' = k'(k'+1)/2$ et $d'' = k''(k''+1)/2$. Avec ces hypothèses, la partition $\mu := \mu' + \mu''$ a la forme suivante :

$$k' + k'', k' + k'' - 2, \cdots, |k' - k''| + 2, |k' - k''|, |k' - k''| - 1, \cdots, 1, 0.$$

En d'autres termes $Jord(O_{\mu',\mu''})$ contient d'une part des éléments de la forme $2i$ pour $i \leq |(k' + (-1)^{k'+k''}k'')|/2$ et d'autre part des éléments de la forme $2j+1$ pour $j \leq (|k' + (-1)^{k'+k''+1}k''| - 1)/2$. On note ζ le signe de $k' - k''$ (si $k' = k''$ ce signe n'intervient pas, donc le mieux est de ne pas le définir). On note $O := O_{\mu',\mu''}$ et $Jord(O)_{pair}$ l'ensemble des blocs de Jordan de O de taille paire.

Remarque 3.1. Avec la définition donnée de $\epsilon_{\mu',\mu''}$ on obtient :

$$\epsilon_{\mu',\mu''}(2i) = (-1)^i; \epsilon_{\mu',\mu''}(2j+1) = (-1)^j\zeta(-1)^{|Jord_{pair}(O)|}.$$

On regarde d'abord la valeur de $\epsilon_{\mu',\mu''}$ sur les blocs de Jordan de taille paire de $O_{\mu',\mu''}$. Soit $2i$ l'un de ces blocs. Ou bien il existe $t \in [0, inf(k', k'')[$ tel que $2i = k' + k'' - 2t$ ou bien il existe $s \in [0, |k' - k''|]$ tel que $2i = |k' - k''| - s$. Dans tous les cas, on écrit $2i = \mu'(2i) + \mu''(2i)$ où $\mu'(2i), \mu''(2i)$ est de la forme $\mu'_?, \mu''_?$ pour ? convenable. On remarque que pour $2i \leq |k' - k''|$, $\mu''(2i) = 0$ si $k'' \leq k'$ et $\mu''(2i) = 2i$ si $k'' \geq k'$. De plus le nombre d'éléments impairs plus petits que $2i$ dans la partition $\mu' + \mu''$ est exactement i. D'où la remarque dans ce cas. Supposons maintenant que $2i > |k' - k''|$; cette possibilité n'est effective que si $k' + k''$ est pair et, alors, avec les notations ci-dessus, $\mu''(2i) = k'' - t$ avec $i = (k' + k'')/2 - t$; c'est-à-dire que $\mu''(2i) = (k'' - k')/2 + i$. Le nombre d'éléments impairs de la partition $\mu' + \mu''$ plus petits que $2i$ est $|k' - k''|/2$. Donc, d'après la définition $\epsilon_{\mu',\mu''} = (-1)^{i+(k''-k')/2+(k'-k'')/2} = (-1)^i$ comme annoncé.

Regardons maintenant le cas des blocs de Jordan de taille impaire, c'est-à-dire $2j+1$. Comme ci-dessus soit $2j+1 \leq |k' - k''|$ avec $\mu''(2j+1) = 0$ si $k'' \leq k'$ et $\mu''(2j+1) = 2j+1$ si $k'' \geq k'$, soit $2j+1 > |k' - k''|$ avec l'existence d'un entier t (éventuellement 0) tel que $\mu'(2j+1) = k' - t$, $\mu''(2j+1) = k'' - t$ et $2j+1 = k'+k''-2t$. Cette dernière possibilité ne peut se produire que si $k'+k''$ est impair. Revenons au cas où $2j+1 \leq |k' - k''|$; le nombre d'éléments de la partition $\mu' + \mu''$ qui sont pairs et plus petits que $2j+1$ est exactement j ; et $(-1)^{\mu''(2j+1)} = \zeta$ d'après la définition de ζ. D'où $\epsilon_{\mu',\mu''} = \zeta(-1)^{j+|Jord_{pair}(O)}$. Considérons maintenant le cas où $k' + k''$ est impair et $2j+1 > |k' - k''|$. On a $\mu''(2j+1) = k'' - (k' + k'' - 1)/2 + j = (k'' - k' + 1)/2 + j$ et le nombre d'éléments de $\mu' + \mu''$ pairs et plus grands que $2j+1$ est ici 0. D'où $\epsilon_{\mu',\mu''}(2j+1) = (-1)^{j+(k''-k'+1)/2} = \zeta(-1)^{j+(|k''-k'|-1)/2}$ (on a ici utilisé le fait que $k' - k''$ est impair). De plus $|Jord_{pair}(O)| = (|(k'-k'')| - 1)/2$ et on trouve le résultat annoncé.

3.6. Restriction au parahorique Soit π une représentation irréductible (lisse) de $U(n, E/F)_\sharp$ de niveau 0 et de réduction unipotente (cf. l'introduction). Pour $(d', d'') \in D(n)$, on fixe un réseau presque autodual $L_{d',d''}$, d'où un parahorique et on note $res^{par}_{d',d''}$ la représentation de $U(d', \mathbb{F}_q) \times U(d'', \mathbb{F}_q)$ dans l'espace des invariants de π sous le radical pro-p-unipotent. Et on pose :

$$res^{par}\pi := \oplus_{(d',d'')\in D(n)} res^{par}_{d',d''}\pi.$$

C'est une somme (formelle) de représentations unipotentes; grâce aux faisceaux caractères, on peut voir cette représentation comme un élément de $\mathbb{C}[\hat{\mathfrak{S}}_{D(n)}]$. Le but (qui ne sera atteint que plus loin) est de caractériser les éléments de $\mathbb{C}[\mathfrak{S}_{D(n)}]$ ainsi obtenus, au moins quand on suppose que π est tempérée versus elliptique. La propriété clé est la suivante, où $\sharp$ n'est pas fixé (comme on le voit ci-dessus, en ajoutant $2x$ on ne change pas la parité de d'' et $\sharp$ n'est pas intervenu) :

pour tout $(d', d'') \in D(n)$ pour tout $x \leq [d'/2]$, les représentations $res^{par}_{d',d''}(\pi)$ et $res^{par}_{d'-2x,d''+2x}(\pi)$, vues comme représentations des groupes $U(d', \mathbb{F}_q) \times U(d'', \mathbb{F}_q)$ d'une part et $U(d'-2x, \mathbb{F}_q) \times U(d''+2x, \mathbb{F}_q)$ d'autre part, ont même module de Jacquet pour le groupe $GL(x, \mathbb{F}_{q^2}) \times U(d'-2x, \mathbb{F}_q) \times U(d'', \mathbb{F}_q)$.

Une telle propriété est claire par construction, c'est la transitivité des restrictions. Mais elle justifie la définition suivante :

Définition Un élément de $\mathbb{C}[\hat{\mathfrak{S}}_{D(n)}]$ que l'on écrit sous la forme $\oplus_{(d',d'')\in D(n)}\pi_{d',d''}$ (où $\pi_{d',d''}$ est une représentation du groupe $U(d', \mathbb{F}_q) \times U(d'', \mathbb{F}_q)$ pour tout $(d', d'') \in D(n)$), est dit de nature globale si pour tout $(d', d'') \in D(n)$ et pour tout $x \leq [d'/2]$, les représentations $\pi_{d',d''}$ et $\pi_{d'-2x,d''+2x}$ ont des modules de Jacquet isomorphes pour les sous-groupes paraboliques de Levi $GL(x, \mathbb{F}_{q^2}) \times U(d'-2x, \mathbb{F}_q) \times U(d'', \mathbb{F}_q)$. On note $\mathbb{C}[\hat{\mathfrak{S}}_{D(n)}]_{global}$ l'ensemble de ces éléments.

3.7. Description des éléments de nature globale

3.7.1. Calcul de la dimension

Proposition 3.2. *La dimension de* $\mathbb{C}[\hat{\mathfrak{S}}_{D(n)}]_{global}$ *est le coefficient de* x^n *dans la série* $\prod_{i\in\mathbb{N}}(1+x^i)/(1-x^i)$.

Pour H un groupe réductif sur $\mathbb{F}_q$, on note $Unip^H_{Ell}$ l'espace des caractères (i.e. fonctions H-invariantes sur H) qui s'annulent hors des éléments elliptiques et qui s'expriment comme trace de représentations unipotentes. Et on rappelle que si H est un groupe général linéaire alors $Unip^H_{Ell}$ est de dimension 1. Pour H de nouveau un groupe réductif arbitraire, on obtient une base du groupe de Grothendieck des représentations unipotentes en prenant les caractères induits à partir d'une base de $Unip^M_{Ell}$ quand M parcourt l'ensemble des sous-groupes de Levi de paraboliques de H pris à conjugaison près . En appliquant cela à chaque $H := U(d',\mathbb{F}_q)\times U(d'',\mathbb{F}_q)$ pour $(d',d'')\in D(n)$, on en déduit une description de $\mathbb{C}[\hat{\mathfrak{S}}_{D(n)}]_{global}$; on induit les caractères de chaque $Unip_{Ell}^{H_{n_1\geq\cdots\geq n_k;n',n''}}$ où $n=2(n_1+\cdots+n_k)+n'+n''$ et où $H_{n_1\geq\cdots\geq n_k;n',n''}$ est le groupe $\prod_{i\in[1,k]}GL(n_i,\mathbb{F}_{q^2})\times U(n',\mathbb{F}_q)\times U(n'',\mathbb{F}_q)$. On connait la dimension de l'espace des traces de représentations unipotentes pour un groupe unitaire (de type $U(n',\mathbb{F}_q)$) ; c'est la dimension de l'espace des représentations de $\mathfrak{S}_{n'}$, c'est-à-dire le coefficient de $x^{n'}$ dans la série génératrice $\prod_{i\in\mathbb{N}}(1-x^i)^{-1}$. Avec le même argument, on en déduit que la dimension de $Unip_{Ell}^{U(n',\mathbb{F}_q)}$ est le coefficient de $x^{n'}$ dans la série $\prod_{i\in\mathbb{N}}(1-x^{2i})/(1-x^i)$ c'est-à-dire dans $\prod_{i\in\mathbb{N}}(1+x^i)$. Finalement la dimension de $\mathbb{C}[\hat{\mathfrak{S}}_{D(n)}]_{global}$ est le coefficient de x^n dans la série :

$$\prod_{i\in\mathbb{N}}(1-x^{2i})^{-1}(1+x^i)^2=\prod_{i\in\mathbb{N}}(1+x^i)/(1-x^i)$$

comme annoncé.

3.7.2. Une construction d'éléments de nature globale Comme les caractères des représentations unipotentes se décrivent aisément avec les faisceaux caractères, on va faire de même pour l'espace des caractères de nature globale décrit ci-dessus. Fixons $(d'_0,d''_0)\in D(n)$ et π'_0,π''_0 des représentations de $\mathfrak{S}_{d'_0}$ et $\mathfrak{S}_{d''_0}$ respectivement. On définit l'application $Ind\circ Res$ de $\mathbb{C}[\hat{\mathfrak{S}}_{d'_0}]\otimes\mathbb{C}[\hat{\mathfrak{S}}_{d''_0}]$ dans $\mathbb{C}[\hat{\mathfrak{S}}_{D(n)}]$, en posant :

$$Ind\circ Res\,\pi'_0\otimes\pi''_0 := \sum_{(d'^{,\prime},d'^{,\prime\prime})\in D(d'_0);(d''^{,\prime},d''^{,\prime\prime})\in D(d''_0)}(-1)^{d''^{,\prime\prime}}$$

$$ind_{\mathfrak{S}_{d'^{,\prime}}\times\mathfrak{S}_{d''^{,\prime}}\times\mathfrak{S}_{d'^{,\prime\prime}}\times\mathfrak{S}_{d''^{,\prime\prime}}}^{\mathfrak{S}_{d'^{,\prime}+d''^{,\prime}}\times\mathfrak{S}_{d'^{,\prime\prime}+d''^{,\prime\prime}}}\circ Res_{\mathfrak{S}_{d'^{,\prime}}\times\mathfrak{S}_{d'^{,\prime\prime}}\times\mathfrak{S}_{d''^{,\prime}}\times\mathfrak{S}_{d''^{,\prime}}}\pi'_0\otimes\pi''_0.$$

On voit cet élément comme un élément de $\mathbb{C}[\hat{\mathfrak{S}}_{D(n)}]$. Cette définition à l'air surprenante ; c'est une sorte de globalisation mais où on a caché le Frobenius. Cette définition est la même que celle de **[MW]** 3.1 et 3.2 (du moins quand on se limite aux éléments elliptiques ce que nous ferons le moment venu) mais présentée un peu différemment.

Proposition 3.3. *L'élément* $Ind \circ Res\, \pi_0' \otimes \pi_0''$ *construit ci-dessus est dans* $\mathbb{C}[\hat{\mathfrak{S}}_{D(n)}]_{global}$.

Il faut traduire le module de Jacquet pour un faisceau caractère comme une restriction à un sous-groupe. Précisément fixons $d \in \mathbb{N}$ et $x \leq [d/2]$; soit π une représentation de $U(d, \mathbb{F}_q)$ correspondant à une représention, encore notée π de $\mathfrak{S}_d$. Le module de Jacquet de π pour le parabolique de Levi $GL(x, \mathbb{F}_{q^2}) \times U(d-2x, \mathbb{F}_q)$ est une représentation de ce dernier groupe dont la trace est le faisceau caractère associé à la restriction de π au sous-ensemble $\mathfrak{S}_x \times \mathfrak{S}_{d-2x}$ où $\mathfrak{S}_x \times \mathfrak{S}_{d-2x}$ s'envoie dans $\mathfrak{S}_d$ par

$$\sigma \times w \in \mathfrak{S}_x \times \mathfrak{S}_{d-2x} \mapsto \begin{pmatrix} 0 & 0 & 1 \\ 0 & w & 0 \\ \sigma & 0 & 0 \end{pmatrix} \in \mathfrak{S}_d.$$

Cette application n'est pas une application de groupes mais respecte les classes de conjugaison. Si $\sigma \times w$ est comme ci-dessus, la classe de conjugaison de l'image contient l'élément $\sigma' \times w$ où σ' est une permutation dans $\mathfrak{S}_{2x}$ dont les cycles sont le double des cycles de σ.

Il faut écrire pour $(d', d'') \in D(n)$ fixé et pour $x \leq [d'/2]$,

$$Res_{\mathfrak{S}_x \times \mathfrak{S}_{d'-2x} \times \mathfrak{S}_{d''}} \sum_{(d'^{,\prime}, d'^{,\prime\prime}) \in D(d_0'); (d''^{,\prime}, d''^{,\prime\prime}) \in D(d_0''); d'^{,\prime} + d''^{,\prime} = d'}$$

$$ind_{\mathfrak{S}_{d'^{,\prime}} \times \mathfrak{S}_{d''^{,\prime}} \times \mathfrak{S}_{d'^{,\prime\prime}} \times \mathfrak{S}_{d''^{,\prime\prime}}}^{\mathfrak{S}_{d'} \times \mathfrak{S}_{d''}} Res_{\mathfrak{S}_{d'^{,\prime}} \times \mathfrak{S}_{d'^{,\prime\prime}} \times \mathfrak{S}_{d''^{,\prime}} \times \mathfrak{S}_{d''^{,\prime\prime}}}.$$

Par la théorie de Mackey (c'est assez facile de le redémontrer), on peut commuter la dernière restriction avec l'induction, pour cela il faut faire une double somme ; fixons $d^{i,j}$ pour $i, j \in \{', ''\}$. On regarde les décompositions, pour $i \in \{', ''\}$, $d^{i,\prime} = 2d^{i,\prime,lin} + d^{i,\prime,u}; d^{\prime,\prime,lin} + d^{\prime\prime,\prime,lin} = x$. Et le module de Jacquet ci-dessus s'obtient pour de bonnes applications comme la somme sur toutes les décompositions des opérations :

$$ind_{\mathfrak{S}_{d'^{,\prime,lin}} \times \mathfrak{S}_{d''^{,\prime,lin}} \times \mathfrak{S}_{d'^{,\prime,u}} \times \mathfrak{S}_{d''^{,\prime,u}} \times \mathfrak{S}_{d'^{,\prime\prime}} \times \mathfrak{S}_{d''^{,\prime\prime}}}^{\mathfrak{S}_x \times \mathfrak{S}_{d'-2x} \times \mathfrak{S}_{d''}}$$

$$Res_{\mathfrak{S}_{d'^{,\prime,lin}} \times \mathfrak{S}_{d'^{,\prime,u}} \times \mathfrak{S}_{d'^{,\prime\prime}} \times \mathfrak{S}_{d''^{,\prime,lin}} \times \mathfrak{S}_{d''^{,\prime,u}} \times \mathfrak{S}_{d''^{,\prime\prime}}}$$

Il faut tenir compte du signe mis dans la définition de $Ind \circ Res$ qui est $(-1)^{d^{'',''}}$. On a une formule du même type quand on remplace d' par $d'-2x$ et d'' par $d''+2x$. Et on somme exactement sur les mêmes nombres en utilisant le fait que ci-dessus $d^{',',u}+d^{'',',u}=d'-2x$ et $2d^{',',lin}+2d^{'',',lin}+d^{','' }+d^{'',''}=d''+2x$ le point est que ci-dessus on fait la somme sur les décompositions : $2d^{',',lin}+d^{',',u}+d^{','' }=d'_0; 2d^{'',',lin}+d^{'',',u}+d^{'',''}=d''_0$. Mais le signe est ici $(-1)^{d^{'',''}+2d^{'',',lin}}$ qui est aussi $(-1)^{d^{'',''}}$ comme ci-dessus. Cela termine la preuve.

3.7.3. Variante Pour π une représentation irréductible d'un groupe symétrique, correspondant donc à une partition, on note π^* la représentation irréductible qui correspond à la partition duale. Sur les caractères, le changement se fait par la multiplication par $(-1)^{l(\sigma)}$ où $l(\sigma)$ est la longueur de l'élément σ. Cette dualité se généralise à un produit de groupes symétriques (elle se fait terme à terme)

Corollaire *Soient* $(d'_0, d''_0) \in D(n)$ *et* $\pi'_0 \otimes \pi''_0$ *une représentation irréductible de* $U(d'_0, \mathbb{F}_q) \times U(d''_0, \mathbb{F}_q)$. *Alors* $\left(Ind \circ Res\, \pi'_0 \otimes \pi''_0\right)^*$ *est de nature globale.*

Cela résulte de ce qui précède.

3.7.4. Calcul du gradué Soient encore $(d'_0, d''_0) \in D(n)$ et soient μ'_0, μ''_0 des partitions de d'_0 et d''_0 ; ci-dessous on utilisera la notation $\mu'_0 \cup \mu''_O$ pour la partition de $d'_0 + d''_0$ dont l'ensemble des termes est la réunion des termes de μ'_0 et de μ''_0, avec multiplicité évidemment. On les voit comme des représentations de groupes symétriques convenables ; on utilisera en fait les représentations associées aux partitions duales, $(\mu'_0)^*$ et $(\mu''_0)^*$. On note $O(\mu'_0 \cup \mu''_0)$ l'orbite unipotente de $Gl(n, \mathbb{C})$ correspondant à la partition $\mu'_0 \cup \mu''_0$ et on note $\epsilon_{\mu'_0 \cup \mu''_0}$ l'application de $Jord(O(\mu'_0 \cup \mu''_0))$ dans $\{\pm 1\}$ définie par $\epsilon_{\mu'_0 \cup \mu''_0}(\alpha) = (-1)^{mult_\alpha(\mu''_0)}$ pour tout $\alpha \in \mu'_0 \cup \mu''_0$. On rappelle qu'en 3.3 et suivant on a associé une orbite, $O_{\mu',\mu''}$, à un couple de partitions μ', μ'' de d' et d'' respectivement, avec $(d', d'') \in D(n)$ ainsi qu'une application, $\epsilon_{\mu',\mu''}$ de $Jord(O_{\mu',\mu''})$ dans $\{\pm 1\}$. Pour μ', μ'' tel que $O_{\mu',\mu''} = O_{\mu'_0 \cup \mu''_0}$, on pose :

$$< \epsilon_{\mu',\mu''}, \epsilon_{\mu'_0 \cup \mu''_0} >:= \prod\nolimits_{\alpha \in Jord(O_{\mu',\mu''}); \epsilon_{\mu',\mu''}(\alpha)=-1} \epsilon_{\mu'_0 \cup \mu''_0}(\alpha)$$
$$\sigma_{\mu'_0 \cup \mu''_0} := \prod\nolimits_{\alpha \in Jord(O_{\mu'_0 \cup \mu''_0}); \epsilon_{\mu'_0 \cup \mu''_0}=-1} (-1)^{|Jord_{?\alpha, opp}|},$$

où ? =< si α est pair et ? => si α est impair

Remarque 3.4. $< \epsilon_{\mu',\mu''}, \epsilon_{\mu'_0 \cup \mu''_0} >= \sigma_{\mu'_0 \cup \mu''_0} \prod_{\alpha \in \mu'_0 \cup \mu''_0} (-1)^{\mu''(\alpha) mult_\alpha(\mu''_0)}$.

On pose $O := O_{\mu',\mu''} = O_{\mu'_0 \cup \mu''_0}$. On revient à la définition de 3.4 pour calculer

$$\prod_{\alpha \in Jord(O)} (-1)^{mult_\alpha(\mu''_0)\mu''(\alpha)} = \prod_{\alpha \in Jord(O);(-1)^{mult_\alpha(\mu''_0)}=-1} (-1)^{\mu''(\alpha)} =$$

$$\prod_{\alpha \in Jord(O);\epsilon_{\mu'_0 \cup \mu''_0}(\alpha)=-1} \epsilon_{\mu',\mu''}(\alpha)(-1)^{|Jord_{?\alpha,opp}|},$$

où ? est comme dans ce qui précède la remarque. D'où la remarque.

Dans l'énoncé ci-dessous, la notation $(\dots)^*$ représente la représentation correspondant aux partitions duales comme expliqué dans 3.7.3.

Proposition 3.5. *Avec les notations précédentes,* $\left(Ind \circ Res(\mu'_0)^* \otimes (\mu''_0)^*\right)^* \in \left(\mathbb{C}[\hat{\mathfrak{S}}_{d'}] \otimes \mathbb{C}[\hat{\mathfrak{S}}_{d''}]\right)_{\geq O}$ *et même plus précisément dans cette filtration le terme de plus bas degré est*

$$\sum_{\mu',\mu'' | O_{\mu',\mu''}=O(\mu'_0 \cup \mu''_0)} \sigma_{\mu'_0 \cup \mu''_0} < \epsilon_{\mu'_0 \cup \mu''_0}, \epsilon_{\mu',\mu''} > \mu' \otimes \mu''.$$

Soit $(d', d'') \in D(n)$ et β', β'' des partitions de d' et d'' respectivement. On cherche à calculer quand la représentation associée à β', β'' (notée abusivement $\beta' \otimes \beta''$) intervient dans la représentation $Ind \circ Res(\mu'_0)^* \otimes (\mu''_0)^*$; c'est un problème de théorie de Mackey et il est étudié dans **[W1]** XII.6 (précisément pages 422 à 425). Il y est montré que la multiplicité est non nulle seulement si

$$\beta' \cup \beta'' \leq (\mu'_0)^* + (\mu''_0)^*.$$

Si l'inégalité est une égalité, alors le coefficient est calculé en loc.cite ; ce qui nous intéresse est la formule qui suit (11) dans la démonstration ; cela n'a pas d'intérêt de l'écrire ici car les notations sont vraiment différentes et le plus simple est de refaire le calcul dans ce cas précis ; ce que l'on cherche à calculer est le signe $(-1)^{d'''}$. On note O^* l'orbite unipotente duale de $O_{\mu'_0 \cup \mu''_0}$. Pour $\alpha \in Jord(O_{\mu'_0 \cup \mu''_0})$, on note $\alpha^* := |\{\beta \in Jord(O_{\mu'_0 \cup \mu''_0})); \beta \geq \alpha\}$ où les multiplicités sont prises en compte. On remarque aisément que $Jord(O^*)$ est l'ensemble des α^* précédemment définis avec comme multiplicité $\alpha - \alpha_-$ où α_- est l'élément de $Jord(O_{\mu'_O \cup \mu''_0})$ strictement inférieur à α (éventuellement 0). Pour tout α comme ci-dessus, on écrit $(\mu'_0)^*(\alpha^*)$ et $(\mu''_0)^*(\alpha^*)$ les termes de la partition $(\mu'_0)^*$ et $(\mu''_0)^*$ respectivement, tels que $\alpha^* = (\mu'_0)^*(\alpha^*) + (\mu''_0)^*(\alpha^*)$;

ceci est bien défini. Et on trouve que le signe est

$$\prod_{\alpha \in Jord(O_{\mu'_0 \cup \mu''_0})} (-1)^{mult_{\alpha^*}(\beta'')(\mu''_0)^*(\alpha^*)}.$$

Pour le résultat cherché, il faut dualiser β' et β'' et la condition trouvée s'écrit donc :

$$(\beta')^* + (\beta'')^* \geq \mu'_0 \cup \mu''_0;$$

on a simplement utilisé le fait que la dualité renverse somme et union et inverse l'ordre. On pose maintenant $\mu' = (\beta')^*$ et $\mu'' = (\beta'')^*$ pour retrouver les notations de l'énoncé et il reste à calculer le signe en terme de $\mu', \mu'', \mu'_0, \mu''_0$ dans le cas où l'inégalité est une égalité et on va vérifier que ceci n'est autre que $\prod_{\alpha \in Jord(O_{\mu'_0 \cup \mu''_0})} (-1)^{mult_\alpha(\mu''_0)\mu''(\alpha)}$. En effet, on reprend les notations α^* et α_- introduites ci-dessus et on vérifie que $(\mu''_0)^*(\alpha^*) = \sum_{\beta \geq \alpha} mult_\beta(\mu''_0)$ et $mult_{\alpha^*}(\beta'') = \mu''(\alpha) - \mu''(\alpha_-)$. En remplaçant, on obtient :

$$\prod_{\alpha \in Jord(O_{\mu'_0 \cup \mu''_0})} (-1)^{mult_{\alpha^*}(\beta'')(\mu''_0)^*(\alpha^*)} =$$

$$\prod_{\beta | (-1)^{mult_\beta(\mu''_0)} = -1} \prod_{\alpha \leq \beta} (-1)^{mult_{\alpha^*}(\beta'')} =$$

$$\prod_{\beta | (-1)^{mult_\beta(\mu''_0)} = -1} \prod_{\alpha \leq \beta} (-1)^{\mu''(\alpha) - \mu''(\alpha_-)} =$$

$$\prod_{\beta | (-1)^{mult_\beta(\mu''_0)} = -1} (-1)^{\mu''(\beta)}.$$

Ceci est bien la formule cherchée. La proposition résulte alors de la remarque qui précède.

3.7.5. Remarque sur les signes Il faut maintenant tenir compte du signe qui relie partition et représentation ; précisément, pour μ' une partition de d', on note $\pi(\mu')$ la représentation irréductible de $U(d', \mathbb{F}_q)$ qui lui correspond au signe près et on note $\sigma_{\mu'}$ ce signe. Pour $(d', d'') \in D(n)$, et pour μ', μ'' des partitions de d' et d'' respectivement, on définit donc $\sigma_{\mu'}$ et $\sigma_{\mu''}$. On pose, avec la notation $(-1)^{\binom{D}{2}}$ pour $(-1)^{D(D-1)/2}$:

$$\sigma_{O_{\mu',\mu''}} := (-1)^{\left(\frac{|\{\alpha \in Jord(O_{\mu',\mu''});\alpha \text{ impair}\}|}{2}\right)} \prod_{\alpha\in Jord(O_{\mu',\mu''});\alpha \text{ impair}} (-1)^{|Jord(O_{\mu',\mu''})_{>\alpha,opp}|}.$$

L'ensemble $Jord(O_{\mu',\mu''})$ est considéré avec multiplicité. Toujours en considérant cet ensemble avec multiplicité, on pose : $\sigma^{\epsilon_{\mu',\mu''}} := \prod_{\alpha\in Jord(O_{\mu',\mu''});\alpha\equiv n+1[2]} \epsilon_{\mu',\mu''}(\alpha)$.

Remarque 3.6. *Soient $(d',d'') \in D(n)$ et μ',μ'' des partitions de d' et d'' respectivement. Alors $\sigma_{\mu'}\sigma_{\mu''} = \sigma_{O_{\mu',\mu''}}\sigma^{\epsilon_{\mu',\mu''}}$.*

C'est un calcul sans mystère qui ne fait qu'utiliser le fait que pour tout $\alpha \in Jord(O_{\mu',\mu''})$, $\mu'(\alpha) + \mu''(\alpha) = \alpha$ avec les notations de 3.4. On a donc $\sigma_{\mu'}\sigma_{\mu''}$

$$= (-1)^{d'(d'-1)/2)+d''(d''-1)/2)} \prod_{\alpha\in Jord(O_{\mu',\mu''})} (-1)^{\mu'(\alpha)(\mu'(\alpha)-1)/2+\mu''(\alpha)(\mu''(\alpha)-1)/2}.$$

On remplace partout $\mu'(\alpha)$ par $\alpha - \mu''(\alpha)$ et on trouve que le signe vaut

$$(-1)^{d'(d'-1)/2+d''(d''-1)/2+\sum_\alpha \alpha(\alpha-1)/2+\sum_\alpha \mu''(\alpha)\alpha+\mu''(\alpha)^2}.$$

On remplace encore d' par $n-d''$ et on utilise le fait qu'un nombre et son carré ont même parité. On trouve $(-1)^{n(n-1)/2+d''(n+1)+\sum_\alpha(\alpha(\alpha-1)/2+\mu''(\alpha)(\alpha+1)}$. De plus $n = \sum_\alpha \alpha$ et $d'' = \sum_\alpha \mu''(\alpha)$. En remplaçant le signe cherché est $(-1)^{\sum_{\alpha\neq\beta}\alpha\beta+\sum_\alpha \mu''(\alpha)(n+\alpha)}$. Or $(-1)^{\sum_{\alpha\neq\beta}\alpha\beta} = (-1)^{N_i(N_i-1)/2}$, où N_i est le nombre de blocs de Jordan de taille impaire de $Jord(O_{\mu',\mu''})$ tandis que $(-1)^{\sum_\alpha \mu''(\alpha)(n+\alpha)} = (-1)^{\sum_{\alpha;\alpha\equiv n+1[2]}\mu''(\alpha)}$. On remplace encore, pour α un bloc de Jordan de $O_{\mu',\mu''}$ tel que

$$\alpha \equiv n+1[2], (-1)^{\mu''(\alpha)} = \epsilon_{\mu',\mu''}(\alpha)(-1)^{|Jord_{?\alpha,opp}|}$$

où ? est < si n est impair et ? => si n est pair. Mais on remarque que

$$\prod_{\alpha\in Jord(O_{\mu',\mu''});\alpha \text{ pair}} (-1)^{|Jord_{<\alpha,opp}|} = \prod_{\alpha\in Jord(O_{\mu',\mu''});\alpha \text{ impair}} (-1)^{|Jord_{>\alpha,opp}|}.$$

Cela termine le calcul.

3.7.6. Description complète Corollaire *L'espace vectoriel engendré par les représentations $Ind \circ Res\,\mu_0' \otimes \mu_0''$ où μ_0', μ_0'' parcourt l'ensemble des partitions des éléments de $D(n)$ coïncide avec l'espace $\mathbb{C}[\hat{\mathfrak{S}}_{D(n)}]_{global}$.*

Il suffit de démontrer que la dimension de l'espace engendré par les représentations $Ind \circ Res\mu_0' \otimes \mu_0''$ où μ_0', μ_0'' est supérieure ou égale à la dimension de

$\mathbb{C}[\hat{\mathfrak{S}}_{D(n)}]_{global}$. Grâce au calcul du gradué fait dans 3.7.4 et une inversion de Fourier simple, on voit que cette dimension est supérieure ou égale au cardinal de l'ensemble $\{(O,\epsilon)\}$ où O est une orbite unipotente de $GL(n,\mathbb{C})$ et ϵ est une application de $Jord(O)$ dans $\{\pm 1\}$. Le cardinal de cet ensemble est le coefficient de x^n dans la série $\prod_{i\in\mathbb{N}}(1+\sum_{k>0}2x^{ki})$. Or pour tout $i\in\mathbb{N}$, $(1+x^i)(1+\sum_{k>0}x^k) = 1+\sum_{k>0}2x^{ki}$ d'où la série s'écrit aussi $\prod_{i\in\mathbb{N}}(1+x^i)/(1-x^i)$. C'est le résultat cherché grâce à 3.2.

4. Induction endoscopique

4.1. Définition de l'induction endoscopique Dans **[W1]** toute une combinatoire est développée de façon à calculer le transfert endoscopique de distributions stables à support unipotent. On a remarqué dans **[MW]** que cette combinatoire se retrouve quand on calcule les "K"-types des représentations de niveau 0, c'est-à-dire quand on décrit l'espace des invariants de la représentation sous le radical pro-p-unipotent des parahoriques comme représentation "du" parahorique en réduction. Cela se confirme ici et c'est d'ailleurs plus simple (comme on peut s'y attendre a priori) que dans le cas des groupes orthogonaux impairs traités en **[MW]**. Voici la combinatoire dont nous allons avoir besoin.

côté symplectique : soit O_p une orbite unipotente d'un groupe symplectique et ϵ_p un caractère du groupe des composantes du centralisateur d'un de ses éléments. On pose $n_p := \sum_{\alpha\in Jord(O_p)}\alpha$. Suivant la construction de la représentation de Springer généralisée par Lusztig, on a déjà associé à (O_p,ϵ_p) un entier k_p et par la représentation de Springer une représentation irréductible ρ du groupe de Weyl de type C et de rang $N_p := 1/2(n_p - k_p(k_p+1))$. Suivant Lusztig on associe aussi à (O_p,ϵ_p) un symbole de rang n_p et de défaut $\pm(2[k_p/2]+1)$; le symbole qui est ici, peut être vu comme 2 collections d'entiers tous distincts avec des conditions et en particulier dans chaque collection la différence entre 2 entiers distincts est au moins 2 ; cela permet aussi de l'écrire comme un ensemble de couples (λ,δ_λ) avec $\lambda\in\mathbb{N}$ et $\delta_\lambda = 0$ ou 1, les éléments λ de la première ligne étant exactement les couples $(\lambda,+1)$. On va préciser les signes dans le cas qui nous intéresse c'est-à-dire celui où $Jord(O_p)$ n'a que des entiers pairs. Sous cette hypothèse, on pose $\tilde{Jord}(O_p) := Jord(O_p)$ si cet ensemble a un nombre impair d'éléments et $Jord(O_p)\cup\{0\}$ sinon. On considère $\tilde{Jord}(O_p)$ comme un ensemble de nombres croissants et donc comme un ensemble totalement ordonné, pour ϵ_p donné comme ci-dessus, on prolonge ϵ_p à $\tilde{Jord}(O_p)$ si nécessaire en posant $\epsilon_p(0) = +1$. Le symbole associé à (O_p,ϵ_p) est alors $\{(1/2\alpha + |\tilde{Jord}(O_p)_{<\alpha}|, \delta_\alpha)\}$, où $(-1)^{\delta_\alpha} = \epsilon(\alpha)(-1)^{|\tilde{Jord}(O_p)_{<\alpha}|}$.

Soit maintenant N'_p, N''_p un couple d'entiers tel que $N'_p + N''_p = N_p := 1/2(n_p - k_p(k_p+1))$ et soit ρ'_p, ρ''_p des représentations des groupes $W_{N'_p}$ et $W_{N''_p}$ respectivement ; on note (λ'_p, μ'_p) et (λ''_p, μ''_p) les partitions correspondant à ces représentations, c'est-à-dire que $\lambda'_p = \{\lambda'_{p,1} \geq \cdots \lambda'_{p,t} \geq 0\}$, où t peut être choisi grand comme on le souhaite et de même pour les autres partitions (où on prend le même t). On dit que l'induite endoscopique de ρ'_p, ρ''_p est (O_p, ϵ_p) si l'on a les égalités $\lambda_p = \lambda'_p + \lambda''_p$ et $\mu_p = \mu'_p + \mu''_p$, la somme est la somme des partitions. Une variante utile consiste à remplacer ρ'_p et ρ''_p par des symboles (au sens ordinaire) de défaut impair (positif ou négatif) et à demander que la somme des symboles soit le symbole associé à (O_p, ϵ_p). On décrit précisément la situation dans le cas qui nous est utile où O_p n'a que des blocs de Jordan de taille paire. La condition pour que ρ'_p, ρ''_p ait (O_p, ϵ_p) pour induite endoscopique se traduit par l'égalité des ensembles (avec les notations précédentes) à des termes nuls près :

$$\{\lambda'_{p,i} + \lambda''_{p,i}; i \in [1,t]\} = \{1/2\alpha + |\tilde{Jord}_{<\alpha}| - 2|\{\beta \in \tilde{Jord}_{<\alpha}; \delta_\beta = 0\}|; \delta_\alpha = 0\}$$

$$\{\mu'_{p,i} + \mu''_{p,i}; i \in [1,t]\} = \{1/2\alpha + 1 + |\tilde{Jord}_{<\alpha}| - 2|\{\beta \in \tilde{Jord}_{<\alpha}; \delta_\beta = 1\}|; \delta_\alpha = 1\}.$$

En fait, un entier n sera fixé pour nous (c'est le n du groupe unitaire) et l'induction endoscopique pour le côté symplectique que nous utiliserons n'est l'opération décrite ci-dessus que si n est pair ; sinon, c'est-à-dire quand n est impair, on demande que la somme des symboles associés à ρ'_p et ρ''_p soit le symbole associé à $O_p, -\epsilon_p$, c'est-à-dire que l'on échange les 2 lignes du symbole associé à (O_p, ϵ_p). On parlera donc d'induction n-endoscopique.

côté orthogonal : soit ici O_i, ϵ_i une orbite unipotente d'un groupe orthogonal et un caractère du groupe des composantes de cette orbite. On a encore l'entier k_i mais ici le fait que $N_i := 1/2(n_i - k_i^2)$ soit entier résulte de la construction de la représentation de Springer généralisée. On note encore ρ_i la représentation irréductible associée à (O_i, ϵ_i) et on reprend les notations du cas symplectique en mettant simplement i en indice au lieu de p. On dit ici que (O_i, ϵ_i) est l'induite endoscopique de ρ'_i, ρ''_i si les relations suivantes sont satisfaites : $\lambda''_i + \mu'_i = \lambda_i$ et $\mu''_i + \lambda'_i = \mu_i$, où λ_i, μ_i est la représentation de Springer généralisée associée à (O_i, ϵ_i). En terme de symbole cela revient à dire que le symbole associé à (O_i, ϵ_i) est la somme du symbole associé à ρ'_i et de celui associé à ρ''_i après avoir échangé les 2 collections d'entiers (c'est-à-dire dans la forme classique échanger les 2 lignes) dans le symbole de ρ'_i. On a des formules comme dans le cas symplectique.

induction endoscopique faible : ici on se donne une orbite unipotente O de $GL(n, \mathbb{C})$ et une application ϵ de $Jord(O)$ dans $\{\pm 1\}$. On note O_p l'orbite unipotente dont les blocs de Jordan sont les blocs de Jordan pairs de O et

O_i l'orbite dont les blocs de Jordan sont ceux qui sont impairs ; on définit par restriction de ϵ, les applications ϵ_i et ϵ_p. Un K-type irréductible est la donnée de $d', d'' \in D(n)$ et d'un couple de représentations irréductibles de $U(d', \mathbb{F}_q)$ et $U(d'', \mathbb{F}_q)$. Et comme on l'a déjà expliqué en 3.4, se donner ces représentations irréductibles revient à se donner des entiers k', k'' et des représentations irréductibles, ρ', ρ'', des groupes de Weil de type C, $W_{1/2(d'-1/2(k'(k'+1))}$ et $W_{1/2(d''-1/2(k''(k''+1))}$. On dit que l'induite endoscopique de ρ', ρ'' est $(O_i \cup O_p, \epsilon_i \cup \epsilon_p)$ si en notant $(\lambda'; \mu')$ et $(\lambda''; \mu'')$ les partitions paramétrant ρ' et ρ'' :

il existe des partitions $\lambda'_i, \lambda'_p, \mu'_i, \mu'_p$ telles que $\lambda' = \lambda'_i \cup \lambda'_p$ et $\mu' = \mu'_i \cup \mu'_p$ ainsi que des partitions analogues avec $'$ remplacé par $''$. Et l'induite n-endoscopique de $(\lambda'_p; \mu'_p)$ et $(\lambda''_p; \mu''_p)$ est (O_p, ϵ_p) tandis que l'induite endoscopique de $(\lambda'_i; \mu'_i)$ et $(\lambda''_i; \mu''_i)$ est (O_i, ϵ_i).
Quand on identifie ρ' et ρ'' à des partitions (au signe près), ν', ν'', et quand les conditions ci-dessus sont satisfaites on dira simplement que ν', ν'' est dans l'induite endoscopique faible de $(O_p \cup O_i, \epsilon_p \cup \epsilon_i)$. On ne peut pas le dire dans l'autre sens car il n'y a pas unicité (en général) de $O_p, O_i, \epsilon_p, \epsilon_i$ qui vérifient cette condition quand ν', ν'' sont fixés.

Induction endoscopique : soient ρ', ρ'' et ν', ν'' comme ci-dessus, on dit que ρ', ρ'' ou, suivant le point de vue, ν', ν'', sont dans l'induite endocopique d'un couple (O, ϵ) où O est une orbite unipotente de $GL(n, \mathbb{C})$ et ϵ est une application de $Jord(O)$ dans $\{\pm 1\}$, si $O = O_{\nu', \nu''}$ et $\epsilon = \epsilon_{\nu', \nu''}$.

4.2. Caractérisation de l'induction endoscopique Pour la proposition ci-dessous, on fixe O une orbite unipotente de $GL(n, \mathbb{C})$ et ϵ un morphisme de $Jord(O)$ dans $\{\pm 1\}$. On définit O_p et O_i ainsi que ϵ_p et ϵ_i à partir de O et ϵ comme cela a été fait en 4.1. On note alors $O = O_p \cup O_i$ et $\epsilon = \epsilon_p \cup \epsilon_i$.

Proposition 4.1. *Avec les notations ci-dessus. Pour $d', d'' \in D(n)$ et pour ν', ν'' des partitions de d' et d''.*

(i) On suppose que ν', ν'' est dans l'induite endoscopique faible de $(O_p, \epsilon_p) \cup (O_i, \epsilon_i)$ alors $\overline{O_{\nu', \nu''}} \supset \overline{(O_p \cup O_i)}$. De plus s'il y a égalité, alors $\epsilon_{\nu', \nu''} = \epsilon_p \cup \epsilon_i$.

(ii) On suppose que $O_{\nu', \nu''} = O_p \cup O_i$ et que $\epsilon_{\nu', \nu''} = \epsilon_p \cup \epsilon_i$. Alors ν', ν'' est dans l'induite endoscopique faible de $(O_p, \epsilon_p) \cup (O_i, \epsilon_i)$.

Fixons $d' \in \mathbb{N}$, on a déjà dit qu'il existe une bijection entre l'ensemble des couples (k', ρ') où $k' \in \mathbb{N}$, vérifie $d' - k'(k'+1)/2$ est un nombre positif et pair et ρ' est une représentation irréductible du groupe de Weyl de type C et de rang $1/2(d' - k'(k'+1)/2)$. On a déjà écrit cette bijection en 3.2 mais on la récrit ici sous la forme qui nous intéresse. Précisément, on pose $\tilde{k'} = k'$ si k' est impair et $k'+1$ si k' est pair. On fixe aussi t' grand (qui ne joue absolument aucun rôle) et on note $\tilde{\mathcal{E}}' := \{\tilde{e}'\}$ un ensemble de couples $(e', \delta'_{e'})$ où e' est un entier positif ou nul, $\delta'_{e'} \in \{0,1\}$ et vérifiant, avec la notation, pour $\delta = 0$ ou 1, $\tilde{\mathcal{E}}'_\delta := \{(e', \delta_{e'}) \in \tilde{\mathcal{E}}'; \delta_{e'} = \delta\}$:

$|\tilde{\mathcal{E}}'| = 2t + \tilde{k}'$,

$|\tilde{\mathcal{E}}'_0| - |\tilde{\mathcal{E}}'_1| = \tilde{k}'$ si $\tilde{k}' = k'$ et $-\tilde{k}'$ sinon

et ρ' est paramétré par le couple de partition $\{e'|(e',0) \in \tilde{\mathcal{E}}'\}; \{e'|(e',1) \in \tilde{\mathcal{E}}'\}$;

l'ensemble $\tilde{\mathcal{E}}'$ peut avoir de la multiplicité et cette multiplicité se retrouve alors dans les partitions que l'on vient de définir. On remarque que l'on peut toujours ajouter des 0 aux partitions définissant ρ' de façon à ce que les conditions précédentes soient satisfaites. Mais évidemment l'ensemble $\tilde{\mathcal{E}}'$ n'est pas uniquement défini (à cause d'un ajout de 0 possible). On ordonne totalement les ensembles $\tilde{\mathcal{E}}'_\delta$ pour $\delta = 0$ ou 1 de sorte que $(e', \delta) > (f', \delta)$ entraîne que $e' \geq f'$. On ordonne ensuite totalement l'ensemble $\tilde{\mathcal{E}}'$ lui-même de la façon suivante :

$$(e', \delta_{e'}) > (f', \delta_{f'}) \Leftrightarrow$$

$$2e' + \delta_{e'} + 2|\{(\tilde{e} \in \tilde{\mathcal{E}}'_{\delta_{e'}}; \tilde{e} < (e', \delta_{e'})\}| > 2f' + \delta_{f'} + 2|\{(\tilde{f} \in \tilde{\mathcal{E}}'_{\delta_{f'}}; \tilde{f} < (f', \delta_{f'})\}|.$$

Alors $\nu' =$

$$\{2e' + \delta_{e'} + |\{(\tilde{e} \in \tilde{\mathcal{E}}'_{\delta_{e'}}; \tilde{e} < (e', \delta_{e'})\}| - |\{(\tilde{e} \in \tilde{\mathcal{E}}'_{1-\delta_{e'}}; \tilde{e} < (e', \delta_{e'})\}|; (e', \delta_{e'}) \in \tilde{\mathcal{E}}'\}.$$

L'ordre total sur $\tilde{\mathcal{E}}'$ induit un ordre total sur ν' qui est compatible avec l'ordre usuel. On définit aussi l'application de ν' (comme ensemble ordonné) dans $\{\pm 1\}$ en posant $\epsilon'(\alpha') = (-1)^{\delta_{e'}}$ si α' est le terme correspondant à $(e', \delta_{e'})$ l'ordre compte ici (sinon ceci ne serait pas défini) car $\mathcal{E}'$ n'a pas de raison d'être sans multiplicité. On voit donc ν' comme un ensemble de couple $(\alpha', \epsilon'_{\alpha'})$ avec $\epsilon'_{\alpha'} \in \{\pm 1\}$.

Supposons maintenant que $(d', d'') \in D(n)$ soit donné avec un $U(d', \mathbb{F}_q) \times U(d'', \mathbb{F}_q)$-type, c'est à dire 2 couples (k', ρ') et (k'', ρ'') comme ci-dessus. On suppose que $2t' + \tilde{k}' = 2t'' + \tilde{k}''$ (il suffit de faire des choix convenables). On reprend les notations précédentes en mettant des $''$ chaque fois qu'il le faut. On a donc les ensembles totalement ordonnés $\tilde{\mathcal{E}}'$ et $\tilde{\mathcal{E}}''$ qui permettent de calculer les partitions ν' et ν''. On définit une application δ de $\tilde{\mathcal{E}}' \cup \tilde{\mathcal{E}}''$ dans $\{\pm 1\}$, en écrivant $\delta(\tilde{e}'_j) = \delta(\tilde{e}''_j) = (-1)^{\delta_{e'_j} + \delta_{e''_j}}$ pour tout $j \in [2t' + \tilde{k}']$ et où $\tilde{e}'_j$ sont les éléments de $\tilde{\mathcal{E}}'$ ordonnés de façon décroissante et de même pour $\tilde{e}''_j$. On pose alors

$$\tilde{\mathcal{E}}'_{pair} := \{\tilde{e}' \in \tilde{\mathcal{E}}'; \delta(\tilde{e}') = +\} \qquad \tilde{\mathcal{E}}'_{impair} := \{\tilde{e}' \in \tilde{\mathcal{E}}'; \delta(\tilde{e}') = -\}$$

$$\tilde{\mathcal{E}}''_{pair} := \{\tilde{e}'' \in \tilde{\mathcal{E}}''; \delta(\tilde{e}'') = +\} \qquad \tilde{\mathcal{E}}''_{impair} := \{\tilde{e}'' \in \tilde{\mathcal{E}}''; \delta(\tilde{e}'') = -\}.$$

En revenant à ρ' et ρ'', on définit ρ'_{pair} à l'aide des 2 partitions issues de $\tilde{\mathcal{E}}'_{pair}$ et on définit de même ρ'_{impair}, ρ''_{pair} et ρ''_{impair} ; on pose aussi $k_{pair} := |\tilde{\mathcal{E}}'_{pair} \cap \tilde{\mathcal{E}}'_0| - |\tilde{\mathcal{E}}'_{pair} \cap \tilde{\mathcal{E}}'_1|$ et $k_{impair} := |\tilde{\mathcal{E}}'_{impair} \cap \tilde{\mathcal{E}}'_0| - |\tilde{\mathcal{E}}'_{impair} \cap \tilde{\mathcal{E}}'_1|$. Ces nombres ne sont pas nécessairement positifs mais on a déjà dit que pour nous les symboles n'avaient pas forcément de défaut positif. Le point clé est le suivant ; pour $\tilde{e}' \in \tilde{\mathcal{E}}'$, on pose :

$$\nu'_{\tilde{e}',parite} := 2e' + \delta_{e'} + |\{\tilde{f}' \in \tilde{\mathcal{E}}'; \tilde{f}' < \tilde{e}'; \delta_{f'} = \delta_{e'}; \delta(\tilde{f}') = \delta(\tilde{e}')\}| -$$

$$|\{\tilde{f}' \in \tilde{\mathcal{E}}'; \tilde{f}' < \tilde{e}'; \delta_{f'} = -\delta_{e'}; \delta(\tilde{f}') = \delta(\tilde{e}')\}|.$$

On définit de même $\nu''_{\tilde{e}'',parite}$. On définit alors la partition ν'_{pair} en regardant la collection des $\nu'_{\tilde{e}',parite}$ pour $\tilde{e}'$ tel que $\delta(\tilde{e}') = +$. Et on définit de même les partitions ν''_{pair}, ν'_{impair} pour laquelle $\delta_{\tilde{e}'} = -$ puis ν''_{impair}.

La partition somme $\nu' + \nu''$ est la réunion des 2 partitions $\nu'_{pair} + \nu''_{pair}$ et $\nu'_{impair} + \nu''_{impair}$.

Cela résulte de l'égalité pour tout couple $\tilde{e}'_j, \tilde{e}''_j$ (notations déjà introduites ci-dessus) :

$$|\{\tilde{f}' \in \tilde{\mathcal{E}}'; \tilde{f}' < \tilde{e}'_j; \delta_{f'} = \delta_{e'_j}; \delta(\tilde{f}') = \delta(\tilde{e}')_j\}|$$

$$-|\{\tilde{f}' \in \tilde{\mathcal{E}}'; \tilde{f}' < \tilde{e}'_j; \delta_{f'} = -\delta_{e'_j}; \delta(\tilde{f}') = \delta(\tilde{e}')_j\}|$$

$$+|\{\tilde{f}'' \in \tilde{\mathcal{E}}''; \tilde{f}'' < \tilde{e}''_j; \delta_{f''} = \delta_{e''_j}; \delta(\tilde{f}'') = \delta(\tilde{e}''_j)\}|$$

$$-|\{\tilde{f}'' \in \tilde{\mathcal{E}}''; \tilde{f}'' < \tilde{e}''_j; \delta_{f''} = -\delta_{e''_j}; \delta(\tilde{f}'') = \delta(\tilde{e}''_j)\}|$$

$$= |\{\tilde{f}' \in \tilde{\mathcal{E}}'; \tilde{f}' < \tilde{e}'_j; \delta_{f'} = \delta_{e'_j}\}| - |\{\tilde{f}' \in \tilde{\mathcal{E}}'; \tilde{f}' < \tilde{e}'_j; \delta_{f'} = \delta_{e'_j}\}|$$

$$+|\{\tilde{f}'' \in \tilde{\mathcal{E}}''; \tilde{f}'' < \tilde{e}''_j; \delta_{f''} = \delta_{e''_j}\}| - |\{\tilde{f}'' \in \tilde{\mathcal{E}}''; \tilde{f}'' < \tilde{e}''_j; \delta_{f''} = -\delta_{e''_j}\}|.$$

On pose $O = O_{\nu',\nu''}$ et O_{pair} l'orbite définie par les blocs de Jordan de taille paire ainsi que O_{impair} celle définie par les blocs de Jordan de taille impaire. Avec cela il n'est pas difficile de démontrer que $\rho'_{pair}, \rho''_{pair}$ est dans l'induite endoscopique de $O_{pair}, \epsilon_{pair}$ et le même résultat avec pair remplacé par impair ; on a écrit les formules qu'il fallait dans la définition de l'induite endoscopique. On va plutôt montrer que $\epsilon_{pair}, \epsilon_{impair}$ s'obtient à partir de $\epsilon_{\nu',\nu''}$ comme expliqué avant l'énoncé. Pour cela, on a besoin de la partition ν'' ; elle est indexée par les éléments de $\tilde{\mathcal{E}}''$ et on rappelle que ce dernier ensemble a par hypothèse un nombre impair de termes et le même nombre de terme que $\tilde{\mathcal{E}}'$. Pour calculer $O_{\nu',\nu''}$, on fait la somme terme à terme $\nu' + \nu''$. Ainsi on trouve les blocs de Jordan de $O_{\nu',\nu''}$, dont certains peuvent être nuls car par définition on en a un nombre impair. On note $J\tilde{o}rd(O)$ l'ensemble des blocs de Jordan trouvé; quitte à diminuer t' et t'', on peut s'arranger pour que $J\tilde{o}rd(O)$ ait au plus un terme nul et à ce moment là, on a pour tout $\alpha \in J\tilde{o}rd(O)$ qui correspond, par définition, à 2 éléments $\tilde{e'}_j, \tilde{e}''_j$ (même j), $(-1)^\alpha = (-1)^{\delta_{e'_j}+\delta_{e''_j}}$. On reprend les notations $J\tilde{o}rd(O)_{<\alpha,opp}$ analogues à celles de 3.4 et on introduit de façon similaire $J\tilde{o}rd(O)_{<\alpha,parite}$. Avec les notations, $\alpha, \tilde{e}''_j$ qui viennent d'être introduites, on a :

$$\nu''(\alpha) = 2e''_j + \delta_{e''_j} + 2|\{\tilde{f}'' \in \tilde{\mathcal{E}}''; \tilde{f}'' < \tilde{e}''_j; \delta_{f''} = \delta_{e''_j}\}| - |\tilde{\mathcal{E}}''_{<\tilde{e}''_j}|.$$

On a donc $(-1)^{\nu''(\alpha)} = (-1)^{\delta_{e''_j}}(-1)^{|-|\tilde{\mathcal{E}}''_{<\tilde{e}''_j}|}$. On remplace $| - |\tilde{\mathcal{E}}''_{<\tilde{e}''_j}|$ par $|J\tilde{o}rd(O)_{<\alpha}|$. Et on décompose

$$|J\tilde{o}rd(O)_{<\alpha}| = |J\tilde{o}rd(O)_{<\alpha,parite}| + |J\tilde{o}rd(O)_{<\alpha,opp}|.$$

On écrit donc :

$$(-1)^{\nu''(\alpha)}(-1)^{J\tilde{o}rd(O)_{<\alpha,opp}} = (-1)^{\delta_{e''_j}}(-1)^{J\tilde{o}rd(O)_{<\alpha,parite}}.$$

On a besoin de remarquer que $J\tilde{o}rd(O)_{impair} = Jord(O)_{impair}$ et que le cardinal de cet ensemble a la parité de n. Par contre

$$J\tilde{o}rd(O)_{pair} = \begin{cases} Jord(O)_{pair} \text{ si } Jord(O)_{pair} \not\equiv n[2] , \\ Jord(O)_{pair} \cup \{0\} \text{ sinon} \end{cases}$$

et que dans tous les cas la parité de $|J\tilde{o}rd(O)|$ est l'opposée de celle de n. Il s'en suit que pour α pair $J\tilde{o}rd(O)_{<\alpha,opp} = Jord(O)_{<\alpha,opp}$ et que pour α impair $|J\tilde{o}rd(O)_{<\alpha,opp}| \equiv |Jord(O)_{>\alpha,opp}| + n + 1$. D'où :

$$(-1)^{\nu''(\alpha)}(-1)|J\tilde{o}rd(O)_{<\alpha,opp}| =$$

$$\begin{cases} \epsilon_{\nu',\nu''}(\alpha) \text{ si } \alpha \text{ est pair} \\ \epsilon_{\nu',\nu''}(\alpha)(-1)^{|J\tilde{o}rd(O)_{pair}|} = \epsilon_{\nu',\nu''}(\alpha)(-1)^{n+1} \text{ si } \alpha \text{ est impair.} \end{cases}$$

De plus par définition par définition de l'induction endocopique :

$$(-1)^{\delta_{e''_j}}(-1)^{J\tilde{o}rd(O)_{<\alpha,parite}} = \begin{cases} (-1)^{n+1}\epsilon_{impair}(\alpha) \text{ si } \alpha \text{ est impair,} \\ \epsilon_{pair}(\alpha) \text{ si } \alpha \text{ est pair;} \end{cases}$$

pour l'égalité dans le cas α pair, il y a un $(-1)^n$ qui vient de notre définition et un autre $(-1)^n$ qui vient du fait que le cardinal de $J\tilde{o}rd(O)_{pair}$ est de la parité de $n+1$. D'où $\epsilon_{\nu',\nu''}(\alpha) = \begin{cases} \epsilon_{impair}(\alpha) \text{ si } \alpha \text{ est impair} \\ \epsilon_{pair}(\alpha) \text{ si } \alpha \text{ est pair.} \end{cases}$ Maintenant pour démontrer le lemme, on part d'une partition de ρ' et ρ'' en pair et impair et on fait les constructions en sens inverse; on a donc $\tilde{\mathcal{E}}'_{pair} \ldots$ qui donne des partitions de $\tilde{\mathcal{E}}'$ ... La seule chose qui diffère est que l'ordre sur $\tilde{\mathcal{E}}'$ tel qu'on l'a défini n'est pas nécessairement compatible avec les ordres sur $\tilde{\mathcal{E}}'_{pair}$ ... Il n'est même compatible que quand on a la partition spéciale définie ci-dessus. Et en remontant les calculs, on voit que quand les ordres ne sont pas compatibles alors $O_{\nu',\nu''}$ est strictement plus grande que la réunion de l'orbite paire et impaire définies par la décomposition. Cela termine la preuve.

5. Construction de représentations

5.1. Lien avec les constructions de Lusztig La construction donnée par Lusztig des représentations de réduction unipotentes des groupes unitaires fait partie du cadre général **[L3]**. On revient à la classification de Langlands 2.2. Cette classification se fait à l'aide des couples (O, ϵ) où O est une orbite unipotente de $GL(n, \mathbb{C})$ et où ϵ est une application de $Jord(O)$ dans $\{\pm 1\}$; on pose $n_p := \sum_{\alpha \in Jord(O); (-1)^\alpha = +1} \alpha$ et $n_i := \sum_{\alpha \in Jord(O); (-1)^\alpha = -1} \alpha$. On note aussi O_p et O_i les orbites unipotentes de $GL(n_p, \mathbb{C})$ et $GL(n_i, \mathbb{C})$ dont les blocs de Jordan sont respectivement les blocs de Jordan de O de taille paire et ceux de taille impaire. On note ϵ_p et ϵ_i les restrictions de ϵ à $Jord(O_p)$ et $Jord(O_i)$ respectivement. Pour $m \in \mathbb{N}$, on note W_m le groupe de Weyl de type C et de rang m et $\mathcal{H}(W_m)$ l'algèbre de Hecke affine associée à ce groupe de Weyl. La représentation de Springer généralisée, associe à (O_p, ϵ_p) (resp. (O_i, ϵ_i)) un entier k_p (k_i tel que $n_i - k_i^2$ est pair) et une représentation non irréductible de $W_{(n_p - k_p(k_p+1))/2}$ (resp. $W_{(n_i - k_i^2)/2}$) ; c'est la représentation dans la cohomologie de la variété de Springer généralisée. On utilisera plus tard ces représentations. Pour le moment on n'utilise que k_i et k_p ; on note ζ le signe de $2(k_i - k_p) - 1$ et on pose $A := 2(k_i + k_p) + 1$ et $B := \zeta(2(k_i - k_p) - 1)$. On pose aussi $N := (n - k_i^2 - k_p(k_p + 1))/2$. On forme alors l'algèbre de Hecke $\mathcal{H}_{N,A,B}$ (cf. **[W2]** 1.4). On pose encore $C_p := A - \zeta B$ et $C_i := A + \zeta B$; avec ces nombres on forme les algèbres de Hecke graduées, $\overline{\mathcal{H}}(M, C_?)$ pour tout $M \in \mathbb{N}$. Pour un couple d'entiers N', N'' tels que $N = N' + N''$, Lusztig a construit une équivalence de catégorie entre la catégorie des classes d'isomorphie de représentations de $\mathcal{H}_{N,A,B}$ avec un caractère central fixé par N', N'' et les classes d'isomorphie de représentations de l'algèbre $\mathcal{H}_{N',A,B} \otimes \mathcal{H}_{N'',A,B}$ pour un caractère central correspondant (cf. **[W2]** 1.7), cela dépend d'un certain nombre de choix et en particulier du choix d'un signe que nous prenons égal à notre ζ ci-dessus (les notations sont alors celles de loc.cite). Et toujours grâce aux constructions de Lusztig, cette dernière catégorie est encore isomorphe (par un passage au gradué) à la catégorie des représentations de $\overline{\mathcal{H}}(N', C_i) \otimes \overline{\mathcal{H}}(N'', C_p)$ (cf. **[W2]** 1.9). Le calcul des constantes A, B, C_i, C_p a été fait à partir des tables de Lusztig (**[L3]**) de façon à obtenir le résultat ci-dessous. Aux entiers k_i, k_p et à $\sharp$ fixé, on associe une représentation cuspidale de $U(k_i^2 +$

$k_p(k_p+1), F)_\sharp$ si k_i est impair, aucune si k_i est pair et $\sharp \neq (-1)^{k_p(k_p+1)/2+k_i^2/2}$, 1 si $k_i = 0$ et 2 dans le cas restant $k_i \neq 0$ pair et $\sharp = (-1)^{(k_p(k_p+1)+k_i)/2}$; on renvoit à 3.5 pour la construction ; on écrit ces représentations cuspidales $cusp(k_i, k_p, \eta)$ où η est un signe (on n'a pas besoin de le préciser, il est lié au ζ de 3.5 qui n'est pas le ζ qui est ici). En résumé

La catégorie des représentations de $U(n, F)_\sharp$ de réduction unipotente de type $n_i, n_p, cusp(k_i, k_p, \eta)$ est équivalente à la catégorie des représentations de l'algèbre $\overline{\mathcal{H}}(N_i, C_i) \otimes \overline{\mathcal{H}}(N_p, C_p)$. On note $\pi_{unip}(O, \epsilon)$ la représentation construite à partir des données ci-dessus.

Toutefois il faut faire attention à un point clé, c'est que les constructions donnent naturellement non pas les séries discrètes mais leur image par une involution (c'est-à-dire ce que l'on a appellé ailleurs des représentations quadratiques unipotentes). Cette involution, a été définie entre autre par Aubert dans **[Au]** et par Schneider et Stuhler dans **[SS]**. Il faut s'accorder sur les signes ; ici on note $|D|$ la dualité qui est telle qu'une irréductible soit une irréductible (on l'avait noté D en **[MW]**). On pose $\pi(O, \epsilon) := |D|\pi_{unip}(O, \epsilon)$.

5.2. Calcul du signe introduit par l'involution La propriété (qui résulte facilement des formules de **[Au]** et **[SS]**) est que pour π une représentation irréductible et γ un élément elliptique régulier du groupe, le caractère de π en γ diffère du caractère de $D(\pi)$ en γ par le signe $(-1)^{rgG-rg(P_{cusp})}$ où P_{cusp} est un parabolique minimal avec la propriété que le module de Jacquet de π le long de son radical unipotent soit non nul. En notant D l'involution définie par $D(\pi) := \sum_P (-1)^{rg(G)-rg(P)} Ind_P^G \circ Res_P\, \pi$, on a simplement $D(\pi) = (-1)^{rgG-rg(P_{cusp})}|D|(\pi)$ avec la même définition que ci-dessus. On a besoin de calculer ce signe.

Proposition 5.1. *Soit π une représentation tempérée versus elliptique, autoduale, de $U(n, E/F)_\sharp$ paramétrée par le couple (ψ, ϵ) et soit γ un élément elliptique régulier de ce groupe. Alors le caractère de π et de $|D|\pi$ en γ diffère par un signe qui ne dépend que de ψ et le signe $\prod_{\alpha \in Jord(O); \alpha \equiv 0[2]} \epsilon(\alpha)$.*

C'est en fait un résultat très général pour tous les groupes classiques qui dépasse le niveau 0, mais on n'a pas introduit les notations nécessaires.

On doit calculer $(-1)^{(n-k_p(k_p+1)-k_i^2)/2}$ où k_i et k_p ont été définis en 5.1. On remarque que $(-1)^{(n-k_i^2)/2} = (-1)^{[n/2]}$ par le fait que k_i est un carré de même parité que n. Il s'agit donc de montrer que $(-1)^{k_p(k_p+1)/2} = \prod_{\alpha \equiv 0[2]} \epsilon(\alpha)$. C'est une propriété de la représentation de Springer : on part d'une orbite O_p d'un groupe symplectique avec un morphisme de $Jord(O_p)$ dans ± 1. On cherche à calculer la partie cuspidale de la représentation de Springer ; pour cela on peut enlever de O_p tout couple α, α' de blocs de Jordan consécutifs et tels que $\epsilon(\alpha) = \epsilon(\alpha')$ et on peut répéter cette opération autant de fois que possible et on doit encore enlever le plus petit des blocs de Jordan, α_{min} si $\epsilon(\alpha_{min}) = +1$. A la fin, on se retrouve avec une orbite $O_{pseudo-cusp}$ d'un groupe symplectique plus petit avec $Jord(O_{pseudo-cusp}) \subset Jord(O_p)$ et avec la propriété que pour α', α 2 blocs de Jordan consécutif $\epsilon(\alpha') = -\epsilon(\alpha)$ et $\epsilon(\alpha_{min}) = -1$. Alors $k_p = |Jord(O_{pseudo-cusp})|$. Et avec les propriétés de $\epsilon_{|Jord(O_{pseudo-cusp})}$ que l'on vient de donner, on vérifie que $(-1)^{k_p(k_p+1)/2} = \prod_{\alpha \in Jord(O_{pseudo-cusp})} \epsilon(\alpha)$. Or par construction $\prod_{\alpha \in Jord(O_{pseudo-cusp})} \epsilon(\alpha) = \prod_{\alpha \in Jord(O)} \epsilon(\alpha)$. D'où le résultat annoncé.

5.3. Calcul du gradué des représentations

Proposition 5.2. *Soit (O, ϵ) comme ci-dessus et donc $\pi_{unip}(O, \epsilon)$. Alors $res^{par} \pi_{unip}(O, \epsilon) \in \mathbb{C}[\hat{\mathfrak{S}}_{D(n)}]_{\geq O, global}$. Et les termes de plus bas degré sont exactement les K-types ν', ν'' tels que $(O, \epsilon) = (O_{\nu',\nu''}, \epsilon_{\nu',\nu''})$. Chacun de ces K-types intervenant avec multiplicité 1.*

Le premier point est de se ramener à un problème de représentation de groupe de Weyl de type C. C'est ce qui est fait dans la première partie de la démonstration **[W2]** proposition de 5.2. La démonstration est complètement générale, elle s'inspire d'une idée de Lusztig que les représentations des groupes de Weyl ne se déforment pas continuement. Ensuite ce dont on a besoin est que les K-types qui interviennent ont une décomposition en parties paire et impaire qui provient exactement de la séparation entre la partie symplectique (paire) et la partie orthogonale (impaire). On a donc pour chaque K-type, par induction endoscopique symplectique, resp. orthogonale des orbites O_{pair} resp. O_{impair} dont la fermeture contient O_p et O_i respectivement. A fortiori avec le lemme de 4.2, on obtient que l'orbite associée en faisant la somme des partitions contient dans sa fermeture $O_p \cup O_i$. On voit aussi que quand ces orbites coïncident les caractères aussi coïncident. Quant à la multiplicité 1, il suffit de voir que les K-types pour lesquels il y a coïncidence interviennent avec multiplicité exactement 1 ; c'est une adaptation de la démonstration de **[W1]** XII.6.

6. Transformation

6.1. Définition Soit O une orbite unipotente de $GL(n, \mathbb{C})$. Soient aussi ϵ, ϵ' des applications de $Jord(O)$ dans $\{\pm 1\}$. On pose (les produits ci-dessous sont indexés par $Jord(O)$ considéré sans multiplicité) :

$$< \epsilon', \epsilon > := \prod_{\alpha \in Jord(O); \epsilon'(\alpha)=-1} \epsilon(\alpha),$$
$$\sigma^{\epsilon'} := \prod_{\alpha \in Jord(O); \alpha \equiv n+1} \epsilon'(\alpha)^{mult_\alpha(Jord(O))},$$
$$\sigma_\epsilon := \prod_{\alpha \in Jord(O); \epsilon(\alpha)=-1} (-1)^{|Jord_{?\alpha,opp}(O)|\, mult_\alpha(Jord(O))},$$

où $Jord_{?\alpha,opp}(O)$ est comme en 3.4. On considère l'espace vectoriel $\mathbb{C}[(O,\epsilon)]$ de base l'ensemble des couples (O,ϵ) où O est une orbite unipotente comme ci-dessus et ϵ est une application de $Jord(O)$ dans $\{\pm 1\}$. On définit un endomorphisme linéaire de $\mathbb{C}[(O,\epsilon)]$, en posant pour tout (O,ϵ) comme ci-dessus,

$$\mathcal{F}(O,\epsilon) := \sum_{\epsilon' : Jord(O) \to \{\pm 1\}} \sigma^{\epsilon'} \sigma_\epsilon < \epsilon, \epsilon' > (O, \epsilon').$$

On remarque que le facteur $\sigma^{\epsilon'} \sigma_\epsilon < \epsilon, \epsilon' >$ n'est pas symétrique en ϵ et ϵ'. On peut toutefois définir un autre endomorphisme de $\mathbb{C}[(O,\epsilon)]$, en posant avec les mêmes notations que ci-dessus :

$$\mathcal{F}'(O,\epsilon) := \sum_{\epsilon' : Jord(O) \to \{\pm 1\}} \sigma^{\epsilon} \sigma_{\epsilon'} < \epsilon, \epsilon' > (O, \epsilon').$$

On vérifie alors sans difficultés que le composé $\mathcal{F}'\mathcal{F}(O,\epsilon) = 2^{|Jord(0)|}(O,\epsilon)$. Cela prouve que $\mathcal{F}$ est bijectif et donne un inverse explicite.

Comme dans [MW], on veut que $\mathcal{F}$ laisse stable ce qui correspond aux représentations elliptiques. Ici c'est beaucoup plus simple que dans loc. cite mais cela nécessite des notations.

6.2. Ellipticité La partie "discrète" est, elle, facile à décrire, c'est le sous-espace notée $\mathbb{C}[(O,\epsilon)]_{disc}$ qui est engendré par les couples (O,ϵ) tel que $Jord(O)$ soit sans multiplicité.

La partie "elliptique" notée $\mathbb{C}[(O,\epsilon_{ell})]$ est plus difficile à décrire. La notation est volontairement différente de ci-dessus. La partie elliptique est un sous-espace vectoriel du sous-espace de base les couples (O,ϵ) où O est tel que dans $Jord(O)$ la multiplicité est au plus 2 ; une telle orbite O est dite pseudo-elliptique. Maintenant on décrit une base de $\mathbb{C}[(O,\epsilon_{ell})]$, en considérant les couples (O,ϵ_{ell}) où O est pseudo-elliptique et où ϵ_{ell} est une application du sous-ensemble de $Jord(O)$ formé des éléments n'ayant que multiplicité 1 dans $\{\pm 1\}$. Et l'élément de la base correspondant est

$$2^{-|\{\alpha \in Jord(O); mult_\alpha(Jord(O))=2\}|} \sum_{\epsilon : Jord(O) \to \{\pm 1\}} \prod_{\substack{\alpha \in Jord(O); \\ mult_\alpha(Jord(O)) = 2}} \epsilon(\alpha)(O,\epsilon),$$

où la somme ne porte que sur les ϵ dont la restriction aux éléments de $Jord(O)$ de multiplicité 1 est le ϵ_{ell} fixé. Attention ici, dans le produit chaque α n'intervient qu'une fois (sinon le signe serait constant égal à 1). Avec les notations ci-dessus, on pose encore $\epsilon_{ell,-}$ l'application de $Jord(O)$ dans $\{\pm 1\}$ qui prolonge ϵ_{ell} par $\epsilon(\alpha) = -$ pour tout α intervenant avec multiplicité 2 dans $Jord(O)$.

Remarque 6.1. L'application $\mathcal{F}$ laisse stable $\mathbb{C}[(O, \epsilon)]_{disc}$ et induit une bijection de $\mathbb{C}[(O, \epsilon_{ell}]$ sur $\mathbb{C}[(O, \epsilon_{ell,-})]$.

L'assertion relative à *disc* est claire et nous démontrons l'autre. On est libre et nous le ferons donc de fixer O une orbite unipotente pseudo-elliptique. Fixons aussi ϵ_{ell} comme dans ce qui précède l'énoncé. L'élément de la base correspond à (O, ϵ_{ell}) est une combinaison linéaire des (O, ϵ) pour ϵ étendant ϵ_{ell} à tout $Jord(O)$. Pour cette preuve, on note $[\epsilon_{ell}]$ l'ensemble des applications ϵ ayant cette propriété. Pour toute application ϵ'_{ell} de même nature que ϵ_{ell}, on définit de la même façon $[\epsilon'_{ell}]$. On note $< (O, \epsilon'_{ell}) >$ l'élément de la base qui lui correspond. On remarque que σ_ϵ ne dépend pas du choix de $\epsilon \in [\epsilon_{ell}]$ et de même $\sigma^{\epsilon'}$ ne dépend pas du choix de ϵ' dans $[\epsilon'_{ell}]$; en effet les α sur lesquels ϵ ou ϵ' peuvent varier interviennent avec multiplicité exactement 2 et $(-1)^2 = 1$. On note alors ces signes $\sigma_{[\epsilon_{ell}]}$ et $\sigma^{[\epsilon'_{ell}]}$. Cette indépendance ne vaut pas pour le signe $< \epsilon', \epsilon >$, mais en séparant dans la formule qui définit ce signe la contribution des α qui interviennent avec multiplicité 1 dans $Jord(O)$ de celle des α qui interviennent avec multiplicité 2 sous la forme :

$$< \epsilon, \epsilon' >=< [\epsilon'_{ell}], [\epsilon_{ell}] > \prod_{\substack{\alpha \in Jord(O); \\ mult_\alpha(Jord(O)) = 2, \\ \epsilon'(\alpha) = -1}} \epsilon(\alpha).$$

Montrons alors la formule : $\mathcal{F}(< (O, \epsilon_{ell}) >) = \sum_{\epsilon'_{ell}} \sigma^{[\epsilon'_{ell}]} \sigma_{[\epsilon_{ell}]} < [\epsilon'_{ell}], [\epsilon_{ell}] >$ $(O, \epsilon'_{ell,-})$. En effet, en appliquant la définition, on obtient :

$$\mathcal{F}(< (O, \epsilon_{ell}) >) = 2^{-|\{\alpha \in Jord(O); mult_\alpha(Jord(O))=2\}|}$$

$$\sum_{\epsilon'_{ell}} \sum_{\epsilon' \in [\epsilon'_{ell}], \epsilon \in [\epsilon_{ell}]} \sigma^{[\epsilon'_{ell}]} \sigma_{[\epsilon_{ell}]} < [\epsilon'_{ell}], [\epsilon_{ell}] > \prod_{\substack{\alpha \in Jord(O); \\ mult_\alpha(Jord(O)) = 2, \\ \epsilon'(\alpha) = -1}} \epsilon(\alpha)$$

$$\prod_{\substack{\alpha \in Jord(O); \\ mult_\alpha(Jord(O)) = 2}} \epsilon(\alpha)(O, \epsilon').$$

On remarque que

$$\prod_{\substack{\alpha \in Jord(O);\\ mult_\alpha(Jord(O))=2,\\ \epsilon'(\alpha)=-1}} \epsilon(\alpha) \prod_{\substack{\alpha \in Jord(O);\\ mult_\alpha(Jord(O))=2}} \epsilon(\alpha) = \prod_{\alpha; mult_\alpha(Jord(O))=2, \epsilon'(\alpha)=1} \epsilon(\alpha)$$

et on a à sommer ces signes, pour ϵ' fixé, sur $\epsilon \in [\epsilon_{ell}]$. Cette somme est nulle sauf si $\epsilon' = \epsilon'_{ell,-}$ auquel cas elle vaut $2^{|\{\alpha \in Jord(O); mult_\alpha(Jord(O))=2\}|}$.

6.3. Produit scalaire elliptique L'ensemble des représentations elliptiques (pour un groupe réductif p-adique très général) a été muni d'un produit scalaire par Arthur ; c'est ce que nous reprenons comme produit scalaire. On obtient une base orthonormée pour ce produit scalaire en considérant l'ensemble des représentations $\pi(O, \epsilon_{ell})$ (cf ci-dessus) où O parcourt l'ensemble des orbites pour lesquelles $Jord(O)$ est de multiplicité au plus 2 et où ϵ_{ell} est une application de $\{\alpha \in Jord(O); mult_\alpha(Jord(O)) = 1\}$ dans $\{\pm 1\}$.

On a aussi dans ce travail considéré les représentations des groupes finis $\mathfrak{S}_{d'} \times \mathfrak{S}_{d''}$; le produit scalaire elliptique pour un groupe fini devrait être l'intégration des caractères sur les éléments elliptiques. C'est ce que l'on fait ici mais en considérant qu'un élément de $\mathfrak{S}_d$ (pour $d = d'$ ou d'') est elliptique s'il n'a pas de cycle de longueur paire ; ces éléments sont dits U-elliptiques. C'est ce qui est utilisé dans [**W1**] et [**MW**] et justifié par le paramétrage des tores elliptiques des groupes unitaires.

Définition On appelle représentation U-elliptique d'un groupe symétrique toute représentation irréductible de ce groupe paramétré par une partition sans multiplicité.

Cette définition est justifiée par la proposition ci-dessous.

Proposition 6.2. *Fixons $d \in \mathbb{N}$. Les caractères des représentations irréductibles U-elliptiques de $\mathfrak{S}_d$ forment une base d'un supplémentaire dans l'ensemble des fonctions invariantes sur $\mathfrak{S}_d$ du noyau du produit scalaire elliptique.*

On montre d'abord qu'il y a le même nombre de partitions de d sans multiplicité que de partitions de d ne contenant que des nombres impairs. Pour cela, on interprète le nombre de partitions de d sans multiplicité comme le coefficient de x^d dans la série $\prod_{i \in \mathbb{N}_{>0}}(1 + x^i)$ et le nombre de partitions de d ne contenant que des nombres impairs comme le coefficient de x^d dans la série $\prod_{j \in \mathbb{N}} 1/(1 - x^{2j+1})$. Pour démontrer que ces nombres sont égaux, on démontre que les séries sont égales. Pour cela, il suffit de démontrer qu'on obtient la même série en les multipliant par la série $\prod_{i\in \mathbb{N}_{>0}} 1/(1-x^{2i})$. Dans les 2 cas, le résultat est $\prod_{i\in \mathbb{N}_{>0}} 1/(1-x^{i})$.

La démonstration ci-dessous m'a été communiquée par J.-L. Waldspurger. On pose $\tilde{\mathfrak{S}}_d$ le produit semi-direct de $\mathfrak{S}_d$ avec le groupe à 2 éléments $\{1, F\}$ où F agit sur $\mathfrak{S}_d$ par $F.\sigma = w_d \sigma w_d$ où w_d est la permutation qui envoie $j \in [1, d]$ sur $d-j$. On appelle Levi de $\tilde{\mathfrak{S}}_d$ les sous-groupes

$$\times_{d_1,\cdots,d_r;d_0 | \sum 2d_j+d_0=d} \mathfrak{S}_{d_1} \times \cdots \times \mathfrak{S}_{d_r} \times \mathfrak{S}_{d_0} \times \mathfrak{S}_{d_1} \times \cdots \times \mathfrak{S}_{d_r} \ltimes \{1, F\}.$$

Au lieu de regarder les fonctions sur $\mathfrak{S}_d$ on regarde les fonctions sur le sous-ensemble $\mathfrak{S}_d F$ de $\tilde{\mathfrak{S}}_d$, quand on regarde les fonctions invariantes, cela ne change pas grand chose sauf que cela permet d'interpréter les fonctions à support dans l'ensemble des éléments de $\mathfrak{S}_d$ ayant des cycles tous de longueur impaire comme les fonctions cuspidales. On remarque maintenant que pour π une représentation irréductible de $\mathfrak{S}_d$, il y a exactement 2 façons de la remonter en une représentation irréductible de $\tilde{\mathfrak{S}}_d$ et quand on restreint le caractère de ces 2 prolongements à $\mathfrak{S}_d F$, on réobtient une fonction invariante, essentiellement le caractère de la représentation de départ. Pour démontrer notre assertion, il suffit donc de démontrer que les traces des représentations irréductibles de $\mathfrak{S}_d$ vues comme fonctions sur $\mathfrak{S}_d F$ forment un supplémentaire des fonctions obtenues comme trace des induites à partir des Lévis décrits ci-dessus. Ceci est clair, c'est en fait essentiellement l'interprétation de l'égalité :

$$\prod_{i\in \mathbb{N}} 1/(1-x^{2i+1}) \times \prod_{i\in \mathbb{N}_{>0}} 1/(1-x^{2i}) = \prod_{i\in \mathbb{N}_{>0}} 1/(1-x^{i})$$

qui est évidente.

6.4. Isométrie On a défini l'application $Ind \circ Res$ de $\mathbb{C}[\hat{\mathfrak{S}}_{D(n)}]$ dans lui-même. On munit $\mathbb{C}[\hat{\mathfrak{S}}_{D(n)}]$ du produit scalaire elliptique comme décrit dans 6.3.

Lemme 6.3. *L'application $Ind \circ Res$ est une isométrie.*

Avec les définitions données ici, c'est loin d'être clair, il faut en revenir à la définition de [**MW**] 3.1 et 3.2. On note W_n le groupe de Weyl de type C et de rang n. On note p_n la projection de W_n sur $\mathfrak{S}_n$ évidente et on note sgn_{CD} le signe sur W_n dont le noyau est le groupe de Weyl de type D. Toutes ces définitions se généralisent pour d un entier quelconque. Pour $(d_1, d_2) \in D(n)$, on note ρ_{d_1,d_2} l'application de $\mathfrak{S}_{d_1} \times \mathfrak{S}_{d_2}$ dans W_n qui à w', w'' associe l'élément w vérifiant :

$$w(i) = \begin{cases} w'(i), \text{ si } i \in \{1, \cdots, d_1\}, \\ -d_1 - w''(i-d_1), \text{ si } i \in \{d_1+1, \cdots, n\}. \end{cases}$$

Ce n'est donc pas un isomorphisme de groupe mais une application qui respecte les classes de conjugaison. En faisant la somme sur tout $(d_1, d_2) \in D(n)$, on peut donc définir une application $\rho^* : \mathbb{C}[\hat{W}_n] \to \mathbb{C}[\hat{\mathfrak{S}}_{D(n)}]$.

Pour $(d', d'') \in D(n)$ et pour $f' \in \mathbb{C}[\hat{\mathfrak{S}}_{d'}]$ et $f'' \in \mathbb{C}[\hat{\mathfrak{S}}_{d''}]$, on définit d'abord f_0 une fonction sur $W_{d'} \times W_{d''}$, en posant $f_0(w', w'') := f' \circ \rho_{d'}(w') f'' \circ \rho_{d''}(w'') sgn_{CD}(w'')$. Puis on pose $\iota_{d',d''}(f', f'')$ l'induite de f_0 à W_d (c'est-à-dire la fonction f_0 rendue invariante). En sommant sur tous les (d', d''), on définit $\iota : \mathbb{C}[\hat{\mathfrak{S}}_{D(n)}] \to \mathbb{C}[\hat{W}_n]$. Et

$$Ind \circ Res = \rho^* \circ \iota.$$

Il est rappelé dans loc.cite que ι est un isomorphisme ; ce point est facile et c'est en fait une tautologie car c'est la description des représentations irréductibles de W_n (à l'aide de 2 partitions). Il est aussi démontré dans loc. cite, que ρ^* est un isomorphisme quand on se limite aux éléments U-elliptiques (un élément de W_n est dit U-elliptique si son image par p_n l'est). L'argument ici est que pour $w \in W_n$ que l'on suppose U-elliptique il existe exactement un couple $(d_1, d_2) \in D(n)$ et une seule classe de conjugaison C_1 dans $\mathfrak{S}_{d_1}$ et une C_2 dans $\mathfrak{S}_{d_2}$ tels que $\rho_{d_1,d_2}(C_1, C_2) \in C$, où C est la classe de conjugaison de w.

C'est cela qui prouve que $\rho^* \circ \iota$ est une isométrie pour le produit scalaire elliptique.

6.5. Définition de Sp_{ell} Soit (O, ϵ_{ell}) une orbite unipotente de $GL(n, \mathbb{C})$ et ϵ_{ell} une application de $\{\alpha \in Jord(O); mult_\alpha(Jord(O)) = 1\}$ dans $\{\pm 1\}$. On suppose que les multiplicités dans $Jord(O)$ sont au plus 2 et on prolonge ϵ_{ell} à tout $Jord(O)$ en posant $\epsilon_{ell}(\alpha) = -1$ pour tout α intervenant avec multiplicité 2 dans $Jord(O)$; c'est ce que l'on a noté $\epsilon_{ell,-}$ en 6.2. On décompose $Jord(O) = Jord_+(O) \cup Jord_-(O)$ de telle sorte que $(-1)^{mult_\alpha(Jord_-(O))} = \epsilon_{ell,-}(\alpha)$. On pose encore, pour $\eta = \pm$, $n_\eta := \sum_{\alpha \in Jord_\eta(O)} \alpha$. Grâce aux travaux de Springer, pour $\eta = \pm$, on sait associer une représentation irréductible $Sp(O_\eta)$ de $\mathfrak{S}_{n_\eta}$ à l'orbite O_η. Et on pose $Sp(O, \epsilon_{ell}) := Sp(O_+) \otimes Sp(O_-)$, c'est une représentation de $\mathfrak{S}_{n_+} \times \mathfrak{S}_{n_-}$, elliptique d'ailleurs au sens de 6.3.

Définition Soit (O, ϵ_{ell}) comme ci-dessus. On note $Sp_{ell}(O, \epsilon)$ l'unique représentation de $\mathfrak{S}_{n_+} \times \mathfrak{S}_{n_-}$ avec les propriétés suivantes :

$$Sp_{ell}(O, \epsilon_{ell}) \in Sp(O, \epsilon_{ell}) + \sum_{(O', \epsilon'_{ell}); \overline{O'} \supsetneq \overline{O}} \mathbb{C} Sp(O', \epsilon'_{ell})$$

$$< Sp_{ell}(O, \epsilon_{ell}), Sp(O', \epsilon'_{ell} >_{ell} = 0, \forall (O', \epsilon'_{ell}) \text{ elliptique avec } \overline{O'} \supsetneq \overline{O},$$

la somme ci-dessus ne fait intervenir que les données elliptiques (O', ϵ'_{ell}).

On a vu en 6.3 que cela était bien défini et que l'on construisait ainsi une base pour le produit scalaire elliptique. En fait on a un peu plus concernant $Sp_{ell}(O, \epsilon_{ell})$ avec la même référence :

Remarque 6.4. Dans les notations ci-dessus, $Sp_{ell}(O, \epsilon_{ell}) - Sp(O, \epsilon_{ell}) \in \sum_{O'_+, O'_-} \mathbb{C}(Sp(O'_+) \otimes Sp(O'_-))$, où la somme porte sur les orbites unipotentes O'_+ de $GL(n_+, \mathbb{C})$ et O'_- de $GL(n_-, \mathbb{C})$ qui vérifient $\overline{O'_\eta} \supset O_\eta$ et en notant O' l'orbite de $GL(n, \mathbb{C})$ dont les blocs de Jordan sont la réunion de ceux de O'_+ et O'_-, alors $\overline{O}' \underset{\neq}{\supset} \overline{O}$.

7. Calcul de la restriction des représentations aux parahoriques

On a défini $\mathcal{F}$ en 6 et on utilise la notation k de 3.1. Soient (O, ϵ_{ell}) une orbite de $GL(n, \mathbb{C})$ et ϵ_{ell} comme dans 6.5 ; en particulier O est supposée pseudo elliptique. On reprend des notations déjà introduites dans 3.7.5,

$$\sigma_O := \prod_{\alpha \in Jord(O); \alpha \text{ impair}} (-1)^{|\{\beta > \alpha; \beta \text{ pair}\}|} (-1)^{N_i(N_i - 1)/2},$$

où $N_i := |\{\alpha \in Jord(O), \alpha \text{ impair}\}|$. On pose

$$\mathcal{F}|D|\pi(O, \epsilon) := \sum_{\epsilon'} \sigma_O \sigma^{\epsilon'} \sigma_\epsilon < \epsilon, \epsilon' > |D|\pi(O, \epsilon').$$

Théorème 7.1. *Avec les notations ci-dessus :*

$$res^{par} \mathcal{F}|D|\pi(O, \epsilon_{ell}) =_{ell} \sigma_O \big(k \circ Ind \circ Res\, Sp_{ell}(O, \epsilon_{ell})^*\big)^*,$$

*c'est-à-dire que l'égalité ci-dessus est l'égalité des caractères des représentations écrites sur les éléments elliptiques et la dernière * est la dualité pour les représentations.*

Le lecteur aura vu que le signe σ_O ne sert à rien ici mais je crois qu'il a un rôle à jouer quand on passe du cas de niveau 0 au cas de réduction unipotente, de toute façon il ne gêne pas beaucoup. Avant de faire la preuve écrivons plus explicitement $\mathcal{F}$, c'est-à-dire que la formule du théorème est (les α ci-dessous sont tous dans $Jord(O)$) :

$$res^{par} \sum_{\epsilon'_{ell}} \prod_{\alpha \equiv n+1[2]} \epsilon'_{ell}(\alpha) \prod_{\alpha \equiv 0[2]} \epsilon'_{ell}(\alpha) \prod_{\alpha; \epsilon_{ell}(\alpha) = -1} \epsilon'_{ell}(\alpha) \pi_{unip}(O, \epsilon'_{ell})$$
$$=_{ell} \sigma_O \sigma_{\epsilon_{ell}} \big(k \circ Ind \circ Res\, Sp_{ell}(O, \epsilon_{ell})^*\big)^*.$$

En fait on suit la démonstration de [MW]. On vérifie d'abord que chaque côté de l'égalité écrite se trouve dans $\mathbb{C}[\hat{\mathfrak{S}}_{D(n)}]_{\geq O,global}$ et on calcul le terme de plus bas degré ; cela résulte de 3.7.4 et 5.2. La définition de $\mathcal{F}$ donnée en 6 est exactement faite pour que ces 2 côtés aient même gradué (on utilise d'abord la remarque de 6.2). On sait aussi que $\mathbb{C}[\hat{\mathfrak{S}}_{D(n)}]_{>O,global}$ est engendré par les $res^{par}\pi_{unip}(O',\epsilon')$ pour $\overline{O'} \supset O$. Mais pour le produit scalaire elliptique on peut se limiter aux $\pi_{unip}(O',\epsilon'_{ell})$ pour (O',ϵ'_{ell}) elliptique, c'est-à-dire démontrer que chacun des 2 côtés de l'égalité du théorème a un produit scalaire elliptique nul contre les $res^{par}\pi_{unip}(O',\epsilon'_{ell})$ pour O' comme ci-dessus. C'est clair pour le terme de gauche. En admettant le résultat du théorème par récurrence descendante, on remplace $res^{par}\pi_{unip}(O',\epsilon'_{ell})$ par $\left(k \circ Ind \circ Res\, Sp_{ell}(O',\epsilon'_{ell})^*\right)^*$. La dualité est la tensorisation par un signe et est donc une isométrie (pour n'importe quoi) ; on a vu que $Ind \circ Res$ est une isométrie pour le produit scalaire elliptique (6.4). L'application k est aussi une isométrie ; cela fait partie des propriétés des faisceaux caractères. L'orthogonalité cherchée résulte alors de la construction même de Sp_{ell} donnée ci-dessus.

Corollaire *Soient O,ϵ comme ci-dessus. Alors, si n est pair (et en identifiant encore $\sharp$ à un élément de $\{\pm 1\}$) :*

$$\left(k \circ Ind \circ Res\, Sp_{ell}(O,\epsilon_{ell})^*\right)^* =_{ell} \sharp \sigma_O \sigma_{\epsilon_{ell}} res^{par} \sum_{\epsilon'_{ell}} \prod_{\alpha;\epsilon_{ell}(\alpha)=-1} \epsilon'_{ell}(\alpha)\pi(O,\epsilon'_{ell})$$

tandis que si n est impair :

$$\left(k \circ Ind \circ Res\, Sp_{ell}(O,\epsilon_{ell})^*\right)^* =_{ell} \sigma_O \sigma_{\epsilon_{ell}} res^{par} \sum_{\epsilon'_{ell}} \prod_{\alpha;\epsilon_{ell}(\alpha)=-1} \epsilon'_{ell}(\alpha)\pi(O,\epsilon'_{ell}).$$

Cela résulte des formules écrites ci-dessus.

8. Stabilité pour les représentations de réduction unipotente

8.1. Enoncé et commentaires Si l'on a fait tout ce travail c'est pour obtenir le théorème de 7 parce qu'il permet de terminer modulo un travail en cours d'Aubert, Kutzko et Morris, la description des paquets stables "de niveau 0" pour les groupes orthogonaux impairs. Toutefois on peut obtenir comme corollaire de ce résultat la description des paquets stables pour les représentations tempérées des groupes unitaires de réduction unipotente (ce sont les représentations que l'on a étudiées ici). A partir de maintenant on suppose que p est grand pour pouvoir utiliser **[MW]** 3.4. Dans le théorème ci-dessous, on utilise la notation suivante : pour O une orbite de $GL(n,\mathbb{C})$ et pour $\epsilon : Jord(O) \to \{\pm 1\}$, on note $\epsilon_O := \prod_{\alpha \in Jord(O)} \epsilon(\alpha)^{mult_\alpha(Jord(O))}$ et on dit que $\epsilon_O = \sharp$ si $\epsilon_O = +$ si $\sharp = iso$ et $\epsilon_O = -$ si $\sharp = -$. En outre on se limite aux représentations tempérées que l'on a appelées versus elliptique en 2.3. Ce cas contient le cas des représentations elliptiques et ce qui manque est donc obtenu par induction irréductible. Pour $\sharp = iso$ ou an et pour (O, ϵ) un couple formé d'une orbite unipotente de $GL(n,\mathbb{C})$ et d'une application ϵ de $Jord(O)$ dans $\{\pm 1\}$, on pose : $\Pi(\sharp, O) := \sum_{\epsilon : Jord(O) \to \{\pm 1\}; \epsilon_O = \sharp} \pi(O,\epsilon)$. Comme O détermine un paramètre ψ (cf 2.2), cette notation est similaire à celle de l'introduction (mais plus commode).

Théorème 8.1. *(p est supposé grand) Soit $\sharp = iso$ ou an et $G_\sharp := U(n,F)_\sharp$.*

(i) Soit O une orbite unipotente de $GL(n,\mathbb{C})$ discrète (c'est-à-dire que $Jord(O)$ est sans multiplicité); la distribution $\Pi(\sharp, O)$ est une distribution stable pour $G_\sharp$.

(ii) L'ensemble des combinaisons linéaires stables de représentations elliptiques pour $G_\sharp$ sont exactement les $\Pi(\sharp, O)$, quand O varie parmi les orbites discrètes de $GL(n,\mathbb{C})$.

(iii)Dans le relèvement de G_{an} vers G_{iso}, si n est pair, $\Pi_{iso,O}$ se tranfert en $-\Pi_{an,O}$. Si n est impair $\Pi_{iso,O}$ et $\Pi_{an,O}$ coïncident.

Avant de faire la preuve, faisons quelques remarques. En (ii) on n'a pas traité le cas tempéré général; c'est juste une question d'écriture (cf 2.4) car l'orbite ne détermine plus le paramètre.

La preuve de ce théorème suit **[MW]** avec des simplifications notables dûes à l'absence de faisceaux caractères cuspidaux (et au fait qu'un certain nombre de résultats de **[MW]** s'appliquent tels quels). Elle fait l'objet des paragraphes ci-dessous.

Rappelons les considérations générales; les travaux d'Arthur et tout particulièrement **[A]** ramènent l'étude des distributions tempérées stables à l'étude des distributions elliptiques stables. Et pour celles-là, il montre qu'il suffit de montrer l'invariance sur les classes de conjugaison stable pour les éléments elliptiques. L'utilisation des faisceaux caractères telle qu'elle a été faite dès **[W1]** est la clé de la preuve; c'est 7 qui permet de s'y ramener ici; on va montrer que les intégrales du terme de droite de ce théorème sont invariantes sur les classes de conjugaison stable si et seulement si ϵ_ℓ est trivial et $Jord(O)$ est sans multiplicité et ensuite on interprétera alors facilement le terme de gauche comme (à un signe près) la somme sur les séries discrètes du paquet associé à ψ.

On fixe donc g un élément elliptique et (O, ϵ) un couple formé d'une orbite unipotente de $Gl(n, \mathbb{C})$ et d'une application de $Jord(O)$ dans $\{\pm 1\}$. On fixe donc $(d', d'') \in D(n)$, des partitions ν', ν'' de d' et d'' respectivement (à un signe près) cela donne un élément de $\mathbb{C}[\hat{\mathfrak{S}}_{D(n)}]$. On cherche à calculer l'intégrale orbitale en g de la fonction $k\, Ind \circ Res\, \nu' \otimes \nu''$; on a ici identifié partitions et représentations, le signe n'a pas d'importance (puisqu'il est indépendant de g). Elle est clairement nulle si g n'est pas compact, ce que nous supposerons donc et l'on écrit $g = g_s g_u$ où g_u est un élément topologiquement unipotent commutant à g_s et où g_s a toutes ses valeurs propres des racines de l'unité d'ordre premier à p.

8.2. Localisation auprès de g_s Le but est de ramener le calcul de l'intégrale d'un faisceau caractère sur l'orbite de g en une intégrale dans le commutant de g_s sur l'orbite de g_u. Il faut d'abord avoir une description du commutant de g_s ; on considère les classes d'équivalence de valeurs propres de g_s où 2 valeurs propres λ et λ' sont équivalentes s'il existe un entier a tel que $\lambda^{(-q)^a} = \lambda'$. On note $[VP(g_s)]$ ces classes d'équivalence et on note, pour $[\lambda] \in [VP(g_s)]$, $m[\lambda]$ la multiplicité de l'une des valeurs propres de g_s dans $[\lambda]$ en tant que valeur propre (ce nombre est bien défini) et on note $q_{[\lambda]} := q^{|[\lambda]|}$. Le commutant de g_s est alors isomorphe à un produit de groupes unitaires $\times_{[\lambda] \in [VP(g_s)]} U(m([\lambda]), \mathbb{F}_{q_{[\lambda]}})_{\sharp(g_s)_{[\lambda]}}$ où $\sharp(g_s)_{[\lambda]}$ indique de quelle forme unitaire il s'agit ; cette forme ne dépend que de la classe de conjugaison de g_s. Quand g_s varie dans sa classe de conjugaison stable, les $\sharp(g_s)_{[\lambda]}$ varient soumis seulement à la condition $\prod_{[\lambda] \in [VP(g_s)]} \sharp(g_s)_{[\lambda]} = \sharp$ (avec la convention identifiant un $\sharp$... avec un signe). On pose $\mathbb{C}[\hat{\mathfrak{S}}_{D(g_s)}] = \otimes_{[\lambda] \in [VP(g_s)]} \mathbb{C}[\hat{\mathfrak{S}}_{D(m([\lambda]))}]$. On définit une application $loc_{g_s} : \mathbb{C}[\hat{\mathfrak{S}}_{D(n)}]$ dans $\mathbb{C}[\hat{\mathfrak{S}}_{D(g_s)}]$, en posant pour $(d', d'') \in D(n)$ et pour ρ', ρ'' des représentations de $\mathfrak{S}_{d'}$ et $\mathfrak{S}_{d''}$ respectivement :

$$loc_{g_s}(\rho' \otimes \rho'') := \sum_{d^j_{[\lambda]};[\lambda] \in [VP(g_s)], j \in \{', ''\}} Res_{\times_{[\lambda],j} \mathfrak{S}_{d^j_{[\lambda]}}} \rho' \otimes \rho'',$$

où les $d^j_{[\lambda]}$ parcourt les collections d'entiers satisfaisant à

$$\sum_{[\lambda]} d'_{[\lambda]} = d', \sum_{[\lambda]} d''_{[\lambda]} = d''$$

et où les inclusions de $\mathfrak{S}_{d'_{[\lambda]}}$ dans $\mathfrak{S}_{d'}$ sont les inclusions naturelles. Les inclusions de $\mathfrak{S}_{d''_{[\lambda]}}$ dans $\mathfrak{S}_{d''}$ sont en fait l'inclusion naturelle de la classe $\times_{[\lambda]} \mathfrak{S}_{d''_{[\lambda]}} F$ dans $\mathfrak{S}_{d''} F$ (où F est le Frobenius).

Proposition 8.2. *Avec les notations qui précèdent, pour $(d', d'') \in D(n)$ et ν', ν'' des partitions de d' et d'' respectivement, on a l'égalité des intégrales* *orbitales (pour de bonnes mesures) :*

$$I_g^G(k\, Ind \circ Res \nu' \otimes \nu'') = I_{g_u}^{Cent_G(g_s)}(Q\, Res \circ Ind\, loc_{g_s}(\nu' \otimes \nu'')),$$

où, suivant la tradition on note Q les fonctions de Green.

C'est l'analogue de la propostion de **[MW]** 3.17, en beaucoup plus simple puisqu'il n'y a pas de partie cuspidale et que le centralisateur est connexe. Mais à ces simplifications près, la démonstration s'applique telle quelle. Avouons que l'on ne s'est pas fatiguée à recalculer les constantes, ici elles ne viennent que de volumes, elles ne jouent aucun rôle.

8.3. Fin de la preuve Comme il n'y a pas de signe venant de partie cuspidale, la démonstration est une conséquence directe de **[MW]** 3.4. Le résultat en question est le suivant : on appelle fonction stable sur un groupe unitaire $U(m, F)$ une fonction dont les intégrales sur les classes de conjugaison ordinaire ne dépendent que de la classe de conjugaison stable ; on dit que la fonction est semi-stable si elle ne dépend que de la classe de conjugaison stable et de la forme unitaire (c'est-à-dire de $\sharp$) et que la dépendance en $\sharp$ se fait précisément par le signe $\sharp$ avec la convention du début de ce papier et on dit que la fonction est instable si la somme des intégrales sur les classes de conjugaison ordinaire à l'intérieur d'une classe de conjugaison stable, pour $\sharp$ fixé, est nulle. Dans loc. cite il est montré que les fonctions du type $Q\, Ind \circ Res$ sont stables exactement quand $d'' = 0$, semi-stables quand $d' = 0$ et $d'' \neq 0$ et instable dans les autres cas. On peut appliquer tel quel en tenant compte de la description du centralisateur de g_s ; la notation ci-dessus d', d'' est à changer car dans le calcul de loc_{g_s} il est à remplacer par $d'_{[\lambda]}, d''_{[\lambda]}$; on réalise ici loc_{g_s} comme une somme et on regarde chaque terme de la somme séparément ; un tel terme est donc indexé par une collection de couples d'entiers $\{(d'_{[\lambda]}, d''_{[\lambda]}); [\lambda] \in [VP(g_s)]\}$. Le terme correspond est certainement instable si pour une valeur de $[\lambda]$ au moins $d'_{[\lambda]} d''_{[\lambda]} \neq 0$. Dans le cas où ce produit est nul pour tout $[\lambda]$, l'intégrale ne dépend que de sa classe de conjugaison stable et du produit $\times_{[\lambda]; d''_{[\lambda]} \neq 0} \sharp_{[\lambda]}(g_s)$. Quand on fait varier g_s dans sa classe de conjugaison stable le signe varie sauf si $d''_{[\lambda]} = 0$ pour tout $[\lambda]$. Dans ce dernier cas l'intégrale ne dépend donc que de la classe de conjugaison stable de g_s ; le signe ne dépend que de $\sharp$ (la forme du groupe auquel appartient le g_s fixé) si pour tout $[\lambda]$, $d''_{[\lambda]} \neq 0$ et dans le cas restant où il existe $[\lambda] \neq [\lambda]'$ tel que $d'_{[\lambda]}, d''_{[\lambda]} \neq 0$, la somme des intégrales sur la classe de conjugaison stable de g_s à l'intérieur de $U(n, G)_\sharp$ est nulle, i.e. instabilité. On se rappelle que

$$d' = \sum_{[\lambda]} d'_{[\lambda]}, d'' = \sum_{[\lambda]} d''_{[\lambda]}.$$

Ainsi on a stabilité si $d'' = 0$, semi-stabilité si $d' = 0, d'' \neq 0$ et instabilité si $d'd'' \neq 0$.

Il ne reste plus qu'à appliquer le résultat précédent à $Sp_{ell}(O, \epsilon_{ell})$ avec les notations de 7 ; fixons donc (O, ϵ_{ell}) une donnée elliptique c'est-à-dire une orbite unipotente O de $GL(n, \mathbb{C})$ telle que $Jord(O)$ a au plus de la multiplicité 2 et où ϵ_{ell} est une application de $Jord(O)$ dans $\{\pm 1\}$ qui vaut -1 sur tous les blocs de Jordan intervenant avec multiplicité 2. On rappelle que Sp_{ell} n'est pas une représentation irréductible mais l'analogue de d', d'' ci-dessus est bien défini et vaut

$$d' = \sum_{\substack{\alpha \in Jord(O) \\ mult_\alpha(Jord(O)) = 2}} \alpha + \sum_{\substack{\alpha \in Jord(O) \\ \epsilon(\alpha) = 1}} \alpha,$$

$$d'' = \sum_{\substack{\alpha \in Jord(O) \\ mult_\alpha(Jord(O)) = 2}} \alpha + \sum_{\substack{\alpha \in Jord(O) \\ \epsilon(\alpha) = -1}} \alpha.$$

On voit (comme on s'y attend) qu'il ne peut y avoir stabilité ou semi-stabilité pour $k\, Ind \circ Res Sp_{ell}(O, \epsilon_{ell})$ que si $Jord(O)$ est sans multiplicité, c'est le cas discret. De plus, dans ce cas, on a stabilité si pour tout α, $\epsilon(\alpha) = 1$, semi-stabilité si pour tout α, $\epsilon(\alpha) = -1$ et instabilité sinon.

Reprenons la notation $\Pi(\sharp, O)$ introduite en 8.1 ; ici, on ne s'intéresse qu'aux orbites discrètes, c'est-à-dire telles que $Jord(O)$ n'a que de la multiplicité 1. La définition de $\Pi(\sharp, O)$ est donc simplement

$$\prod_{\alpha \in Jord(O); \alpha \text{ impair}} \epsilon'(\alpha) \pi(O, \epsilon').$$

On vérifie que si $\epsilon \equiv 1$, alors

$$\mathcal{F}\pi(O, \epsilon) = \prod_{\alpha \in Jord(O); \alpha \equiv n+1} \epsilon'(\alpha) \pi(O, \epsilon').$$

La définition de $\mathcal{F}$ mélange les 2 groupes $G_\sharp$ (même si ils sont isomorphes). En séparant dans la somme, les termes pour lesquels $\prod_\alpha \epsilon(\alpha) = 1$ de ceux où cela vaut -1, on récrit

$$\prod_{\alpha \in Jord(O); \alpha \equiv n+1} \epsilon'(\alpha) = (\sharp)^{n+1} \prod_{\alpha \in Jord(O); \alpha \text{ pair}} \epsilon'(\alpha)$$

et on trouve donc que pour $\epsilon \equiv 1$, $\mathcal{F}\pi(O, \epsilon) = \Pi(iso, O) + (-1)^{n+1} \Pi(an, O)$. Un calcul analogue montre que pour $\epsilon \equiv -1$, et avec σ_O un signe qui ne dépend que de O, $\mathcal{F}\pi(O, \epsilon) = \sigma_O(\Pi(iso, O) + (-1)^n \Pi(an, O))$.

On obtient un supplémentaire de ces combinaisons linéaires (pour le produit scalaire elliptique) en considérant les autres possibilités pour ϵ. Cela prouve que les $\Pi(\sharp, O)$ sont les combinaisons linéraires stables sur les elliptiques pour $\sharp$ fixé. Le transfert se calcule en utilisant le fait que l'on a montré que $\Pi(iso, O) + (-1)^{n+1}\Pi(an, O)$ est stable.

Pour obtenir un résultat général de stabilité, on utilise les résultats généraux qui assure que les distributions tempérées stables sont les induites des distributions elliptiques stables que l'on vient de décrire. Cela termine la preuve du théorème de 8.1.

8.4. Le cas des représentations (quadratique)-unipotentes Dans cette section, on note D l'involution de [**Au**] et [**SS**], c'est-à-dire que pour π une représentation irréductible d'un groupe réductif, on pose

$$D(\pi) := \sum_P Ind_P^G res_P \pi,$$

où P parcourt un ensemble de paraboliques standard. Il est montré en [**Au**], que $D(\pi)$ est à un signe près une représentation irréductible, ce signe étant $(-1)^{rg_G - rg_{P_{cusp,\pi}}}$, où $P_{cusp,\pi}$ est "le" sous-groupe parabolique de G, minimal avec la propriété que $res_P \pi$ est non nul. Dans le cas que l'on traite ici, le signe a été calculé en 5.2.

On fixe ψ un paramètre discret, d'où une orbite O de $GL(n, \mathbb{C})$; pour tout $\epsilon : Jord(O) \to \{\pm 1\}$, on a défini $\pi(O, \epsilon)$ et aussi $\pi_{unip}(O, \epsilon)$. Par définition $\pi(O, \epsilon) = |D|\pi_{unip}(O, \epsilon)$.

Théorème 8.3. *On fixe ψ un paramètre discret.*

(i) pour $\sharp = iso$ ou an, la combinaison linéaire de représentations

$$\Pi_{unip}(\sharp, \psi) := \sum_{\epsilon; \epsilon_Z = \sharp} \prod_{\alpha \in Jord(O); \alpha \equiv n+1[2]} \epsilon(\alpha)\, \pi_{unip}(O, \epsilon)$$

est stable.

(ii)dans le transfert de $U(n, E/F)_{iso}$ à $U(n, E/F)_{an}$, les distribution $\Pi_{unip}(iso, \psi)$ et $\Pi_{unip}(an, \psi)$ se correspondent.

Puisque induction et restriction respectent la stabilité, il suffit d'appliquer D au théorème de 8.1. L'astuce ici est d'étendre la propriété de stabilité que l'on connait sur les elliptiques à tous les éléments semi-simples en utilisant les travaux d'Arthur sur les représentations elliptiques (ce qui ne peut se faire avec les représentations unipotentes). En tenant compte de 5.2, on obtient exactement que la distribution

$$\Pi'_{unip}(\sharp, unip) := \sum_{\epsilon} \prod_{\alpha \in Jord(O), \alpha \equiv 0[2]} \pi_{unip}(O, \epsilon)$$

est stable et quand $\sharp$ varie, les distributions se transfèrent l'une sur l'autre avec un signe $-$ dans le cas n pair (et pas de signe dans le cas n impair). Dans le cas n impair, on a directement $\Pi'_{unip}(\sharp, \psi) = \Pi_{unip}(\sharp, \psi)$, tandis que dans le cas de n pair, $\Pi'_{unip}(\sharp, \psi) = \sharp\Pi_{unip}(\sharp, \psi)$. D'où le résultat. Il y a de bonnes raisons pour avoir choisi de définir $\Pi_{unip}(\sharp, \psi)$ plutôt que $\Pi'_{unip}(\sharp, \psi)$ (cf. ci-dessous).

8.5. Interprétation du signe On a défini l'inclusion de $GL(n, \mathbb{C}) \rtimes W_F/W_E$ dans $GL(2n, \mathbb{C})$ en 2.1. On vérifie sur les formules que l'image de cette inclusion est incluse dans l'ensemble des matrices inversibles X, vérifiant ${}^tXJX = J$ où J est la matrice $\begin{pmatrix} 0 & I_n \\ (-1)^{n-1}I_n & 0 \end{pmatrix}$. Ainsi la représentation définie est symplectique si n est impair et orthogonale si n est pair.

Soit ψ, ϵ un paramètre discret de niveau zéro. En composant $inc \circ \psi$, on obtient une représentation de $W_F \times SL(2, \mathbb{C})$ dans $GL(2n, \mathbb{C})$. Pour $a \in \mathbb{N}$, on note $inc\psi_a$ la restriction de $inc \circ \psi$ à la composante isotypique pour la représentation de dimension a de $SL(2, \mathbb{C})$. On pose $inc\psi_{ortho} := \times_{a \equiv n[2]} inc\psi_a$ et $inc\psi_{symp} := \times_{a \equiv n+1[2]} inc\psi_a$. La restriction de $inc\psi_{ortho}$ à W_F est une représentation orthogonale tandis que la restriction de $inc\psi_{symp}$ à W_F est symplectique ; cela vient du fait qu'une représentation de $SL(2, \mathbb{C})$ irréductible est orthogonale si sa dimension est impaire et symplectique si sa dimension est paire et aussi des propriétés écrites ci-dessus de la représentation définie par *inc*. Il est clair qu'il existe des morphismes ψ_{ortho} (resp. ψ_{symp}) de $W_F \times SL(2, \mathbb{C})$ dans un sous-groupe convenable, $GL(n_{ortho}, \mathbb{C}) \rtimes W_F$ (resp. $GL(n_{symp}, \mathbb{C}) \rtimes W_F$) de $GL(n, \mathbb{C}) \rtimes W_F$, tels que $inc\psi_{ortho}$ (resp. $inc\psi_{symp}$) soit le composé de ψ_{ortho} (resp. ψ_{symp}) avec *inc*. On retrouve ainsi les notations de l'introduction. On note, s_{symp} l'élément non trivial du centre de $GL(n_{symp}, \mathbb{C}) \rtimes W_F$. Il est clairement dans le commutant de ψ et on peut donc évaluer $\epsilon(s_{symp})$. Il est clair que $\epsilon(s_{symp}) = \prod_{\alpha \equiv n+1[2]} \epsilon(\alpha)$. D'où (en écrivant $\pi_{unip}(\psi, \epsilon)$ plutôt que $\pi_{unip}(O, \epsilon)$) :

Remarque 8.4. On fixe ψ, ϵ un paramètre discret de niveau zéro. La seule combinaison linéaire stable combinaison linéaire de représentation $\pi_{unip}(\psi, \epsilon)$ pour ψ fixé discret et ϵ variable est : $\sum_\epsilon \epsilon(s_{symp})\pi_{unip}(\psi, \epsilon)$.

Ceci est conforme aux conjectures d'Arthur.

Références

[A] ARTHUR J. : *On local character relations*, Selecta Math. 2, 1996, pp. 501-579

[Au] AUBERT A.-M. : *Dualité dans le groupe de Grothendieck de la catégorie des représentations lisses de longueur finie d'un groupe réductif p-adique*, TAMS, 347, 1995, pp. 2179-2189 avec l'erratum publié dans TAMS, 348, 1996, pp. 4687-4690

[AMR] AUBERT A.-M., MICHEL J., ROUQUIER R. : *Correspondance de Howe pour les groupes réductifs sur les corps finis*, Duke, 83, 1996, pp. 353-397

[DM] DIGNE F., MICHEL J. : *Foncteurs de Lusztig et caractères des groupes linéaires et unitaires sur un corps fini*, J. of Algebra, 107, 1987, pp. 217-255

[L1] LUSZTIG G. : *Intersection cohomology complexes on a reductive group*, Inventiones 75, 1984, pp. 205-272

[L2] LUSZTIG G. : *Green functions and character sheaves*, Ann of Math 131, 1990, pp. 355-408

[L3] LUSZTIG G. : Classification of Unipotent Representations of Simple p-adic Groups, II, arXiv :math. RT/0111248, 2001

[L4] LUSZTIG G. : *Character scheaves II* Adv in Math 57, 1985, 226-265

[M] MŒGLIN C. : *Stabilité en niveau 0 pour les groupes orthogonaux impairs* prépublication, 2003.

[MW] MŒGLIN C., WALDSPURGER J.-L. : *Paquets stables de représentations tempérées et de réduction unipotente pour $SO(2n+1)$* à paraître à Inv.

[MP] MOY A., PRASAD G. : *Jacquet functors and unrefined minimal K-types*, Comment. Math. Helvetici 71, 1996, pp. 98-121

[SS] SCHNEIDER M., STUHLER U. : *Representation theory and sheaves on the Bruhat-Tits building* Publ. Math. IHES 85, 1997, pp. 97-191

[W1] WALDSPURGER J.-L. : *Intégrales orbitales unipotentes et endoscopie pour les groupes classiques non ramifiés* Astérisque 269, 2001

[W2] WALDSPURGER J.-L. : *Représentations de réduction unipotente pour $SO(2n+1)$: quelques conséquences d'un article de Lusztig* prépublication Inst. Math. Jussieu 298, 2001, à paraître dans recueil en l'honneur de Shalika.

[W3] WALDSPURGER J.-L. : *Une conjecture de Lusztig pour les groupes classiques* prépublication, 2002

Colette Mœglin
Institut de Mathématiques de Jussieu, CNRS
175 rue du Chevaleret, 75013 Paris
moeglin@math.jussieu.fr

Global method in relative trace identities

Zhengyu Mao and Stephen Rallis

Abstract. We explain a way to establish the relative trace identity through comparison of global distributions, instead of through comparing local orbital integrals.

1991 Mathematics Subject Classification: primary 11F70

1. Introduction

An important method in number theory and Langlands program is the theory of the trace formula. Let F be a number field, $\mathbf{A}$ its adele ring. We use v to denote a place of F. Let G be a reductive group. In studying the trace formula, one considers a distribution of the following type: for $f \in S(G(\mathbf{A}))$ (the space of Schwartz functions on $G(\mathbf{A})$), let

$$I_G(f : H_1, \chi_1, H_2, \chi_2)$$

$$= \int_{H_1(F)\backslash H_1(\mathbf{A})} \int_{H_2(F)\backslash H_2(\mathbf{A})} K_f(h_1, h_2)\chi_1(h_1)\chi_2(h_2)dh_1 dh_2. \qquad (1)$$

Here H_1, H_2 are two closed subgroups of G, χ_i $(i = 1, 2)$ is a character (or an automorphic form) on $H_i(\mathbf{A})$, and $K_f(x, y)$ is the kernel function for the representation $\rho(f)$ acting on $L^2(G(F)\backslash G(\mathbf{A}))$; more explicitly

$$K_f(x, y) = \sum_{\gamma \in G(F)} f(x^{-1}\gamma y). \qquad (2)$$

The distribution I_G is a relative trace formula; the usual trace formula can be shown as a special case of this distribution.

Let G and G' be two reductive groups. Assume there is a homomorphism between the L-groups of G and G'. A *relative trace identity* is a relation between the two distributions (1) on G and G'. Explicitly, we say there is relative trace identity

$$I_G(f : H_1, \chi_1, H_2, \chi_2) = I_{G'}(f' : H_1', \chi_1', H_2', \chi_2') \qquad (3)$$

if the following is true:

(a) There exists maps ϵ_v between sufficiently large subsets of $\mathcal{S}(G_v)$ and $\mathcal{S}(G'_v)$ for all places v of F,

(b) There is a finite set S_0 of bad places, such that for any S a finite set of places containing S_0, we have for any $f = \otimes_{v\in S} f_v \otimes_{v\notin S} f_v$ with $f_v \in \mathcal{S}(G_v)$ when $v \in S$ and f_v a Hecke function when $v \notin S$, the equation (3) holds for $f' = \otimes_{v\in S}\epsilon_v(f_v) \otimes_{v\notin S} \lambda_v(f_v)$. Here λ_v is the local Hecke algebra homomorphism between G_v and G'_v given by the Satake isomorphism and the homomorphism between the $L-$groups of G and G'.

The readers interested in more background on relative trace formula should consult Jacquet's contribution to this volume. For some applications of the relative trace identities in number theory and Langlands correspondence, see **[BM]**,**[CJ]**,**[Gu]** and **[J]**.

In this paper we discuss a new method of proving a relative trace identity. We illustrate this method by looking at an example. We consider the case $G = SO(n+1,n)$ the split special orthogonal group, and $G' = \tilde{Sp}_n$ the double cover of the symplectic group Sp_n. In §2, we define two distributions $I_G(f : H_1, \chi_1, H_2, \chi_2)$ and $I_{G'}(f' : H'_1, \chi'_1, H'_2, \chi'_2)$ on the groups. We will prove the relative trace identity between these two distributions using a global method. This identity generalizes an identity considered in **[J]**.

Let us first recall the usual way of proving a trace identity, which is by comparing orbital integrals. Assume $f = \otimes f_v$, one decomposes the distribution $I_G(f : H_1, \chi_1, H_2, \chi_2)$ into sums of orbital integrals:

$$\sum_{\emptyset\in\mathcal{O}} I_G(f, \emptyset : H_1, \chi_1, H_2, \chi_2) = \sum_{\emptyset\in\mathcal{O}} c(\emptyset)\prod_v I_{G,v}(f_v, \emptyset : H_1, \chi_1, H_2, \chi_2).$$

Here $\mathcal{O}$ is the set of representatives of orbits, a subset of $G(F)$; $c(\emptyset)$ a positive coefficient (equals some volume), and $I_{G,v}(f_v, \emptyset : H_1, \chi_1, H_2, \chi_2)$ is the local orbital integral which takes the form:

$$\int_{H_{2,v}\cap\emptyset^{-1}H_{1,v}\emptyset\backslash H_{2,v}} \int_{H_{1,v}} f_v(h_1^{-1}\emptyset h_2)\chi_1(h_1)\chi_2(h_2)dh_1dh_2.$$

Similar decomposition holds for $I_{G'}(f' : H'_1, \chi'_1, H'_2, \chi'_2)$:

$$I_{G'}(f' : H'_1, \chi'_1, H'_2, \chi'_2) = \sum_{\emptyset'\in\mathcal{O}'} c(\emptyset')\prod_v I_{G',v}(f'_v, \emptyset' : H'_1, \chi'_1, H'_2, \chi'_2).$$

To prove the identity (3), one shows:
1) there is a bijection ι between the set of orbits $\mathcal{O}$ and $\mathcal{O}'$;
2) there is a map ϵ_v such that for $f'_v = \epsilon_v(f_v)$, we have an identity of orbital integrals

$$I_{G,v}(f_v, \emptyset : H_1, \chi_1, H_2, \chi_2) = I_{G',v}(f'_v, \iota(\emptyset) : H'_1, \chi'_1, H'_2, \chi'_2)\Delta_v(\emptyset), \quad (4)$$

where $\Delta_v(\emptyset)$ is some transfer factor independent of f_v, satisfying $\prod_v \Delta_v(\emptyset) = c(\emptyset)/c(\iota(\emptyset))$;

3) we have the fundamental lemma, i.e. the identity (4) holds for $f'_v = \lambda(f_v)$ where $v \notin S_0$ and f_v is a Hecke function.

It is easy to see that the above three facts together with the orbital integral decompositions imply the relative trace identity (3). This method is developed in many papers, for example [**FuSh**],[**J**],[**MR1**] and [**MR2**].

The global method involves an introduction of a new distribution, on the group $G \times G'$, which we denote $I_{G\times G'}(\Phi : H_3, \chi_3)$. We establish separately the relation between $I_G(f : H_1, \chi_1, H_2, \chi_2)$ and $I_{G'}(f' : H'_1, \chi'_1, H'_2, \chi'_2)$ with $I_{G\times G'}(\Phi : H_3, \chi_3)$. We put together the two relations using a global identity for $I_{G\times G'}(\Phi : H_3, \chi_3)$.

This paper is part of ours series of works on connecting the dual pair method and relative trace formula. After some early exploration of examples like in [**MR2**], we arrive at a general principle which predicts the relative trace identities for dual pairs G and G'. The principle is stated in [**MR1**] where we conjecture that there is a relative trace identity (3) if over the local places there are G'-module isomorphisms

$$\omega_\psi[H_2, \chi_2^{-1}] \cong Ind_{H'_1}^{G'} \chi'_1 \tag{5}$$

and $G-$module isomorphisms

$$\omega_\psi[H'_2, \chi_2^{'-1}] \cong Ind_{H_1}^{G} \chi_1. \tag{6}$$

Here we use $\omega_\psi[H, \chi]$ to denote the Jacquet module of ω_ψ with respect to the group H and character χ. The motivation behind the conjecture is that we can establish a vector space isomorphism

$$Ind_{H_1}^{G} \chi_1[H_2, \chi_2^{-1}] \cong Ind_{H'_1}^{G'} \chi'_1[H'_2, \chi_2^{'-1}] \tag{7}$$

by applying (5) and (6) to the space $\omega_\psi[H_2 \times H'_2, \chi_2^{-1} \times \chi_2^{'-1}]$. The relative trace identity can be considered as a global analogue of the isomorphism (7).

The global method presented here seems to fit the principle better than the local method used in [**MR1**]. The idea is the introduction of a distribution which can be considered as the global analogue of $\omega_\psi[H_2 \times H'_2, \chi_2^{-1} \times \chi_2^{'-1}]$. The key to the proof is the analogues of the isomorphisms (5) and (6): they are the easily established Propositions 4.1 and 5.1.

We note an interesting feature in the global method, namely it does not require the existence of a map ι matching the orbits $\{\emptyset\}$ and $\{\emptyset'\}$. Though in the example we are considering there is a match of orbits ([**MR1**]), we have found examples where the trace identity holds while there is no match of orbits, (and thus the method of comparing orbital integrals does not apply). We describe such an example.

Consider the case where $G = G' = Sp_2$ be the symplectic group. Let $H_1 = H_1' = SL_2 \times SL_2$ and $\chi_1 = \chi_1'$ be the trivial character of $SL_2 \times SL_2$. Let H_2 be the maximal unipotent subgroup of Sp_2

$$H_2 = \{n(x,y,z,v) = \begin{pmatrix} 1 & x & 0 & 0 \\ & 1 & 0 & 0 \\ & & 1 & -x \\ & & & 1 \end{pmatrix} \begin{pmatrix} 1 & 0 & y & z \\ & 1 & v & -y \\ & & 1 & 0 \\ & & & 1 \end{pmatrix} \},$$

and H_2' be the subgroup of H_2 consisting of the $n(0,y,z,v)$. Let ψ be a nontrivial additive character of $\mathbf{A}/F$; let χ_2' be the character of H_2' with $\chi_2'(n(0,y,z,v)) = \psi(v+z)$; let χ_2 be an automorphic form on H_2:

$$\chi_2(n(x,y,z,v)) = \psi(v)\Theta_\Phi^\psi(n).$$

Here Φ is a Schwartz function of $\mathbf{A}$ and

$$\Theta_\Phi^\psi(n(x,y,z,v)) = \sum_{w \in F} \psi(w^2 v + wy + z)\Phi(w+x).$$

With these data, the relative trace identity (3) holds. One can check the relevant orbits do not match in this case. However following the global work of [**GRS**], it is not difficult to see the identity holds when

$$f(g) = \int_{\mathbf{A}} \Phi(w) f'(gn(w,0,0,0))dw.$$

The paper is organized as follows. In Section 2, we define the distributions on G, G' and state the main result. In Section 3 we introduce a third distribution on $G \times G'$ and prove a global identity. In Section 4, we compare the distributions on G and $G \times G'$. In Section 5, we compare the distributions on G' and $G \times G'$. The proof of the trace identity is given in Section 6.

2. Statement of the result

Let X be a $2n+1$ dimensional space; let $e_1, \ldots, e_{2n+1}$ be the standard basis of $X \cong F^{2n+1}$, Let $SO(n+1,n)$ be the special orthogonal group fixing the symmetric bilinear form $<,>$ given by $< e_i, e_j >= 2$ when i+j=2n+2, $i \neq j$, $< e_{n+1}, e_{n+1} >= 1$, and $< e_i, e_j >= 0$ otherwise. Let V be the subspace of F^{2n+1} spanned by $\{e_1, \ldots, e_{n-1}, e_{n+1}\}$. We define some subgroups of $SO(n+1,n)$.

Let R' be the subgroup of $SO(n+1,n)$ fixing V:

$$R' = \{r \in SO(n+1,n) | r(e_1, \ldots, e_{n-1}, e_{n+1}) = (e_1, \ldots, e_{n-1}, e_{n+1})\}. \quad (8)$$

Let N denote the subgroup of upper triangular matrices with unit diagonal in GL_n which acts on the n-dimensional space V. Define R the subgroup of

$SO(n+1,n)$:

$$R = \{r \in SO(n+1,n) | \exists n \in N, r\mathbf{v} = n\mathbf{v}, \forall \mathbf{v} \in V\}. \tag{9}$$

This is the Bessel group of $SO(n+1,n)$. The group R' is a normal subgroup of R with an isomorphism $\rho : R/R' \cong N$ given by $r \mapsto n$ in equation (9).

Let U be the maximal unipotent subgroup of $SO(n+1,n)$ consisting of the upper triangular matrices with unit diagonal.

Fix ψ a nontrivial additive character of $\mathbf{A}/F$. We define the characters θ on $N(\mathbf{A})$, χ on $R(\mathbf{A})$ and μ on $U(\mathbf{A})$ as follows:

$$\theta(n) = \psi(n_{1,2} + \ldots + n_{n-1,n}), n \in N; \tag{10}$$

$$\chi(r) = \psi^{-1}(r_{1,2} + r_{2,3} + \ldots + r_{n-2,n-1} + r_{n-1,n+1}), r \in R; \tag{11}$$

$$\mu(u) = \psi(u_{1,2} + u_{2,3} + \ldots + u_{n-1,n} + u_{n,n+1}), u \in U. \tag{12}$$

We define the distribution on G as:

$$I_G(f : R, \chi, U, \mu) = \int_{R(F)\backslash R(\mathbf{A})} \int_{U(F)\backslash U(\mathbf{A})} K_f(r,u)\chi(r)\mu(u)drdu. \tag{13}$$

Let Y be a $2n$ dimensional space with a symplectic structure $< *, * >'$, where explicitly with the standard basis $f_1, \ldots, f_{2n}$ of $Y \cong F^{2n}$, $< f_i, f_j >' = 1$ when $i + j = 2n + 1$ and $i \leq n$ and 0 when $i + j \neq 2n + 1$. Let Sp_n be the corresponding symplectic group. Let N' be the group of upper triangular matrices in Sp_n with unit diagonal. Denote an element in the double cover G' of Sp_n by (g, ϵ) with $g \in Sp_n$ and $\epsilon = \pm 1$. If H is a subgroup of $Sp_n(\mathbf{A})$ where the covering splits, we will consider $h \in H$ as an element in G' through the splitting. In particular, the group N' can be considered as a subgroup of G' via the embedding $n \to (n, 1)$, $n \in N'$. Define a character θ' of N' by

$$\theta'(n) = \psi(n_{1,2} + \ldots + n_{n-1,n} + n_{n,n+1}/2) \tag{14}$$

where $n = (n_{i,j}) \in N'$. We define a distribution on G':

$$I_{G'}(\tilde{f} : N', \theta', N', (\theta')^{-1})$$

$$= \int_{N'(F)\backslash N'(\mathbf{A})} \int_{N'(F)\backslash N'(\mathbf{A})} K_{\tilde{f}}(n_1, n_2)\theta'(n_1 n_2^{-1})dn_1 dn_2. \tag{15}$$

Our main result is:

Theorem 2.1. *For $f = \otimes f_v \in C_c^\infty(G(\mathbf{A}))$ there exist the maps $\epsilon_v : f_v \mapsto \tilde{f}_v \in S(G'(F_v))$ such that there is a relative trace identity (in the sense of (3)):*

$$I_G(f : R, \chi, U, \mu) = I_{G'}(\tilde{f} : N', \theta', N', \theta'^{-1}). \tag{16}$$

3. A global identity

To prove the Theorem, we introduce another distribution. Let $Z = X \otimes Y$ be a $2n(2n+1)$ dimensional space, then Z inherits a symplectic structure:

$$< x \otimes y, x' \otimes y' >'' = < x, x' >< y, y' >', \ x, x' \in X, y, y' \in Y.$$

Let $Sp_{n(2n+1)}$ be the corresponding symplectic group. Recall that G and G' consist of a dual pair inside the metaplectic group $\tilde{G} = \tilde{Sp}_{n(2n+1)}$. We will denote by (g, g') the image of $g \in G$ and $g' \in G'$.

Associated to ψ is a Weil representation ω_ψ of $\tilde{G}$, [W]. Choose a maximal isotropic subspace Z^+ of Z, the Weil representation acts on the space $\mathcal{S}(Z^+)$ of Schwartz functions on Z^+. For $\Phi \in \mathcal{S}(Z^+(\mathbf{A}))$, define the Theta function:

$$\Theta_\Phi^\psi(g) = \sum_{A \in Z^+(F)} \omega_\psi(g)\Phi(A), \ g \in \tilde{G}. \tag{17}$$

We now define the distribution

$$I_{\tilde{G}}(\Phi : U \times N', \mu \times (\theta')^{-1})$$

$$= \int_{U(F)\backslash U(\mathbf{A})} \int_{N'(F)\backslash N'(\mathbf{A})} \Theta_\Phi^\psi(u, n)\mu(u)\theta'(n^{-1})dudn. \tag{18}$$

This definition depends on our choice of the maximal isotropic subspace Z^+. Given any two choices Z_1^+ and Z_2^+ of the maximal isotropic subspaces, there is an element $w_0 \in Sp_{n(2n+1)}(F)$ such that $Z_2^+ = w_0 Z_1^+$. We use ω'_ψ to denote the action of the Weil representation on $\mathcal{S}(Z_2^+)$, then for $g \in \tilde{G}(\mathbf{A})$:

$$\omega'_\psi(g)\Phi(A) = \omega_\psi(w_0 g w_0^{-1})\Phi'(w_0^{-1}A), \ A \in Z_2^+(\mathbf{A}), \tag{19}$$

where $\Phi \in \mathcal{S}(Z_2^+)$, and $\Phi'(w_0^{-1}A) = \Phi(A)$ is a Schwartz function on Z_1^+; (we use also the fact that the covering splits over $Sp_{n(2n+1)}(F)$ to consider w_0 as an element in $\tilde{G}(\mathbf{A})$).

The following simple global identity is the key in our proof of Theorem 2.1:

Proposition 3.1. *Let Z_1^+ and Z_2^+ be two maximal isotropic subspaces of Z with $Z_2^+ = w_0 Z_1^+$. Let $\Phi \in \mathcal{S}(Z_2^+)$ and $\Phi' \in \mathcal{S}(Z_1^+)$ be related by $\omega_\psi(w_0)\Phi'(w_0^{-1}A) = \Phi(A)$ with $A \in Z_2^+(\mathbf{A})$. Then*

$$I_{\tilde{G}}(\Phi : U \times N', \mu \times (\theta')^{-1}) = I_{\tilde{G}}(\Phi' : U \times N', \mu \times (\theta')^{-1}). \tag{20}$$

Proof. We only need to show $\Theta_\Phi^\psi(g) = \Theta_{\Phi'}^\psi(g)$. Using the fact that the Theta function is automorphic, (which follows from the Poisson summation formula), we get:

$$\begin{aligned}\Theta^{\psi}_{\Phi'}(g) &= \Theta^{\psi}_{\Phi'}(w_0 g)\\ &= \sum_{A\in Z_1^+(F)} \omega_\psi(w_0 g w_0^{-1})\omega_\psi(w_0)\Phi'(A)\\ &= \sum_{A\in Z_2^+(F)} \omega_\psi(w_0 g w_0^{-1})\omega_\psi(w_0)\Phi'(w_0^{-1}A)\\ &= \sum_{A\in Z_2^+(F)} \omega'_\psi(g)\Phi(A) = \Theta^{\psi}_{\Phi}(g).\end{aligned}$$

□

Two obvious choices of Z^+ are: $Z_1^+ = X\otimes Y^+$ where Y^+ is the maximal isotropic subspace of Y spanned by $\{f_1,\dots,f_n\}$; $Z_2^+ = X^+\otimes Y \oplus X^0\otimes Y^+$ where X^+ is the maximal isotropic subspace of X spanned by $\{e_1,\dots,e_n\}$ and X^0 is the one dimensional subspace generated by e_{n+1}. We will use the choice Z_1^+ when comparing $I_{\tilde{G}}(\Phi : U\times N', \mu\times(\theta')^{-1})$ with the distribution on G, and use the choice Z_2^+ when comparing it with the distribution on G'.

We remark that $I_{\tilde{G}}(\Phi : U\times N', \mu\times(\theta')^{-1})$ can be considered as a relative trace formula on $G\times G'$. Consider the relative trace formula on $G\times G'$:

$$\int\int K_{f\otimes\tilde{f}}((g,h),(u,n))\Theta^{\psi}_{\Phi}(g,h)\mu(u)\theta'(n^{-1})d(u,n)d(g,h) \tag{21}$$

where the first integral is over $G\times G'(F)\backslash G\times G'(\mathbf{A})$, and the second over $U\times N'(F)\backslash U\times N'(\mathbf{A})$. If we define for $A\in Z^+(\mathbf{A})$,

$$\Phi_{f\otimes\tilde{f}}(A) = \int_{G\times G'(\mathbf{A})} f(g^{-1})\tilde{f}(h^{-1})\omega_\psi(g,h)\Phi(A)d(g,h),$$

then the distribution (21) equals $I_{\tilde{G}}(\Phi_{f\otimes\tilde{f}} : U\times N', \mu\times(\theta')^{-1})$.

4. Comparison with $I_G(f : R, \chi, U, \mu)$

We identify $Z_1^+ = X\otimes Y^+$ with $M_{2n+1,n}$ the space of $(2n+1)\times n$ matrices. The action of ω_ψ on $\mathcal{S}(M_{2n+1,n})$ restricted to $G\times G'$ can be described as follows:

$$\omega_\psi(g,\delta(h))\Phi(A) = |\det(h)|^{n+1/2}\frac{\gamma(1,\psi)}{\gamma(\det(h)^{2n+1},\psi)}\Phi(g^{-1}Ah), \tag{22}$$

$$\omega_\psi(1,\sigma'_l)\Phi(A) = \gamma(1,\psi)^{-(2n+1)l/2}\Phi^{\wedge l}(A), \tag{23}$$

$$\omega_\psi(1,(\begin{pmatrix} 1_n & V \\ & 1_n \end{pmatrix},1))\Phi(A) = \psi(tr({}^tA\sigma_{2n+1}AV\sigma_n)/2)\Phi(A). \tag{24}$$

Here $g \in G, h \in GL_n$, σ_l is the matrix in GL_l with 1's on anti-diagonal and 0's elsewhere; $h^* = \sigma_n^t h^{-1}\sigma_n$;

$$\delta(h) = (\begin{pmatrix} h & \\ & h^* \end{pmatrix},1),\ \sigma_l' = (\begin{pmatrix} & & -\sigma_l \\ & 1_{2n-2l} & \\ \sigma_l & & \end{pmatrix},1); \tag{25}$$

$\gamma(a,\psi)$ denotes the Weil constant, and for $A = [A_1, A_2] \in M_{2n+1,n}$, where $A_1 \in M_{2n+1,l}$,

$$\Phi^{\wedge l}(A) = \int_{A' \in M_{2n+1,l}} \psi(tr({}^tA'\sigma_{2n+1}A_1))\Phi([A', A_2])dA'. \tag{26}$$

The distribution $I_{\tilde{G}}(\Phi : U \times N', \mu \times (\theta')^{-1})$ equals:

$$\sum_{A \in M_{2n+1,n}(F)} \int_{U(F)\backslash U(\mathbf{A})} \int_{N'(F)\backslash N'(\mathbf{A})} \omega_\psi(u,n)\Phi(A)\mu(u)\theta'(n^{-1})dudn. \tag{27}$$

Let $E_1 \in M_{2n+1,n}$ be the matrix $[e_1, \ldots, e_{n-1}, e_{n+1}]$.

Proposition 4.1. *With suitable choice of the measure, when for $g \in G(\mathbf{A})$*

$$\int_{R'(\mathbf{A})\backslash R(\mathbf{A})} \omega_\psi(r^{-1}g, 1_{2n})\Phi(E_1)\chi(r)dr = \int_{R(\mathbf{A})} f(r^{-1}g)\chi(r)dr, \tag{28}$$

we have

$$\sum_{A \in M_{2n+1,n}(F)} \int_{N'(F)\backslash N'(\mathbf{A})} \omega_\psi(g,n)\Phi(A)\theta'(n^{-1})dn$$

$$= \int_{R(F)\backslash R(\mathbf{A})} K_f(r,g)\chi(r)dr. \tag{29}$$

Proof. Let S be the variety of $A = [A_1, \ldots, A_n] \in M_{2n+1,n}(F)$ with $< A_i, A_j >= 0$ for all $1 \le i,j \le n$ except $< A_n, A_n >= 1$, and with A being of rank n.

Lemma 4.2. *If $A \notin S$, then*

$$\int_{N'(F)\backslash N'(\mathbf{A})} \omega_\psi(g,n)\Phi(A)\theta'(n^{-1})dn = 0.$$

Proof. If the value of $< A_i, A_j >$ is not as prescribed above, then the integration over the unipotent radical of the Siegel parabolic N_U' of Sp_n (which is a normal subgroup of N') would vanish by formula (24). If A is not of the maximal rank, then there is a $n = \begin{pmatrix} n_1 & \\ & n_1^* \end{pmatrix} \in N'(\mathbf{A})$ with $\theta'(n) \neq 1$ and $An_1 = A$; from formula (22), the integral vanishes. □

From the Witt's Theorem, the map $g \mapsto A_g = g^{-1}E_1$ gives a bijection between $R'(F)\backslash G(F)$ and the variety S. From the above Lemma, we get the left hand side of (29) equals:

$$\sum_{h \in R'(F)\backslash G(F)} \int_{N'(F)\backslash N'(\mathbf{A})} \omega_\psi(g, n)\Phi(A_h)\theta'(n^{-1})dn. \tag{30}$$

Lemma 4.3. *With suitable choice of the measure, the expression (30) equals*

$$\sum_{h \in R(F)\backslash G(F)} \int_{R'(\mathbf{A})\backslash R(\mathbf{A})} \omega_\psi(r^{-1}hg, 1_{2n})\Phi(E_1)\chi(r)dr. \tag{31}$$

Proof. We first integrate out the unipotent radical of the Siegel parabolic subgroup N'_U of Sp_n in (30). From (24), the integration yields a factor which equals the volume of $N'_U(F)\backslash N'_U(\mathbf{A})$; with suitable measure, this volume equals 1. As $N'_U\backslash N' \cong N$, from (22) the expression (30) equals:

$$\sum_{h \in R'(F)\backslash G(F)} \int_{N(F)\backslash N(\mathbf{A})} \omega_\psi(g, 1_{2n})\Phi(A_h n)\theta(n^{-1})dn. \tag{32}$$

Recall there is an isomorphism $\rho : R'\backslash R \cong N$. It is easy to verify that $rE_1 = E_1 n$ when $\rho(r) = n$, thus $A_h n = h^{-1}rE_1$. Also clear is that $\chi(r) = \theta^{-1}(\rho(r))$. From (22) we get (32) equals:

$$\sum_{h \in R'(F)\backslash G(F)} \int_{(R'\backslash R(F))\backslash(R'\backslash R(\mathbf{A}))} \omega_\psi(r^{-1}hg, 1_{2n})\Phi(E_1)\chi(r)dr. \tag{33}$$

A standard unwinding of the summation and integration gives the Lemma. □

On the other hand from (2), the right hand side of (29) equals

$$\sum_{h \in R(F)\backslash G(F)} \int_{R(\mathbf{A})} f(r^{-1}hg)\chi(r)dr. \tag{34}$$

Thus the Proposition 4.1 follows from (31) and (34). □

It follows immediately from Proposition 4.1, (13) and (27) that:

Proposition 4.4. *When Φ and f satisfy (28), we have $I_{\tilde{G}}(\Phi : U \times N', \mu \times (\theta')^{-1}) = I_G(f : R, \chi, U, \mu)$.*

5. Comparison with $I_{G'}(\tilde{f} : N', \theta', N', (\theta')^{-1})$

We use the maximal isotropic subspace $Z_2^+ \cong X^+ \otimes Y \oplus X^0 \otimes Y^+$. We will write an element in this space as (A, B) where $A \in M_{2n,n}$ and $B \in F^n$. Use ω'_ψ to denote the Weil representation of $\tilde{G}$ acting on $\mathcal{S}(Z_2^+)$. We will use the following formulas on this action:

For a function $\Phi \otimes \Phi_0(A, B) = \Phi(A)\Phi_0(B)$ in $\mathcal{S}(Z_2^+)$, for $g' \in Sp_n$:

$$\omega'_\psi(1, (g', 1))\Phi \otimes \Phi_0(A, B) = \Phi(g'^{-1}A)\omega_\psi(g', 1)\Phi_0(B) \tag{35}$$

the second ω_ψ being the Weil representation of G' acting on $\mathcal{S}(Y^+)$. Let U^1, U^2, U^3 be subgroups of U consisting respectively of

$$u_1(n) = \begin{pmatrix} n & & \\ & 1 & \\ & & n^* \end{pmatrix}, u_2(v) = \begin{pmatrix} 1_n & v & -vv^*/2 \\ & 1 & v^* \\ & & 1_n \end{pmatrix}, u_3(V) = \begin{pmatrix} 1_n & 0 & V \\ & 1 & 0 \\ & & 1_n \end{pmatrix}$$

Here in the definition of $u_2(v)$, we understand v as a vector in F^n. Then U^3 and U^2U^3 are normal subgroups of U. The action of U on $\mathcal{S}(Z_2^+)$ can be described as follows:

$$\omega_\psi(u_1(n), 1)\Phi \otimes \Phi_0(A, B) = \Phi \otimes \Phi_0(An, B), \tag{36}$$

$$\omega_\psi(u_3(V), 1)\Phi \otimes \Phi_0(A, B) = \psi(tr({}^tAV\sigma_n A)/2)\Phi \otimes \Phi_0(A, B). \tag{37}$$

We only need to know the value of $\omega_\psi(u_2(v), 1)\Phi \otimes \Phi_0(A, B)$ for some special choice of A:

$$\omega_\psi(u_2(v), 1)\Phi \otimes \Phi_0(A_0, B) = \psi({}^tBv)\Phi \otimes \Phi_0(A_0, B) \tag{38}$$

where $A_0 = [f_1, \ldots, f_n]$ and $v \in F^n$.

For $\Phi \in \mathcal{S}(Z_2^+(\mathbf{A}))$, the distribution $I_{\tilde{G}}(\Phi : U \times N', \mu \times (\theta')^{-1})$ equals:

$$\sum_{A \in M_{2n,n}(F), B \in F^n} \int_{U(F)\backslash U(\mathbf{A})} \int_{N'(F)\backslash N'(\mathbf{A})} \omega'_\psi(u, n)\Phi(A, B)\mu(u)\theta'(n^{-1})du dn. \tag{39}$$

Let $E_2 = (A_0, B_0)$ with $B_0 = e_n \in F^n$. Recall N'_U is the unipotent radical of the Siegel parabolic subgroup of Sp_n.

Proposition 5.1. *With suitable choice of the measure, when for $g \in G'(\mathbf{A})$*

$$\int_{N'_U(\mathbf{A})\backslash N'(\mathbf{A})} \omega'_\psi(1_{2n+1}, ng)\Phi(E_2)\theta'(n^{-1})dn = \int_{N'(\mathbf{A})} \tilde{f}(n^{-1}g)\theta'(n)dn, \tag{40}$$

we have

$$\sum_{A\in M_{2n,n}(F),B\in F^n}\int_{U(F)\backslash U(\mathbf{A})}\omega'_\psi(u,g)\Phi(A,B)\mu(u)du$$

$$=\int_{N'(F)\backslash N'(\mathbf{A})}K_{\bar{f}}(n,g)\theta'(n)dn. \tag{41}$$

Proof. Let S' be the variety of $A = [A_1,\ldots,A_n] \in M_{2n,n}(F)$ with $< A_i, A_j >' = 0$ for all $1 \le i,j \le n$ and with A being of rank n. Similar to Lemma 4.2, using (37) and (36) we get

Lemma 5.2. *If $A \notin S'$, then*

$$\int_{U(F)\backslash U(\mathbf{A})}\omega'_\psi(u,g)\Phi(A,B)\mu(u)du = 0.$$

Thus from Witt's Theorem, there is a bijection from $N'_U(F)\backslash Sp_n(F)$ to S' given by $h \mapsto h^{-1}A_0$; the left hand side of (41) becomes:

$$\sum_{h\in N'_U\backslash Sp_n(F),B\in F^n}\int_{U(F)\backslash U(\mathbf{A})}\omega'_\psi(u,g)\Phi(h^{-1}A_0,B)\mu(u)du. \tag{42}$$

From Lemma 2 of [**Fu**], this expression equals:

$$\sum_{h\in N'_U\backslash Sp_n(F),B\in F^n}\int_{U(F)\backslash U(\mathbf{A})}\omega'_\psi(u,hg)\Phi(A_0,B)\mu(u)du. \tag{43}$$

From (38), we see

Lemma 5.3. *If $B \neq B_0$,*

$$\int_{U(F)\backslash U(\mathbf{A})}\omega'_\psi(u,g)\Phi(A_0,B)\mu(u)du = 0.$$

Thus the left hand side of (41) equals:

$$\sum_{h\in N'_U\backslash Sp_n(F)}\int_{U(F)\backslash U(\mathbf{A})}\omega'_\psi(u,hg)\Phi(A_0,B_0)\mu(u)du. \tag{44}$$

Integrating over the normal subgroup U^2U^3 gives a factor which equals the volume of $U^2U^3(F)\backslash U^2U^3(\mathbf{A})$; with suitable choice of measure, this volume equals 1. As $(U^2U^3\backslash U) \cong U^1 \cong N$, from (36) we get (44) equals:

$$\sum_{h\in N'_U\backslash Sp_n(F)}\int_{N(F)\backslash N(\mathbf{A})}\omega'_\psi(1_{2n+1},hg)\Phi(A_0n,B_0)\theta(n)dn. \tag{45}$$

Here we used the fact $\mu(u_1(n)) = \theta(n)$.

Since $A_0 n = \begin{pmatrix} n & \\ & n^* \end{pmatrix} A_0$, from (35), we see (45) equals:

$$\sum_{h\in N'_U\backslash Sp_n(F)} \int_{N(F)\backslash N(\mathbf{A})} \omega'_\psi(1_{2n+1}, \begin{pmatrix} n & \\ & n^* \end{pmatrix} hg)\Phi(A_0, B_0)\theta(n^{-1})dn. \quad (46)$$

As $\theta'(\begin{pmatrix} n & \\ & n^* \end{pmatrix}) = \theta(n)$ and $N \cong N'_U\backslash N'$, we can unwind the above sum and integral to get

$$\sum_{h\in N'\backslash Sp_n(F)} \int_{N'_U\backslash N'(\mathbf{A})} \omega'_\psi(1_{2n+1}, nhg)\Phi(A_0, B_0)\theta'(n^{-1})dn. \quad (47)$$

On the other hand, from (2) the right hand side of (41) equals:

$$\sum_{h\in N'\backslash Sp_n(F)} \int_{N'(\mathbf{A})} \tilde{f}(n^{-1}hg)\theta'(n)dn. \quad (48)$$

The Proposition 5.1 follows from comparing (47) and (48). □

It follows immediately from Proposition 5.1, (15) and (39) that:

Proposition 5.4. *When Φ and $\tilde{f}$ satisfy (40), we have $I_{\tilde{G}}(\Phi : U \times N', \mu \times (\theta')^{-1}) = I_{G'}(\tilde{f} : N', \theta', N', (\theta')^{-1})$.*

6. Proof of the result

We first show how to define the map ϵ_v that associates to each $f_v \in C_c^\infty(G_v)$ a function $\tilde{f}_v \in \mathcal{S}(G'_v)$. Given $f_v \in C_c^\infty(G_v)$, from Lemma 5.3 in **[MR1]**, there exists a $\Phi_v = \epsilon_{1,v}(f_v) \in \mathcal{S}(Z_1^+(F_v))$ such that:

$$\int_{R'(F_v)\backslash R(F_v)} \omega_\psi(r^{-1}g, 1_{2n})\Phi_v(E_1)\chi(r)dr = \int_{R(F_v)} f_v(r^{-1}g)\chi(r)dr. \quad (49)$$

We remark that this statement is rather easy to prove; the above equation determines the function Φ_v on the closed variety $S(F_v)$, we extend it to a function on $Z_1^+(F_v)$, (and that is the reason for the restriction $f_v \in C_c^\infty(G_v)$).

Let $\Phi'_v \in \mathcal{S}(Z_2^+(F_v))$ be defined by

$$\epsilon'_v : \Phi_v \mapsto \Phi'_v, \ \Phi'_v(A) = \omega_\psi(w_0)\Phi(w_0^{-1}A)$$

for $A \in Z_2^+(F_v)$. From Lemma 5.5 in **[MR1]**, there is a function $\tilde{f}_v = \epsilon_{2,v}(\Phi'_v) \in \mathcal{S}(G'_v)$ such that

$$\int_{N'_U(F_v)\backslash N'(F_v)} \omega'_\psi(1_{2n+1}, ng)\Phi'_v(E_2)\theta'(n^{-1})dn$$

$$= \int_{N'(F_v)} \tilde{f}_v(n^{-1}g)\theta'(n)dn. \tag{50}$$

We remark that this statement is again easy to prove: the left hand side of (50) gives an N'−equivariant function that is a Schwartz function on $N'\backslash G'$, thus the existence of $\tilde{f}_v$.

We will define $\epsilon_v(f_v)$ through this association $f_v \mapsto \Phi_v \mapsto \Phi'_v \mapsto \tilde{f}_v$. Next we prove the equivalent of the fundamental lemma in this case. Let v be a non-archimedean place, let $\Phi_{0,v}$ and $\Phi'_{0,v}$ be the characteristic functions of the integer lattices of $Z_1^+(F_v)$ and $Z_2^+(F_v)$; when f_v and $\tilde{f}_v$ are Hecke functions, let

$$\Phi_{f,v}(A) = \int_{G_v} f_v(g)\omega_\psi(g, 1_{2n})\Phi_{0,v}(A)dg,\ A \in Z_1^+(F_v);$$

$$\Phi'_{\tilde{f},v}(A) = \int_{G'_v} \tilde{f}_v(g)\omega'_\psi(1_{2n+1}, g)\Phi'_{0,v}(A)dg,\ A \in Z_2^+(F_v).$$

Proposition 6.1. *At almost all places v, when f_v and $\tilde{f}_v$ are Hecke functions:*

$$\int_{R'(F_v)\backslash R(F_v)} \omega_\psi(r^{-1}g, 1_{2n})\Phi_{f,v}(E_1)\chi(r)dr = \int_{R(F_v)} f_v(r^{-1}g)\chi(r)dr; \tag{51}$$

$$\int_{N'_U(F_v)\backslash N'(F_v)} \omega'_\psi(1_{2n+1}, ng)\Phi'_{\tilde{f},v}(E_2)\theta'(n^{-1})dn$$

$$= \int_{N'(F_v)} \tilde{f}_v(n^{-1}g)\theta'(n)dn; \tag{52}$$

Moreover when $\tilde{f}_v = \lambda_v(f_v)$ under the Hecke algebra homomorphism, $\epsilon'_v(\Phi_{f,v}) = \Phi'_{\tilde{f},v}$.

Proof. Let $f_{0,v}$ and $\tilde{f}_{0,v}$ be the unit elements of the Hecke algebras. For a general Hecke function f (or $\tilde{f}$), we have $f = f * f_0$ (and $\tilde{f} = \tilde{f} * \tilde{f}_0$). Thus from the definition of $\Phi_{f,v}$ and $\Phi'_{\tilde{f},v}$, we only need to show the identities (51) and (52) when $f_v = f_{0,v}$, $\tilde{f}_v = \tilde{f}_{0,v}$. In these cases $\Phi_{f,v} = \Phi_{0,v}$ and $\Phi'_{\tilde{f},v} = \Phi'_{0,v}$. The proof of the identities in this case can be done by direct computation. For example both sides of (52) gives left N' equivariant and right K' invariant functions, (here K' is the maximal compact subgroup of G'); from Iwasawa decomposition, we only need to check the two functions equal over the diagonal elements. A simple computation shows both functions vanish over diagonal elements not in K' and equals 1 over identity. The detail of the proof is given in Lemmas 7.3 and 7.2 in **[MR1]**.

Over almost all places, w_0 is an integral matrix; we have $\epsilon'_v(\Phi_{0,v}) = \Phi'_{0,v}$. The last statement of the Proposition is just another statement of the unramified Howe duality, see **[H]**,**[R]**,**[Wa]**. □

We are now ready to prove Theorem 2.1.

Proof. Let S_0 be a large enough finite set of places, such that if $v \notin S_0$ the statements in Proposition 6.1 hold. For $f = \otimes_{v \in S_0} f_v \otimes_{v \notin S_0} f_v \in S(G(\mathbf{A}))$ where f_v are Hecke functions when $v \notin S_0$, let $\tilde{f} = \otimes_{v \in S_0} \epsilon_v(f_v) \otimes_{v \notin S_0} \lambda_v(f_v)$. We prove the identity (16). From Proposition 4.4, (49) and (51), we get when $\Phi = \otimes_{v \in S_0} \epsilon_{1,v}(f_v) \otimes_{v \notin S_0} \Phi_{f,v}$,

$$I_G(f : R, \chi, U, \mu) = I_{\tilde{G}}(\Phi : U \times N', \mu \times (\theta')^{-1}). \qquad (53)$$

From Propositions 3.1 and 6.1, we see when $\Phi' = \otimes_{v \in S_0} \epsilon'_v(\Phi_v) \otimes_{v \notin S_0} \Phi'_{\tilde{f},v}$,

$$I_{\tilde{G}}(\Phi : U \times N', \mu \times (\theta')^{-1}) = I_{\tilde{G}}(\Phi' : U \times N', \mu \times (\theta')^{-1}). \qquad (54)$$

From Proposition 5.4, (50) and (52), we get

$$I_{\tilde{G}}(\Phi' : U \times N', \mu \times (\theta')^{-1}) = I_{G'}(\tilde{f} : N', \theta', N', (\theta')^{-1}). \qquad (55)$$

The identity (16) follows. □

We remark that the same argument also shows the relative trace identities between the distribution (21) and the distributions $I_G(f : R, \chi, U, \mu)$ and $I_{G'}(\tilde{f} : N', \theta', N', (\theta')^{-1})$.

References

[BM] E. M. Baruch, Z. Mao, Central values of automorphic L-functions, preprint.

[CJ] N. Chen, H. Jacquet, Positivity of quadratic base change $L-$functions, Bull. Soc. Math. France **129** (2001), 33–90.

[Fu] M. Furusawa, On the theta lift from SO_{2n+1} to $\tilde{Sp}_n$, J. Reine. Angew. Math. **466** (1995), 87–110.

[FuSh] M. Furusawa, J. Shalika, The fundamental lemma for the Bessel and Novodvorsky subgroups of $GSp(4)$, C. R. Acad. Sci. Paris Sér. I Math. **328** (1999), 105–110.

[GRS] D. Ginzburg, S. Rallis, D. Soudry, On a correspondence between cuspidal representations of GL_{2n} and $\widetilde{Sp}_{2n}$, J. Amer. Math. Soc. **12** (1999), 849–907.

[Gu] J. Guo, On the positivity of the central critical varlues of automorphic L-functions for $GL(2)$, Duke Math. J. **83** (1996), 157–190.

[H] R. Howe, Another look at the local θ−correspondence for an unramified dual pair, in Israel Math. Conf. Proc. **2** (1990), 93–124.

[J] H. Jacquet, On the nonvanishing of some L−functions, Proc. Indian Acad. Sci. **97** (1987), 117–155.

[MR1] Z. Mao, S. Rallis, Jacquet module of the oscillator representation and families of relative trace identities, Comp. Math. 140 (2004), 855-886.

[MR2] Z. Mao, S. Rallis, The induced Weil representation and the Shalika period, Pacific J. Math. **101** (1999), 313–327.

[R] S. Rallis, Langlands' functoriality and the Weil representation. Amer. J. Math. **104** (1982), 469–515.

[Wa] J-L. Waldspurger, Demonstration d'une conjecture de duality de Howe dans le case p−adiques, $p \neq 2$, in Israel Math. Conf. Proc. **2** (1990), 267–324.

[W] A. Weil, Sur certains groupes d'opérateurs unitaires, Acta Math. **111** (1964), 143–211.

Zhengyu Mao
Department of Mathematics and Computer Science
Rutgers University at Newark
Newark, NJ 07102, USA
zmao@andromeda.rutgers.edu

Stephen Rallis
Department of Mathematics
The Ohio State University
Columbus, OH 43210, USA
haar@math.ohio-state.edu

Remark on congruences between algebraic automorphic forms

Marie-France Vignéras

Abstract. The purpose of this paper is to present and to extend the method of [9] in order to obtain congruences between algebraic automorphic forms [7]. The congruences are "easy" to obtain; this is "soft" mathematics as Steve Rallis likes to say. However they are quite important; via the "hard" functorialities of Langlands, they allow to get congruences between automorphic representations of $GL(n)$ or Galois representations, as for instance in [5], [6], or in the proof of the Langlands correspondence modulo ℓ [9]. It is a pleasure to dedicate this work to Steve Rallis, congruences are related to his work via the theory of L-functions where his contribution is so fundamental.

Let p be a prime number, F a number field, and G a reductive connected group over F such that $\{1\}$ is an arithmetic subgroup. We denote by $G' = \mathrm{Res}_{F/\mathbf{Q}} G$ the scalar restriction of G to $\mathbf{Q}$ [8] and by $G_A = G'_A = G_\infty \times G^\infty$ the adele group of G or of G'.

Theorem 0.1. *We consider*

- V, V' two absolutely irreducible algebraic representations of G',

- v a finite place of F and K_v an open compact subgroup of $G(F_v)$ of pro-order prime to p,

- $K^{v,p}$ an open compact subgroup of the component $G^{\infty,v,p}$ of G^∞ outside of v, p, and W^p a complex irreducible smooth representation of $K^p = K_v \times K^{v,p}$.

Then we have the following property:

Any irreducible automorphic representation π_A of G_A of algebraic type V which contains (K^p, W^p) is rational over a number field, its component $\pi^{\infty,p}$ outside of ∞, p is p-integral, and is congruent modulo p to an irreducible automorphic representation of G_A of algebraic type V' which contains (K^p, W^p) and has finite slope at p.

The term **finite slope** or **congruent modulo** p used in the theorem are defined as follows. An irreducible automorphic representation π_A of G_A has a **finite slope at** p, if the p-component π_p of π_A contains the trivial representation of a pro-p-Iwahori subgroup Iw_p of $G_p = G'(\mathbf{Q}_p)$.

Two irreducible automorphic representations π_A and π'_A of G_A are called **congruent modulo** p if there exists an isomorphism i_p from $\mathbf{C}$ to an algebraic closure $\overline{\mathbf{Q}}_p$ of $\mathbf{Q}_p$, a finite extension $E/\mathbf{Q}_p$ contained in $\overline{\mathbf{Q}}_p$ and a finite set S of places of F containing the places above ∞ and p, such that

- the components π^S and $\pi'^{,S}$ of π_A and π'_A outside of S are unramified,
- the values of the characters $i_p\lambda^S, i_p\lambda'^{,S} : \mathcal{H}^S \to \mathbf{C} \to \overline{\mathbf{Q}}_p$ of the commutative Satake-Hecke algebra $\mathcal{H}^S$ given by π^S and $\pi'^{,S}$ belong to the ring of integers O_E of E and are congruent $i_p\lambda^S \equiv i_p\lambda'^{,S}$ modulo the maximal ideal of O_E.

The theorem has a converse:

Let W_p be an irreducible complex representation of the pro-p-Iwahori group Iw_p. Any irreducible automorphic representation of G_A of algebraic type V', of finite slope at p, which contains (K^p, W^p), is congruent modulo p to an irreducible automorphic representation of G_A of algebraic type V, which contains $(Iw_p \times K^p, W_p \times W^p)$.

When $p \neq 2$, we obtain a slightly stronger form of the theorem and of its converse in §20.

One gets also congruences modulo $p^n, n > 1$, between automorphic forms instead of representations, in §17.

The method of the proof is a combination of arguments given in **[9]** only for automorphic forms trivial at ∞, and in the excellent introduction to algebraic automorphic forms **[7]**.

The beautiful recent results of Buzzard, Chenevier, Emerton and Kisin on p-adic interpolation of automorphic forms make this paper look out of date, but their results do not imply our theorem. The interest of Barry Mazur when I sent him these congruences in July 2002, motivated me to write their proof. It is a pleasure to thank him and Steve Rallis for many friendly discussions.

Proof of the theorem (see §19)

1 *We consider a number field F and a reductive connected group G over F such that $G(F)$ is discrete in the group G^∞ of finite adeles.*

Equivalent properties are given in [7]. One of them is that the arithmetic subgroups are finite. This condition is very restrictive. There are only three possibilities:

- *$F = \mathbf{Q}$ and the split centre S of G is a maximal $\mathbf{R}$-split torus of G,*
- *F is an imaginary quadratic field and G is a F-split torus,*

- *F is totally real and the infinite part G_∞ of G is compact (in particular the split centre of G is trivial).*

The group G_∞ has a unique maximal compact subgroup K_∞. The connected components G_∞^o and K_∞^o are related to the split and anisotropic centres S and C of G' as follows. We have $K_\infty^o = (C(G', G'))_\infty$ and $G_\infty^o = S_\infty^o \times K_\infty^o$. We denote by S_1 the finite algebraic $\mathbf{Q}$-group of $x \in S$ with $x^2 = 1$. The groups G_∞, K_∞ are equal to $G_\infty = S_\infty K_\infty = S_{1,\infty} G_\infty^o$ and $K_\infty = S_{1,\infty} K_\infty^o$. They have the same group of components $\pi_{G,o} = G_\infty / G_\infty^o = K_\infty / K_\infty^o$, isomorphic to the 2-group $S_{1,\infty}/(S_{1,\infty} \cap G_\infty^o)$. We have $S_{1,\infty} = S_1(\mathbf{Q})$.

2 The representations are always left representations (unless right is specified).

An irreducible automorphic representation of G_A of dimension 1 is a Hecke character, i.e. a continuous morphism $G(F)\backslash G_A \rightarrow \mathbf{C}^*$ (for any reductive connected group over F). We denote by Ω the set of Hecke characters ω of $S_A = S(\mathbf{Q}) \times S_\infty^o \times \prod_p S(\mathbf{Z}_p)$.

An automorphic form on G_A is a finite sum $f = \sum_{\omega \in \Omega} f_\omega$ of functions $f_\omega : G(F)\backslash G_A \rightarrow \mathbf{C}$ which are smooth (continuous and K_∞-finite) on G_∞, locally constant on G^∞, and $f_\omega(sg) = \omega(s) f_\omega(g)$ for all $s \in S_A, g \in G_A$. The action of G_A by right translations on the space $\mathcal{S}_{\mathbf{C}} = \oplus_{\omega \in \Omega} \mathcal{S}_{\mathbf{C},\omega}$ of automorphic forms is semi-simple. The irreducible constituents are the irreducible automorphic representations of G_A.

A non-degenerate bilinear symmetric form $< f_{\omega,1}, f_{\omega,2} >$ on $\mathcal{S}_{\mathbf{C},\omega}$ is the integral of $f_{\omega,1} \overline{f}_{\omega,2} \eta^{-1}$ on the compact space $G(F) S_A \backslash G_A$, for an invariant Haar measure, where η is the positive Hecke character of G_A equal to $\overline{\omega}\omega$ on S_A.

3 Let (R, H, G, M, X) be any commutative ring R, a subgroup H of a group G, an R-module M with an action of H, and a space X with a right action of G. We denote by $C(X, R)$ the R-module of functions $f : X \rightarrow R$ with the natural action of G and by $\tilde{M}$ the R-module of linear forms on M with the contragredient action of H, such that $< \tilde{m}, m > = < h\tilde{m}, hm >$ for all $h \in H, m \in M, \tilde{m} \in \tilde{M}$. Then

$$\mathrm{Hom}_{RH}(M, C(X, R))$$

and the R-module $\mathcal{A}_H(X, M)$ of functions on X of type (H, M)

$$\phi : X \rightarrow \tilde{M}, \quad \phi(xh) = h^{-1}\phi(x) \text{ for all } h \in H,$$

are isomorphic by the map associating to a RH-morphism $m \rightarrow f_m$ the function ϕ such that $< \phi(x), m > = f_m(x)$. The relations $f_{hm}(x) = f_m(xh)$ and $< \phi(xh), m > = < \phi(x), hm > = < h^{-1}\phi(x), m >$ are equivalent.

When the groups and spaces have a topology, the representations and the functions ϕ are supposed continuous.

4 We decompose $\mathcal{S}_{\mathbf{C}}$ according to the set $\hat{G}^o_\infty$ of isomorphism classes of irreducible representations of G^o_∞,

$$\mathcal{S}_{\mathbf{C}} \simeq \oplus_{\pi^o_\infty \in \hat{G}^o_\infty} \pi^o_\infty \otimes \rho_{\pi^o_\infty}.$$

For each $(\pi^o_\infty, V_\infty) \in \hat{G}^o_\infty$, the complex representation $\rho_{\pi^o_\infty}$ of the locally profinite group $G^o_\infty \backslash G_A$ is smooth, semisimple and admissible.

The semisimplicity results from §2 because G_∞ / G^o_∞ is finite, the admissibility results from the finiteness of $G(F)G_\infty \backslash G_A / K$ for any open compact subgroup K of G^∞, for any reductive connected group over F [2] 5.1.

The model $\mathcal{A}_{G^o_\infty}(G(F)\backslash G_A, V_\infty)$ of $\rho_{\pi^o_\infty} \simeq \mathrm{Hom}_{\mathbf{C}G^o_\infty}(V_\infty, \mathcal{S}_{\mathbf{C}})$ obtained by applying §3, is the complex space of functions

$$\phi : G(F)\backslash G_A \to \tilde{V}_\infty, \quad \phi(g g_\infty) = g_\infty^{-1}\phi(g) \text{ for all } g_\infty \in G^o_\infty,$$

locally constant on G^∞, with $G^o_\infty \backslash G_A$ acting by right translations.

We can extend the representation π^o_∞ of G^o_∞ to G_∞ because $\pi_{G,o}$ is a finite 2-group. We choose an extension π_∞. Then we set

$$\phi(g) = g_\infty^{-1}\psi(g),$$

where g_∞ is the infinite component of $g \in G_A$. Now ψ is invariant by right translation by G^o_∞ and we have, for $\gamma \in G(F)$,

$$\psi(\gamma g) = \gamma g_\infty \phi(\gamma g) = \gamma\psi(g).$$

The natural action of $G^o_\infty \backslash G_A$ by right translations on the space $\mathcal{A}(V_\infty)$ of locally constant functions

$$\psi : G^o_\infty \backslash G_A \to \tilde{V}_\infty, \quad \psi(\gamma g) = \gamma\psi(g) \quad \text{for all} \quad \gamma \in G(F),$$

is a new model of $\rho_{\pi^o_\infty}$, depending on the choice of π_∞. We could replace $G^o_\infty \backslash G_A$ by $\pi_{G,o} \times G^\infty$ of course.

5 When π^o_∞ is trivial, we deduce that $\rho_{\pi^o_\infty}$ is isomorphic to the natural representation of $G^o_\infty \backslash G_A$ on the space of locally constant complex functions on $G(F)G^o_\infty \backslash G_A$. The functions with values in $\mathbf{Z}$ form a $G^o_\infty \backslash G_A$-stable $\mathbf{Z}$-lattice. An irreducible automorphic representation of G_A trivial on G^o_∞ is **defined over the ring of integers of a number field** [9].

6 *We consider*
- *a number field L,*

- *a place* $\lambda_\infty \in \mathrm{Hom}(L, \mathbf{C})$,
- *a finite dimensional* L*-vector space* V *with an absolutely irreducible algebraic action of* G'.

In §4, we fix π_∞, resp. π^o_∞, to be the representation of G_∞, resp. G^o_∞, equal to the restriction of the algebraic action of $G'(\mathbf{C})$ on $V_\infty = V \otimes_{L,\lambda_\infty} \mathbf{C}$.

The representation $\pi_\infty \otimes \varepsilon$ *of* G_∞, *twist of* π_∞ *by a character* ε *of the component group* $\pi_{G,o}$, *is called of* **algebraic type** (V, ε) *or* V. *An irreducible automorphic representation of* G_A *of infinite component* $\pi_\infty \otimes \varepsilon$ *is also called of algebraic type* (V, ε) *or* V.

The irreducible complex representations of G_∞ of algebraic type are characterized by the property that their restriction to the component G^o_∞ is algebraic.
Any continuous irreducible complex representation to K^o_∞ is algebraic, and any continuous irreducible complex representation of G^o_∞ is the twist by a character of an algebraic representation because $G^o_\infty = S^o_\infty \times K^o_\infty$ (see §1).

The L-vector space $\mathcal{A}(V)$ is a $G^o_\infty\backslash G_A$-stable L-structure of the semi-simple admissible representation $\mathcal{A}(V_\infty)$. The finite part of an irreducible automorphic representation of G_A of algebraic type is **defined over a number field.**

7 In order to obtain an integral structure when V is not trivial (§5), we will work p-adically. We choose

- *an isomorphism* $i_p : \mathbf{C} \to \overline{\mathbf{Q}}_p$.

The place λ_∞ and i_p give a p-adic place $\lambda_p = i_p\lambda_\infty \in \mathrm{Hom}(L, \overline{\mathbf{Q}}_p)$ which allows to identify the complex space V_∞ defined in §6 with the $\overline{\mathbf{Q}}_p$-vector space $V_p = V \otimes_{L,\lambda_p} \overline{\mathbf{Q}}_p$.
The interest is that there is a natural action of the p-adic group $G_p = G'(\mathbf{Q}_p) \subset G'(\overline{\mathbf{Q}}_p)$ on V_p, such that the actions of $G(F)$ on V_p as a subgroup of G_p and on V_∞ as a subgroup of G_∞ coincide via the isomorphism $V_\infty \to V_p$. One identifies $\mathcal{A}_{G^o_\infty}(G(F)\backslash G_A, V_\infty)$ and $\mathcal{A}_{G^o_\infty}(G(F)\backslash G_A, V_p)$ and the trick is to set

$$f(g) = g_p^{-1}\psi(g) = (g_\infty, g_p^{-1})\phi(g)$$

where $\phi \in \mathcal{A}_{G^o_\infty}(G(F)\backslash G_A, V_\infty)$ and (g_∞, g_p) is the (∞, p)-component of $g \in G_A$. There exists an open compact subgroup $K = K_p \times K^p$ of $G^\infty = G_p \times G^{\infty,p}$ (depending on ϕ) such that ϕ, ψ are right invariant by K. The relation $\psi(\gamma g k) = \gamma\psi(g)$ for $(\gamma, k) \in G(F) \times K$ is equivalent to

$$f(\gamma g k) = (\gamma g_p k_p)^{-1}\psi(\gamma g k) = (\gamma g_p k_p)^{-1}\gamma\psi(g) = k_p^{-1} f(g).$$

These are the functions studied in §3 when

$$(R,H,G,M,X) = (\overline{\mathbf{Q}}_p, K, G^\infty, V_p, G(F)G^o_\infty\backslash G_A).$$

the action of K on V_p is the inflation of the action of K_p (trivial on K^p).

8 *The $\overline{\mathbf{Q}}_p$-vector space $\mathcal{A}_K(G(F)G^o_\infty\backslash G_A, V_p)$ of p-adic automorphic forms on G_A of locally algebraic type (K, V_p) is*

$$\{f : G(F)G^o_\infty\backslash G_A \to \tilde{V}_p, \quad f(gk) = k_p^{-1}f(g) \text{ for all } k \in K\},$$

with $G^o_\infty\backslash G^p_A$ acting by right translations.

The inductive limit

$$\mathcal{A}(G(F)G^o_\infty\backslash G_A, V_p) = \lim_K \mathcal{A}_K(G(F)G^o_\infty\backslash G_A, V_p)$$

over all open compact subgroups $K = K_p \times K^p$ of G^∞, is the space of p-adic automorphic forms on G_A of locally algebraic type V_p. The complex and p-adic spaces

$$\mathcal{A}_{G^o_\infty}(G(F)\backslash G_A, V_\infty), \quad \mathcal{A}(G(F)G^o_\infty\backslash G_A, V_p)$$

are isomorphic via i_p. The isomorphism is $\pi_{G,o} \times G^{\infty,p}$-equivariant only. The action of G_p is locally algebraic on $\mathcal{A}(G(F)G^o_\infty\backslash G_A, V_p)$ and is smooth on $\mathcal{A}_{G^o_\infty}(G(F)\backslash G_A, V_\infty)$.

9 We denote by $E = L_{\lambda_p}$, O_E the ring of integers, p_E an uniformizer, k_E the residual field. We choose:

- *an open compact subgroup K^o_p of G_p,*
- *an O_E-structure $\mathcal{L}(V_p)$ of V_p stable by K^o_p.*

The linear O_E-dual $\tilde{\mathcal{L}}(V_p) = \mathcal{L}(\tilde{V}_p)$ is an O_E-structure of the contragredient $\tilde{V}_p$. In §8, we can suppose that the groups K_p are contained in K^o_p.
The O_E-module $\mathcal{A}(G(F)G^o_\infty\backslash G^p_A, \mathcal{L}(V_p))$ is free, $G^o_\infty\backslash G_A$-stable, and generates the $\overline{\mathbf{Q}}_p$-vector space $\mathcal{A}(G(F)G^o_\infty\backslash G_A, V_p)$. The finite part outside of p of an irreducible automorphic representation of G_A of algebraic type is **defined over the ring of integers of a finite extension of $\mathbf{Q}_p$.**

We could not do this in the complex realization of §4, because the representation V_∞ of G_∞, when it is not trivial, has no integral structure stable by $G(F)$, by density.

10 We return to the generalities of §3. We suppose that G is a locally profinite group acting continuously on a locally profinite space X, and that $H = K$ is an open subgroup of G acting smoothly on an R-module M. We consider the compact induction from the R-representations of K to the R-representations of G

$$\operatorname{ind}_K^G M = \{f : G \to M,\ f(kg) = kf(g),\ (k,g) \in K \times G,$$

$$f \text{ with compact support}\},$$

the restriction res_K^G in the other direction, and the Hecke R-algebra

$$\mathcal{H}(G,K,M) = \operatorname{End}_{RG} \operatorname{ind}_K^G M.$$

These functors respect smoothness. In the abelian category of smooth R-representations, the induction from K to G is left adjoint to the restriction from G to K [**10**],

$$\operatorname{Hom}_{RG}(\operatorname{ind}_K^G M, V) \simeq \operatorname{Hom}_{RK}(M, \operatorname{res}_K^G V).$$

Hence $\operatorname{Hom}_{RK}(M, \operatorname{res}_K^G \cdot)$ simply denoted by $\operatorname{Hom}_{RK}(M,\cdot)$ is a functor from the smooth R-representations of K to the right $\mathcal{H}(G,K,M)$-modules.

The sum V^{sm} of the smooth subrepresentations of an R-representation V of G is a smooth subrepresentation of V and for any smooth representation M of K

$$Hom_{RK}(M,V) = \operatorname{Hom}_{RK}(M, V^{sm})$$

because K is open. One then deduces that $\mathcal{A}_K(X,M)$ *is naturally a right* $\mathcal{H}(G,K,M)$*-module.*

This remains true when the action of K on M is locally algebraic. The proof follows the same pattern.

11 We can add to the machine:

- *a finite dimensional L-vector space W with an absolutely irreducible smooth action of an open compact subgroup K of G^∞.*

The functor $V \to \operatorname{Hom}_{RK}(W,V)$ (see §10) gives a bijection from the isomorphism classes of the irreducible smooth representations of G^∞ which contain (K,W) onto the isomorphism classes of the simple right modules of the Hecke algebra $\mathcal{H}(G^\infty,K,W)$ (consider the idempotent of the global Hecke algebra of G^∞ associated to (K,W)).

The representation $\rho_{\pi^o_\infty}$ of $\pi_{G,o} \times G^\infty$ is admissible and semisimple. The isomorphism class of the subrepresentation of $\rho_{\pi^o_\infty}$ generated by the irreducible subrepresentations which contain (K,W), is identified with (see §3, §4):

The right $\mathcal{H}(G^\infty,K,W)$-module $\mathcal{A}_{G^o_\infty \times K}(G(F)\backslash G_A, V_\infty \otimes_{L,\lambda_\infty} W)$ of automorphic forms on G^∞ of algebraic type V and smooth type (K,W) .

This semi-simple module is isomorphic to

$$\mathcal{A}(V_\infty \otimes_{L,\lambda_\infty} W) = \{\psi : G^o_\infty \backslash G_A \to \tilde{V}_\infty \otimes_{L,\lambda_\infty} \tilde{W},$$
$$\psi(\gamma g k) = (\gamma, k^{-1})\psi(g), \text{ for all } (\gamma, k) \in G(F) \times K\}.$$

The finite set $G(F)G^o_\infty \backslash G_A/K$ has a system of representatives $g_1, \ldots, g_n$ in G^∞ because $G(F)$ is dense in G_∞ (for any reductive connected group) [2]. The values $\psi(g_i)$ are invariant by the finite arithmetic group $\Gamma_i = G(F) \cap (G^o_\infty \times g_i K g_i^{-1})$. As a complex vector space,

$$\mathcal{A}(V_\infty \otimes_{L,\lambda_\infty} W) \simeq \oplus_{i=1}^n (\tilde{V}_\infty \otimes_{L,\lambda_\infty} \tilde{W})^{\Gamma_i}, \quad \psi \mapsto (\psi(g_i))_{1 \le i \le n}.$$

It is easy to choose K such that $G(F) \cap K$ is trivial, or even such that all the finite groups Γ_i are trivial, noting that if $K = \prod_v K_v$ is factorizable, then the order of Γ_i divides the pro-order of K_v for any finite place v of F.

12 One deduces from §11 a formula for the finite dimension of the space of automorphic representations with given algebraic and smooth types. In particular:

For any algebraic absolutely irreducible representation V of G', there exists an irreducible automorphic representation of algebraic type V. If K is an open compact subgroup K of G^∞ such that $K \cap G(F) = \{1\}$, for any irreducible complex representation W of K, there exists an irreducible automorphic representation of algebraic type V which contains (K, W).

13 There exists a finite set S of places of F which contains the infinite places, such that $K = K_S \times K^S$ (see §11) where $K_S \subset G_S$ and outside of S, K^S is a good maximal open compact subgroup of G^S, and W^S is trivial.

The Hecke $\mathbf{Z}$-algebra $\mathcal{H}^S = \mathcal{H}(G^S, K^S)$ is commutative and acts semi-simply on $\mathcal{A}(V_\infty \otimes_{L,\lambda_\infty} W)$ with eigenvalues λ_{π^S} where λ_{π^S} is the character of $\mathcal{H}^S$ acting on the one dimensional space of K^S-invariants of π^S, for all irreducible automorphic representations $\pi_A = \pi_S \otimes \pi^S$ of G_A of algebraic type V which contain (K, W).

14 We suppose in §11 that $K = K_p \times K^p$ with $K_p \subset K_p^o$ (see §9) and $W = W_p \otimes_L W^p$ has a factorization compatible with the factorization of $G^\infty = G_p \times G^{\infty,p}$. The group K acts on $V_p \otimes_{L,\lambda_p} W$ by the diagonal locally algebraic irreducible action of K_p and by the smooth action of K^p natural on W and trivial on V_p. We choose

- *a K-stable O_L-structure $\mathcal{L}(W) = \mathcal{L}(W_p) \otimes_{O_L} \mathcal{L}(W^p)$ of W over the ring of integers O_L of L* (see §11).

The linear O_L-dual $\tilde{\mathcal{L}}(W) = \mathcal{L}(\tilde{W})$ is an O_L-structure of the contragredient $\tilde{W}$. As in §8,

15 *The $\overline{\mathbf{Q}}_p$-space of p-adic automorphic forms of locally algebraic type* $(K, V_p \otimes_{L,\lambda_p} W)$ *is*

$$\mathcal{A}_K(G(F)G^o_\infty\backslash G_A, V_p \otimes_{L,\lambda_p} W) = \{f : G(F)G^o_\infty\backslash G_A \to \tilde{V}_p \otimes_L \tilde{W},\\ f(gk) = k^{-1}f(g) \text{ for all } k \in K\}.$$

This is a right $\pi_{G,o} \times \mathcal{H}(G^\infty, K, V_p \otimes_{L,\lambda_p} W)$-module as in §10, but the proof of the theorem uses only its $\pi_{G,o} \times \mathcal{H}(G^{\infty,p}, K^p, W^p)$-module structure. The natural isomorphism

$$\mathcal{A}_{G^o_\infty \times K}(G(F)\backslash G_A, V_\infty \otimes_{L,\lambda_\infty} W) \to \mathcal{A}_K(G(F)G^o_\infty\backslash G_A, V_p \otimes_{L,\lambda_p} W)$$

is $\pi_{G,o} \times \mathcal{H}(G^{\infty,p}, K^p, W^p)$-equivariant.

16 The integral structures chosen in §9 and in §14, give an O_E-structure $\mathcal{L} = \mathcal{L}(V_p) \otimes_{O_L} \mathcal{L}(W)$ of the locally algebraic representation of K on $V_p \otimes_L W$. As in §11, the O_E-module $\mathcal{A}_K(G(F)G^o_\infty\backslash G_A, \mathcal{L})$ is isomorphic to

$$\oplus_{i=1}^n \tilde{\mathcal{L}}^{\Gamma_i},$$

is O_E-free, stable by $\pi_{G,o} \times \mathcal{H}(G^{\infty,p}, K^p, \mathcal{L})$, and generates the $\overline{\mathbf{Q}}_p$-vector space $\mathcal{A}_K(G(F)G^o_\infty\backslash G_A, V_p \otimes_{L,\lambda_p} W)$.

17 We suppose that **the orders of the finite arithmetic groups Γ_i in §11, are prime to p.** The congruences between automorphic forms rely on the fact that the natural $\pi_{G,o} \times \mathcal{H}(G^{\infty,p}, K^p, \mathcal{L})$-equivariant map between

$$\mathcal{A}_K(G(F)G^o_\infty\backslash G_A, \mathcal{L}) \;/\; p_E^n \mathcal{A}_K(G(F)G^o_\infty\backslash G_A, \mathcal{L})$$

and

$$\mathcal{A}_K(G(F)G^o_\infty\backslash G_A, \mathcal{L}/p_E^n\mathcal{L})$$

is an isomorphism for any integer $n \geq 1$.

The congruences between automorphic representations when $n = 1$, rely on the Deligne-Serre lemma [4]. Let $\mathcal{X}$ be any commutative subalgebra of $\pi_{G,o} \times \mathcal{H}(G^{\infty,p}, K^p, \mathcal{L})$. For any eigenvalue μ of $\mathcal{X}$ on the reduction

$$\mathcal{A}_K(G(F)G^o_\infty\backslash G_A, \mathcal{L}/p_E\mathcal{L})$$

there exists a finite extension E'/E and an eigenvalue λ of $\mathcal{X}$ acting on

$$\mathcal{A}_K(G(F)G^o_\infty\backslash G_A, \mathcal{L} \otimes_{O_E} O_{E'}) = \mathcal{A}_K(G(F)G^o_\infty\backslash G_A, \mathcal{L}) \otimes_{O_E} O_{E'},$$

congruent modulo p to μ, i.e. $\lambda \equiv \mu \mod p_{E'}$.

18 The pro-p-radical Iw_p of an Iwahori subgroup Iw of G_p, called a **pro-p-Iwahori subgroup** of G_p, should be considered as the pro-p-Sylow of G_p

because it is a pro-p-group, uniquely defined modulo conjugation, which contains the pro-p-radical of any parahoric subgroup of G_p containing Iw. As Iw_p is a pro-p-group, it satisfies the following properties:

1) An irreducible locally algebraic representation of a pro-p-group over a field of characteristic p is trivial.

2) For any locally algebraic E-representation W of a pro-p-group I_p, there exists a finite extension E'/E and an $O_{E'}$-structure of $W \otimes_E E'$ with trivial reduction modulo $p_{E'}$ (for the action of I_p) **[3]**.

19 We are now ready to prove the theorem and its converse. Let π_A be an irreducible automorphic representation of G_A of algebraic type V and containing $(K^p = K_v \times K^{p,v}, W^p)$, let V' be as in the theorem, and let W_p be an irreducible smooth complex representation of Iw_p.

In §11, we take $K = Iw_p \times K^p$. The orders of the arithmetic groups Γ_i in §11 are trivial because the pro-orders of K_v and of Iw_p are relatively prime. We choose the number field L and the place λ_∞ in §6, and the isomorphism i_p in §7, such that:

a) The algebraic representations V, V' of G' and the smooth representation $W_p \otimes_{\mathbf{C}} W^p$ of K are defined over L embedded in $\mathbf{C}$ via λ_∞.

b) The finite dimensional $\overline{\mathbf{Q}}_p$-vector spaces $U_p = V_p, V_p', W_p \otimes_{\mathbf{C},i_p} \overline{\mathbf{Q}}_p$ have Iw_p-stable O_E-structures $\mathcal{L}(U_p)$ of trivial reduction modulo p_E (for the action of Iw_p), using §18 2).

c) The finite components prime to p in the finite set of irreducible automorphic representations of G_A

- of algebraic type V and containing $(K, W_p \otimes_{\mathbf{C}} W^p)$, or
- of algebraic type V' and containing $(K, \mathrm{id}_p \otimes_{\mathbf{C}} W^p)$ (trivial on Iw_p),

are defined over O_E.

In §11, the L-representation of K is either W' equal to a K^p-stable L-structure of W^p with K acting by inflation (trivial on Iw_p), or $W = W_{p,L} \otimes_L W'$ for a K_p-stable L-structure $W_{p,L}$ of W_p.

We choose a K^p-stable O_E-structure $\mathcal{L}(W^p)$ of $W^p \otimes_{\mathbf{C},i_p} \overline{\mathbf{Q}}_p$. With b), we obtain the O_E-structures in §9, §14 and in §16, $\mathcal{L}$ for $V_p \otimes_{L,\lambda_p} W$ and $\mathcal{L}'$ for $V_p' \otimes_{L,\lambda_p} W'$. We denote by d the complex dimension of W_p.

There is a K-isomorphism

$$\mathcal{L}/p_E\mathcal{L} \simeq \oplus^d \mathcal{L}'/p_E\mathcal{L}'.$$

The same is true for the contragredients. We deduce as in §17 that the reduction modulo p_E of

$$\mathcal{A}_K(G(F)G_\infty^o \backslash G_A, \mathcal{L}) \quad \text{and} \quad \oplus^d \mathcal{A}_K(G(F)G_\infty^o \backslash G_A, \mathcal{L}')$$

are $\pi_{G,o} \times \mathcal{H}(G^{\infty,p}, K^p, \mathcal{L}(W^p))$-isomorphic.

We suppose in §13 that S contains p. Then $\pi_{G,o} \times \mathcal{H}^S$ has the same eigenvalues on the reduction modulo p_E of

$$\mathcal{A}_K(G(F)G^o_\infty\backslash G_A, \mathcal{L}) \quad \text{and} \quad \oplus^d \mathcal{A}_K(G(F)G^o_\infty\backslash G_A, \mathcal{L}').$$

For $* = \mathcal{L}$ or $\mathcal{L}'$, we deduce from c) that for any finite extension E'/E, the eigenvalues of $\mathcal{H}^S$ on

$$\mathcal{A}_K(G(F)G^o_\infty\backslash G_A, *) \quad \text{or on} \quad \mathcal{A}_K(G(F)G^o_\infty\backslash G_A, * \otimes_{O_E} O_{E'})$$

are the same. Remembering from §1 that $\pi_{G,o}$ is a finite 2-group, we deduce as in §17:

When $p \neq 2$, the reduction modulo p_E of the eigenvalues of $\pi_{G,o} \times \mathcal{H}^S$ on

$$\mathcal{A}_K(G(F)G^o_\infty\backslash G_A, \mathcal{L}) \text{ and on } \mathcal{A}_K(G(F)G^o_\infty\backslash G_A, \mathcal{L}')$$

are the same. If $p = 2$, this remains true without $\pi_{G,o}$.

20 Using §13, we obtain a slightly stronger version of the theorem and of its converse.

With the notations of the theorem and of its converse, let ε be a character of $\pi_{G,o}$. If $p \neq 2$, any automorphic irreducible representation of G_A of algebraic type (V, ε), and containing $(Iw_p \times K^p, W_p \otimes W^p)$, is congruent to an automorphic irreducible representation of G_A of algebraic type (V', ε) and containing $(Iw_p \times K^p, \mathrm{id}_p \otimes W^p)$, and conversely.
If $p = 2$, this remains true without the character ε.

References

[1] Nicolas Bourbaki, Algèbre I. Hermann (1970).

[2] Armand Borel, Some finiteness properties of adele groups over number fields. Publ. Math., Inst. Hautes Etud. Sci. 16 (1963) 5-30.

[3] Jean-Francois Dat, ν-tempered representations of p-adic groups. I : ℓ-adic case. Preprint 2003.

[4] Pierre Deligne et Jean-Pierre Serre, Formes modulaires de poids 1. Ann. Sci. E.N.S. (iv) 7 (1974), 507-530.

[5] Fred Diamond and Richard Taylor, Lifting modular mod ℓ representations. Duke Math. J., Vol. 74, No 2, (1994), 253-269.

[6] Chandrasekhar Khare, On the local correspondence mod ℓ. J. Number Theory 88 (2001), no. 2, 357–365.

[7] Benedict H. Gross, Algebraic Modular Forms. Israel Journal of Mathematics 113 (1999), 61-93.

[8] T.A. Springer, Linear algebraic groups. Second Edition. Birkhäuser. PM 9 (1998).

[9] Marie-France Vignéras, Correspondance de Langlands semi-simple pour $GL(n, F)$ modulo $\ell \neq p$. Invent. math. 144, 177-223 (2001).

[10] Marie-France Vignéras, Représentations ℓ-modulaires d'un groupe réductif p-adique avec $\ell \neq p$. Birkhaüser PM 137 (1996).

Marie-France Vignéras
Institut de Mathematiques de Jussieu
175 rue du Chevaleret
Paris 75013, France
vigneras@math.jussieu.fr

Bei Fragen zur Produktsicherheit wenden Sie sich bitte an:
If you have any questions regarding product safety,
please contact:

Walter de Gruyter GmbH
Genthiner Straße 13
10785 Berlin
productsafety@degruyterbrill.com